FORMULATING, PACKAGING, AND MARKETING OF NATURAL COSMETIC PRODUCTS

FORMULATING, PACKAGING, AND MARKETING OF NATURAL COSMETIC PRODUCTS

Edited By

NAVA DAYAN
Lipo Chemicals, Inc.

LAMBROS KROMIDAS
OnPoint Scientific Solutions, LLC.

WILEY
A JOHN WILEY & SONS, INC., PUBLICATION

Published by John Wiley & Sons, Inc., Hoboken, New Jersey.
Published simultaneously in Canada

Library of Congress Cataloging-in-Publication Data:

Formulating, packaging, and marketing of natural cosmetic products / edited by Nava Dayan, Lambros Kromidas C.
p. cm.
ISBN 978-0-470-48408-1 (cloth)
1. Cosmetics–Composition. 2. Cosmetics containers. 3. Organic compounds.
I. Dayan, Nava. II. Kromidas, Lambros.
TP983.F595 2011
668′.55–dc22

2010054045

Printed in Singapore

oBook ISBN: 9781118056806
ePDF ISBN: 9781118056783
ePub ISBN: 9781118056790

10 9 8 7 6 5 4 3 2 1

CONTENTS

Preface ix
Contributors xi

PART I
MARKET AND TRENDS

CHAPTER 1 *THE NATURAL PERSONAL CARE MARKET* **3**
Darrin C. Duber-Smith

CHAPTER 2 *CONSUMER AND FORMULATOR OF NATURAL COSMETICS: UNDERSTANDING AND INTEGRATING EACH OTHER'S NEEDS* **15**
Yi Hsin Chang

CHAPTER 3 *THE MECHANICS OF DEVELOPING A NATURAL/ORGANIC NONGOVERNMENTAL COSMETIC STANDARD* **27**
Amy B. Olson

PART II
REGULATORY ASPECTS

CHAPTER 4 *REGULATORY PERSPECTIVE OF NATURAL AND ORGANIC COSMETIC PRODUCTS* **51**
Azalea P. Rosholt

CHAPTER 5 *THE U.S. LEGAL PERSPECTIVE ON MAKING ORGANIC AND NATURAL CLAIMS FOR COSMETIC PRODUCTS* **67**
Miriam J. Guggenheim and MaryJoy Ballantyne

CHAPTER 6 *REGULATORY AND SAFETY ASPECTS OF NATURAL FRAGRANCE INGREDIENTS* **89**
Anne Marie Api and Matthias Vey

CHAPTER 7 *ADVERTISING SELF-REGULATION: A REVIEW OF COSMETIC CLAIMS AND NATURAL/ORGANIC CLAIMS* **107**

Annie M. Ugurlayan

PART III
SAFETY ASPECTS

CHAPTER 8 *THE SAFETY ASSESSMENT OF PLANT-DERIVED INGREDIENTS IN COSMETICS* **125**

Viny Srinivasan and Eric Antignac

CHAPTER 9 *APPROACH TO ASSESSING CONSUMER SAFETY OF BOTANICAL INGREDIENTS WITH EMPHASIS TO TYPE I ALLERGY* **149**

Esperanza Troyano, Donna Mcmillan, Katherine Sarlo, Lijuan Li, and Rohan Wimalasena

CHAPTER 10 *PRESERVING COSMETICS WITH NATURAL PRESERVATIVES AND PRESERVING NATURAL COSMETICS* **169**

James Flanagan

CHAPTER 11 *MICROBIAL RISKS AND ECO-FRIENDLY PACKAGING* **179**

John I. Yablonski and Sharon E. Mancuso

PART IV
USE OF NATURAL INGREDIENTS

CHAPTER 12 *FORMULATING COSMETICS WITH NATURAL OILS, FATS, BUTTERS, AND WAXES* **215**

Mark Garrison and Nava Dayan

CHAPTER 13 *NATURAL ANTIOXIDANTS AND THEIR EFFECTS ON THE SKIN* **239**

Anne Pouillot, Luigi L. Polla, Philippe Tacchini, Alice Neequaye, Ada Polla, and Barbara Polla

CHAPTER 14 *THE USE OF QUERCETIN AND CURCUMIN IN SKIN CARE CONSUMER PRODUCTS* **259**

Brian E. Kilfoyle, Diksha Kaushik, Jenna L. Terebetski, Sonali Bose, and Bozena B. Michniak-Kohn

CHAPTER 15 *AYURVEDIC INGREDIENTS IN COSMETICS* **287**

Rajarajeswari Sivalenka and Mangathayaru Putrevu

CHAPTER 16 *FORMULATING WITH TRADITIONAL CHINESE MEDICINE HERBS* **303**

Ina Schlenoff and Robin Choi

CHAPTER 17 *THE INSIDE-OUT CONCEPT AS COMPLEMENT TO THE USE OF TOPICAL SUNSCREEN: THE CASE FOR ENDOGENOUS SKIN PHOTOPROTECTION FROM SUNLIGHT BY NATURAL DIETARY ACTIVES SUCH AS TOMATO CAROTENOIDS* **313**

Joseph Levy and Yoav Sharoni

PART V

ANALYSIS OF NATURALS

CHAPTER 18 *CHROMATOGRAPHIC TECHNIQUES FOR THE ANALYSIS OF NATURAL PRODUCTS IN COSMETICS* **331**

Stefan Gafner and Jacquelyn R. Villinski

CHAPTER 19 *THE USE OF NUCLEAR MAGNETIC RESONANCE SPECTROSCOPY FOR THE IDENTIFICATION OF BIOMARKERS FOR QUALITY CONTROL OF PLANT EXTRACTS* **361**

Kan He and Marc Roller

PART VI

BIODEGRADATION

CHAPTER 20 *BIODEGRADABILITY EVALUATION FOR COSMETIC INGREDIENTS AND FINISHED PRODUCTS* **389**

Jennifer K. Saxe

CHAPTER 21 *OVERVIEW OF BIODEGRADABLE PACKAGING, METHODS, AND CURRENT TRENDS* **411**

Gaurav Kale

Index 421

PREFACE

From an early age, we learn to discern what is natural and what is not. A child knows that a deer is part of nature and a car is not. He may not be able to explain exactly why, but he has, even at this early age, an innate understanding of natural things. The deer is not manmade and cannot be controlled or designed, while the car can be designed by man to assume many shapes and functions. The same can be said about cosmetic chemist, manufacturers, marketers, and consumers of natural cosmetics. The issue with natural cosmetics is that unlike trees, deer, fish, flowers, and so on they are not found in or made by nature. They are manufactured like a car, chair, fork, and a multitude of other items. So how can a manufactured cosmetic product claim to be natural? Some may claim that cosmetics made of all natural products are natural. That sounds reasonable. Does that mean then that a fork made of unprocessed wood is natural? A child may not think so because even though the wood is natural, the finished product does not look natural and involved extensive human intervention. Likewise, many do not think cosmetics are natural even if they are made of natural components. If there is a natural cosmetic, it will be one of the many fruits or berries found in nature that may be used to color ones skin by using their substance. If used in such a manner and for the purpose of coloring the skin, one can argue that berry is a natural cosmetic. For many of us, it may be the packaging and manufacturing parts that make it difficult to accept the concept that a cosmetic on the drug store shelf could be natural. Be that as it may, one cannot argue with something made of all natural components even if manufactured by big noisy steel machines and packaged in an artificial container (i.e., a container not made of natural components). But for argument's sake, let us accept the paradigm that such a product is a natural product. Can a cosmetic then be made of all natural components? Some hardcore "natural" cosmetic chemists may think so, but for the majority the lack of a variety of natural functional ingredients is a rate-limiting step in the development. What percentage of natural components included in the formula will qualify it to claim that it is natural? Furthermore, how would one classify natural ingredients? For some, it matters how they were gathered and processed before use. Regardless, even the hardcore "natural" cosmetic chemist will have to admit that the palate, aesthetics, and stability of cosmetics made of all natural products will be severely limited and compromised. However, the trends of the past two decades clearly indicate that a large part of the population is striving to "get back to nature," as they say. As such, natural has become a created homeostasis outside God's or evolution's responsibility. While it is the nature of life in many poor third world countries, in the Western developed world, the "getting back to nature" trend is mostly driven by the wealthier population that can afford shopping for organic and unprocessed food, vitamins and supplements, and alternative medicinal practices.

When it comes to cosmetics and personal care, this sense of "natural" confusion and created homeostasis has penetrated almost every aspect of product development: from the sourcing and definitions of raw materials used, to the design of the formulations, and to the testing requirements, regulations, and their enforcement. Many of us in the cosmetic industry did not and do not know how to digest this new paradigm of "natural cosmetics," and how to translate it to practicality. This book is an attempt to mitigate the confusion over "natural cosmetics" and is designed to provide the reader with the basic tools and ideas to develop naturally derived formulations. This book is neither developed by hardcore natural cosmetic scientists, manufacturers, or marketers who want to push their own unique agenda. Nor is it designed to champion any special interest group's agenda. It is tailored by conventional cosmetic scientists whose agenda is to merely make sense of this business opportunity. We specifically chose contributors associated with our industry but from varied walks of life, backgrounds, and specialties to give a fresh outlook on the subject matter. Our contributors had no preconceived interests in "natural cosmetics" but were eager to advise, enlighten, and guide the reader whose interest is to take the primary steps toward formulating, packaging, and marketing of natural cosmetic products.

Working with these contributors was a wonderful learning experience and we thank them sincerely.

Lambros Kromidas
Nava Dayan

CONTRIBUTORS

Eric Antignac, L'oreal R&D, Worldwide Safety Evaluation, Asnieres, France

Anne Marie Api, Research Institute for Fragrance Materials, Inc., Woodcliff Lake, NJ

MaryJoy Ballantyne, Covington & Burling LLP, Washington, DC

Sonali Bose, Department of Pharmaceutics, Ernest Mario School of Pharmacy, New Jersey Center for Biomaterials, Rutgers, The State University of New Jersey, Piscataway, NJ; Novartis Pharmaceuticals Corporation, East Hanover, NJ

Yi Hsin Chang, Global Ingredient Solutions, North Tustin, CA

Robin Choi, New York College of Health Professions, Syosset, NY

Nava Dayan, Lipo Chemicals Inc., Paterson, NJ

Darrin C. Duber-Smith, Green Marketing, Inc., Nederland, CO

James Flanagan, Microbiologist, Coty Inc., New York, NY

Stefan Gafner, Tom's of Maine, Kennebunk, ME

Mark Garrison, Avon Products, Suffern, NY

Miriam J. Guggenheim, Covington & Burling LLP, Washington, DC

Kan He, Naturex, Inc., South Hackensack, NJ

Gaurav Kale, Chiquita Brands International, Franklin Park, IL

Diksha Kaushik, Department of Pharmaceutics, Ernest Mario School of Pharmacy, New Jersey Center for Biomaterials, Rutgers, The State University of New Jersey, Piscataway, NJ

Brian E. Kilfoyle, Department of Pharmaceutics, Ernest Mario School of Pharmacy, New Jersey Center for Biomaterials, Rutgers, The State University of New Jersey, Piscataway, NJ

Joseph Levy, Department of Clinical Biochemistry, Faculty of Health Sciences, Ben-Gurion University of the Negev, Israel; Soroka Medical Center of Kupat Holim, Beer-Sheva, Israel

Lijuan Li, The Procter & Gamble Company, Mason Business Center, Mason, OH

Sharon E. Mancuso, Bio-Control Consultants, Inc., Westfield, NY

Donna Mcmillan, The Procter & Gamble Company, Mason Business Center, Mason, OH

Bozena B. Michniak-Kohn, Department of Pharmaceutics, Ernest Mario School of Pharmacy, New Jersey Center for Biomaterials, Rutgers, The State University of New Jersey, Piscataway, NJ

Alice Neequaye, EDEL Therapeutics SA, Lausanne, Switzerland

Amy B. Olson, Washingtonville, NY

Ada Polla, Alchimie Forever LLC, Washington, DC

Barbara Polla, Alchimie Forever Sàrl, Geneva, Switzerland

Luigi L. Polla, Alchimie Forever Sàrl, Geneva, Switzerland

Anne Pouillot, Alchimie Forever Sàrl, Geneva, Switzerland

Mangathayaru Putrevu, Celgene Cellular Therapeutics, Avenel, NJ

Marc Roller, Naturex SA, Avignon, France

Azalea P. Rosholt, Customs and International Trade, Ernst & Young LLP, Washington, DC

Katherine Sarlo, The Procter & Gamble Company, Miami Valley Innovation Center, Cincinnati, OH

Jennifer K. Saxe, EcoSafety Sciences, Natick, MA

Ina Schlenoff, Estee Lauder Companies, Syosset, NY

Yoav Sharoni, Department of Clinical Biochemistry, Faculty of Health Sciences, Ben-Gurion University of the Negev, Israel; Soroka Medical Center of Kupat Holim, Beer-Sheva, Israel

Rajarajeswari Sivalenka, Celgene Cellular Therapeutics, Avenel, NJ

Viny Srinivasan, L'Oreal USA, Inc., New York, NY

Philippe Tacchini, EDEL Therapeutics SA, Lausanne, Switzerland

Jenna L. Terebetski, Department of Pharmaceutics, Ernest Mario School of Pharmacy, New Jersey Center for Biomaterials, Rutgers, The State University of New Jersey, Piscataway, NJ

Esperanza Troyano, The Procter & Gamble Company, Sharon Woods Innovation Center, Cincinnati, OH

Annie M. Ugurlayan, National Advertising Division, New York, NY

Matthias Vey, International Fragrance Association, Brussels, Belgium

Jacquelyn R. Villinski, Tom's of Maine, Kennebunk, ME

Rohan Wimalasena, The Procter & Gamble Company, Mason Business Center, Mason, OH

John I. Yablonski, Bio-Control Consultants, Inc., Westfield, NJ

PART I

MARKET AND TRENDS

CHAPTER 1

THE NATURAL PERSONAL CARE MARKET

Darrin C. Duber-Smith[1,2]

[1]*Green Marketing, Inc., Denver, CO*
[2]*Metropolitan State College of Denver, Denver, CO*

1.1 HISTORY'S "NATURAL" PROGRESSION

"Natural" and "certified organic." For many years, the very words were enough to make a cynical cosmetic chemist scoff and furrow a marketer's brow. Are natural cosmetics a passing fad or an established long-term trend? Can natural cosmetic ingredients ever be as effective as their synthetic counterparts? Will governments ever define the term natural? Will "mainstream" companies ever properly address this opportunity? Well, the jury is in. After two decades, the natural and organic personal care industry continues to grow at double-digit rates, barring the recessionary dip in 2008 and 2009 where rates dropped into the high single digits, and larger cosmetic companies have finally figured out that high growth rates and return on investment for shareholders go hand in hand. Thus, formulating with natural and organic ingredients has moved from a novelty to an imperative for many cosmetic companies.

The origin of the natural and organic products industry, with its roots in healthy foods and beverages, can be traced as far back as the beginning of the last century in recognition of the efforts of the famous health enthusiast Will Kellogg and a handful of early pioneering companies such as lozenge maker Thayer's and vitamin provider KAL. These early visionaries can at least be partially credited with the initial commercialization of natural products on a large scale. However, decades of "better living through chemistry" resulted in the development of tens of thousands of effective, synthetic ingredients that could be easily standardized and manufactured on a large scale. Food and personal care companies were the major beneficiaries of such technological advancements, and consumers appreciated the proliferation of effective product options. Natural products were to take a back seat, at least temporarily.

A peppering of small food and nutritional supplement companies dominated the niche's natural products landscape, as few people were willing to compromise on effectiveness, taste, price, convenience, and other variables, very much preferring the widely available synthetic offerings. The zeitgeist of the 1960s illustrated by books

Formulating, Packaging, and Marketing of Natural Cosmetic Products, First Edition.
Edited by Nava Dayan and Lambros Kromidas.

such as Rachel Carson's environmental bombshell, *Silent Spring*, along with the civil rights movement, helped fuel an increasing consumer demand for effective products that were healthier for people and better for the environment. At that time, natural personal care represented only a small handful of these businesses and the concept of organic agriculture was in its infancy.

A result of this "hippie"-inspired movement was the proliferation of the natural channel, a loose network of independent "health food" stores found in every region of the country, and an important step in getting natural products into the hands of the consumers who needed them. Poorly capitalized entrepreneurs struggled to make effective products for this growing consumer segment, but at least now they had a place to sell them. Most of these stores sold bulk grains and whatever brands they could get their hands on. Personal care items were limited to a variety of salves and balms, but nothing close to the range of sophisticated product offerings we see as of this printing.

The 1970s and 1980s saw huge increases at every stage of the value chain. Since demand for all products and services is ultimately derived from the end user, retailers required more and increasingly effective branded products and manufacturers required the ingredients to produce them. The loose network of suppliers, manufacturers, and intermediaries was transforming into a sophisticated supply chain. More importantly, people were starting to prosper, and early entrepreneurs cashed out their investments to start second-generation firms. Personal care became a bona fide section of the store with the emergence of "supernatural" retailers such as Wild Oats Markets, Whole Foods Markets, Fresh Fields, Bread and Circus, Boney's, Trader Joe's, and dozens of others. Recognized brands such as Aveda and the Body Shop, companies that have traditionally used some natural ingredients, were what passed for natural products in the mainstream channels during that time.

The year 1990 ushered in a new decade of double-digit growth in the food, beverage, functional food, and nutritional supplement segments and saw the beginning of industry consolidation as entrepreneurs implemented their exit strategies and large portfolios of brands were compiled by companies such as the Hain Celestial Group. By the late 1990s, several conferences produced by Natural Business Communications were dedicated to the financial side of the business (as private and public equity players scoped out the opportunity), in addition to a huge proliferation of events all over the world. At that time, the personal care segment, although still dominated by a dozen or so companies generating several million in sales each as well as thousands of crude products made in kitchens and bathtubs, began a sustained growth spurt that continues to this day.

At the end of the first decade in this new century, many of the pioneering brands have been purchased by larger players, as Colgate now owns Tom's of Maine and Clorox owns industry leader Burt's Bees. Organic regulations that were written for food, now for the most part, apply to personal care, although there are still some points of contention. There are no regulations per se regarding use of the word natural in marketing, so there is much liberal use of the term. The practice of "dusting" products with inadequate amounts of active botanicals, for example, is well known. Consumers obviously expect that a product marketed as "natural" contains at least a majority of natural ingredients, but this is often not the case. The question of how to define a natural ingredient is now compounded by how to define and label a natural product, which is supposed to contain some amount of natural ingredients. The industry has come a long way in the decade prior to this writing, and third-party certifications abound causing much confusion in the marketplace.

The remainder of this chapter will be dedicated to providing the reader with an understanding of the current business environment as well as what to expect for the near future in natural cosmetics.

1.2 THE NATURE OF NATURAL

That the current situation around defining what constitutes a natural cosmetic product is unclear is a gross understatement. Any effort to define what natural really means must begin at the root problem. In the case of a natural personal care product, one must first decide what constitutes a "natural" ingredient. Once that has been accomplished, one can set standards as to what percentage of natural ingredients must be present in a product as well as what types of ingredients must never be included in a natural product for it to be called "natural."

Would it be helpful if there were such an accepted standard? Unfortunately this is fantasy, as there exist few government regulations in the area of natural ingredients (except for fragrances and a very loose definition of natural in the United States) and therefore a lack of standards for finished products themselves. As we know, in the absence of government regulations, organizations are "supervised" by a patchwork of stakeholder groups including competitors, supply chain members, the media, non-government organizations (nonprofit organizations, etc.), customers, shareholders, and others. This is generally a model that leads to corporate improvisation and consumer confusion in the marketplace since terminology is not commonly understood and there are numerous opportunities to engage in ethically questionable behavior with regard to product claims.

In the case of natural personal care, it is my opinion that it should be generally understood that a natural ingredient should be neither synthetically derived nor synthetically processed. This means that a natural ingredient must be "found in nature" *and* be free of synthetic additives in processing as well as "overprocessing." Most definitions of natural mention the idea of "minimal processing," but this is vague. Abuses are numerous in this area as well because of the highly interpretive nature of processing. Add this to the traditional unwillingness by the U.S. Food and Drug Administration and other government bodies to address this growing issue and the result is a proverbial "elephant in the living room." Nevertheless, there are a growing number of third-party organizations and industry associations that have developed and marketed various models, certifications, and labeling requirements for natural products. Some of these models are more stringent than others, and this is indeed better than nothing at all, but unfortunately none of these programs is consistent with each other and none has emerged as a standard with a high degree of awareness by industry or consumers. With regard to the "natural" distinction, the most promising as of this writing appear to be the Natural Products Association standard in the United States and the much older BDIH European standard for natural, which among others are listed below.

The following general guidelines may be of preference regardless of the labeling certification used. Any certification should involve the following:

1. A stringent definition of what constitutes a natural ingredient.
2. A list of synthetically and naturally derived ingredients that may not be included in a natural product under any circumstance (examples include many of those identified by various government and third-party organizations, such as the European Union and the California Green Chemistry Initiative, as potentially harmful or actually toxic).
3. A labeling standard that recognizes *four* different levels of a natural product. Water cannot and therefore does not count as ingredient when calculating percentage of the whole numbers since the result would be skewed in favor of a more "natural" product.
    a. A "100% natural" product claim would include no synthetic or banned ingredients whatsoever, and only 100% natural ingredients are allowed.
    b. A "natural" product would include 95% or higher natural ingredients and none of the banned ingredients. The remaining 5% could be synthetically derived and/or processed but not any as discussed in item 2 above.
    c. A "made with natural ingredients" product would contain between 70% and 95% natural ingredients and no banned ingredients.
    d. A product that has fewer than 70% natural ingredients would be allowed to emphasize such ingredients denoting them as natural on the label but would not be allowed to make any other natural claim on the primary display panel (unlike the previous three scenarios), and would allow even banned ingredients.
4. A standard that has government oversight (both United States and European Union for a start) as well as enforcement.

It seems logical that any future natural regulation would mirror the existing U.S. Department of Agriculture (USDA) Certified Organic regulations as far as percentage content and what claims can be made are concerned. Therefore, a product with 100% natural ingredients could be marketed as "100% natural." A product with 95% or higher natural ingredients could denote "natural." A product with 70% or higher natural ingredients could claim "made with natural ingredients." And the product that has less than 70% natural ingredients could only denote an ingredient as natural in the ingredient panel. These are the rules for USDA Certified Organic and could be used as a framework for natural marketing guidelines.

1.3 THE ORGANIC PANIC

It is crucial to understand the difference between "natural" and "certified organic," the latter of which is largely defined and regulated within the United States. When the regulations for "certified organic" were created for foods in the mid-1990s and the USDA tasked with oversight of the third-party certification process, the framers of this groundbreaking legislation did not address personal care. The regulations were intended only for ingestible and not topical ingredients. Personal care was not addressed until a few years later when larger players began to penetrate the natural space and realized that they had great flexibility with regard to formulation and product claims in the area of organic as well. Companies that had traditionally operated within the organic channel scrambled to address the issue.

As a result, the USDA disallowed the use of the USDA Certified Organic seal on cosmetics for a brief period, reinstated it, and is as of this writing still in the process of finalizing the requirements for such products. The outcome would have great industry impact since a generally accepted standard in the United States is likely to have very large global impact. In Europe, the Cosmetics Organic and Natural Standard (COSMOS) program was instituted in 2008 to harmonize some of the various certifications. These standards, however harmonized, enjoy very little awareness in the United States, and it will be interesting to follow future actions as to which certifications will win the hearts and minds of the U.S. market. It may be that the USDA backing combined with the strength and size of the U.S. market for personal care products gives USDA Certified Organic a competitive advantage.

A certified organic product by definition must be natural, but also much, much more. For USDA Organic certification, the ingredients (which must also be certified in order for the finished product to display the seal) must adhere to the requirements outlined in the USDA's National Organic Program, which include agricultural restrictions such as water usage, crop rotation, use of pesticides and herbicides, and so on. These requirements can be found on the organization's web site, and approved third-party certifiers such as Quality Assurance International provide the actual certification services. The USDA acts as overseer. It is all very neat in theory, and the pending clarification by the USDA and others should rocket this area into the stratosphere as it has for USDA Certified Organic foods.

1.4 LABELING AND CERTIFICATIONS

There are dozens of certifications related to eco-friendly and wholesome products and services. However, the "eco-labels" we are most concerned with specifically address natural and/or organic offerings. Below is a listing and description of the major natural and/or organic certifications as of this writing:

- *United States* (nonharmonized standards from four different entities)
    1. USDA Organic
    2. OASIS (Organic and Sustainability Industry Standards)
    3. National Sanitation Foundation (NSF)'s Made with Organic Personal Care Standard
    4. Natural Products Association (NPA)

These standards are very different from one another. For example, USDA Organic is for organic products only while the Natural Products Association standards are intended for natural cosmetics, and not organic ones.

- *Europe*

  COSMOS (formed of the below listed separate organizations for the purpose of developing one harmonized European standard)

  1. ECOCERT (France)
  2. Soil Association (UK)
  3. Bundesverband Deutscher Industrie- und Handelsunternehmen (BDIH) (Germany)
  4. L'Associazione Italiana per l'Agricoltura Biologica (AIAB) and Istituto per la Certificazione Etica e Ambientale (ICEA) (Italy)
  5. BioForum (Belgium)
  6. Cosmebio (France)

Although it is impressive that six different organizations were both willing and able to harmonize their standards, it would appear that greater adoption of one or more of these standards could be accomplished through merging the six under one European Union moniker. The likelihood of this happening is improbable or it would already have been done when the standards were identified in the first place, so some of the confusion that such harmonization is intended to alleviate remains present. Other European certifications include Organic Farmers and Growers (UK), The International Federation of Organic Agriculture Movements, IBD (Instituto Biodinâmico) Certified Organic (Latin America), and IBD Natural Ingredients. It is interesting to note that IDB has separate certifications for natural and organic, and this could be a good model for certifying organizations to follow, as it is easier to gain awareness of one brand name as opposed to more than one.

If the reader is confused, it is because of the current state of affairs in marketing natural/organic cosmetics. As of this writing, other than the COSMOS effort for a single European organic standard (albeit under six different certifying names), there has not been much in the way of advancing the harmonization of natural/organic cosmetics. The USDA organic standards for food, which may be applied to cosmetics and will experience several future adjustments, continue to be the predominant organic standard in the United States. ECOCERT is perhaps the most recognized standard for both natural and organic in Europe. It is likely that a major marketing effort funded by cosmetic companies would dramatically increase awareness, and perhaps demand, among the general population for any certification the organizations choose. Such a program could be facilitated by government and the certifiers themselves.

1.5 THE INDUSTRY

For many years, numerous marketing research companies have produced reports based on limited understanding of the natural and organic products industry. Such is the case with most of the larger, global organizations that analyze multiple industries such as Packaged Facts and Datamonitor. Two research companies that specialize only in natural products and have both produced numerous reports since the late 1990s are therefore chosen for discussion. The consumer market for natural products will be described later with special recognition given to the Natural Marketing Institute (NMI) of Harleysville, PA. The purpose of this section, however, is to describe the supply chain and provide some very revealing industry data provided by *Nutrition Business Journal*, a publication of New Hope Natural Media, the largest and perhaps the most important organization in the natural products industry due to its dominance in print, in person, and online media and communication in the natural and organic space.

The natural and organic personal care industry is served by a wide variety of generalized and highly specialized suppliers and service companies. These organizations are responsible for the ingredients and components of the products consumers buy and range from large global chemical companies that offer a few natural and/or certified organic ingredients to very small regional players that may offer only one or two niche ingredients. Service companies range from insurance and legal to strategic planning consultants and associations. Natural ingredients are usually found in good supply, but certified organic ingredients are less available. Traditionally, demand for many of these ingredients tends to be greater than the supply, which can be a major barrier to adoption on a large scale.

Branded product manufacturers, the most visible of all supply chain members, are now dominated by a small number of larger companies. Many branded product manufacturers have been acquired by either large personal care companies or smaller investment interests and have grown exponentially since the infusion of capital and expansion of distribution have allowed them to do so. A huge number of smaller organizations are still able to compete adequately, enjoying relatively large profit margins, healthy growth rates, and all of the entrepreneurial perks (and headaches) that come with private ownership. The highly publicized acquisition of industry leader Burt's Bees by Clorox and the Colgate-Palmolive's purchase of strongly branded Tom's of Maine in the latter part of the last decade epitomize the recognition by all stakeholders that the health and wellness trend is not merely a fad and will continue to affect the ways by which business operates.

Table 1.1 shows a 2008 list of branded product *natural* personal care manufacturers, their estimated sizes, and estimated share of certified organic sales.

It is obvious that a smaller revenue base is most certainly correlated with the likelihood of offering certified organic products as part of a brand's natural offerings, and the reasons for this are unclear. It should also be noted that the list in Table 1.1 does not include the many hundreds of smaller players offering products through a variety of distribution channels. So, the question remains, as we make our way downstream through the supply chain, how do branded product manufacturers deliver their products to the growing base of natural and organic cosmetics consumers.

Natural and organic products are still offered in the original "natural channel" consisting of literally 10,000 plus independent health food stores, pill shops, and small chains. These stores, which source products directly from manufacturers and from small wholesalers, are present in most communities in the United States and areas of Europe, and tend to be the primary channel for the smaller branded product manufacturers who cannot handle large distribution and who serve the core natural products user. This channel is still very strong, and the criticism with the natural products industry that smaller stores would lose out as stronger distribution channels enter the marketplace has been dramatically overstated.

TABLE 1.1 Largest Natural and Organic Personal Care Companies [1]

Brand/company	Size (millions of dollars)	Certified organic sales (%)
Burt's Bees (Clorox)	250–350	0
Aveda (Estee Lauder)	250–350	0
Aveeno (Johnson & Johnson)	150–200	0
The Body Shop (L'Oreal)	150–200	20
Nu Skin	150–200	0
Hain Celestial Group	150–200	5
N.V. Dr. Perricone, MD Cosm	150–200	0
Bare Escentuals	100–150	5
Jurlique	100–150	0
Tom's of Maine (Colgate-Palmolive)	50–100	0
Dr. Hauschka	30–50	20
Dr. Bronner's	30–50	40
Yes to Carrots	30–50	15
Nature's Gate (Levlad)	30–50	20
Avon	30–50	0
Weleda	30–50	20
Giovanni	30–50	60
Aubrey Organics	30–50	100

Whole Foods Markets remains the leader in the "supernatural channel" consisting of larger supermarket-style "gourmet crossover" format stores that buy products from larger manufacturers and serve a more mainstream and higher income natural products consumer. Other examples include Trader Joe's and Larry's Markets on the west coast.

Larger manufacturers now offer products through big discount retailers such as Target and Wal-Mart and mainstream grocery stores, and smaller ones through spas, salons, gift shops, and other specialty stores. Examples of direct channels include television, radio, Internet, and even the controversial multilevel/network marketing channel that popularized companies such as Avon, Mary Kay, and Amway. A myriad of opportunities await the well-funded brand!

Table 1.2 shows a breakdown of the 2008 U.S. natural and organic personal care market by sales channel in 2008. The six categories described in Table 1.2 are very

TABLE 1.2 Natural and Organic Personal Care Channel Breakdown [2]

Sales channel	2008 sales (%)
Natural/specialty retail	30
Network marketing	21
Mass market/beauty	20
Other direct (TV, Internet)	11
Dept store/salon/spa	10
Specialty personal care	8

TABLE 1.3 Natural and Organic Personal Care Channel Breakdown [2]

Product category	Total conversion sales	NOG sales/penetration	NOG growth (%)
Cosmetics	$15.1 billion	$419 million/2.8%	10.1
Feminine hygiene	$2.76 billion	$100 million/3.6%	15.1
Hair care/color	$10.01 billion	$1.72 billion/17%	11.3
Baby care	$623 million	$159 million/25.5%	18.4
Nail care	$730 million	$24 million/3.2%	9.2
Oral hygiene	$4.91 billion	$605 million/12.3%	7.3
Bath items	$659 million	$147 million/22.2%	6.3
Deodorants	$1.61 billion	$217 million/13.5%	9.5
Shaving	$1.81 billion	$138 million/7.6%	8.7
Skincare	$9.20 billion	$3.05 billion/33.1%	6.6
Soap	$3.53 billion	$998 million/28.2%	7.8
Fragrances/aromas	$5.91 billion	$352 million/6%	6
Total personal care	$56.98 billion	$7.92 billion/13.9%	8.4

broad but provide the reader with an idea as to how and where these products are sold. Of particular note is the continued dominance of the natural channel and the surprising share the multilevel/network marketing category enjoys in the United States.

The growth rates shown in Table 1.3 suggest growing revenue and penetration rates for natural and organic products into the conventional area. Also notable are the high natural penetration rates in skincare, bath items, soap, hair care, and baby care suggesting that a growing number of consumers prefer natural/organic options to conventional offerings. Clearly, there remain myriad opportunities for growth in this not yet mature market.

1.6 THE NATURAL AND ORGANIC CONSUMER

As there are many marketing research companies that study the industry trends, many as has been said from the outside looking in, there are a number of companies that specialize in tracking the trends of consumer demographics (describing consumers in terms of size and structure), geographic location, and psychographics (behaviors and attitudes). The most prolific such study, the LOHAS segmentation (as described below), is one that is developed and fielded by the NMI and now has over 10 years of history in identifying and tracking consumer preferences and characteristics through market segmentation studies.

The first such study was conducted by Natural Business Communications, a producer of industry news and conferences, in 1999 and spawned an annual convention, quarterly, publication, web site, and the important study mentioned above. The pioneering work (yours truly was on the development team) was institutionalized by NMI, while the former parent of Natural Business Communications now operates the event, publication, and web site. It is important that research continues year after year so that outlying data and fads can be separated from the statistically important information, so this study is the one the author has chosen.

LOHAS stands for lifestyles of health and sustainability, and the study is commissioned primarily to identify appropriate target markets for natural, organic, and environment-friendly products and services. Natural personal care represents one of the fastest growing segments of this marketplace. Natural Marketing Institute considers a natural personal care user to be one who has purchased at least one product during the previous 6 months. In 2007, according to NMI 38% of general population consumers have classified themselves this way.

The LOHAS segment breaks down the U.S. population into five market segment categories based on demographic, geographic, and psychographic characteristics reflecting attitudes and behaviors with regard to natural products.

The primary target for most natural products is the aptly named LOHAS segment, consisting of 19% of the general population. Consumers in this segment tend to live a healthy lifestyle and will go out of their way to purchase natural and organic products. They are early adopters, opinion leaders, and avid users of green products. They aren't necessarily high-income group and tend to be higher educated. While concentrated more heavily on the coasts, they are actually located throughout the United States.

Naturalites, another attractive market segment, representing 19% of the general consumer population, are more interested in health than they are in the environment, and they are primary users of natural and organic personal care. They are higher income group and tend to be better educated than the general population. So far in our discussion, the target market for natural and organic personal care products stands at 38% of U.S. adults, but we have an even larger segment to include in our marketing planning.

DRIFTERS, the largest market segment at 25% of the general population, tend to be much younger and have yet to develop their own values structures and ingrained purchase patterns. They are motivated by the latest trends and shift on commitments to many issues, but it is probable that the majority of these drifters will become LOHAS or Naturalites as they get older.

The other two segments, *Conventionals* (lower income folks motivated by price and practicality) and *Unconcerned* (the nomenclature reflects the general attitude), are not primary targets, but the ongoing proliferation of natural/organic cosmetic product availability coupled with the growing health and wellness trend might change this. These segments are ignored in the targeting process for now, but must still be monitored as the growing popularity and availability of natural products will surely bring some of these consumers from these two segments into one of the other three segments. The bottom line is that about 50% of adults are in the primary target for natural and organic personal care products through the LOHAS segmentation model.

Again, this is only one example of many studies dedicated to studying this field, and the author's own personal bias coupled with the consistent findings of the study over a long period of time makes it a very appropriate example for this book [4].

1.7 A LOOK AHEAD

The natural and organic personal care industry itself, together with the behavior of consumers, represents a large part of strategic marketing planning, but marketers must also consider other factors in the external business environment all of which will affect the nature of the product, its pricing, the optimal channels of distribution, and the most effective methods and messaging with regard to promotion.

For instance, the economic environment, a factor completely out of the control of the marketer, affects all aspects of business, from the upstream members of the supply chain down to the consumer at the very end. The question has always been, "To what degree does a recession affect the sales of personal care products?" The so-called lipstick effect is a phenomenon identified decades ago when researchers discovered that sales of personal care products actually ticked up when there was a recession. The reasons for this have been a subject of great debate for many years, ranging from an increase in "pampering" behavior due to economic and personal malaise to a desire to look your best when unemployment is high. In the most recent recession (2007–2009), the results indicate that the conventional category of personal care products took a pretty big hit as sales and revenues dropped off at industry giants such as L'Oreal. The natural personal care industry, on the other hand, dropped from growth rates in the high teens to growth rates around 8–10% annually. This is a true indicator of the power of this market and the necessity to meet demand by formulating appropriate products.

The legal, regulatory, and political environments, also out of the control of the industry, must be monitored by the marketers as they are ever changing and not always favorable to the industry. In the case of natural and organic personal care products, it has been said that few regulations are in place. Rather, industry is a patchwork of association regulations, third-party certifications, and haphazard government attempts at clearing the confusion. Some organizations take advantage of this with misrepresentations, since most consumers cannot differentiate between natural and organic products. Consequently, the organizations that are truly addressing the needs of this market are losing their competitive advantage. Perhaps the most important threat to this industry is the lack of guidelines and oversight.

There is little doubt that the decades long social trend of consumer attitudes and behavior toward health and wellness bodes increasingly well for ethical companies that are producing truly natural products and therefore addressing the basic consumer need for products that are healthier for them and the environment. The target markets will continue to grow, the penetration rates of natural and organic into conventional will increase, the number of players will proliferate, and numerous opportunities for long-term success can be enacted. "Green washers" (those who position products as "green," but whose commitment to social and environmental responsibility is questioned by stakeholder groups) will be fingered by competitors, activist consumers, supply chain members, the media, and so on, so misrepresentation of products as natural or organic is never an effective long-term strategy.

Government regulation may eventually occur in the natural area, an area that is still much larger than the better defined and regulated organic space. It will take time, but the industry is light years ahead of where it was 10 years ago. In general, any uniform standard for a natural product most likely will require at least 95% natural ingredients for a natural product claim to be made. Obviously, with the notable exception of an all-organic product, a 100% natural product, labeled as such and protected by regulations, will enjoy a competitive advantage over all others.

The challenges inherent in working with these interesting ingredients are indeed surmountable, as many cosmetic chemists have recently chosen to specialize in this area as a way to differentiate themselves from other, less progressive chemists. The question is no longer "why" one should produce natural products, but rather "how." Your organization will reap myriad benefits through addressing the growing need in the marketplace with ingredients and finished product. Failure to recognize this important shift in the personal care marketplace will result in repercussions that will last a very long time.

REFERENCES

1. *Nutr. Bus. J.* 2009;XIV(2):7.
2. *Nutr. Bus. J.* 2009;XIV(2):5.
3. *Nutr. Bus. J.* 2009;XIV(2):1.
4. *LOHAS J.* 2009;(Spring):17.

CHAPTER 2

CONSUMER AND FORMULATOR OF NATURAL COSMETICS: UNDERSTANDING AND INTEGRATING EACH OTHER'S NEEDS

Yi Hsin Chang

Global Ingredient Solutions, North Tustin, CA

2.1 NATURAL MARKET TRENDS

2.1.1 Personal Care Market Trends in Europe

Natural and organic cosmetic sales continue to increase at a healthy rate in Europe, according to Organic Monitor, with revenues projected to increase by 13% in 2009. Organic Monitor's 2009 edition of *The European Natural Cosmetics Market* report [1] finds that natural and organic cosmetic sales have been affected by the financial crisis. However, steady growth is continuing because of increasing distribution and high consumer demand for "chemically clean" cosmetics, and revenues are forecast to approach €2 billion in 2010 [2].

The UK market has been the most adversely affected by the economic recession. The market growth rate has dropped from 16% in 2008 to 8% in 2009 because sales have been hit in organic food shops and health food retailers. However, natural and organic cosmetic sales continue to expand at double-digit rates in other European countries [1].

Natural and organic products now comprise 3% of total cosmetic sales in Europe. The market share is above 4% in Germany, Austria, and Switzerland. High growth in the German market is projected to raise the market share of natural and organic cosmetics to 10% in the coming years.

Natural cosmetics are the most successful in Germany, where they comprise 5% of total cosmetic sales. Although the UK market has been showing high growth rates, the market share of total cosmetic sales remains below 3%.

Formulating, Packaging, and Marketing of Natural Cosmetic Products, First Edition.
Edited by Nava Dayan and Lambros Kromidas.

The French and German markets are the fastest growing in Europe. High growth rates are occurring as product penetration increases in mainstream retailers. Both countries have observed a large number of retailer private label launches in recent years. Certified natural and organic cosmetics have been introduced under private labels of supermarkets, drugstores, perfumeries, organic food shops, and discounters in these countries. Private labels are so successful in Germany that they comprise over 10% of total sales. Alverde has become the leading retailer of private label natural cosmetics in the world, having more than 100 certified products across categories.

The UK and German markets have witnessed the largest number of natural cosmetic product launches under private labels. Natural and organic products have been launched under own brands of supermarkets, drugstores, organic food retailers, and even discounters. Private labels are the most successful in the German market, where they comprise more than 10% of total natural cosmetic sales.

Apart from private labels, a major retailing trend is the growing number of concept stores for natural and organic beauty products. Dedicated stores are being opened by leading natural cosmetic brands, large retail groups, and private investors. Most developments are in Paris where highly established brands such as Dr. Hauschka as well as large retailers such as Monoprix are opening concept stores. Established brands such as Korres and Aveda are opening concept stores across Europe. Dedicated beauty retailers such as The Organic Pharmacy and Naturalia are expanding their chained outlets. Most developments are in France where international brands such as Weleda and Natura have opened flagship stores in Paris.

The Organic Monitor report, produced by continuously tracking the European natural and organic cosmetics market since 2003, finds that the competitive landscape has changed significantly. High growth rates have been responsible for mergers, acquisitions, and an influx of new entrants. Rising competition is leading natural and organic cosmetic brands to expand distribution outside traditional sales channels. This strategy has enabled the Swiss company Weleda to consolidate its position as market leader.

With competition for retail shelf space stepping up, Organic Monitor expects market winners to be companies that can successfully differentiate their products from competitors. New entrants and existing participants are advised to consider strategies based on product differentiation and market segmentation: companies should target specific consumer groups or sales channels.

2.1.2 Personal Care Trends in North America

The natural and organic sector is the fastest growing in the North American cosmetics and toiletries industry, with sales estimated at $7.8 billion in 2009 and increasing by 20% a year [1], outpacing that of the traditional market with double-digit growth annually projected to generate $11.7 billion in sales by 2013 representing a growing opportunity for all cosmetic and skincare companies. Organic Monitor projects strong future growth with natural and organic increasing its share of the total U.S. personal care products market growing from 8% in 2009 to 15% in the coming years.

In the United States, product penetration of natural and organic products is increasing in all channels of distribution. Specialty personal care stores and

TABLE 2.1 Market Segment Breakdown

Market segment	Total (in billion dollars)	Natural/organic (in billion dollars)	Growth (%)
Color cosmetics	14.7	0.4	12.1
Skincare	8.8	2.9	12.1
Hair care/coloring	10.2	1.5	22.2
Fragrance/aromatherapy	6.3	0.3	10.5
Oral care	4.9	0.6	11.7
Bath/toilet soap	3.9	0.9	14.3

department stores are introducing premium natural and/or organic brands (e.g., Sephora). High growth is also coming from spas, salons, and beauty centers. But most sales, however, remain from natural food shops that have 45% market share. Leading chains such as Trader Joe's and Wild Oats have developed private labels for natural beauty products, while independents are giving greater shelf space to these products.

Table 2.1 provides a breakdown of each market segment in the beauty category and how much of each market segment is comprised of natural and organic shares and the estimated growth in the natural organic sector by market segment [3].

As of March 2009, there were a total of 138 products certified with organic labels. Brands such as Aubrey Organics, Boom!, Burt's Bees, Highland Labs, J.R. Watkins, Yes to Carrots, S&V, Xango, and CamoCare are the top organic brands. Also as of March 2009, there were over 120 products in the queue awaiting organic certification.

2.2 THE CONSUMER

Although natural beauty products are nothing new, the interest in natural cosmetic products came back in 2007 due to product safety and environmental scares along with a movement to live a more stress-free, holistic lifestyle [4]. The Natural Marketing Institute reported in 2004 that "63% of U.S. consumers have a preference for natural foods and beverages" while Harris Interactive survey in 2006 found that 83% of consumers "sought a government definition of the term."

According to the Benchmarking Survey conducted in 2007, the majority of women (61%) find it difficult to tell which brands are natural and which are not. Furthermore,

- 78% think natural personal care is currently regulated while 97% think it should be;
- 83% think there should be one meaning for "natural" and 86% agree that there should be a symbol to certify a natural personal care product;
- 65% think a product labeled "natural" must be at least 95% natural;

- 86% don't know about the potential risks of parabens, 92% of oxybenzone, 88% of sodium lauryl sulfate, and 89% of glycol;
- 71% wish there were more natural personal care products available in stores and 73% think stores do not do a good job educating consumers on natural personal care products.

Fifty-six percent of all women believe that the U.S. government regulates beauty products claiming to be 100% organic and know that products claiming to be botanical, natural, or pure have less U.S. regulation.

Although most women understand that the U.S. government does not regulate claims other than "organic," 64% of all women believe a beauty product claiming to be "all-natural" must contain 80% or more natural ingredients.

Seventy percent of women who buy natural beauty products and 31% of those who buy traditional beauty products claim to read labels carefully before buying beauty products. There will be a large increase in the number of women who actively read and scrutinize labels in the years to come and therefore it is critical that formulators know what the beauty customer is looking for.

Forty-five percent of all women buy natural/organic beauty products because of their fear of chemicals, but once they purchase the natural/organic brand, they still expect spectacular results. Over 40% of all women surveyed expect to see results in 2–4 weeks.

Most women surveyed in the study also believe that natural/organic products are better for their skin and make them look their best. The reason not every woman buys natural/organic cosmetic products is simply because there is not enough supply to meet the demand resulting in higher prices. Sixty-five percent of the women who did not buy natural/organic products indicated that price was the key reason of not purchasing. Forty-six percent of women bought natural/organic beauty products for the first time just out of curiosity. Curiosity leads to cosmetic use and sooner or later women will try a certain type of natural brand. Women want to try something new, to make them look even better and younger, and the new hope is in natural and organic beauty products. Sixty-eight percent of women keep buying natural beauty brands because they believe that they are doing what's best for their skin.

As much as the consumer fears harmful ingredients, they are willing to give traditional beauty brands another try if they remove the questionable chemicals. The bottom line for the formulation chemists is to make the labels clear and compelling.

In the same survey, women claimed that they do not want synthetic chemicals but do want to hear what nature can do for them. The strength of medical credentials and language that evokes emotion and nature's innate healing powers will entice your buyer.

Nearly half of all women surveyed buy natural/organic beauty products, but certainly not all the time. The women who usually buy traditionally made cosmetics don't buy more natural/organic beauty products because they simply have not though about it yet; with thousand of brands on the market clamoring for their attention and disposable income, 50% of nonnatural beauty buyers have not given it a thought yet and 49% don't know enough about them [4].

It is also important to note that women who normally buy natural/organic beauty products have much higher makeup and skincare aspirations than traditional beauty brand users, and higher expectations for product efficacy. One interesting note is that women who normally purchase natural/organic beauty products have the same attitude and practices toward their personal wellness and these consumers are more likely to believe in the concept of inside/out, nutritional beauty products.

Current natural/organic brand users wished there were more buying options; 69% of respondents wished they were less expensive, while 35% wish they were easier to find; only 14% of the respondents wanted the product to work more effectively. And when it comes to pricing, 57% of women don't always buy their favorite brand due to price [4].

When it comes to color cosmetics, 55% percent of women who buy natural beauty products use mineral makeup and half of the women who use mineral makeup do so because they feel it works and is safe and natural. Eleven percent of women who identified themselves as "not buying natural/organic beauty products" use mineral makeup. Using fewer products and saving time and money are important factors for 41% of respondents who use mineral makeup.

Corporate social responsibility is also extremely important to the beauty consumer. Nearly 70% of all women believe it is somewhat to extremely important that companies refrain from testing on animals, 65% find it equally important that companies refrain from using any animal part in products or packaging, and 65% feel strongly that a company not pollute the environment.

There is a new ethical group of consumers that cares about health and sustainability as part of their purchase decision. This group represents about 16% of the population but is growing in importance and makes purchase decisions based on those companies that adhere to an ethical code of producing healthier products that protect the environment. Even among these consumers, there are varying degrees of adherence from hardcore "Naturalites" to "Drifters" who will purchase both traditional and natural products. The mainstream customer will adopt the values of this new ethical group of consumers as long as she doesn't have to forgo the other benefits she derives from using traditional products, such as esthetics, performance, femininity, and pleasure.

According to the Benchmarking Company, natural and organic beauty products are in their infancy; many women consider them a natural extension of the holistic lifestyle they are already living or want to lead. Marketers should expect greater consumer education about them: more detailed labeling that explains how or why the product will help them live a more holistic lifestyle, and a stronger desire for proven beauty brands.

In summary, customers demand the same product performance, a competitive price, and an organic seal—expectations often beyond current economic or technical means of today. As raw material suppliers work hard to introduce new "natural/organic" ingredients to the market, there will be more availability and options for the chemists to formulate and achieve their customers' demand.

2.3 THE STATE OF REGULATION

There are numerous nongovernmental guidelines to "natural" or "naturally based" formulations. As such, the biggest challenge for the formulators is to determine which of the guidelines to follow.

Traditional cosmetics, toiletries, and fragrances are regulated by the Food and Drug Administration (FDA), but the FDA and the EU's Cosmetics Directive do not have an official, enforceable definition for "natural" or "organic" products, and any references to natural through the FDA are in direct relation to food, not cosmetics. In addition, there is no specific certification or regulation regarding the use of the term "natural" [5].

Furthermore, unlike organic claims, there are no state regulations regarding the claim natural. California is the only state that regulates organic claims for all products, including cosmetics. "Natural" claims are regulated just like any other claim. They must be truthful and not misleading.

The closest governmental definition for natural personal care products was established in Canada by Health Canada as a regulated drug category called Natural Health Products (NHPs) [6] and went into effect in 2004. According to the Canadian definition, Natural Health Products consist of vitamins and minerals, herbal remedies, homeopathic medicines, traditional medicine such as traditional Chinese medicine, probiotics, and other products such as amino acids and essential fatty acids. Health Canada also accepts synthetic duplicates and semisynthetic substances as naturals, provided that these substances share identical chemical structure and pharmacological properties with their natural counterparts.

Although nongovernmental agencies, it will be informative for the reader to become aware of some of the most prominent organizations that give guidance to developing natural/organic cosmetic products.

The Natural Products Association (NPA, United States) [7] was founded in 1936 and was principally concerned with dietary supplements. The group represents more than 10,000 retailers, manufacturers, wholesalers, and distributors of natural products, including foods, dietary supplements, and health and beauty aids. On May 1, 2008, the organization issued its certification program for personal care products.

In order to display the NPA seal, a product must meet the following requirements:

1. contain at least 95% natural ingredients (truly natural ingredients or ingredients that are derived from natural sources);
2. contain no ingredients linked with potentially suspected human health risks;
3. not be processed in ways that significantly or adversely alter the purity of its natural ingredients;
4. include ingredients derived from a purposeful, renewable source found in nature;
5. be minimally processed and avoid use of synthetic or harsh chemicals so as not to dilute the materials' purity;
6. contain nonnatural ingredients only where viable natural alternatives are unavailable, and only when they pose absolutely no potentially suspected human health risks.

The Natural Products Association has also published [8] a list of 839 ingredients that it considers to meet these requirements.

Cosmetics Organic and Natural Standard (COSMOS, EU) [9] has published guidelines in 2008, in an effort to outline organic and natural standards for the European Union markets.

This standard was developed from collaborations between working groups including the Instituto per la Certificazione Etica e Ambientale (ICEA in Italy); the Federation of German Industries and Trading Firms for pharmaceuticals, health care goods, dietary supplements, and personal hygiene products (BDIH in Germany); BioForum in Belgium; the French Professional Association for the Ecological and Organic Cosmetics and a French certification organization (Cosmebio/ECOCERT in France); and an environmental charity promoting sustainable, organic farming and championing human health (the Soil Association in the United Kingdom).

To achieve COSMOS Organic certification, a product must have at least 20% organic content but to put the term organic in the product name the standard requires 95% organic content. Water that is added during manufacturing is not considered organic. Products with high water concentrations such as body lotions or shampoos will therefore struggle to reach the required levels of organic content. A common method of overcoming this clause is to use aloe concentrates or flower water. However, the COSMOS standard cuts off this shortcut stating that water cannot be calculated as organic irrespective of whether it is added directly or indirectly as mixtures with or components of other ingredients.

The COSMOS standards describe five categories of ingredients:

1. water;
2. minerals;
3. physically processed agro ingredients;
4. chemically processed agro ingredients;
5. synthetic materials.

The COSMOS standard also allows the following physical processes for obtaining or treating natural/organic ingredients:

- Extractions must use natural materials with any form of water or with a third solvent of plant origin such as ethyl alcohol, glycerin, vegetable oils, and CO_2 absorption (on an inert support that conforms to these standards).
- Bleaching or deodorization (on an inert support that conforms to these standards).
- Grinding or centrifuging (solid–liquid separation, spin drying).
- Setting, decanting, desiccation, or drying (progressive or not by evaporation/ natural under sun).
- Deterpenation (if fractional distillation with steam).
- Distillation, expression, or extraction (steam).
- Filtration and purification (ultrafiltration, dialysis, crystallization, and ion exchange).
- Lyophilization, blending, percolation, cold pressure, and hot pressure (depending on the fluidity of the fatty acids to be extracted).

- Sterilization with thermal treatments (according to a temperature respectful of the active substances).
- Sifting, maceration, and ultrasound.

The following chemical processes are also allowed by COSMOS for obtaining or treating natural/organic ingredients:

- Alkylation, amidation, calcination of plant residues, and carbonization (resins, fatty organic oils).
- Condensation/addition, esterification, etherification, and fermentation (natural/biotechnological).
- Hydration, hydrogenation, hydrolysis, and neutralization (to obtain sodium, calcium, magnesium, and potassium).
- Oxidation/reduction processes for the manufacture of amphoterics.
- Saponification, sulfation, and roasting.

Any process not listed above, including but not limited to the below, is not allowed:

- Bleaching or deodorization (on a support of animal origin).
- Use of enzymes derived from GMOs.
- Deterpenation (other than with beam).
- Ethoxylation, irradiation, and sulfonation (as the main reaction).
- Techniques employing genetic engineering.
- Treatment with ethylene oxide or using mercury (mercurial soda).
- Use of petrochemical solvents (hexane, toluene, benzene, etc.).
- Propoxylation.

According to Appendix II of the COSMOS standard, the following synthetic ingredients are allowed: benzoic acid, benzyl alcohol, dehydroacetic acid, denatonium benzoate, heliotropine, salicylic acid, sorbic acid, and tetrasodium glutamate diacetate.

Also, natural minerals are allowed as colorants but they cannot be processed; as a minor point, this means that with the exception of mica, none of these natural minerals would be permitted in cosmetics.

Natural iron oxides, for example, would be in violation of FDA, EU, and Japanese standards since ground iron oxide ores have enough lead, mercury, arsenic, cadmium, and so on in them to keep Proposition 65 lawyers in California busy filing lawsuits forever. Natural does not mean safe.

Synthetic duplicates are substances that share identical chemical structures and pharmacological properties with their natural counterparts; an example of such substances is vitamin E and DL-alpha-tocopherol.

A semisynthetic substance may also be acceptable as an NHP, provided that it shares identical chemical structures and pharmacological properties with its natural counterpart. Semisynthetic substances are produced by processes that chemically change a related starting material that has been extracted or isolated from a plant or a plant material, an alga, a fungus, or a nonhuman animal material. An example of such substances is ginsenosides, which are produced from the starting compound betulafolienetriol.

Natural Standards (NAFTA) Reference: All products that are labeled or branded "natural" *must*

- be made with *at least 95%* natural ingredients;
- contain only synthetic ingredients specifically allowed under these standards and environment-friendly products that are nurturing to us and as harmless as possible to the earth:
  - Positive list of natural Ingredients.
  - List of prohibited ingredients.

Other requirements:

- For each ingredient, the substance is listed as generally recognized as safe (GRAS) by FDA when used in accordance with FDA's good manufacturing practices (GMP) and contains no residues of heavy metals or other contaminants in excess of tolerances set by FDA or EPA or has been reviewed using criteria in this standard.
- Companies must be transparent, fully disclosing their ingredients accurately and truthfully.
- Companies should strive to maximize their use of recyclable and post-consumer recycled content in packaging.
- No animal testing of ingredients or products.

In the next major phase of implementation, currently planned for May 1, 2010, the Natural Standard will look to, among other things,

- Disallow the temporary initial phase allowance of synthetic ingredients, including synthetic fragrance.
- Require formal INCI nomenclature to describe all cosmetic ingredients.
- Have on-site auditing capabilities.
- Harmonize with other global standards.

Below are some additional organizations worth mentioning and I encourage the interested reader to become familiar with

Whole Foods: http://www.wholefoodsmarket.com/products/premium-body-care.php.

Oasis: http://www.oasisseal.org.

BDIH: http://www.kontrollierte-naturkosmetik.de/e/index_e.htm.

ECOCERT **:**http://www.ecocert.com.

Soil Association: http://www.soilassociation.org/certification.

ICEA: http://www.icea.info.

NaTrue: http://www.natrue.org.

2.4 THE FORMULATOR

The challenge in formulating natural products is to increase the level of green ingredients in formulations without compromising performance. Often, making the formulation greener means the performance may be negatively influenced because the necessary ingredients are not available yet.

Faced with these challenges, it is very crucial that formulators educate their marketing team on how natural green product can actually be made since there is a lack of specific ingredients such as UV protection agents and cationic conditioning agents. Marketers should be aware of the fact that some products with a high level of natural ingredients cannot deliver the same performance. The question formulators should impose when formulating is "How natural do you want your product to be and more importantly to be clear on how this formulation will impact the performance of your final product." Theoretically, it is possible to produce a completely natural product, at least in some product categories. But in practice, the challenge to deliver the aesthetic and functional performance on economically justifiable terms still remains.

Marketers will need to identify which market segment they would like to offer their product to: consumers not willing to make any sacrifice when living their green lifestyle, and do not mind compromising on high-class product performance, or consumers who do not take greenness so seriously and their priority lies on good product performance.

Most ingredient suppliers generally conduct the research on the botanical or herbal ingredients they are representing. As a formulator, one might want to carefully select the botanical ingredient for their specific, desired beneficial effect based on scientific research and traditional medical knowledge founded on ethnobotany. A number of resources exist that can assist in determining what activities the primary ingredients may have, such as the Natural Medicines Comprehensive Database.

Formulators should support only those products that have renewable ingredients with sustained harvesting since some of the natural ingredients are obtained by wild crafting, that is, harvesting wild plants for commercial use. Thus, skincare professionals should support only those products that have renewable ingredients sourced from sustainable harvesting to prevent the plant species from becoming endangered due to unregulated overharvesting [10, 11]. A listing of these can be found with the Convention on International Trade in Endangered Species of Fauna and Flora at www.cites.org.

Botanical extracts are also much more susceptible to quality variations than synthetic products due to more factors affecting solubility, stability, pharmacokinetics, pharmacology, and toxicity of the active ingredients. Some of these factors include

- growing conditions of the plant;
- harvesting time and care of botanical products during transport and storage;
- method of processing the plant for extraction of active ingredients;
- in order to be of the highest quality, a plant must be healthy and disease-free [12].

A natural or organic product does not always implicitly mean a safe product, either. Botanical-based products usually provide multiple functionalities and stable formulations of highly reactive ingredients. While it can be notoriously difficult to create efficacious and stable commercial formulations, botanicals containing multiples of these ingredients already exist in nature in stable mixtures that can be extracted.

For example, one of the world's most deadly bioterrorist products is ricin. It only takes about a teaspoon of ricin to kill up to 40,000 people if ingested via a city's water supply or inhaled through a heating and cooling system. However, ricin is actually an extract from the castor bean, and another extract from the castor bean is azelaic acid, which is a top ingredient for treating rosacea. So this one plant produces two very different extractable compounds with vastly dissimilar functionalities and safety. This example illustrates why skincare and spa professionals need to be aware of and knowledgeable about what ingredients are present in their products, as well as the ingredients' safety and efficacy.

Lubrizol's range of vegetable-derived methyl glucoside derivatives offers natural ingredient choices for the discriminating formulator. These ingredients offer formulation enhancements such as enhanced emolliency, improved emulsification, humectancy, moisturization, and thickening.

Chattem's Hamposyl product line of sarcosinates is positioned to address the market requirements, as the range is nontoxic, readily biodegradable, and sulfate-free.

Marketers want to get away from petro-derived and sulfated surfactants; however, they want the replacements at the same cost per actives, which is hard to achieve. RITA Corporation offers lactylate and coco-glucoside technologies that are eco-certified, natural-based surfactants.

Many natural colorants have been around for a very long time and are quite commonly used in cosmetics and personal care products. But with the industry moving toward the natural category, the list that includes annatto, beta-carotene, carmine, chlorophyll, and a few others is quite small to produce the wide range of colors that are required for the color cosmetic market segment.

Several companies, such as LCW/Sensient Technologies and Alban Muller, claim to have color products that are all-natural and can be used in a variety of cosmetic applications. Some other companies are now becoming very active in the search for all-natural colorants that really have a wide pallet and go far beyond what is currently available and permitted. BioOrganic Concepts has acquired representation for a line of natural colorants that rival the synthetic colorants in terms of color variety and function. The problem is that all these new ingredients are not permitted in the United States as color additives. The formulator will again have to be creative and use the various extracts (many of these with exotic names) into the formulation as an added extract since the amount used is usually quite small to produce the desired color, but for color cosmetics we would have to await the permission to use them.

Products made of 100% minerals are natural but not if they contain iron oxides, titanium dioxide, bismuth oxychloride, zinc oxide, and so on as minerals. These ingredients although made from natural starting material are synthetic and therefore not natural. Of the iron oxides, only hematite is natural but one cannot use it as a colorant because it does not meet the Code of Federal Regulations (CFR) criteria.

2.4.1 Choosing a Contract Manufacturer for Organics

Choosing the right contract manufacturer always hinges on the same fundamentals: assessing development and production capabilities, capacity, market savvy, and human "chemistry." Experience with a particular class of ingredients is generally not at the top of the list. But in the natural and organic product categories and during this economy, brand managers would be wise to reconsider their priorities.

"The soaring popularity of organic personal products looks a lot like other market trends we have all seen over the years," says Lynn Betz, cofounder of Sensibility Soaps. "But it's not. Organics require a fundamentally different approach to formulation and manufacturing. It's more intuitive and experience-based. This shift is often quite hard for internal R&D and production managers to accept—especially among chemists who have worked only with conventional products. So, the first step is to start thinking unconventionally."

Below are Betz's recommendations in searching for a contract manufacturer for organic products. Her order of priorities reflects her perception of the special needs of organic product development and production.

1. Assess the contract manufacturer's expertise and depth of experience working with natural and organic ingredients.
2. Meet the team in person. Meet all the key people you'll be partnering with, and make sure they share your passion for organics and for satisfying the needs and desires of your customers.
3. Evaluate the company's understanding of the organic market in your target demographics, as well as the whole competitive landscape—with special emphasis on the mass retail channels where sales, due to current economic realities, will likely be strongest.
4. Make a site visit, and be sure to check out the equipment in the laboratories and manufacturing plants. Look for signs of innovative thinking geared toward fast production, superior product quality, and competitive strength at retail.
5. Evaluate the company's quality program from top to bottom.

REFERENCES

1. Organic Monitor. www.organicmonitor.com.
2. Increased market segmentation in European natural cosmetics market, GCI, October 2009.
3. Natural and Organic Personal Care Report 2008. *Nutr. Bus. J.* 2009.
4. Beyer A. *The Age of Naturals*. The Benchmarking Company, 2009.
5. Steinberg DC. Ingredient review: animal vs. vegetable, a continuing controversy. *Skin Inc.* 1999;11 (3):58–62.
6. Natural Health Products Regulations, Health Canada, 2009. www.hc-sc.gc.ca/dhp-mps/prodnatur/legislation/acts-lois/prodnatur/index-eng.php
7. Natural Products Association. www.naturalproductsassoc.org
8. Illustrative "Positive List" of Ingredients, Natural Products Association, 2009. www.naturalproductsassoc.org/site/DocServer/Natural_Ingredients_List.pdf?docID=7341
9. Cosmetics Organic and Natural Standard. www.cosmos-standard.org, http://www.cosmos-standard.org/docs/Cosmetics_Organic_Standard_Consultation.pdf
10. Cornuelle T, Lenhart J. Topical botanicals. In: Draelos ZD, Thaman LA, editors. *Cosmetic Formulations of Skin Care Products*. New York: Taylor & Francis, 2006. pp. 299–308.
11. Chamberlain J, Bush R, Hammett A. Non-timber forest products. *Forest Products J.* 1998;10–19.
12. Skidmore-Roth L. *Herbs and Natural Supplements*, 3rd ed. Philadelphia, PA: Elsevier Mosby, 2006. pp. xiii, xvii.

CHAPTER 3

THE MECHANICS OF DEVELOPING A NATURAL/ ORGANIC NONGOVERNMENTAL COSMETIC STANDARD

Amy B. Olson, PhD

Washingtonville, NY

3.1 THE RISE OF NONGOVERNMENTAL STANDARDS

Over the past several decades, there has been an increased influence that led to the use of nongovernmental standards to establish credibility for products marketed with a variety of environmental and social attributes. These standards seek to establish criteria to define a given claim within the marketplace. Nongovernmental standards have emerged from nonprofits, nongovernmental organizations (NGOs), industry, or trade groups [1]. These standards have emerged as a result of a combination of factors, including a lack of state authority, consumer expectation, and industry drive for legitimacy.

As the world increasingly transitions to a global economy, the jurisdiction of national laws is diminished [2]. Nongovernmental standards present a mechanism for establishing common rules of marketing across borders. Such standards have emerged for sectors involved in global trade, such as forestry, fishing, ecotourism, fair labor, and fair trade commodities. Organic and natural personal care products are another sector that has entered the realm of nongovernmental standards. With an increase in the development, evolution, and application of these standards, it is important to understand the underlying factors that can contribute to or detract from their success in the marketplace. This chapter will seek to identify and provide context for these factors in order to assist the marketer and standard making body in avoiding some of the challenges that can detract from a standard's success.

There is a long history of standard development across a variety of disciplines from product manufacturing to analytical methodology. Certain commonly recognized standardization bodies, such as the International Standardization Organization (ISO), European Committee for Standardization (CEN), and American National

Formulating, Packaging, and Marketing of Natural Cosmetic Products, First Edition. Edited by Nava Dayan and Lambros Kromidas.

Standards Institute (ANSI) have been developing voluntary standards for a range of processes and products for some time [3, 4]. These are well recognized as industry standards in many regulated and self-regulated industries [5]. As an international organization, the ISO is perhaps the most well-known example of establishing voluntary standards [6]. ISO competes with national standardizing bodies such as CEN and ANSI to gain international acceptance of their many industry technical standards and protocols. These organizations have achieved legitimacy and are recognized as authoritative standard making bodies.

However, the market is demanding standards related to more social or environmental-based issues, areas in which standard making bodies such as ISO, CEN, and ANSI may have less experience. Until recently, these standard making bodies had remained relatively free from competition [3]. Now, industry-specific initiatives are emerging to address various social and environmental issues.

3.1.1 Theory of Nonstate Market-Driven Governance

Cashore studied the phenomenon of nongovernmental standards driven by market forces and termed them "nonstate market-driven" standards or NSMD [7]. Cashore's studies in the forestry sector investigated how market forces and other factors contribute to the success of a standard. In 2004, Cashore, Auld, and Newsom (CAN) authored a framework entitled the CAN Explanatory Framework. This framework sought to identify causal factors and criteria impacting the success of NSMD initiatives. CAN's work originated from observation and study of the forestry sector. In practice, the framework provides a tool to help manage the challenges related to implementation of a successful NSMD initiative in any social or environmental area [7].

The CAN Explanatory Framework is the only framework that comprehensively attempts to explain how NSMD governance systems gain authority in the marketplace [7]. The framework evolved from several academic disciplines. The authors drew from political science theories and literature [8]. One framework considered by CAN is the advocacy coalition framework [9]. This framework is used by policy makers to resolve issues related to technical disputes and goal conflicts in environmental policy. Although the principles of this framework are relevant in resolving conflicts between competing NSMD standards, the advocacy coalition framework relies too heavily on public precepts to be applied to NSMD governance.

In developing a mechanism to explain NSMD, Cashore drew heavily on Suchman's legitimacy research from organizational sociology [10]. The CAN Explanatory Framework uses concepts of pragmatic, moral, and cognitive legitimacy to lay the groundwork for explaining how NSMD governance gains legitimacy. With a heavy focus on sociology, the CAN Explanatory Framework is ideal for application to resolving environmental and social issues on a global scale [8].

3.1.2 CAN Explanatory Framework

Cashore et al. developed 10 hypotheses based on their extensive literature review (Table 3.1). The hypotheses predicted factors that impact a standard's likelihood of

TABLE 3.1 CAN Explanatory Framework Hypotheses [7]

CAN Factor Themes	CAN Hypotheses	Whether Promotes or Hinders the Success of a Standard
Sector's place in the global economy	H1: reliance on foreign imports	Promotes
	H2: reliance on foreign exports	Promotes
Structure of the sector	H3: presence of large concentrated companies	Promotes
	H4: unfragmented industry	Promotes
	H5: presence of cohesive trade associations	Hinders
History on the public policy agenda	H6: presence of public dissatisfaction with industry practices	Promotes
	H7: shared access to relevant agencies between industry and nonindustry stakeholders	Promotes
Additional themes	H8: cohesion of the supply chain	Promotes
	H9: independence of companies from one another	Hinders
	H10: presence of competitor standards	Hinders

gaining acceptance and legitimacy in the marketplace. CAN grouped the 10 hypotheses into three major categories [8]. The first category contains factors related to the sector's place in the global economy. These hypotheses proposed that a reliance on imports and exports of a particular product makes it more likely for a standard to be adopted. The second category of factors is the structure of the sector. The hypotheses within this group proposed that independent behavior of individual companies and cohesion of trade associations can have a role in facilitating or debilitating the acceptance of a standard. The third group of hypotheses deals with the history of forestry on the public policy agenda. These hypotheses relate to whether there was public dissatisfaction with the way the forestry sector was being managed and what role the government played. Three additional hypotheses were added as a result of the authors' data collection (Table 3.1) [8]. These hypotheses delve into the role of the supply chain and the competition between emerging standards. In their original framework, Cashore et al. tested each of these hypotheses related to the forestry sector in Canada, the United States, the United Kingdom, Germany, and Sweden [8]. They later tested the hypotheses in Finland and made some revisions to the framework based on their findings [11].

In addition to the 10 factors, the framework delineates three discrete phases of NSMD governance [11]. Phase I is characterized by initiation of an NSMD system. Phase II begins with the proliferation of alternative standards and ends with widespread support of one or more approaches. Phase II is characterized by a standardizing body attempting to inform potential supporters about the standard and gain their support. During this time, if informing and educating about standard do not lead to support, the standardizing body may lower the criteria within the standard to entice additional supporters [12]. Phase III is referred to as political legitimacy, in which a standard is fully recognized as authoritative within the marketplace. Phase III is the ultimate goal of NSMD initiatives within the CAN Explanatory Framework [13]. Natural and organic cosmetic standards are currently in Phase II, with multiple competing standards vying for legitimacy in the marketplace.

3.2 NONGOVERNMENTAL STANDARDS FOR NATURAL/ORGANIC COSMETICS

The organic and natural cosmetic sector has become a forum for the promotion of nongovernmental standards in the past decade. This is due in part to the growing global market in organic and natural consumer products. The leading organic and natural markets are the United States and Europe, although there is a growing interest in Latin America and Asia. The trend initiated in the food industry and has spread to various other agricultural-based sectors.

The growth of organic farming practices has been an important movement toward the development of a sustainable future in agriculture. Consumers are also increasingly demanding more sustainable agriculture with less pesticide use [14, 15]. Researchers have shown that organic foods are more nutritious, containing higher levels of vitamin C, iron, magnesium, and phosphorus and 15% less nitrates than conventional crops [16, 17]. National organic programs have been established in some markets, including the United States and Europe, for foods. While laws exist for production standards exclusive to organic food, other downstream industries are seeking organic options for consumers.

NSMD systems are increasingly prevalent in circumstances in which the government chooses not to take a formal position on a particular issue [7]. This has been the case with organic and natural personal care products, leading to the development of multiple voluntary standards and certification schemes for organic, natural, and a combination of both [2]. To date in the United States, no governmental agency has been willing to regulate the terms *organic* or *natural* for personal care products. Organic food products have been regulated by the U.S. Department of Agriculture (USDA) since 2002 [18]. From that time on, market interest has grown significantly for organic products. Other industries, such as personal care, have started to incorporate organic claims into their products. This is in response to consumers' growing concern regarding the chemicals used in cosmetic products [19].

Consumers' desire for organic and natural cosmetic products is the basis for development of the latest NSMD initiatives. Consumers who have embraced an organic lifestyle are looking to use organic and natural products in all areas of their lives [14]. The trend in NSMD standards for consumer products appears to be growing in niche marketing sectors in which companies need to ensure that environmental product claims are substantiated in a transparent and legitimate way. Adoption of voluntary standards is a way to substantiate those claims.

The status as an NSMD initiative is warranted for these organic and natural standards because those standards have been developed for third-party certification, independent of any governing body or authority [2]. Furthermore, as consumers have decision-making power in the purchase of a certified organic or natural cosmetic product, this constitutes a market-based initiative that may ultimately gain momentum and legitimacy through consumer understanding and acceptance.

3.2.1 The Five Key Features of NSMD for Organic and Natural Personal Care Products

Cashore et al. identified five key features of NSMD governance. These key features are nonstate authority, the role of stakeholders, social domain, the role of the market, and legitimacy or enforcement. This section briefly reviews each of the features in the context of organic and natural personal care.

The situation surrounding the emerging standards for organic and natural personal care products fits within the NSMD paradigm by originating from nonstate authority. The standardization bodies for each of the proposed organic and natural standards are nonstate actors. None of the standards are legally binding. The FDA in the United States and European Commission in Europe that have jurisdiction over personal care products have no definitions of the terms organic and natural [20]. No public agency in any market currently has the authority to enforce any one of the nongovernmental personal care standards.

The second feature of NSMD governance focuses on the role stakeholders play in developing standards. A number of stakeholders are involved in the debate over organic and natural cosmetic standardization globally. These stakeholders range in perspective based on their affiliations. Those involved in the NSMD process for organic and natural standards include industry leaders, trade associations, third-party independent labelers, consumer advocacy groups, and organic ingredient suppliers [2].

The social domain is important in the context of organic and natural cosmetics. Several companies were engaged in a lawsuit with Dr. Bronner's Magic Soaps. The public perception of how an organic and natural personal care product is defined could have played a significant role in the decisions made by the judges as the court case progressed. It appeared that Dr. Bronner's and similar supporters were attempting to cajole and coerce industry through litigation into complying with the NSF or USDA organic standard. Neither the NSF nor the USDA NOP are currently legally binding for cosmetics.

The role of the market is important to the situation with organic and natural cosmetics. Currently, there are products on store shelves that have several different symbols identifying them as certified organic or natural. In the United States, these are the USDA, NSF, NPA, OASIS (Organic and Sustainable Industry Standards), and EcoCert. In Europe, they are EcoCert, CosmeBio, NaTrue, ICEA, UK Soil Association, BDIH, and BioGaranti. The public has the opportunity to "vote" for a particular standard with their buying power. Those standards that are viewed as the most desirable by consumers may have a stronger chance of gaining long-term legitimacy. However, it is also possible that the quantity of symbols could lead to symbol fatigue, decreasing the credibility of any one standard.

Legitimacy plays a huge role in the organic and natural personal care product arena. The current lawsuit between Dr. Bronner's and other certifiers and the media efforts by the Organic Consumers Association (OCA) are a form of public shaming. In theory, each voluntary standard is considered legitimate in its own right, regardless of the differences between them. However, Dr. Bronner's and the OCA have been attempting to obstruct legitimacy of all other certifying bodies that do not comply with the NSF or USDA requirements.

3.2.2 Overview of the Existing Standards Across United States and Europe

The personal care industry has responded to growing consumer interest by introducing cosmetics making organic and natural claims into the marketplace. Some of these products claim to be 100% organic, organic, made with organic or natural, while others contain only small quantities of organic and natural ingredients and make similar claims [21, 22]. Some are certified to standards and some are not.

Several stakeholder companies in the United States that were using high percentages of organic ingredients decided to take action to protect their organic market share by developing their own standards. This action came from the desire to effectively substantiate their organic personal care product claims in the absence of formal regulation [21]. The situation is an example of NSMD governance, whereas Cashore (2002) asserted that the industry "derived their policy-making authority not from the state, but from manipulation of global markets and attention to consumer preferences" ([7], p. 504). Two competing groups in the United States simultaneously developed contradictory organic standards for personal care products: one from the NSF, entitled "NSF Standard for Organic Personal Care Products" [23] and the other from OASIS entitled "Organic Production Standards (for Health and Beauty Products). A Voluntary Standard" [24]. The Natural Products Association also produced the Natural Cosmetics Standard [25].

The only legally binding legislation directly related to organic personal care products is found in California's Health and Safety Code, Article 7, The California Organic Products Act of 2003 [26]. This regulation stipulates guidelines for organic cosmetics. Specifically, the regulation requires a minimum of 70% organically produced ingredients for cosmetics to make a claim of organic or made with organic [26]. This differs from the requirement of 95% content required by the USDA NOP [18]. The regulation also defines how the total calculation of the organic percentage is to be determined. However, the regulation does specify the same parameters as the USDA, prohibiting any reference to organic when the total percentage of the product is less than 70% except in the ingredient listing.

Europe and the United States have different positioning of natural and organic products. In the United States, natural and organic are seen as separate claims, with different formulation philosophies. In Europe, organic and natural standards are interwoven. Many of the European standards present a combination of organic and natural ingredients and percentages, such as EcoCert, CosmeBio, COSMOS, and NaTrue. In Europe, consumers see organic and natural as conjoined, and therefore almost all the European standards combine organic and natural into the same standard [27, 28]. In contrast, in the United States, organic and natural products each have their own discrete standards [23–25].

There are numerous organic and natural NSMD standards across Europe. These include EcoCert, BDIH, CosmeBio, COSMOS, EcoGarantie, NaTrue, ICEA, and UK Soil Association. Furthermore, Colipa, the European Cosmetic Trade Association, proposed definitions for the terms organic, natural, and naturally derived to be applicable to cosmetics [2]. These definitions were presented to the ISO in 2009 and at the time of publication are under consideration for global adoption [29].

3.2.2.1 USDA National Organic Program Organic products are clearly regulated for foods. In 2002, the USDA Organic Foods Standard was published, passing organic standards for foods into law. This NOP is a marketing standard that established minimum content labeling requirements as well as definitions and processes for organically farmed agricultural products [18]. Only agricultural foods fall within the scope of the program.

The USDA NOP for foods specifies that products claiming organic must have a minimum content of 95% organic. To claim 100% organic, there cannot be any nonorganic materials used. To make a claim that a product is made with organic, the product must contain a minimum of 70% organic materials. Any use of the claim organic in food products with less than 70% content is prohibited, with the exception of listing the organic ingredient as such in the ingredient list [18].

To date, the USDA has not regulated the term organic for personal care cosmetics in the United States. The USDA, which holds jurisdiction over the term organic, has been unwilling to enfold cosmetics officially within the agricultural standard [30]. The Food and Drug Administration (FDA) holds jurisdiction over cosmetics. However, the FDA has declined to participate in a rulemaking process on organic or natural cosmetics.

Dr. Bronner's Magic Soaps, along with other cosmetic product manufacturers, began developing organic personal care products compliant with the USDA standard, even though it was not directly applicable to personal care products. However, these products could not originally bear the USDA seal, as they did not fall within the scope and definition of the jurisdiction of NOP or USDA. In 2004, the USDA stated that personal care products were not under their jurisdiction [31]. Following this statement, Bronner, president of Dr. Bronner's Magic Soaps, sued the USDA to allow use of the USDA seal on the products Bronner was producing that legitimately met the stringent requirements of the USDA NOP. The USDA conceded in response to the lawsuit, admitting that the USDA seal could be used if the standard and certification requirements were met. This did not constitute an inclusion of personal care products in the scope of the regulation, as some have argued. The USDA clearly issued a disclaimer that although personal care products that met the USDA NOP could bear the USDA label, they fall under the jurisdiction of the FDA and would not be formally included in the USDA standard [30]. In this statement, the USDA (2008) affirmed that

> *USDA has no authority over the production and labeling of cosmetics, body care products, and personal care products that are not made up of agricultural ingredients or do not make any claims to meeting the USDA organic standards. Cosmetics, body care products, and personal care products may be certified to other, private standards and be*

> *marketed to those private standards in the United States. These standards might include foreign organic standards, eco-labels, earth friendly, etc. USDA's NOP does not regulate these labels at this time.* (*Source*: Ref. 27, p. 2.)

To use the USDA organic seal on cosmetics, the product must contain 95% organic ingredients for the claim organic or 100% organic in which only organic ingredients are used. The organic percentage of a product excludes water and salts from the calculation. Despite the seeming acceptance of personal care products into the USDA NOP by allowing use of the seal, the USDA has no statutory authority for enforcement of claims on personal care products [30]. This asserts the use of and certification to the USDA organic standard as another NSMD option for voluntary certification.

This statement was issued in April 2008, following the publication of two other organic personal care standards. The USDA's comment confirmed the industry's perception that the USDA would continue to decline formally including personal care products in the NOP. On the basis of the communication from the USDA, only personal care products that bear the USDA organic seal fall within their jurisdiction. The USDA publicly confirmed that industry was free to reference other private organic personal care standards [30].

In 2009, the Certification, Accreditation, and Compliance Committee of the National Organics Standards Board (NOSB) that advises the USDA published a discussion document recommending that the USDA NOP fold personal care products into the existing food standard [32]. The suggestion came from an assertion that products on the market are mislabeled as organic [43]. The USDA considered the proposal at a public meeting in 2009 [33]. To date, the USDA has not chosen to follow the recommendations of the NOSB and regulate the term organic for personal care products [33]. If it chooses to do so in the future, this sector will cease to allow NSMD organic standards in the United States.

3.2.2.2 NSF/ANSI Standard for Organic Personal Care Products In lieu of formal regulations, the organic personal care market has seen the emergence of products ranging from 0.01% organic to 100% organic making the claim [34]. Furthermore, the USDA has continued to resist acceptance of personal care products within the NOP for food. This led to the publication of an organic personal care standard by the NSF in 2007. NSF is an accredited third-party certification body. NSF standards are a precursor to the ANSI standards [31]. This certification body was chosen to develop an organic personal care standard because of its strong reputation in consensus-based standards development. NSF/ANSI standards follow protocols to ensure due process, balance of interests, and consensus as well as an appeals process [31]. Stakeholders that participated in the standard making process with NSF include Dr. Bronner's, Aveda, and several other small marketers of organic personal care products.

The original standard prescribed that to make a product claiming 100% organic, it must contain 100% organic agricultural ingredients, after subtracting the water and salts from the calculation. To claim organic, the product must contain at least 95% organic agricultural ingredients [23]. Products claiming made with organic must meet a minimum content of 70% organic. NSF further changed the standard midway through 2008, removing the provisions for 100% organic and organic claims. The standard now covers the made with organic product claim, requiring a minimum content of 70% organic ingredients. The revised NSF standard defers to the USDA NOP for all higher level claims of organic (95% minimum) or 100% organic [34]. NSF (2009) officially announced that ANSI had adopted the NSF/ANSI 305: Made with Organic Personal Care Products standard, identifying itself as the only consensus-based standard for organic cosmetics.

3.2.2.3 Organic and Sustainable Industry Standards In March 2008, another group issued an organic personal care standard. OASIS is a nonprofit mutual benefit organization, also known as a trade association. The bylaws require that the members participate, vote, and direct the goals of the group. Founding members include Esteé Lauder Companies, Inc., L'Oreal, and many other smaller organic personal care manufacturers and suppliers of organic raw materials.

OASIS is an organization whose purpose is to support standard development for organic and sustainable health and beauty consumer goods. OASIS has identified the concept of sustainable as covering a broader spectrum, in which organic is one measurement. OASIS is working to develop standards that will measure different types of sustainable practices, including organic production and sustainable packaging for the personal care category, and intends to eventually include household cleaners and other products [24].

The OASIS standard calls for an incremental approach to the organic claim with at least 85% organic content from allowed chemical processes. The 85% content was allowed until January 2010, at which point certified products were expected to increase the organic content to 90%. A 95% minimum content is expected in 2012 [24]. This standard allows certain chemical processes, such as hydrogenation and sulfation deemed important to product aesthetics and stability in personal care products. Unlike food, which is expected to have a short shelf life (1–2 weeks for most perishables), personal care products are expected to maintain integrity for months, if not years, of use. For this reason, the OASIS group has developed a standard that takes such technical formulation challenges into account [35].

3.2.2.4 Natural Products Association The NPA was founded in 1936 as a nonprofit organization dedicated to supporting the natural products industry [36]. A third-party survey conducted revealed that consumers were very confused about the regulation of the term natural. A total of 78% of respondents thought the term was regulated for personal care products, and 83% believed there ought to be only one meaning for the term natural [37]. The NPA decided to develop a natural standard with the assistance of its member companies and provide certification services to ensure that natural product consumers could easily identify natural products that conformed to a defined standard [37].

NPA defines natural as "ingredients that come from or are made from a renewable resource found in nature (flora, fauna, mineral), with absolutely no petroleum compounds" ([25], p. 1). The standard requires a minimum content of 95% natural ingredients. The other 5% cannot be composed of any ingredients specifically disallowed under the standard. The disallowed ingredients include a number of ingredient categories suspected to have potential health or environmental risks, such as parabens, phthalates, and ethoxylated materials [25]. NPA has established an incremental standard, hoping to ratchet up the requirement to eliminate temporarily allowed synthetics by 2010.

3.2.2.5 COSMOS Standard The COSMOS standard represents the most recent attempt at harmonization of organic and natural standards in Europe. The standard was developed as a collaboration of a number of natural and organic standards that had been leading in various member states. These include BDIH (Germany), BioForum (Belgium), CosmeBio and EcoCert (France), ICEA (Italy), and UK Soil Association [27]. The COSMOS standard identified the purpose of the harmonized effort as "to define minimum requirements and common definitions for organic and/or natural cosmetics" ([27], p. 1).

A new seal was not created by COSMOS for the new harmonized requirements. Instead, each of the EU member state-based certifying bodies will continue to certify the product and use existing seals [38]. This allows the continued recognition of the seals that have had longstanding respect in each country, yet cross-regional understanding of a common set of requirements. To date, COSMOS is the only organization to successfully harmonize requirements of multiple standardizing bodies.

3.2.2.6 NaTrue Standard The NaTrue standard was developed in 2008 based on recognition of the complexity of cosmetic products. The NaTrue standard points out that unlike foods, cosmetic products are made up of many complex substances that are impractical in their unmodified form [28]. It is essential to chemically or physically modify natural starting materials to achieve a desired aesthetic [28]. Therefore, the standard clearly defines the terms natural, nature-identical, and nearly natural raw materials as acceptable within NaTrue certified products.

NaTrue is the first third-party labeling organization to establish overseas negotiations with U.S. standards bodies [28]. At present, NaTrue has negotiated a mutual recognition of symbols with the NPA Natural Standard and the NSF organic standard [39]. The mutual recognition will allow market access for products certified to either standard in the United States or Europe.

3.2.3 Role of the Federal Trade Commission and Green Guides

All consumer products marketed in the United States fall under the jurisdiction of the FTC. The FTC ensures that consumer products are marketed in a truthful and nonmisleading manner. In the early 1990s, the FTC developed the Green Guides, which are environmental marketing guidelines. The guides cover how to substantiate an environmental claim and discuss more generally how to avoid overstatement of environmental attributes. Guidance is provided on how to properly qualify claims to avoid deception and how to handle comparative claims for a product that has been improved with an environmental attribute [40]. FTC requires that all claims be properly qualified and substantiated. FTC states that

> *All marketers making express or implied claims about the attributes of their product, package or service must have substantiation, that is, a reasonable basis for their claims. When it comes to environmental claims, a reasonable basis often may require competent and reliable scientific evidence, which is defined as tests, analyses, research, studies or other evidence based on the expertise of professionals in the relevant area conducted and evaluated in an objective way by qualified people using procedures generally accepted in the profession to yield accurate and reliable results.* (*Source*: Ref. 42.)

Under the various proposed NSMD standards, no products meeting a standard could be considered negligent of this FTC requirement to substantiate the organic or natural claim. The adherence to one or more standards presents substantiation. The purpose of having a standard is to establish credibility and product parameters against which consumers can compare products and standards. If a standard existed that consumers could reference prior to purchasing a product, a company making such a certified claim would meet its obligations under FTC to adequately substantiate [2].

In 2010, the FTC reissued the Green Guides to address concerns related to new environmental claims that have surfaced since the last update. In this reissue, FTC proposed a position on the NSMD standards to quality environmental claims. FTC (2010) advised marketers

> *. . . not to use unqualified seals or certifications. Marketers should accompany seals or certifications with clear and prominent language limiting the general environmental benefit claim to the particular attribute or attributes for which they have substantiation.* (*Source*: Ref. 42, p. 60.)

Furthermore, FTC specifies that in the event a third-party certification is used, the marketer must follow the FTC Endorsement Guide and specifies that certifications by trade associations which the marketer is a member of may be deceptive. FTC stated that

> *. . . as explained by the Endorsement Guides, marketers are required to disclose a "material connection" or a "connection between the endorser and the seller of the advertised product that might materially affect the weight or credibility of the endorsement." [41]. Accordingly, this example makes clear that the marketer's failure to disclose its material connection with the endorsing association, i.e., that it is a dues-paying member of the endorsing association, is deceptive.* (*Source*: Ref. 42, p. 60.)

FTC did assert that third-party certifications can be an effective means of environmental claim substantiation provided they have adequate scientific evidence to prove the claim. Despite comments requesting further FTC oversight on developing and evaluating third-party certifications, the revised draft of the Green Guides does not present any more active role in the NSMD process than previously described [42].

3.3 HOW TO BUILD AND EFFECTIVELY MARKET PRODUCTS UNDER NONGOVERNMENTAL STANDARDS

The factors identified by Cashore et al. presented a useful tool for the planning, development, and implementation of a nongovernmental standard [7]. This work was added to by Olson in evaluating the CAN Explanatory Framework against the organic and natural personal care sector [2]. The following section presents a practical tool for implementing the findings in the organic and natural cosmetic sector.

3.3.1 Inputs for Success

There are various factors that can contribute to the success of an organic or natural personal care products standard. Understanding and applying the inputs for success can help to increase a standard's legitimacy in the marketplace. The CAN Explanatory Framework identified those factors that would contribute to the success of the standard [7]. Olson evaluated whether the factors were relevant to the organic and natural cosmetic sector and suggested additional factors specific to the sector [2]. The following section presents those factors that were shown to contribute to the success of a standard.

3.3.1.1 Imports and Exports The personal care industry is extremely dynamic and continues to increase marketing products across international borders. The CAN Explanatory Framework identified that reliance on imports and exports facilitates the acceptance of a standard. This was hypothesized based on a reliance on the imported or exported product seen in the forestry sector [7]. The international movement of organic and natural personal care products is still growing, and as such there is not yet a reliance on imports and exports by industry. For personal care products, the driver for certification seems to be more related to a company's desire to increase the marketability and legitimacy of its organic or natural products. The current exports and imports are a marketing expansion. Despite this difference from what CAN hypothesized, imports and exports do play a role in standard legitimacy.

Some companies may experience greater market acceptance by certifying to a standard. Manufacturers may choose to certify their products to a standard of notoriety in hopes of gaining support for their brand in a new region. A standard used on products in multiple international markets is more likely to become recognized and accepted by consumers than a relatively unknown standard completely new to the market. For standard making bodies, it is important to promote a standard across multiple countries to increase global recognition. The standard making body should not rely only on the appearance of its seal on certified products to educate consumers.

It is possible that some companies that choose to align with a single standard may face market resistance if the standard they chose is not in line with national expectations. Before choosing a standard, it is important for the marketer to understand the local standards and market expectations in each country where a product is to be sold. For example, in the United States, a company choosing a standard must be aware of the legal and political risks before aligning with a particular standard. In Europe, it is important to understand the history of organic and natural personal care standards as many countries have a nationally recognized standard that has already gained considerable market recognition. Entering the market with a foreign standard may present obstacles. It has been shown that increasing imports and exports are leading to discussions between standard making bodies to seek reciprocity instead of trying to claim market presence for national-based standards.

In summary, imports and exports can play a positive role in both growing the acceptance of an NSMD standard and establishing brand credibility [2]. Marketers who choose to certify products to a standard with global recognition and credibility is more likely to find market acceptance than an uncertified product. It is up to the marketer to consider the local impressions about existing standards, or in some cases a lack of standards, before deciding what approach to take.

3.3.1.2 State Disinterest in Regulation One positive input for success is state disinterest in regulation. When it is clear that there is no intention by state authorities to regulate, there is a clear path for NSMD initiatives to proliferate and evolve without threat of regulation. In the case of the EU, cosmetics were clearly excluded from the organic food regulations, making a clear path for the national standards, such as EcoCert and BDIH that have had a long uncontested standing in the European marketplace.

In the United States, despite public pressure for the NOP to include personal care products, it does not appear that personal care will be regulated in the near future. Miles McEvoy, the deputy administrator of the NOP, concluded the NOSB discussions on organic personal care by stating

> *I think it is a very important issue. It is important to protect the organic label. It is a very complex issue and it has some jurisdictional issues that we need to work out... It is not on the top 10 in terms of our priority list, it is certainly really important. We will certainly get around to addressing it sometime.* (*Source*: Ref. 34.)

While it presently seems unlikely to occur soon, if the USDA does decide to regulate the term organic for personal care products, the category will cease to have opportunities in NSMD governance. This will jeopardize the marketing status of existing organic products that carry a seal from a nongovernmental certifying body. It is important for marketers of organic products in the United States to understand the risk of certifying to an NSMD standard. In the event organic products are regulated under the USDA Organic Food Standard, products certified to EcoCert, NaTrue, and other private standards will not be in compliance and will face reformulation, relabeling, or market withdrawal. As the environment currently stands, market-based nongovernmental standards are safe to proceed in both the United States and Europe.

3.3.1.3 Cohesion Along the Supply Chain Cohesion along the supply chain is an important factor in the successful promotion of a standard. CAN's hypothesis suggested that concentrated ownership would increase the success of a standard. In the organic and natural personal care sector, there is diffuse ownership of raw material suppliers, manufacturers, and distributors, each with their own ideals about product and ingredient certification. Raw material suppliers certify ingredients to many different standards to maximize their business opportunities. This corroborates the CAN hypothesis, but for an expanded reason that CAN did not explicitly address. CAN suggested that the concentrated ownership along the supply chain would increase support for an NSMD approach. It has been observed that raw material suppliers are adopting and certifying to multiple standards, thereby creating the possibility for success of multiple NSMD standards.

The resulting menu of materials across multiple NSMD approaches sets the stage for firm adoption of any standard. Many raw material suppliers for natural and organic ingredients understand the diverse nature of cosmetic manufacturers. While some may be seeking a national-based standard, others may need the same materials to be marketed in products that will be sold in multiple countries, requiring multiple certifications. Suppliers who certify materials to multiple standards facilitate a marketing company to formulate with a single ingredient that can be marketed under multiple certification seals across multiple countries.

3.3.1.4 Transparency Olson showed transparency to be an extremely significant factor in the success of an organic and natural standard [2]. If consumers and manufacturers are not able to easily understand what the standard means, then it is unlikely to be successfully adopted by manufacturers. A standard must be transparent and objective. It must encapsulate the consumer's expectations and communicate them in a truthful and accurate way. Furthermore, the standard making body must make readily available to the public the criteria in advertising, labeling, and communications. NSMD initiatives will be beneficial only to their environmental or social cause with a clear, transparent, and well-marketed standard. Potential critics of the organic and natural NSMD initiatives would ask if the standards were transparent, if the information about the organization and origin of the standard is transparent, and if the board of directors is transparent. Consumers and consumer advocates want to ensure that the standard was developed with integrity for the cause. If the standard is not transparent, and cannot be measured by the consumer, it detracts from the ultimate cause of preserving the integrity of organic and natural products. In general, the consumer is unable to measure the percentage of organic or natural ingredients in the product. They must trust the label and the standard behind the label.

Not all the current organic and natural standards are found to be transparent, in that the criteria were not readily accessible or understandable by the lay consumer. Certain past standards maintained proprietary elements that prevented consumers from verifying the standard's criteria themselves. In some cases, understanding the criteria required purchase of the standard. Removing such obstacles and promoting transparency is essential to the success of any standard.

Another important factor is the use of strict, yet feasible criteria. The standard must be attainable for manufacturers of varying types of organic and natural products.

If the criteria are so strict that they are limited to only a few product forms or if attaining the criteria is financially prohibitive, very few companies will adopt it. Ultimately, if very few products can be certified to the standard, it is detrimental to the standard making body and the marketing of the product.

During the evolution process of a standard, there is a certain amount of converting and conforming that happens. The converting portion involves the recruiting of supporters and manufacturers to align with the standard. Conforming efforts are related to modifying the standard to reach a more attainable or market acceptable set of criteria. However, when standards are conforming to encourage further acceptance by a broader audience of manufacturers, it is important that criteria remain strict enough that the standard continues to be meaningful to the environmental or social cause. Standards that are relaxed too much compromise the integrity of the terms organic and natural. Furthermore, it is important that standard making bodies continuously strive to understand the market expectations and attempt to increase the stringency of standards over time as warranted. An example of this is the NPA standard that entered the market with an understanding that certain ingredients would be phased out at a certain point as industry was able to find technically feasible alternatives. Similarly, the OASIS standard set to increase the percentage of organic ingredients within 2 years of the standard's inception. These measures show that the standard making bodies recognize the technical challenges faced by industry to meet the criteria but promise consumers a stronger standard as manufacturers evolve formulation technology.

3.3.2 Factors Presenting Obstacles

In addition to observing which factors contribute to the success of the standard, it is important to understand what factors detract from the success of a standard. In many cases, the obstacles to success were the lack of the items previously identified as promoting a standard's acceptance in the market (e.g., transparency, consensus, attainable rules). However, there are several other factors related to the structure of the sector that have proved to detract from the success of any one standard in the organic and natural personal care sector. These include fragmentation among the industry, involvement of industry associations, independence of organic and natural product manufacturers, and competition between standards. The following section elaborates on each of these obstacles for success.

3.3.2.1 Fragmentation A challenge present in the organic and natural personal care sector is that significant fragmentation has driven stakeholders into different camps. Fragmentation has manifested in the evolution of varying viewpoints on the philosophy of what specific criteria and processes should be allowed. This fragmentation delays the process of any standard gaining success. With a divide in consensus about the criteria for organic and natural products, an environment has been created where there is an opportunity for a variety of alternate approaches and innovative solutions.

A major source of the fragmentation in the organic and natural personal care sector is the controversy between the USDA NOP certified products and alternate approaches. As discussed earlier, supporters of the NOP argued that organic consumers expect their products to meet the USDA criteria under all circumstances in which the term organic is used. OASIS and alternative NSMD approach supporters suggested that non-USDA NOP processes are required to achieve various complex emulsions and surfactant systems that are pivotal to creating a variety of personal care products. This philosophical difference of opinion continues to be a technical obstacle and source of fragmentation in the U.S. organic personal care sector. This issue is unlikely to be resolved effectively until the USDA takes a definitive stand as to whether personal care products will be folded into the USDA NOP.

In both the United States and Europe, the fragmentation has manifested in the creation of alternate approaches. With multiple approaches, there is now competition between NSMD initiatives that perpetuate industry fragmentation. The fragmentation has led to discussions among competitor standards and industry, which is a natural part of the NSMD process described by Cashore et al. Phase II of NSMD governance is depicted by a proliferation of alternative approaches that eventually convert and conform over time to evolve into a more broadly accepted consensus standard.

The organic and natural standardization process is behaving very typically as a Phase II NSMD situation. As such, it must be acknowledged that fragmentation has delayed the process and is likely to further delay the movement of any standard into the final NSMD stage of political legitimacy. Phase II will continue until there is more industry solidarity or state direction. This may mean that certain standards work toward merging with one another over time or seeking mutual international recognition.

For marketing companies, the best approach to manage fragmentation in the sector is to commit to certifying products to a standard that is most aligned with the company's needs. Sitting on the fence and avoiding commitment only perpetuates fragmentation.

3.3.2.2 Industry Associations Industry associations play a significant role in both supporting and hindering the success of standards. While the previous factor discussed the fragmentation in the organic and natural personal care sector, there is also a strong presence of associations, polarized into various camps. There is significant cohesion within associational groups in the organic and natural personal care sector. The cohesion is presently least strong among the large traditional industry trade associations due to diversity in membership. These associations have refrained from taking a formal position and in doing so allow market forces to take their course. This has had a facilitating effect on allowing multiple NSMD approached to proliferate in the sector.

The strongest associations in the United States that have presented an obstacle to NSMD approaches are the consumer advocacy groups that are extremely cohesive around the application of the USDA NOP food standard to cosmetics. These associations facilitate the acceptance of the USDA NOP, but debilitate the entry of any other NSMD approach. In the EU, where the regulation specifically excludes cosmetics, NSMD approaches are seen as the only viable mechanism for verifying the legitimacy of the organic and natural claim and few associations, if any, oppose the approach. This has resulted in industry and standard making organizations working together toward acceptable market-based standards. In the United States, procedural neutrality on the part of USDA has left the door open for various associations to oppose any approach that is not aligned with the organic food standard.

As such, significant challenges are likely to be seen by NSMD approaches attempting to gain acceptance in the U.S. market unless the consumer advocacy groups can be convinced of the legitimacy of such approaches. In contrast, the EU's only cohesive associations are those surrounding particular standards. These are debilitating to external NSMD approaches, but promote success with strength and backing to the represented standard.

Furthermore, with the reissue of the FTC Green Guides, marketers of organic and natural products certified by a trade association or mutual benefit association must be certain to follow FTC Endorsement Guides to verify that any material relationships are disclosed to the consumer [42]. Adherence to this principle is required for advertising compliance in the United States.

3.3.2.3 Independence In general, organic and natural personal care companies operate very independently when it comes to deciding what standard to align with. CAN (2004) proposed that independence would be a hindrance, as companies may "place a higher value on independence, rendering certification programs in which they feel they do not play a key role in standard development less likely to gain support" (p. 235). Many companies, although they are seemingly independent, are working closely with competitors to develop common standards. In these cases, the dependence is facilitating consensus and success for a standard. When companies work together toward a mutually acceptable standard, it can help facilitate the legitimacy of that standard.

In the discussion on supply chain cohesion, it was observed that the raw material suppliers are choosing to certify to multiple standards to appeal to a wide variety of customers. This is likely to create successful adoption of multiple NSMD approaches as manufacturers independently choose to adopt various standards. Those marketers remaining independent and not aligning with a particular NSMD approach detract from the success of any standard.

Furthermore, those companies who choose to market organic and natural products without aligning with any standard take the risk of marketing without the transparency and qualification that standards provide. This can lead to consumer confusion and disillusion about the organic or natural claim in general, which will detract from the overall goal of the organic and natural movement.

3.3.2.4 Competition Between Standards The presence of competition is found to be a hindrance to success for standard making bodies in the United States and Europe. In both regions, this is shown to be an intersecting factor in other CAN factors, such as fragmentation [2]. Much of the reason for fragmentation relates to differing beliefs around standards and their associated criteria. This difference of opinion led to the creation of alternative approaches as the CAN Explanatory Framework would expect in Phase II [7].

While competition has slowed down the NSMD evolution process, it is also a critical step in a market-based initiative. Competition ensures that standards will continuously work to improve and evolve to meet the changing market needs. Considering such, it is likely that the existing standards will compete against each other in the market and gain success, fall away, or evolve through conforming efforts.

The main source of competition in the United States has been the creation of the NSF/ANSI standard and the OASIS standard for organic. With the vehement opposition that has been seen in the United States, it is expected that there will be significant challenges for any NSMD approach to win acceptance with the NGO community in the United States. In the event efforts are successful in convincing the USDA to regulate, no other NSMD approaches to organic personal care will be allowed. This will effectively bar European national standard approaches, such as COSMOS or NaTrue, from certifying products to be sold in the United States. Similar consumer advocacy groups for organic and natural products were not identified in Europe. A challenge that will continue in the United States is the competitive stalemate between the USDA NOP supporters and alternative approaches, such as OASIS. This stall in standard development is likely to continue until the USDA establishes a definitive position on whether it intends to regulate organic personal care products.

A major obstacle presented by the competition between standards is the public denigration of the criteria by competitor's approaches. This competitive behavior can lead to consumer and state distrust of the NSMD approach. Competition is a normal and expected process in a NSMD standard's evolution. However, competition that denigrates other transparent and potentially legitimate standards is nonproductive to the process.

3.3.2.5 Presence of a Regulation in a Sister Category A major debilitating factor to NSMD systems in the U.S. organic sector is the presence of a regulatory framework in a nearby sector. The USDA NOP has proven to be the largest detriment to NSMD approaches in the United States. The USDA NOP itself in the personal care arena is an NSMD initiative, as it is not formally supported by the USDA. This program has become the gold standard in the United States for what organic personal care should be across many small organic manufacturers despite technical challenges in formulating diverse product types. As discussed previously, supporters of the USDA NOP want to see the standard made into law for personal care products. The vehement support of the USDA NOP by consumer advocacy groups and manufacturers has effectively prevented other NSMD approaches from seeking legitimacy in the United States. This immensely debilitating effect suggested an additional hypothesis to CAN framework related to the history of the issue on the public policy agenda.

3.4 CONCLUSIONS AROUND IMPLEMENTING NONGOVERNMENTAL STANDARDS

In marketing a personal care product containing organic or natural ingredients, it is important to understand all the regulatory and market parameters for the desired product claims. This chapter has presented an overview of the factors that can facilitate and hinder market acceptance of a standard. Increasing imports and exports of certified products is seen as a way that a marketer can increase awareness and acceptance of an NSMD standard. It was also observed that the best environment to promote a standard is one where there is little or no state interest in regulating. Having a regulation in a sister category, such as food, had shown to be a hindrance to the proliferation of NSMD approaches in the United States.

The transparency of the standard is also a critical factor in promoting market acceptability. NSMD approaches are more likely to gain acceptance if they are transparent to the manufacturer and the consumer and if the standard is made up of workable, yet strict criteria. Furthermore, any material associations between the marketer and the certifying body should be made transparent as well.

It is important for marketers and third-party certifiers to be cognizant of the associational factors that can help or hinder a standard. Fragmentation and independence are factors that detract from the success of a standard. Associations can be beneficial to the standard around which the associated is based, but can be damaging to external NSMD approaches. The recommendation for marketers is to be aware of various associations' positions and prepared to manage potential issues. Ultimately, it is better to make a decision about what standard to align with than to remain independent and uncertified. Companies should choose a standard with criteria that best represent the marketing approach desired.

When choosing to develop an organic or natural personal care standard, it is important to be aware of and follow all laws and regulations in the intended market. When marketing in the United States, this includes the California Health and Safety Code and FTC Green Guides for using environmental marketing claims. Once all legal obligations have been met, it is the responsibility of the marketer to ensure the integrity of the product by certifying it to an appropriate standard or making claims that do not overstate the benefit. If a product does not contain enough organic or natural ingredients to meet the criteria of an existing standard, the marketer has the option to establish a new standard or to limit market claims to be truthful and qualified around the content of the ingredients.

Once strong NSMD initiatives are in place for globally traded products, there will be a limited market for those products obtained through unsustainable or unjust means. NSMD governance may drive down the demand for goods obtained through exploitation of workers, environmental devastation, and sweatshops. Consumers will now have a mechanism to gauge such products and can drive out those who do not measure up.

REFERENCES

1. Cashore B, Auld G, Levin K. The emergence of NSMD systems across sectors. Paper presented at the *Annual Meeting of the International Studies Association*, San Diego, CA, 2006.
2. Olson A. Non-State Market Driven Governance: An International Case Study in Organic and Natural Personal Care Products. Ph.D. dissertation, Walden University, 2010.
3. Bernstein S, Hannah E. Non-state global standard setting and the WTO: legitimacy and the need for regulatory space. *J. Int. Econ. Law* 2008;11(3):575.
4. Wood S. The International Organization for Standardization, Non-State Market-Driven Governance and the Struggle for the Soul of Global Business. Paper presented at the *Annual Meeting of the International Studies Association*, San Diego, CA, 2006.

5. Campins-Eritja M, Gupta J. Non-state actors and sustainability labelling schemes: implications for international law. *Non-State Actors Int. Law* 2002;2(3):213–240.
6. Wood S. Forecasting the market for authority: what can ISO social responsibility standards tell us about the future of non-state market-driven governance? Paper presented at the *Annual Meeting of the International Studies Association 48th Annual Convention*, Hilton Chicago, Chicago, IL, 2007.
7. Cashore B. Legitimacy and the privatization of environmental governance: how non-state market-driven (NSMD) governance systems gain rule-making authority. *Governance* 2002;15(4):503–529.
8. Cashore B, Auld G, Newsom D. *Governing Through Markets: Forest Certification and the Emergence of Non-State Authority*. New Haven, CT: Yale University Press, 2004.
9. Sabatier PA. *Theories of the Policy Process*, 2nd ed. Boulder, CO: Westview Press, 2007.
10. Suchman MC. Managing legitimacy: strategic and institutional approaches. *Acad. Manage. Rev.* 1995;20(3):571–610.
11. Cashore B, Auld G, Bernstein S, McDermott C. Can non-state governance "ratchet up" global environmental standards? Lessons from the forest sector. *Rev. Eur. Commun. Int. Environ. Law* 2007;16(2):158–172.
12. Cashore B, Egan E, Auld G, Newsom D. Revising theories of nonstate market-driven (NSMD) governance: lessons from the Finnish forest certification experience. *Global Environ. Polit.* 2007;7 (1):1–44.
13. Bernstein S, Cashore B. Can non-state global governance be legitimate? A theoretical framework. Paper presented at the *International Studies Association 48th Annual Convention*, Chicago, IL, 2007.
14. Hughner RS, McDonagh P, Prothero A, Shultz CJ, Stanton J. Who are organic food consumers? A compilation and review of why people purchase organic food. *J. Consum. Behav.* 2007;6(2/3):94–110.
15. Kortbech-Olesen, R. Growing taste for organic products in the United States. *Intl. Trade Forum* 2002;2: 27.
16. Shan Y. Going organic: is it nutritionally better? *Primary Health Care* 2006;16(3):37–40.
17. Worthington V. Nutritional quality of organic versus conventional fruits, vegetables, and grains. *J. Altern. Complement. Med.* 2001;7(2):161–173.
18. Snowe O. USDA Organic Food Standards. *FDCH Press Releases*, 2002.
19. Sahota A. Global & North American market for natural & organic cosmetics. Paper presented at the *Natural Beauty Summit*, New York, 2008.
20. Wasserman I. Natural, organic and green: ensuring FDA and USDA compliance when "eco-labeling." *Thompson Interactive Webinar*, 2008.
21. OCA. Coming Clean. A Campaign to Establish Strict Organic Standards for Body Care Products, 2003.
22. OCA. Complaint Against Handling Operations Selling and Labeling Personal Care Products as "Organic" In Violation of OFPA and NOP Regulations, 2010. www.organicconsumers.org/bodycare/documents/Complaint.pdf (retrieved January 22, 2010).
23. NSF., NSF, Standard for Organic Personal Care Products 305 2007. www.nsf.org/business/newsroom/articles/0903_n3_nsf305.asp (retrieved January 5, 2010).
24. OASIS. Organic and Sustainable Industry Standards, Inc. Health & Beauty Products, Beta Version #4, 2008. www.oasisseal.org/organic_production_standard.htm (retrieved September 12, 2008).
25. NPA. Natural Products Association Standard and Certification for Personal Care Products, 2008. www.npainfo.org/index.php?src=gendocs&ref=NaturalStandard_new&category=NaturalStandard (retrieved January 15, 2009).
26. California. California Health and Safety Code—The California Organic Products Act of 2003 (Vol. Division 104, Part 5, Article 7), 2003.
27. COSMOS. Cosmetics organic and natural standards, 2009. www.cosmos-standard.org/docs/COSMOS-standard_mai_09.pdf (retrieved May 25, 2009).
28. NaTrue. Requirements to be met by natural and organic cosmetics, Version 1.6. www.natrue-label.de/fileadmin/General_Mediapool/Label/Certifiers/Criteria/Criteria_EN.pdf (retrieved January 28, 2010).
29. ISO. Cosmetics—technical definitions and criteria for "natural" and "organic" ingredients and products; ISO/NP 16128, 2009. www.iso.org/iso/catalogue_detail.htm?csnumber=55725 (retrieved March, 15, 2010).
30. USDA. National Organic Program; Cosmetics, Body Care Products and Personal Care, 2008. www.ams.usda.gov/AMSv1.0/getfile?dDocName=STELPRDC5068442.

31. Smillie, J. Update on the American NSF Standard. Paper presented at the *Natural Beauty Summit*, New York, 2008.
32. NOSB. Solving the Problem of Mislabeled Organic Cosmetics & Personal Care Products, 2009. www.organicconsumers.org/bodycare/documents/CACCNOSBOrganicPersonalCareAug312009.pdf (retrieved September 5, 2009).
33. NOSB. National Organic Standards Board meeting transcript, November 3–5, 2009. www.ams.usda.gov/AMSv1.0/ams.fetchTemplateData.do?template=TemplateN&navID=NationalOrganicProgram&leftNav=NationalOrganicProgram& page=Nov2009Transcripts&description=NOSB%20Meeting%20Transcripts:%20November%203-5,%202009 (retrieved January 5, 2010).
34. Bronner D. Applying the USDA NOP standard to cosmetics. Paper presented at the *Natural Beauty Summit*, New York, 2008.
35. Kapsner T. Organic Cosmetics: US and International Standards. Paper presented at a *Personal Care Products Council sponsored webinar*, 2008.
36. NPA. *About Natural Products Association,* 2009. www.npainfo.org/index.php?submenu=About&src=gendocs&ref=AboutNPA&category=About (retrieved January 15, 2009).
37. Indursky, M.Build a global naturals brand. Paper presented at the *Natural Beauty Summit,* New York, 2008.
38. Ditmar, H.COSMOS/BDIH. Paper presented at the *Natural Beauty Summit,* New York, 2009.
39. OrganicMonitor. Europe: New Harmonised Natural & Organic Cosmetic Standards Launched, 2008. www.organicmonitor.com/r1411.htm (retrieved January 2009).
40. FTC Guides for the Use of Environmental Marketing Claims, 15 U.S.C. § 41–58, 1991.
41. FTC. Complying with the Environmental Marketing Guides, 2000. www.lawpublish.com/ftcenvironmental-marketing-guides.html (retrieved February 2, 2010).
42. FTC. Proposed revisions to the green guides. FTC File No. P954501, 2010.
43. CACC. Solving the Problem of Mislabeled Organic Cosmetics & Personal Care Products, 2009. www.organicconsumers.org/bodycare/documents/CACCNOSBOrganicPersonalCareAug312009.pdf (retrieved September 20, 2009).

PART II

REGULATORY ASPECTS

CHAPTER 4

REGULATORY PERSPECTIVE OF NATURAL AND ORGANIC COSMETIC PRODUCTS[1]

Azalea P. Rosholt

Ernst & Young LLP, Washington, DC

4.1 INTRODUCTION

The role of the regulator in issues concerning consumer health and welfare has evolved to a formidable magnitude over the last century and a half. Food and drugs are subject to complex regulatory control in almost every country of the world and while the regulation of products defined as cosmetics[2] has been relatively flexible, this is rapidly changing as well. The manufacture and sale of cosmetics is subject to two main forms of regulation: for health protection and for consumer fraud prevention. From a health protection perspective, the regulator's concern is focused on consumers' perception that so-called "natural" ingredients or even finished products are safe. From a consumer fraud prevention perspective, the concern focuses on the indiscriminate and potentially deceptive use of "natural" claims to take advantage of current fashion trends. Nevertheless, the U.S. government has been reluctant to define "natural" and related claims, specifically with regard to food, which makes it apparent that regulation of "natural" claims for cosmetics is even more remote. To fill the gap, a number of private organizations have stepped in to formulate their own definitions and standards and their own certification programs for cosmetics. Most of these programs,

[1] The views expressed here are those of the author and do not necessarily reflect the views of Ernst & Young LLP or any other member firm of Ernst & Young Global Limited. (Ms. Rosholt is a former senior director, International Regulatory Affairs, The Cosmetic, Toiletry, and Fragrance Association, currently renamed the Personal Care Products Council, Washington, DC.)

[2] This chapter uses the terms "cosmetic" and "personal care product" interchangeably. The Federal Food, Drug, and Cosmetic Act (FDCA) codified in 21 U.S.C. § 301 et seq. defines "cosmetic" as (1) articles intended to be rubbed, poured, sprinkled, or sprayed on, introduced into, or otherwise applied to the human body or any part thereof for cleansing, beautifying, promoting attractiveness, or altering the appearance and (2) articles intended for use as a component of any such articles.... The FDCA does not define "personal care product"; however, it is industry practice to refer to most cosmetics as "personal care products."

Formulating, Packaging, and Marketing of Natural Cosmetic Products, First Edition.
Edited by Nava Dayan and Lambros Kromidas.

however, go well beyond defining the term "natural" and impose many and diverse requirements for certification.

This chapter will describe regulatory attempts to define "natural" and other related terms, such as "organic" and "artificial" in the United States. It will also outline regulations in the European Union (EU) and Canada on the subject. Finally, it will describe some private certification programs and ways they have approached the definition of "natural cosmetics."

4.2 REGULATORY ATTEMPTS TO DEFINE "NATURAL" AND OTHER RELATED CLAIMS IN THE UNITED STATES

4.2.1 Definition of "Natural" for Food Claims

The U.S. Food and Drug Administration (FDA) has not so far specifically defined the term "natural" with regard to food product claims, although the issue has come up and has been addressed on a number of occasions in the context of food labeling. In 1991, the FDA initiated an inquiry into whether to develop a definition for the term "natural."[3] The FDA noted that the term has been used to describe a variety of products and although it could be taken to mean a variety of things, most notably it is perceived as the absence of substances that are man-made.

The FDA reviewed definitions of the term "natural" used by other government agencies and observed that the U.S. Department of Agriculture (USDA) permits use of the term "natural" on the labeling of meat and poultry products if

1. they contain no artificial flavor or flavoring, coloring ingredient, chemical preservative, or any other artificial or synthetic ingredient, and
2. they and their ingredients are not more than minimally processed.[4]

"Minimally processed," according to the USDA definition, may include traditional food processes such as smoking, roasting, freezing, drying, and fermenting. It may also include those processes that do not fundamentally alter the raw product and that only separate a whole, intact food into component parts such as grinding meat or pressing fruits to produce juices. Solvent extraction, acid hydrolysis, chemical bleaching, and other such relatively complex processes do not meet the criteria for minimal processing, and thus if they have occurred, the product would not be allowed by USDA to be labeled as "natural."[5] In addition, all labels of meat and poultry products bearing the term "natural" must be accompanied by a brief statement informing consumers that the product is natural because it contains no artificial ingredients and is only minimally processed.[6]

[3] 56 Fed. Reg. 60421, November 27, 1991.

[4] United States Department of Agriculture, FSIS, Policy Memo 055, November 12, 1983, cited in 56 Fed. Reg. 60421, November 27, 1991.

[5] Ibid.

[6] Ibid. See also www.fsis.usda.gov/FactSheets/Meat_&_Poultry_Labeling_Terms/index.a.

The FDA also referred to an earlier attempt by the Federal Trade Commission (FTC) to define the term "natural" for food claims in 1983. According to the FTC's proposed rule, "'natural' foods are those with no artificial ingredients and only minimal processing."[7] The FTC noted that the main problem with defining "natural" is that the meaning changes depending on the context in which it is used, and therefore no single definition would be appropriate, nor is such definition expected by consumers.[8] The FTC reviewed numerous comments and terminated its rulemaking proceeding without formulating a definition and decided to continue treating the issue on a case-by-case basis.

Similarly, the FDA received numerous comments to its proposed rulemaking. The comments were so diverse that the FDA could not reasonably decide on any one direction to follow for developing a definition of the term "natural."[9] Given the great ambiguity surrounding the use of the term and because of resource limitations and other priorities for the agency, the FDA decided not to undertake rulemaking to establish such definition, but to follow its established policy not to restrict the use of the term "natural" except for added color, synthetic substances, and flavors. Furthermore, the FDA decided to maintain its existing policy of interpreting "natural" to mean that "nothing artificial or synthetic has been added to the product that would not normally be expected to be there."[10] The FDA did not include a condition of minimal processing because such definition would likely conflict with its current definitions of "natural" and "artificial" flavors as discussed below.[11]

4.2.2 Regulation of Flavors and Color Additives

4.2.2.1 Flavors The FDA's definitions for "natural" and "artificial" flavor are as follows:

> "The term 'natural flavor' or 'natural flavoring' means the essential oil, oleoresin, essence or extractive, protein hydrolysate, distillate, or any product of roasting, heating enzymolysis, which contains the flavoring constituents derived from a spice, fruit or fruit juice, vegetable or vegetable juice, edible yeast, herb, bark, bud, root, leaf or similar plant material, meat, seafood, poultry, eggs, dairy products, or fermentation products thereof, whose significant function in food is flavoring rather than nutritional."[12]

It is important to note that under this definition, there is significant amount of processing that may be performed on a substance and still be permitted to claim that it is "natural." Another interesting aspect of the definition of "natural flavor" is that it does not cover inorganic substances. This is probably logical for food because most inorganic substances that could be considered "natural" are added to food products,

[7] Termination of Proposed Trade Regulation; Rule on Food Advertising, 48 Fed. Reg. 23270, May 24, 1983.

[8] Ibid.

[9] 58 Fed. Reg. 2302, January 6, 1993.

[10] Ibid. at 2408.

[11] Ibid.

[12] 21 C.F.R. 101.22 (a)(3).

such as sodium chloride, are also nutrients, and as such are excluded from the definition of "natural flavor." This is consistent with a definition of a "natural" product that was provided by the FDA in its *Guidance for Industry: Changes in an Approved NDA or ANDA*,[13] which states "For the purposes of this guidance, 'natural product' refers to materials (e.g., drug substance, excipients) that are derived from plants, animals, or microorganisms, and that are subject to approval under section 505[14] of the Act. The specific recommendations for natural products are not applicable to inorganic compounds (e.g., salts, minerals)."[15] Thus, if a definition of natural were to be developed with regard to cosmetics that would be consistent with the aforementioned definitions of a "natural product," it would have to exclude inorganic substances. This is not likely to be received well by manufacturers of certain cosmetics who are currently advertising mineral makeup as "natural."

Taking a look from the other side, the FDA's definition of "artificial flavor" covers any flavoring substance not covered by the definition of a natural flavor as follows: "The term 'artificial flavor' or 'artificial flavoring' means any substance, the function of which is to impart flavor, which is not derived from a spice, fruit or fruit juice, vegetable or vegetable juice, edible yeast, herb, bark, bud, root, leaf or similar plant material, meat, fish, poultry, eggs, dairy products, or fermentation products thereof."[16]

This appears to be fairly simple and straightforward; however, in practice, the FDA has given broad application of the term "artificial flavor" by considering not only the origin, but also the intended end use of a flavor additive. Thus, not only synthetic source substances must be labeled as artificial, but also substances that otherwise fall under the definition of "natural flavor" that are used in a food product to simulate the flavor of a food other than the one from which the flavor is derived. Thus, to claim that a flavor is "natural," the flavor must not only be derived from the plant or animal sources enumerated in the definition, but also be the flavor associated with the food in which it is used. The FDA provides an example: where a lemon-type pie is made with flavors derived from plant material sources, but where these sources are actually various other citrus fruit rather than lemons, such lemon-type pie would have to be labeled "artificially flavored lemon pie" because none of the flavoring is derived from lemons as would be implied had the product been simply labeled "lemon pie."[17] The policy of considering both the origin and the intended use of a substance to determine its "natural" or "artificial" status for flavors is paralleled in the regulation of color additives discussed below.

[13] *Guidance for Industry: Changes in an Approved NDA or ANDA*, U.S. Department of Health and Human Services, Food and Drug Administration, Center for Drug Evaluation and Research (CDER), Fn. 16. p. 13, April 2004. This guidance provides recommendations to holders of new drug applications (NDAs) and abbreviated new drug applications (ANDAs) who intend to make postapproval changes in accordance with section 506A of the Federal Food, Drug, and Cosmetic Act and 21 C.F.R. § 314.70.

[14] Section 505 of the Food, Drug, and Cosmetics Act covers the drug approval process.

[15] Ibid.

[16] 21 C.F.R. § 101.22 (a)(1).

[17] See 61 Fed. Reg. 29701, June 12, 1996, Food Standards; Reinvention of Regulations Needing Revision; Request for Comments on Certain Existing Regulations, at 29704.

4.2.2.2 Color Additives There is a long history in the United States of color additive regulation. It started in 1886, when Congress passed an act regulating the use of a synthetic organic color additive in butter.[18] In the early 1900s, the use of color additives in other types of food, and later in drugs and cosmetics, was brought under federal supervision. The Pure Food and Drugs Act of 1906[19] prohibited the use of poisonous or deleterious colors in confectionary and the coloring or staining of food to conceal damage or inferiority. Subsequently, the government (the FDA was not established until 1930) started a voluntary color certification program that later became mandatory. This color certification program continues to the present time.

The Federal Food, Drug, and Cosmetics Act of 1938 (FDCA)[20] required the FDA to list color additives that were "harmless and suitable" for use in foods, drugs, and cosmetics and to certify all batches of synthetic listed (i.e., approved) colors. In 1960, Congress passed amendments to the FDCA that implemented a complex certification system for synthetic organic color additives and included provisions for a separate listing of certain color additives obtained from plant, animal, or mineral sources as exempt from certification.

The color additive amendments also established the definition of a color additive as "any material...that is a dye, pigment, or other substance made by a process of synthesis of similar artifice, or extracted, isolated, or otherwise derived, with or without intermediate or final change of identity, from a vegetable, animal, mineral or other source and that, when added or applied to a food, drug, or cosmetic or to the human body or any part thereof, is capable (alone or through reaction with another substance) of imparting a color thereto."[21]

Color additives not subject to certification by the FDA include a variety of colors that some in the cosmetic industry consider "natural," such as henna, carmine, ultramarines, and metallic powders such as aluminum powder, copper powder, iron oxides, and titanium dioxide. Although they are not required to be certified by the FDA, their production is subject to good manufacturing practices, and they must meet detailed identity and specification requirements.[22]

Similar to the FDA policy for "artificial flavors," any substance added, regardless of source, to a product for the purpose of imparting color to that product is considered to be an artificial color. The reason for designating it as "artificial" is because it is not normally present or expected to be in the product. Thus, adding beet juice to lemonade to make it pink results in a product with a color additive that cannot be called "natural," regardless of the fact that beet juice is a simple aqueous extraction from a plant (the beet root) that occurs in nature.[23]

As in the regulations on flavor additives, the regulations on color additives consider both the origin and the intended use of a given color additive to determine

[18] Act of August 2, 1886, 24 Stat. at L. 209, amended by Act of May 9, 1902, 32 Stat. at L. 194.

[19] Pure Food and Drugs Act of 1906, United States Statutes at Large (59th Cong., Sess. I, Chp. 3915, pp. 768–772).

[20] 21 U.S.C. 301 et seq.

[21] 21 C.F.R. § 70.3(f).

[22] 21 C.F.R. § 73.

[23] 56 Fed. Reg. 60421, November 27, 1991.

whether it is artificial or not. This policy originated in the latter part of the nineteenth century when the government adopted a statute[24] to regulate the manufacture of oleomargarine, a much less expensive substitute for butter at the time. The statute enumerated acceptable ingredients for oleomargarine, which included both synthetic and natural source color additives and imposed a hefty tax of 10 cents per pound ($1.45 in today's dollars) on oleomargarine that was artificially colored to look like butter. The tax was only a quarter of 1 cent (3.6 cents in today's dollars) per pound if the oleomargarine was free from artificial coloration that "causes it to look like butter of any shade of yellow." The purpose of the tax was to discourage the deceptive practice of selling oleomargarine as butter. In *Cliff v. United States*, [25] a case that worked its way to the Supreme Court in 1904, the Court faced the issue whether palm oil, which was in fact one of the statutorily approved ingredients, when mixed in small quantities with the other approved oleomargarine ingredients for the purpose of imparting a light yellow color to the finished product would cause the resultant oleomargarine to be taxed at the higher rate. After all, palm oil is a vegetable oil and a natural (i.e., approved) ingredient of oleomargarine. The Court held that "when any substance, although named as a possible ingredient of oleomargarine, substantially serves only the function of coloring the mass and so as to cause the product to 'look like butter of any shade of yellow,' it is an artificial coloration."[26] In a similar case, *Moxley v. Hertz*, [27] the Court noted that the only natural colorant in oleomargarine to impart the color of butter is butter itself, thus only butter being added to oleomargarine would make the oleomargarine free from artificial coloration.

Thus, any substance added to a food product that is not normally present in that product for the purpose of imparting color or flavor to it would be considered an artificial flavor or color and cannot, at the same time, be considered a "natural" flavor or color of that product.

4.2.3 Applying FDA Policy for "Natural" Food Claims to Cosmetics

To what extent can the FDA policy be applied to cosmetic ingredients? Certainly many cosmetic ingredients are derived from "a spice, fruit or fruit juice, vegetable or vegetable juice, edible yeast, herb, bark, bud, root, leaf or similar plant material, meat, seafood, poultry, eggs, dairy products, or fermentation products thereof" and as such would meet a definition similar to the definition of "natural flavor." Conceivably, a definition for a "natural" cosmetic ingredient can be expanded to include inorganic substances found in nature to the extent these are used in cosmetic products. It would be difficult, however, to formulate a definition for a "natural" cosmetic ingredient that is consistent with the FDA policy to factor in the intended use of the ingredient into the definition. The fundamental difference between cosmetics and food would come into play; that is, unlike many foods, there are no finished cosmetic products that occur in

[24] Act of August 2, 1886, 24 Stat. at L. 209, amended by Act of May 9, 1902, 32 Stat.at L. 194.

[25] Cliff v. United States, 195 U.S. 159, 25 S. Ct. 1 (1904).

[26] Ibid., 165.

[27] Moxley v. Hertz, 216 U.S. 344, 30 S. Ct. 305 (1910).

nature. This, of course, refers to commercialized cosmetic products and not the cosmetic uses of certain foods, for example, raw cucumber slices or mashed banana applied to the skin as a facial mask. Under the current FDA policy, any color additive added to a cosmetic product for the purpose of imparting color to it would be an "artificial" color and any flavor, or by analogy, any fragrance added to such product would also be artificial. It follows that if the FDA were to stay consistent with its policy for food, manufacturers would be allowed to make only "natural" claims for some of the ingredients used in a product and not for the finished product. In some way, this is parallel to the USDA policy used for organic claims discussed below.

4.3 REGULATORY TREATMENT OF "ORGANIC" CLAIMS

4.3.1 Background

Organic claims and certification are regulated by the USDA. The USDA administers the National Organic Program (NOP), which was created under the Organic Foods Production Act (OFPA) of 1990. The OFPA[28] and the regulations issued under its authority[29] provide for production, handling, and labeling standards, in addition to accrediting private businesses, organizations, and state agencies to certify producers and handlers of agricultural products. Operations that produce or handle agricultural products intended to be sold, labeled, or represented as "100% organic," "organic," or "made with organic ingredients" or food groups need to be certified before such claims may be made. To become certified and allowed to make organic claims, the product must

1. have been produced and handled without the use of synthetic chemicals,
2. not be produced on land to which any prohibited substances, including synthetic chemicals, have been applied during the 3 years immediately preceding the harvest of the agricultural products, and
3. be produced and handled in compliance with an organic plan agreed to by the producer and handler of such product and the certifying agent.[30]

The term "organic," as defined in the regulations, is a labeling term that refers to an agricultural product produced in accordance with the OFPA and its regulations.[31] Furthermore, the term, "organic" may be used only on labels of raw or processed agricultural products, including ingredients that have been produced and handled in accordance with the regulations.[32] Certain methods are specifically excluded,[33] namely, methods used to genetically modify organisms or influence their growth

[28] Organic Foods Production Act of 1990, Public Law 101-624 as amended through Public Law 109-97, November 10, 2005, codified in 7 U.S.C. § 6501 et seq. (2009).

[29] 7 C.F.R. § 205.

[30] 7 U.S.C. § 6504.

[31] 7 C.F.R. § 205.2.

[32] 7 C.F.R. § 205.300.

[33] 7 C.F.R. §205.105.

and development by means that are not possible under natural conditions or processes and are not considered compatible with organic production.[34] The regulations distinguish between modern genetic modification (genetic engineering) and traditional genetic modification, such as by selective breeding.

4.3.2 Certification Procedure

The USDA certification requirements are complex and the certification process covers all aspects of production. Specific certification requirements exist for farms, livestock operations, and processing and manufacturing operations. The objective is to provide assurance to the consumer that production systems are managed in accordance with the OFPA and its regulations to respond to site-specific conditions by integrating cultural, biological, and mechanical practices that foster cycling of resources, promote ecological balance, and conserve biodiversity.[35] In turn, the ability to use "organic" claims and the USDA seal on products provides significant market advantage.

4.3.3 Organic Claims

Under the OFPA and its regulations, products made entirely with certified organic ingredients and methods can be labeled "100% organic."[36] Products with at least 95% organically produced raw or processed agricultural products may be labeled as "organic."[37] Products meeting the requirements for "100% organic" and "organic" may display these terms and the percentage of organic content on their principal display panel. The USDA seal may appear on product packages and in advertisements.

A third category, containing a minimum of 70% (by weight or fluid volume, excluding water and salt) organically produced, may be labeled "made with organic ingredients"[38] and list up to three of the organic ingredients or food groups on the principal display panel.

Products with less than 70% organically produced ingredients may use the "organic" claim, identify each organically produced ingredient in the ingredient statement with the word "organic" or with an asterisk or other reference mark to indicate the ingredient is organically produced, and display the product's percentage of organic contents on the information panel. Nevertheless, products with less than 70% organically produced ingredients must not display the USDA seal and any

[34] 7 C.F.R. § 201.2. Methods not compatible with organic certification include cell fusion, microencapsulation and macroencapsulation, and recombinant DNA technology (including gene deletion, gene doubling, introducing a foreign gene, and changing the positions of genes when achieved by recombinant DNA technology). The exclusion does not apply to breeding, conjugation, fermentation, hybridization, *in vitro* fertilization, or tissue culture. Id.

[35] 7 C.F.R. § 205.2.

[36] 7 C.F.R. § 205.301(a).

[37] 7 C.F.R. § 205.301(b).

[38] 7 C.F.R. § 205.301(c).

certifying agent seal, logo, or other identifying mark that represents organic certification of a product or product ingredients.[39]

4.3.4 Natural Claims

The OFPA and NOP also provide for "natural" claims. According to the USDA *Food Standards and Labeling Policy Book*,[40] the term "natural" may be used so long as the manufacturer is able to demonstrate that the product does not contain any artificial flavor or flavoring, coloring ingredient, chemical preservative, or any other artificial or synthetic ingredient and the product and its ingredients are not more than minimally processed. "Minimal processing" includes all "traditional processes used to make food edible or to preserve it or make it safe for human consumption." These processes include smoking, roasting, freezing, drying, and fermenting, while any more severe processes such as acid hydrolysis, solvent extraction, and chemical bleaching would not be considered minimal processing. Nevertheless, where the manufacturer can demonstrate that the use of a more than minimally processed ingredient would not significantly change the character of the product to the point that it could no longer be considered a natural product, a "natural" claim may still be made. In these cases, the USDA may grant an exception on a case-by-case basis.

Recognizing that consumers may interpret "natural" claims in a variety of ways, the USDA policy also requires manufacturers who make "natural" claims to provide a brief statement on the product label to explain what is meant by the term "natural." Acceptable reasons for use of the "natural" claim are that the product contains no artificial ingredients and is only minimally processed.

It is interesting to note that under the USDA OFPA and NOP rules, technically only 100% organic products may not have genetically engineered organisms. Products with "organic" and "made with organic ingredients" claims must be 95% and 70% organic, respectively. The rest can be anything. However, note that a "natural" claim, if used, needs to be qualified to make clear what the manufacturer means by "natural" as described above.

4.3.5 Organic Claims for Cosmetic Products

The USDA does not have authority to regulate cosmetic products; however, in 2005 after being sued by an organic soap company and a consumer group for not extending its regulatory authority to personal care products,[41] the USDA issued a memorandum[42] to all USDA accredited certifying agents. The memorandum noted that certain personal care products, which by virtue of their organic agricultural product content meet the NOP standards, may be labeled as "100% organic," organic," or "made with organic"

[39] 7 C.F.R. § 205.305.

[40] USDA Food Standards and Labeling Policy Book, August 2005.

[41] USDA Reverses Itself on Organic Seals, Libby Quaid, August 24, 2005, available at www.mindfully.org/Health/2005/USDA-Organic-Cosmetics24aug05.htm.

[42] Memorandum from Barbara C. Robinson to: All USDA Accredited Certifying Agents, dated August 23, 2005, available at www.nal.usda.gov/afsic/pubs/OAP/OAPGuide1.shtml.

pursuant to the NOP regulations. Businesses that manufacture and distribute such products may be certified under the NOP, and such products may be labeled as "100% organic," "organic," or "made with organic" so long as they meet NOP requirements. In addition, products that may be labeled "100% organic" or "organic" may also carry the USDA organic seal. Contrary to what has been expected and notwithstanding various communications and guidance documents, the USDA has not been enforcing its policy[43] regarding personal care products and has not implemented regulations to formally cover cosmetic products under the NOP regulations.[44]

The FDA, which is the agency that does in fact have authority to regulate cosmetics, has noted that neither the Food, Drug, and Cosmetics Act[45] nor the Fair Packaging and Labeling Act[46] defines the term "organic" or gives the FDA authority to regulate such claims. Furthermore, the USDA requirements for the use of the term "organic" are separate from the laws and regulations that the FDA enforces for cosmetics. Thus, cosmetic products labeled with organic claims must comply with both USDA regulations for the organic claim and FDA regulations for the labeling and safety of cosmetics. Finally, the FDA makes clear that the source of cosmetic ingredients is not determinative of safety; for example, extracts from certain plants, whether produced organically or not, contain toxic substances (consider, for example, strychnine or digitalis) that may be harmful for human health.[47]

How is USDA's experience with defining and enforcing use of the term "organic" instructive for any possible future regulatory endeavors to define "natural" for cosmetic products? In the first place, it is likely to take a very long time and in the second place, it will involve a significant investment of resources. Considering that the USDA spent about 12 years and reviewed more than 230,000 comments before issuing a final rule on "organic" claims,[48] any rulemaking on a definition of "natural" for cosmetic claims would represent a formidable task for the FDA. Judging from the attempts to regulate "natural" claims for food and other higher priority commitments, the FDA is unlikely to devote any significant time or resources to the definition of "natural" for cosmetic claims in the foreseeable future.

4.4 REGULATION OF "NATURAL" IN OTHER COUNTRIES

4.4.1 European Union

The EU regulates extensively in the area of "organic" production methods with regard to food and is beginning to do so in the area of cosmetics. The objective of "organic"

[43] Controversy Continues Over Regulation of Organic Cosmetics and Personal Care Products, Sarah A. Sunday, Esq.,West, Andrews Litigation Reporter, November 9, 2009.

[44] The NOP regulations are found in 7 C.F.R. pt. 205.

[45] 21 U.S.C. § 301 et seq.

[46] 15 U.S.C. § 1451 et seq.

[47] "Organic" Cosmetics, March 10, 2010, available at www.fda.gov/Cosmetics/ProductandIngredientSafety/ProductInformation/ucm203078.htm.

[48] See The National Organic Program (NOP), available at filebox.vt.edu/cals/cses/chagedor/USDAorganic.html.

farming in general is environmental conservation and protection. For this reason, what may be called in the U.S. "organic" farming may be described as "ecological" in the EU and elsewhere. A recent Commission Decision[49] establishes the ecological criteria for the award of the Community eco-label to soaps, shampoos, and hair conditioners. The decision also covers related assessment and verification requirements. It follows Regulation (EC) 1980/2000 of the European Parliament and of the Council[50] and allows the Community eco-label to be awarded to a product possessing characteristics that enable it to contribute significantly to improvements in relation to key environmental aspects. The eco-label program or "scheme" is voluntary and has as its main purpose to "promote products that have the potential to reduce negative environmental impacts, compared to other products in the same product group, thus contributing to the efficient use of resources and a high level of environmental protection. This objective, according to the Regulation, is to be pursued by providing guidance and accurate, nondeceptive, and scientifically based information to consumers on such products.[51]

The selection of shampoos, conditioners, and soap for this program is logical as these are personal care products that are released in the environment in greater quantities than other personal care products. Accordingly, the criteria applied under the Regulation include the level of toxicity to aquatic organisms, aerobic biodegradability, anaerobic biodegradability, selection of packaging, fragrances, color additives, preservatives and list of environmentally hazardous ingredients, and others. Consumer testing is required and must show that 80% of the test subjects (the number of test subjects must be at least 10) find the eco-label product as satisfactory as the market-leading product. Finally, products that are approved must provide the following statements on the product label: "minimal impact on aquatic ecosystems; fulfills strict biodegradability requirements; limits packaging waste."[52]

The aforementioned regulations do not define "natural" in the context of the Community eco-label scheme.

4.4.2 Canada

Canada also regulates extensively organic farming and has a Canada Organic logo that appears on certified food products. The Organic Products Regulations[53] apply only to food products, animal feed, and products used for the cultivation of plants; however, cosmetics and other products not included may make organic claims.

An interesting development in Canada occurred several years ago when Health Canada adopted, the Natural Health Product (NHP) regulations.[54] The regulations were intended to cover dietary supplements; however, the definition included a number of products that are also regulated as cosmetics and nonprescription drugs.

[49] Commission Decision of 21 June 2007, 20007/506/EC, OJ (L 186) 36, July 18, 2007.

[50] Regulation (EC) 1980/2000 of the European Parliament and of the Council of July 17, 2000 on a revised Community eco-label award scheme, OJ (L 237) 1, September 21, 2000.

[51] Ibid. Article 1(1).

[52] Commission Decision of 21 June 2007, Annex (11).

[53] Organic Product Regulations, P.C. 2009-944, June 11, 2009.

[54] Natural Health Products Regulations, P.C. 2003-847, June 5, 2003.

A natural health product under the NHP Regulations "means a substance set out in Schedule 1 (see below) or a combination of substances in which all the medicinal ingredients are substances set out in Schedule 1, a homeopathic medicine or a traditional medicine, that is manufactured, sold or represented for use in

1. the diagnosis, treatment, mitigation or prevention of a disease, disorder or abnormal physical state or its symptoms in humans;
2. restoring or correcting organic functions in humans; or
3. modifying organic functions in humans, such as modifying those functions in a manner that maintains or promotes health.

However, a natural health product does not include a substance set out in Schedule 2 (see below), any combination of substances that includes a substance set out in Schedule 2 or a homeopathic medicine or a traditional medicine that is or includes a substance set out in Schedule 2."

"Schedule 1" is appended to the Regulations and contains a list of natural health product substances as follows: a plant or a plant material, an alga, a bacterium, a fungus or a nonhuman animal material; an extract or isolate of a substance described above, the primary molecular structure of which is identical to the one that it had prior to its extraction or isolation; any of the vitamins biotin, folate, niacin, pantothenic acid, riboflavin, thiamine, vitamin A, vitamin B_6, vitamin B_{12}, vitamin C, vitamin D, vitamin E; an amino acid, essential fatty acid, or a synthetic duplicate thereof; a mineral; and a probiotic. "Schedule 2" of the NHP Regulations contains a list of excluded natural health product substances and these include prescription drug actives, controlled substances, pharmaceutical biologics, antibiotics, and others.

The NHP Regulations apply to a number of personal care products, such as toothpaste, antiperspirants, and sunscreens. Interestingly, the definition allows "synthetic duplicates thereof" to be used in NHP. In addition, the presence of an active that is an inorganic substance in a nonprescription topical drug that also falls under the definition of a cosmetic, for example, sodium fluoride in toothpaste, would make the product an NHP regardless of the "natural" or "synthetic duplicate" status of the sodium fluoride. The regulations designate certain products as "natural" but do not provide the rationale for doing so.

4.5 OTHER CERTIFICATION AND SEALS

A number of private organizations have developed programs that contain certification requirements and standards that apply to cosmetic products. Two of the most recognized are outlined below.

4.5.1 ECOCERT

ECOCERT is a nongovernmental organization with headquarters in France and representation in many countries. As one of the most popular eco-seals in Europe, it is awarded to companies who follow some very strict requirements. The stated objectives of the program in the Standards for Natural and Organic Cosmetics[55] are as follows:

- To define a quality level superior to the one defined by the French and European legislation on cosmetic products that will safeguard a real enhanced value of the natural substances used, a real practice of the respect of the environment throughout the production process, and a real respect for the consumer.
- To establish a link between certain cosmetic products and Organic Agriculture in promoting the use of Organic Agriculture plant products.
- To establish a link between certain cosmetic products and the respect for the environment.

Products that qualify are designated as a "natural cosmetic" or a "natural and organic cosmetic." A "natural cosmetic" contains at least 50% (by weight) of plant ingredients issued from Organic Agriculture (as defined in EU regulations). A "natural and organic cosmetic" is one that contains at least 95% (by weight) of plant ingredients issued from Organic Agriculture. Of the total ingredients, a "natural cosmetic" contains at least 5% of all ingredients, including water, issued from Organic Agriculture, and a "natural and organic cosmetic" contains at least 10% by weight of all ingredients, including water, issued from Organic Agriculture. Plant and animal ingredients that are processed according to specific standards are not subject to a positive list (i.e., an ingredient may only be used if it is on the list); however, ingredients of mineral (inorganic) origin are subjected to a positive list. Ingredients produced from biotechnologies, such as *in vitro* cultures, cloning, cell cultures, and fermentation with microorganisms, are allowed so long as there is no intervention or use of genetically modified organisms. Synthetic ingredients, such as synthetic color additives, fragrances, antioxidants, and emollients, produced by pure chemical synthesis are not allowed at all in certifiable products. In addition, the following are prohibited: treatment with ionizing radiation, processes involving genetically modified organisms, nitrosamine-forming substances, and animal testing.

Manufacturing processes must be simple and nonpolluting and the products must be biodegradable. Production processes of certifiable products must be separated from production of products not destined for certification. Chlorinated gas or any chlorine gas by-product may not be used in manufacturing. Similarly, primary packaging must be recyclable and may not be made of PVC or polystyrene foam. Storage of raw materials used in certifiable products and the finished products must be separated from storage of other raw materials and finished products.

The Appendix to the Standards contains a list of permitted processes, subdivided into a list of permitted physical and a list of permitted chemical processes. The Standards list a number of prohibited processes as well, such as bleaching, treatments with ethylene oxide or mercury, and others. The Standards list a number of approved

[55] Standards for Natural and Organic Cosmetics, ECOCERT, January 2003.

synthesized ingredients, including approved preservatives, a list of "natural" ingredients, such as beeswax, milk, and honey, and a list of ingredients of natural origin, subdivided into lists of ingredients of mineral and marine origin that are approved for use in certified products.

The certification process is complex and involves a detailed inspection of the manufacturing facilities, including records. Approved products must follow specific requirements on how to use the ECOCERT logo. Moreover, the Standards provide consumer information and note that the ECOCERT requirements are in addition to the regulatory requirements for cosmetics in effect in the European Union, namely, that the product must be safe for its intended purpose.

4.5.2 BDIH

BDIH (*Bundesverband deutscher Industrie- und Handelsunternehmen*) is the Association of German Industries and Trading Firms[56] for pharmaceuticals, health care products, food supplements, and personal hygiene products. The association certifies cosmetics making "natural" but not "organic" claims. To be eligible for certification, certain restrictions on ingredients apply as follows: raw materials obtained from plants must be from controlled biological cultivation or controlled biological wild collections and mineral raw materials with certain exceptions are allowed as well as components extracted through hydrolysis, hydrogenation, esterification, transesterification, or other crackings and condensations from the natural materials such as fats, oils, and waxes; lecithins, lanolin; monosaccharides, oligosaccharides, and polysaccharides; proteins and lipoproteins. Ingredients that are not permitted include organic synthetic dyes, synthetic fragrances, ethoxylated raw materials, silicones, paraffin, and other petroleum products. Processes that are not permitted include treatment with ionizing radiation, animal testing, and use of raw materials derived from dead vertebrates (raw materials that would require slaughter of the animal). Stated additional goals[57] include use of raw materials with traceable production and clear processes, education of consumers, and clear opposition to raw materials extracted from genetically modified plants or animals. The association also strives toward ecological compatibility and natural sources of raw materials, certification under the European Community eco-label scheme,[58] environmental-friendly manufacturing processes ensuring optimal degradability of raw materials and finished products, and economical and environmental-friendly recyclable packaging. The association further encourages use of raw materials from fair trading and Third World projects. In many ways, the criteria are similar to other certification programs in that they establish some relatively specific definitions and guidelines for manufacture and labeling.

[56] BDIH web site: www.bdih.de.

[57] Guidelines, available at www.kontrolliertenaturkosmetik.de/e/guideline_natural_cosmetics.htm.

[58] Regulation (EC) 1980/2000 of the European Parliament and of the Council of July 17, 2000 on a revised Community eco-label award scheme, OJ (L 237) 1, September 21, 2000.

4.6 CONCLUSION

As consumer demand for "natural" products increases, there will be more pressure for regulators to come up with an acceptable definition of "natural" that can be enforced in a reasonable and uniform manner. Regulation of "organic" claims in the United States and elsewhere provides some insight. Regulators are faced with the complex task of balancing a variety of expectations on many levels, including the task of ensuring that definitions remain consistent with the regulatory framework that has already been established for cosmetic products and other similarly regulated products. On the other hand, the availability of private certification, which has emerged recently, with specific definitions and criteria of eligibility, is a workable alternative for companies willing to deal with the sometimes formidable certification requirements.

CONCLUSION

CHAPTER 5

THE U.S. LEGAL PERSPECTIVE ON MAKING ORGANIC AND NATURAL CLAIMS FOR COSMETIC PRODUCTS

Miriam J. Guggenheim and MaryJoy Ballantyne

Covington & Burling LLP, Washington, DC

5.1 INTRODUCTION

This chapter provides an overview of the U.S. legal perspective on making "organic" and "natural" claims for cosmetic products. Also briefly addressed are other types of environmental or "green" claims. In response to the recent rise in consumer preferences for natural and environmental-friendly goods, more and more products are being identified with these claims. Despite the rise in consumer demand for such goods, the current federal standards do not specifically mandate how and when cosmetic products lawfully can make claims such as "organic," "natural," and "green." Recently, the federal government was asked to prioritize the creation of guidance for "organic" cosmetic claims, but no cosmetic-specific guidance is underway for "natural" and "green" cosmetic claims. Nevertheless, industry can rely on general federal and state antideception and consumer protection standards to ensure the lawfulness of such claims when used for cosmetic products.

In the United States, the federal regulation of "natural," "organic," and other "green" claims for cosmetics is largely shared among three agencies, the Food and Drug Administration (FDA), the Federal Trade Commission (FTC), and the U.S. Department of Agriculture (USDA). In addition, a few states have taken the initiative to regulate certain aspects of cosmetic products more stringently than the current federal standards. Industry and consumer groups also play a role in the informal regulation of cosmetics through industry-sponsored standards and self-regulation of advertising and through consumer advocacy aimed at developing standards for cosmetic claims.

Formulating, Packaging, and Marketing of Natural Cosmetic Products, First Edition.
Edited by Nava Dayan and Lambros Kromidas.

5.2 FEDERAL REGULATION OF COSMETIC CLAIMS

The three federal agencies, FDA, FTC, and USDA, have somewhat overlapping roles in the regulation of claims for cosmetics. In general, FDA has primary jurisdiction over the labeling of cosmetics, including the prevention of misbranding.[1] The FTC has primary jurisdiction over cosmetic advertising, including the authority to take enforcement action against advertising claims that are false or unsubstantiated.[2] Whether FDA or FTC has primary jurisdiction over a misleading cosmetic claim depends on whether the claim is made in advertising or in labeling. Both agencies can take simultaneous enforcement actions if the claim is made in an advertisement and in labeling. The USDA's involvement in the regulation of cosmetic claims has recently become a priority for the agency, and the USDA standards for "organic" claims, while not formally mandatory at this time, provide a "safe harbor" from regulatory enforcement action for cosmetic products making "organic" claims.

5.2.1 FDA Regulation of Cosmetic Claims in Product Labeling

FDA's jurisdiction to regulate cosmetic claims arises under the Federal Food, Drug, and Cosmetic Act (FDCA) and the Fair Packaging and Labeling Act (FPLA). FDA's regulatory authority over cosmetics is somewhat different from the authority it has to regulate other products under its jurisdiction. Unlike its premarket authority to regulate the safety and effectiveness of drugs and devices or to preapprove food additives, FDA's authority over cosmetics begins only after the products reach the marketplace. FDA does not have authority to require a safety assessment or premarket approval of cosmetics and cosmetic ingredients, with the exception of color additives.[3] Therefore, FDA does not review cosmetic products, ingredients, or labeling before the products reach the market.[4] Rather, the safety of cosmetic ingredients is addressed through an industry-sponsored yet

[1] See 36 Fed. Reg. 18539, FTC, Memorandum of Understanding Between FTC and the FDA (September 16, 1971).

[2] See ibid.

[3] Cosmetics companies are responsible for substantiating the safety of their products and ingredients before marketing. Failure to adequately substantiate the safety of a cosmetic product or its ingredients prior to marketing causes the product to be misbranded unless the following warning statement appears conspicuously on the principal display panel of the product's label, "*Warning*—The safety of this product has not been determined." 21 C.F.R. § 740.10. In addition, regulations prohibit or restrict the use of several ingredients in cosmetic products and require warning statements on labels of certain types of cosmetics. See 21 C.F.R. §§ 700.11–.27, 21 C.F.R. Part 740.

[4] In general, except for color additives and those ingredients that are prohibited or restricted from use in cosmetics by regulation, a manufacturer may use any ingredient in the formulation of a cosmetic provided that the ingredient and the finished cosmetic are safe, the product is properly labeled, and the use of the ingredient does not otherwise cause the cosmetic to be adulterated or misbranded.

industry-independent voluntary review process conducted by the Cosmetic Ingredient Review (CIR) Panel.[5]

As defined by the FDCA, cosmetic products include " (1) articles intended to be rubbed, poured, sprinkled, or sprayed on, introduced into, or otherwise applied to the human body or any part thereof for cleansing, beautifying, promoting attractiveness, or altering the appearance, and (2) articles intended for use as a component of any such articles; except that such term shall not include soap."[6] Pursuant to the FDCA, a company is prohibited from marketing adulterated[7] or misbranded[8] cosmetics in interstate commerce.

FDA's regulation of cosmetic claims generally arises under the misbranding provisions in sections 602 and 201(n) of the FDCA. FDA will find a cosmetic product misbranded if (1) its labeling is "false or misleading in any particular," (2) its label fails to provide required information, (3) its required label information is not properly displayed, or (4) its packaging or labeling violates requirements of the Poison Prevention Packaging Act of 1970.[9] Cosmetic labeling also may be deemed misleading if it fails to reveal facts that would be material to consumers in light of other representations made on the label.[10]

FDA reviews the labeling to determine whether a cosmetic claim is "false or misleading in any particular." "Labeling"[11] includes all labels and other written, printed, or graphic matters on or "accompanying" a product. For purposes of the FPLA, FDA has also defined the principal display panel (PDP) of retail packages; this may be on the actual container or on the box used in retail display. The PDP

[5] CIR assesses ingredient safety data by gathering information on its own and by seeking industry input. According to CIR's mission statement, CIR "thoroughly reviews and assesses the safety of ingredients used in cosmetics in an open, unbiased, and expert manner, and publishes the results in the peer-reviewed scientific literature." See CIR, Mission, www.cir-safety.org/info.shtml. CIR makes final safety assessments in the following categories: (1) safe ingredients—ingredients that are safe in the practices of use (product categories) and concentrations of use for each product category as documented in the safety assessment; (2) unsafe ingredients—ingredients with specific adverse effects that make them unsuitable for use in cosmetics; (3) safe ingredients, with qualifications—ingredient can be used safely, but only under certain conditions (e.g., qualifications relate to maximum concentration, rinse-off versus leave-on uses, and other restrictions); and (4) ingredients for which the data are insufficient—ingredient has not been found to be safe or unsafe, and the available data do not support safety. CIR safety assessments are made available as monographs that are compiled in a comprehensive compendium. See ibid.

[6] FDCA § 201(i) (21 U.S.C. § 321(i)).

[7] Under section 601 of the FDCA (21 U.S.C. § 361), a cosmetic is adulterated and therefore subject to FDA enforcement if (1) it contains a substance that may make the product harmful to consumers under customary conditions of use; (2) it contains a filthy, putrid, or decomposed substance; (3) it is manufactured or held under insanitary conditions whereby it may have become contaminated with filth, or whereby it may have been rendered injurious to health; (4) its container contains a substance that may render the contents injurious to health; or (5) it contains a color additive (except for hair dyes) that is unsafe.

[8] Under section 602 of the FDCA (21 U.S.C. § 362), improperly labeled or deceptively packaged products are considered misbranded and subject to regulatory action.

[9] See FDCA § 602(a) (21 U.S.C. § 362(a)); see also 21 C.F.R. Parts 701, 740.

[10] FDCA § 201(n) (21 U.S.C. § 321(n)).

[11] See ibid. § 201(m) (21 U.S.C. § 321(m)).

determines the location and size of the quantity declaration in the retail package. The PDP[12] includes the part of the label most likely displayed or examined under customary conditions of display for sale.

FDA also used its regulatory authority under the FPLA to require cosmetics to include an ingredient declaration on every product offered for sale to consumers[13] and to list the ingredients in descending order of quantity. Cosmetics that fail to comply with the FPLA-based requirements are considered misbranded under the FDCA.

Under the misbranding provisions of the FDCA, FDA may pursue enforcement action against violative products or against firms or individuals who violate the law. The FDCA misbranding provisions would apply to cosmetic claims such as "organic," "natural," and "green" if use of such claims results in labeling that is "false or misleading in any particular." To date, however, FDA has not brought an enforcement action against a cosmetic product due to misbranding caused by any such claim.

5.2.2 FTC Regulation of Cosmetic Claims in Advertising

Under section 5 of the Federal Trade Commission Act (FTCA), the FTC has authority to bring enforcement actions against false or misleading marketing claims. The FTC's authority applies to all forms of marketing including advertisements, labels, package inserts, promotional materials, words, symbols, logos, product brand names, and marketing through digital or electronic media such as the Internet or email.

The FTC reviews marketing claims, both express and implied, to evaluate the message conveyed to the consumer. As a result, the FTC attributes to marketing claims the meaning that consumers give them, which is not necessarily the technical or scientific meaning of the claims. The FTC also requires substantiation for all reasonable interpretations of an advertisement. Marketers making express or implied claims about the attributes of a product must be able to substantiate their marketing claims with competent and reliable evidence. The FTC has defined "competent and scientific reliable evidence" to mean "tests, analyses, research, studies, or other evidence based on the expertise of professionals in the relevant area, that have been conducted and evaluated in an objective manner by persons qualified to do so, using procedures generally accepted in the profession to yield accurate and reliable results."[14]

5.2.3 Nongovernmental Challenges to Cosmetic Claims

Cosmetic companies should also be aware that interested parties may bring legal challenges to a company's use of the terms "organic" and "natural" or "green"-type claims. For example, consumers may bring lawsuits for deceptive claims under state

[12] See 21 C.F.R. § 701.10.

[13] See ibid. § 701.3. The ingredient declaration applies only to retail products intended for home use. Cosmetic samples and products used exclusively by beauticians in salons are not required to include the ingredient declaration but must include other information including any required warning statements.

[14] See, for example, FTC, Dietary Supplements: An Advertising Guide for Industry 9, available at www.ftc.gov/bcp/edu/pubs/business/adv/bus09.pdf.

consumer protection statutes. In addition, competitors may bring challenges in court under section 43(a) of the Lanham Act[15] or bring disputes before the National Advertising Division of the Council of Better Business Bureaus (NAD).[16] Moreover, "organic," "natural," and "green"-type claims in advertising also may be subject to challenge by the NAD on its own accord.

5.3 "ORGANIC" CLAIMS ON COSMETIC PRODUCTS

There are currently no mandatory federal standards for "organic" claims for cosmetic products. Companies and manufacturers can choose to certify their cosmetic products under a variety of standards, including federal organic standards, private industry standards, or a combination. Cosmetics sold in the state of California, however, may bear "organic" claims only if they comply with that state's organic body care standards (discussed below). Finally, cosmetic companies that sell their products outside the United States should be mindful of foreign standards that may apply to "organic" claims on cosmetics.

5.3.1 USDA's "Organic" Standards

USDA's involvement with cosmetic claims solely arises under the Organic Foods Production Act (OFPA) that is administered through the National Organic Program (NOP). The OFPA gives USDA the authority to regulate the use of the term "organic" and the USDA Organic seal in the marketing of food and agricultural products. Under the NOP, "organic" is a labeling term that denotes products produced according to the provisions of the OFPA. The principal guidelines for organic production are to use materials and practices that enhance the ecological balance of natural systems and integrate the parts of the farming system into an ecological whole.[17]

While the extent of USDA's authority over cosmetic claims is not concretely established and does not appear to be formally mandatory, the agency allows cosmetic products to be certified under the NOP. Notably, the National Organic Standards Board (NOSB), the citizen advisory board that guides USDA's thinking on many subjects, recently recommended that the agency make cosmetic "organic" claims a priority.[18] Although the NOP originally did not consider its jurisdiction to extend

[15] 15 U.S.C. § 1125(a)(1)(B).

[16] See National Advertising Division, www.nadreview.org.

[17] See definition of "organic," USDA National Organic Standards Board (April 1995).

[18] On December 10, 2009, by a 12-1 vote, the NOSB formally recommended to the NOP "that organic personal care products be recognized explicitly by the National Organic Program to ensure consumers and businesses alike that the products have an unquestioned home in the USDA National Organic Program." Formal recommendations by the NOSB to the NOP (December 10, 2009), available at www.ams.usda.gov/AMSv1.0/getfile?dDocName=STELPRDC5081493&acct=nosb. In addition, in July 2009, the USDA posted for public comment a draft document discussing certification and labeling of soap products made from organic agricultural ingredients. See Agricultural Marketing Service, USDA, Draft for Comment Only: Certification and Labeling of Soap Products Made from Agricultural Ingredients (July 2009), available at www.ams.usda.gov/AMSv1.0/getfile?dDocName=STELPRDC5078399.

beyond organic food, it clarified in 2005 that all agricultural products, including personal care products such as cosmetics, may be certified "organic" under the NOP as long as they meet NOP requirements.[19] Thus, the current NOP certification requirements, while specific to food products, can be used as a safe harbor for cosmetic products making "organic" claims and cosmetic products seeking certification under the program.[20] The NOP develops, implements, and administers the national production, handling, and labeling standards for organic agricultural products certified under the USDA organic standard. In addition, the NOP accredits certifying agents (foreign and domestic) who inspect organic production and handling operations to certify that they meet USDA standards. The NOP organic certification process is based on the content of the product and how it is made. The NOP does not address safety issues. Products that are sold, labeled, or represented as "USDA organic" must be produced and processed in accordance with NOP standards.

The NOP permits four main labeling categories based on the percentage of organic ingredients in the product: "100% organic," "organic," "made with organic ingredients," and less than 70% organic.[21] The four terms and their criteria were originally drafted primarily for food products, but now may be used for cosmetics meeting the NOP criteria.

5.3.1.1 "100% Organic" or "Organic" USDA products and cosmetics seeking USDA certification, labeled as "100% organic," must contain (excluding water and salt) only organically produced ingredients and be processed using only organically produced processing aids. Products labeled "organic" must consist of at least 95% organically produced ingredients (excluding water and salt). Any remaining product ingredients must consist of nonagricultural substances approved for use on the National List of Allowed and Prohibited Substances (National List)[22] or specific nonorganically produced agricultural products not commercially available in organic form. Each organically produced ingredient must be identified as "organic" in the ingredient statement on the information panel, as part of the name of the ingredient, that is, "organic chamomile" or "chamomile (organic)," or through an asterisk or other reference mark linking to a footnote identifying the ingredient as "organic." Agricultural products labeled "100% organic" and "organic" cannot be produced or composed of components using excluded methods, sewage sludge, or ionizing radiation.

[19] See Memorandum from Barbara C. Robinson, Deputy Administrator, USDA Agricultural Marketing Service, to all USDA accredited certifying agents (August 23, 2005), available at http://www.purefood.org/bodycare/nop-memo.pdf.

[20] See Agricultural Marketing Service, USDA: Cosmetics, Body Care Products, and Personal Care Products (April 2008), available at www.ams.usda.gov/AMSv1.0/getfile?dDocName=STELPRDC5068442 (providing informal guidance on how cosmetic products can be certified under the program). To obtain USDA certification, the operations that produce the organic agricultural ingredients used in cosmetic products, the handlers of these agricultural ingredients, and the manufacturer of the final product must all be certified by an NOP accredited organic certifying agent.

[21] See 7 C.F.R. §§ 205.300–.305.

[22] See ibid. §§ 205.600–.606.

Products meeting the requirements for "100% organic" and "organic" may display these terms and the percentage of organic content on their PDP. Below the name and address of the handler of the finished product, the following statement or comparable phrase must appear: "certified organic by (name of the certifying agent)." The USDA Organic seal and the seal or mark of the certifying agents also may appear on product packages or in advertisements.

5.3.1.2 "Made with Organic" (Ingredients or Ingredient Categories)

Processed products that contain at least 70% organic ingredients can use the phrase "made with organic . . ." and then list up to three of the organic ingredients or ingredient categories on the PDP. For example, a qualifying body lotion made with at least 70% organic ingredients (excluding water and salt) and only organic herbs may be labeled "body lotion made with organic lavender, rosemary, and chamomile" or "body lotion made with organic herbs." Processed products labeled "made with organic . . ." cannot be produced using excluded methods, sewage sludge, or ionizing radiation.

As with "100% organic" and "organic" products, "made with organic . . ." products must identify each organically produced ingredient as such in the ingredient statement and include the "certified organic by . . ." statement. The percentage of organic content and the certifying agent seal or mark may be used on the PDP, along with claims that the organic ingredients are certified to USDA standards, but the USDA seal may not be used anywhere on the package.

5.3.1.3 Less than 70% Organic Under the NOP regulations, processed products containing less than 70% USDA organic ingredients cannot use the term "organic" anywhere on the PDP. They may, however, identify the specific ingredients that are USDA certified as being organically produced in the ingredient statement on the information panel. Products in this category may include claims that specific ingredients are USDA certified, but may not display the USDA Organic seal or include the name, address, or seal of a certifying agent.

Pursuant to the NOP regulations, organic products that fail to meet the requirements for one labeling category may be eligible for a lower labeling category. For example, if a product contains wholly organic ingredients but the product formulation requires a processing aid or less than 5% of a minor ingredient that does not exist in organic form, the product cannot be labeled "100% organic" and must be labeled as "organic." Similarly, if a multi-ingredient product is 95% or more organic but contains a prohibited substance in the remaining 5%, the product cannot be labeled as "organic" because of the presence of the prohibited substance but may be labeled as a "made with organic" product. Furthermore, under the NOP regulations, a handler who produces a "100% organic" or "organic" product but chooses not to be certified under this program may display only the organic percentage on the information panel and label the ingredients as "organic" on the ingredient statement in the manner discussed above. As stated previously, at this time, the NOP requirements provide a safe harbor for personal care products such as cosmetics that make "organic" claims and wish to be certified under the program. As discussed in more detail below, the USDA recognizes that cosmetic products may be certified to other private standards, but unless the cosmetic product fulfills the NOP requirements, it could not claim USDA organic certification.

As noted previously, the NOSB has recommended that USDA make cosmetic "organic" claims a priority. On December 10, 2009, the NOSB by a 12-1 vote formally recommended to the NOP that "organic personal care products be recognized explicitly by the National Organic Program."[23] The recommendation included proposed changes to the regulations, in which "personal care product" would be added to the scope of the regulations and would be defined similarly to FDA's statutory definition of cosmetic.[24]

5.3.2 Relationship Between NOP, FDA, and FTC Standards for "Organic" Cosmetics

As discussed previously, the NOP standards are currently voluntary and act as a safe harbor against regulatory enforcement action for cosmetics bearing "organic" claims. FDA retains primary jurisdiction over the labeling of cosmetic products and even the use of the term "organic" in cosmetic labeling, for the OFPA expressly states that "Nothing in this title shall alter ... any of the authorities of the [FDA] under the Federal Food, Drug, and Cosmetic Act."[25] Compliance with the NOP standards can, however, help ensure that "organic" claims in cosmetic labeling will be deemed truthful and nonmisleading by FDA and that such claims in advertising will be considered truthful and substantiated by the FTC.

Because for cosmetic products the NOP organic standards are currently voluntary, cosmetic products may be certified under other private standards and may be marketed according to those standards in the United States. The USDA specifically recognizes that cosmetics could be certified to "other private standards" and marketed in the United States pursuant to the private certification.[26] Given the presence of the NOP standards and the USDA Organic seal in the organic marketplace, however, adherence to the NOP standards in labeling and marketing a cosmetic product in the United States generally will provide security against FDA and FTC enforcement actions for false or misleading "organic" cosmetic claims. To date, FDA has declined to define the term "organic" for cosmetic products[27] and neither FDA nor the FTC has

[23] See supra, n. 18, Formal Recommendations by the NOSB to the NOP, at 1 (December 10, 2009).

[24] See ibid. at 2 (defining "cosmetic" as "An article intended to be rubbed, poured, sprinkled, or sprayed on, introduced into, or otherwise applied to the human body or any part thereof for cleansing, beautifying, promoting attractiveness, or altering the appearance").

[25] 7 U.S.C. § 6519(f).

[26] See supra, n. 18, Formal recommendations by the NOSB to the NOP, at 5 (April 2008) (stating "Cosmetics, body care products, and personal care products may be certified to other, private standards and be marketed to those private standards in the United States. These standards might include foreign organic standards, eco-labels, earth friendly, etc. USDA's NOP does not regulate these labels at this time.").

[27] See FDA Response to Aubrey Organics, Inc. Citizen Petition, Docket No. 96P-0267/CP 1 (July 11, 1997) (responding to Aubrey's request that FDA issue regulations setting forth the circumstances under which a cosmetic product will be misbranded for the false and/or misleading use of the term "organic" in label claims and stating that the agency had been unable to reach a decision on the petition).

brought an action against a cosmetic product due to a deceptive "organic" claim based on failure to meet USDA criteria.

5.3.3 State Standards for "Organic" Cosmetic Claims

Under the NOP, a state can establish organic certification standards, which must be approved by and meet the requirements of the NOP.[28] The state program can be more restrictive, however, because of environmental conditions or the necessity of specific production or handling practices particular to a state or region of the United States.[29] Several states have adopted organic standards under the NOP, but only California has enacted specific legislation directed to organic cosmetics.[30]

California has established its "Organic Body Care Standards" under the California Organic Products Act. The standards require a minimum of 70% organic content by weight for personal care products, including cosmetics, sold in the state that are "labeled or represented as organic or made with organic ingredients."[31] In addition, multi-ingredient cosmetic products sold as "organic" in California with less than 70% organically produced ingredients, excluding water and salt, may only identify the organic content by (1) identifying each organically produced ingredient in the ingredient statement with the word "organic" or with a reference mark to indicate the ingredient is organically produced, or (2) displaying the product's percentage of organic content on an information panel if the organically produced ingredients are identified in the ingredient statement. Thus, a cosmetic product sold in California that makes an "organic" claim must follow the California "organic" standards.

5.3.4 Private Standards for "Organic" Cosmetic Claims

In addition to the NOP organic certification standards, several organizations have developed private standards to certify cosmetic products as "organic." This chapter identifies two of the many private certification standards as examples of the types of standards that are available to companies and manufacturers who wish to market "certified" (but not USDA/NOP certified) organic cosmetic products. Currently, if a company certifies a cosmetic product under a non-USDA organic standard, the company can expect that it will not be subject to USDA enforcement unless the standard used conflicts with USDA policy or misleads consumers as to the relationship between the two standards. Any marketing indication that the product meets USDA's existing certification standards if it does not, or if it misuses the USDA

[28] See 7 C.F.R. §§ 205.620–.621.

[29] See ibid. § 205.620(c).

[30] California was also the first state to pass legislation governing the safety and reporting of cosmetic ingredients. The California Safe Cosmetics Act requires manufacturers to disclose to the state any product ingredient that is on the state or federal lists of chemicals that cause cancer, reproductive harm, or birth defects. In addition, Washington State in 2008 enacted its Children's Safe Products Act, which bans phthalates from personal care products marketed to or used by children.

[31] See CAL. HEALTH & SAFETY CODE § 110838(a).

Organic seal, will clearly fall outside of any safe harbor that USDA intended to create with its 2008 Guidance.

5.3.4.1 NSF/ANSI 305: Made with Organic Personal Care Products Standard NSF International through Quality Assurance International (QAI) provides a voluntary certification standard for personal care products pursuant to NSF/ANSI (American National Standards Institute) 305: Personal Care Products Containing Organic Ingredients (the NSF standard).[32] Products covered by the standard include rinse-off and leave-on personal care and cosmetic products, as well as oral care and personal hygiene products. The standard applies only to "made with organic . . ." products that contain from 70% up to 95% organic content. The NSF standard defers to the NOP standards for products containing 95% or greater organic content and labeled "organic" or "100% organic." The standard prohibits all nonorganic ingredients from being produced using excluded methods, sewage sludge, ionizing radiation, or genetically engineered organisms and from containing any petroleum compounds except as listed. It allows, however, certain chemical processes and ingredients common in the personal care industry that are not permitted in the NOP, such as mineral acid-catalyzed hydrolysis, esterification, or transesterification; glucosidation; sulfation; hydrogenation of oils; and protein fragment (nonpetroleum) acylation.[33] This voluntary standard specifies materials, processes, production criteria, and conditions that must be met in order for "made with organic . . ." personal care products to make "organic" labeling and marketing claims. The standard also includes requirements for agricultural ingredients, synthetic ingredients, and methods of extraction based on the final product's label claim.

5.3.4.2 The OASIS for Certifying Organic Cosmetic Products Organic and Sustainability Industry Standards (OASIS) is an industry developed and sponsored organic certification standard that was issued in March 2008. The OASIS standard is less stringent than other standards and permits certified "organic" cosmetic products to include cleansing agents made from nonorganic material including petrochemicals. In addition, a cosmetic product certified as "organic" under OASIS needs to include only a minimum of 85% organic content, which, unlike the NOP standard, can include organic water extracts and aloe vera.

In April 2008, Dr. Bronner's Magic Soaps filed a lawsuit against OASIS and other defendants for alleged deceptive "organic" marketing claims. In its second amended complaint, Dr. Bronner's asserted that the defendants' "organic" claims for their cosmetic products that would be certified under the OASIS standard were false and misleading to consumers because the cleansing agents in the products were made from conventional, as opposed to organic, agricultural materials including

[32] See NSF Int'l, NSF Standard for Organic Personal Care Products, Issue 1, Draft 7 (August 2008), available at http://standards.nsf.org/apps/group_public/download.php/2790/305i1r7.pdf. The American National Standards Institute (ANSI) adopted the NSF Standard on February 18, 2009. Copies of the NSF Standard are available for purchase from ANSI at http://webstore.ansi.org/RecordDetail.aspx?sku=NSF%2FANSI + 305-2009).

[33] See supra, n. 33, NSF, Int'l, NSF Standard for Organic Personal Care Products, Issue 1, Draft 7 (August 2008) at 12.

petrochemicals and synthetic preservatives as reflected in the NOP standard.[34] A federal district court dismissed the case in December 2009 on the grounds that if it proceeded with the case, the court would have to interpret the NOP standards and enforce the standards against the defendants. The court concluded that to do so in this case was outside its jurisdiction because only USDA had the authority to interpret and enforce the NOP standards in the first instance.[35]

Based on the court's dismissal of Dr. Bronner's lawsuit, companies may have more confidence that certification under non-USDA private standards will protect a cosmetic manufacturer against a challenge that its "organic" cosmetic claims are false or misleading. The NSF standard appears to be more reliable than the OASIS standard and is perhaps more reflective of the realities of formulating "organic" personal care products than are NOP organic requirements. Nevertheless, the safest approach for avoiding allegations of false or misleading "organic" cosmetic claims is to formulate and label cosmetic products in accordance with NOP standards. Given USDA's recent receipt of the NOSB's recommendation regarding organic cosmetics, following the NOP standard likely offers a manufacturer the greatest assurance that its "organic" claims will be deemed truthful, nonmisleading, and substantiated under FDA and FTC standards.

5.3.5 Foreign "Organic" Standards That May Apply to "Organic" Cosmetic Claims

In addition to the organic standards that may apply to organic cosmetics sold in the United States, cosmetic companies that market cosmetic products abroad should be aware that various foreign organic standards may apply to cosmetics making "organic" claims. Currently, there are a number of foreign standards that are at various stages of development. This chapter describes two examples: the European Union's (EU) COSMOS Standard (Cosmetics Organic and Natural Standard (European)) and the Canadian Organic Standards (COS). While these standards are not mandatory for cosmetics at this time, they are addressed here to illustrate the certification standards that may apply in the near future to cosmetics sold abroad that make "organic" claims. Cosmetic companies operating in foreign markets should monitor the development of organic standards in the countries in which they distribute products.

5.3.5.1 EU COSMOS Standard, Cosmetics Organic, and Natural Standard

The EU standard, known as the COSMOS standard, is a harmonization of several private European standards for natural and organic cosmetics, including those established by the Soil Association (UK), BDiH (Germany), Ecocert (France),

[34] See Second Amended Complaint for Damages and Injunctive Relief Against False Advertising in Violation of Section 43(A) of the Lanham Act, *Dr. Brommer's Magic Soaps v. The Hain Celestial Group, Inc.* (Cal. Super. Ct., San Francisco County, July 1, 2009) (No. 474707), available at www.drbronner.com/pdf/drbronner_second_amended_complaint.pdf.

[35] See *All One God Faith, Inc. v. The Hain Celestial Group, Inc.*, No. 09-03517 (N.D. Cal., December 14, 2009), order granting defendants' motions to dismiss, slip op. ("The Court could not grant such relief without reference to and interpretation of the NOP, which would be in excess of its jurisdiction." Id. at 14).

Cosmebio (France), and ICEA (Italy). The COSMOS standard, proposed in November 2008 and effective as of January 2010, sets minimum standards for natural and organic cosmetics. Among the requirements for COSMOS certification of organic cosmetics are

- at least 20% of the total product must be organic.
- at least 95% of the physically processed agro-ingredients must be organically produced, and by 36 months after January 2010, the remaining physically processed agro-ingredients must be organic if they are available (in sufficient quantity and quality).
- by 60 months after January 2010, at least 30% of the chemically processed agricultural ingredients must be of organic origin.[36]

The EU also abides by a Regulation (EC) No 1223/2009, which prohibits specific ingredients from being included in cosmetic products sold in the EU. The Regulation defines products that are considered as cosmetics for purposes of the Regulation and lists the substances that are prohibited from being used as ingredients in cosmetics sold in the EU. The Regulation bans from use in cosmetics over 1300 substances that are linked to cancer, genetic mutation, reproductive harm, or birth defects—regardless of the concentration of the substance in the product.

5.3.5.2 Canada's Organic Products Regulations The Canadian organic certification process is administered pursuant to the Organic Products Regulations (OPR). Canada recently amended the OPR, effective June 30, 2009, to mandate certification pursuant to the COS for organic products that are traded across provincial or international borders or that use the Canada Organic Logo. Because the regulations embodied in the OPR expressly apply only to organic food and agricultural products, the regulations are similar to the U.S. NOP standards for cosmetics in that they may be helpful in formulating and promoting cosmetic products in Canada with lawful "organic" claims.

On June 17, 2009, Agriculture Deputy Secretary Kathleen Merrigan announced that the United States and Canada have reached agreement for organic trade equivalence.[37] Under an equivalence determination, producers and processors that are certified by a USDA accredited certifying agent to the NOP standards do not have to become certified under the Canada OPR standards in order for their products to be represented as organic in Canada.[38] In addition, organic products certified to OPR

[36] COSMOS-Standard, Cosmetics Organic and Natural Standard, European Cosmetics Standards Working Group, (January 31, 2010), available at www.cosmos-standard.org/docs/cosmos-standard-final-jan-10.pdf.

[37] News Release, USDA, Agriculture Deputy Secretary Merrigan Announces U.S.–Canada Agreement for Organic Trade Equivalence (June 17, 2009), available at www.usda.gov/wps/portal/?contentidonly=true&contentid=2009/06/0212.xml.

[38] Agriculture Marketing Service, USDA, U.S.–Canada Determination of Equivalence Questions and Answers, available at www.ams.usda.gov/AMSv1.0/getfile?dDocName=STELPRDC5077525&acct=nopgeninfo.

standards may be sold or labeled as organically produced in the United States. Certified products from both countries may use the USDA Organic seal and the Canada Organic Biologique logo.

Despite this equivalence, there are some differences between the Canadian and U.S. standards. For example, Canada does not have a 100% organic claim for products.[39] In addition, both the Canada and U.S. letters noting the countries' respective equivalency determinations include some exceptions. For example, the Canada determination provides that agricultural products produced with the use of sodium nitrate or by hydroponic or aeroponic production methods shall not be sold or marketed as organic in Canada and animal-derived products must be produced in accordance with specified livestock stocking rates.[40] The U.S. determination provides that agricultural products derived from antibiotic-treated animals shall not be marked as organic in the United States.[41]

While the mandatory organic certification requirements under the OPR for food and agricultural products are not enforceable for cosmetics, Canada has indicated that it is developing "organic" standards for personal care products. Until the government issues its standard, companies can certify cosmetic products as "organic" under private standards. As in the United States, Canada requires all claims on cosmetic products sold in Canada to be truthful, and a company may be subject to enforcement for a false or misleading claim that its cosmetic product is "organic."[42]

Canada has also launched a Cosmetic Ingredient Hotlist that includes a list of chemicals that are prohibited or restricted for use in cosmetic products.[43] Cosmetics manufacturers are required to register their products and disclose a list of ingredients and the concentration of each ingredient used. Canadian labeling requirements mandate that ingredient lists appear on all cosmetics labels. Similar to the U.S. antideception laws, Canadian law prohibits cosmetic labeling that is deceptive and misleading.

5.4 "NATURAL" CLAIMS FOR COSMETIC PRODUCTS

As is the case for "organic" cosmetic claims, no U.S. government standards have been formally established for the use of "natural" claims on cosmetics or other personal

[39] Ibid. at 1.

[40] Letter from Karen McIntyre, Executive Director, Agrifood, Meat, Seafood Safety Directorate, Canadian Food Inspection Agency, to Barbara C. Robinson, Deputy Administrator, Transportation and Marketing Programs, Agriculture Marketing Service, USDA (June 17, 2009), available at www.ams.usda.gov/AMSv1.0/getfile?dDocName=STELPRDC5077628&acct=nopgeninfo.

[41] Letter from Barbara C. Robinson, Deputy Administrator, Transportation and Marketing Programs, Agriculture Marketing Service, United States Department of Agriculture, to Karen McIntyre, Executive Director, Agrifood, Meat, Seafood Safety Directorate, Canadian Food Inspection Agency, available at www.ams.usda.gov/AMSv1.0/getfile?dDocName=STELPRDC5077629&acct=nopgeninfo.

[42] See Canada Consumer Packaging and Labeling Act, R.S., 1985, c. C.-38, s.7.

[43] See Health Canada, List of Prohibited and Restricted Cosmetic Ingredients (The Cosmetic Ingredient Hotlist) (June 2010), www.hc-sc.gc.ca/cps-spc/alt_formats/hec-sesc/pdf/cosmet-person/indust/hot-list-critique/hotlist-liste_2010-eng.pdf.

care products at the federal or state levels. In the absence of a specific "natural" standard or definition for these products, the lawfulness of a "natural" claim is determined under the general antideception provisions of the FDCA and other federal and state consumer protection laws. The legal issue is whether such claims are false or misleading in any way, including whether the claims are deceptive in light of any failure to disclose information that would also be material to consumers in evaluating "natural" claim.

As discussed above, FDA has primary jurisdiction over the labeling of cosmetics, and false or misleading statements or material omissions in labeling would render the products misbranded under the FDCA.[44] The FTC has jurisdiction to take enforcement action against "natural" claims in advertising and promotion (including labels) that are false because they are "misleading in a material respect," including where the advertisement fails to reveal facts material in light of representations made.[45]

Without an objective or recognized standard for "natural," however, it can be difficult to determine what it means for a "natural" claim to be deceptive. Despite the tremendous consumer interest in products labeled as "natural," no comprehensive research has demonstrated any widely accepted consensus as to what "natural" means when used in cosmetic labeling. There are, however, a few benchmarks that reasonably may guide marketers considering the use of "natural" claims for cosmetics.

5.4.1 Guiding Standards for "Natural" Claims

5.4.1.1 Organic Labeling Standards Because less stringent standards apply to "natural" claims than apply to "organic" claims, cosmetic ingredients and products that satisfy NOP organic standards reasonably may be labeled as "natural" in most instances. Accordingly, individual ingredients that are certified organic may be generally labeled as "natural," and finished products that meet NOP criteria for "organic" or "100% organic" also are reasonably termed "natural." In addition, the inclusion of an ingredient on the NOP National List as a "nonsynthetic allowed" in products labeled as organic[46] can provide reasonable support for calling that ingredient "natural."

5.4.1.2 Regulatory Policies Regarding the Use of the Term "Natural" in Food Labeling Another helpful touchstone for determining whether "natural" claims for cosmetics are truthful and nonmisleading is FDA's policy regarding the use of the claim "natural" in food labeling. FDA has never defined the term "natural" by

[44] See FDCA § 602(a) (21 U.S.C. § 361(a)); FDCA § 201(n) (21 U.S.C. § 321(n)).

[45] See Federal Trade Commission Act § 15, 15 U.S.C. § 55.

[46] See 7. C.F.R. § 205.605(a).

regulation. The agency once considered doing so,[47] but ultimately declined, advising that FDA would continue its current policy regarding "natural" claims:

> The agency will . . . not . . . restrict the use of the term "natural" except for added color, synthetic substances, and flavors . . . . Additionally, the agency will maintain its policy . . . regarding the use of "natural," as meaning that nothing artificial or synthetic (including all color additives regardless of source) has been included in, or has been added to, a food that would not normally be expected to be in the food.[48]

FDA has restated this core "natural" policy for over a decade.[49]

In the food context, FDA considers all colorants to be "artificial color" because they impart a color that the food would not have naturally, and therefore FDA does not permit a "natural" claim on foods colored even with natural substances such as beet juice. In the context of cosmetics that are intended to impart color, however, such as lipsticks and other makeup, the presence of such colorants arguably should not preclude "natural" claims for such products.

FDA's "natural" policy is focused on the source from which a substance is derived, rather than the means through which it is produced. For example, FDA's "natural flavor" definition includes the "essence or extractive, protein hydrolysate, distillate, or any product of roasting, heating or enzymolysis, which contains the flavoring constituents derived from" basic plant- or animal-derived foods, as well as "fermentation products thereof."[50] "Artificial flavor," by contrast, is defined to mean any flavoring substance that is not derived from animal or plant sources.[51] In promulgating these regulations, FDA reiterated its focus on the source of the ingredients, stating that the "term 'artificial' connotes a synthetic source."[52] In addition, in a consumer-directed brochure published jointly with the International Food Information Council (IFIC), FDA explained that the difference between natural and artificial ingredients is that "[n]atural ingredients are derived from natural sources."[53]

Occasionally, however, FDA has considered processing methods in determining whether a product may bear a "natural" claim. The USDA imposes a condition of "minimal processing" for "natural" claims in the labeling of meat and poultry products.[54] Minimal processing is generally considered to include the types of

[47] See 56 Fed. Reg. 60421, 60466-67 (November 27, 1991); 58 Fed. Reg. 2302, 2407 (January 6, 1993).

[48] 58 Fed. Reg. at 2407.

[49] See, for example, Letter from Margaret O'K. Glavin, Assoc. Comm'r for Regulatory Affairs, FDA, to Antonio Zamora, at 2 (December 12, 2005), available at www.fda.gov/ohrms/dockets/dockets/04p0009/04p-0009-pdn0001-vol1.pdf; 62 Fed. Reg. 49826, 49841 (September 23, 1997).

[50] 21 C.F.R. § 101.22(a)(3).

[51] Id. § 101.22(a)(1).

[52] 38 Fed. Reg. 20718, 20719 (August 2, 1973).

[53] International Food Information Council (IFIC) and FDA, "Food Ingredients & Colors" (April 2010), available at www.cfsan.fda.gov/~dms/foodic.html#qanatural.

[54] See Food Safety and Inspection Service, USDA, "Food Standards and Labeling Policy Book" (August 2005), available at www.fsis.usda.gov/OPPDE/larc/Policies/Labeling_Policy_Book_082005.pdf.

processing that could be done in a home kitchen. FDA at times has referred to this element in evaluating whether "natural" claims may be false or misleading. Processing steps also may be relevant in determining whether a substance is "synthetic" because it has been "synthesized." The NOP definition of "synthetic process" may be instructive in this regard, focusing on whether the process "chemically changes a substance extracted from" a natural source.[55] A "natural" claim for a substance that has undergone such a synthetic process could be deemed false or misleading even on products not labeled as "organic."

By contrast, the lawful processing steps enumerated in FDA's "natural flavor" regulation cited above would not disqualify a substance or product from bearing a "natural" claim. Also potentially relevant is the one-page legislative history of the Dietary Supplement Health and Education Act (DSHEA) of 1994, in which Congress clarified that the term "chemically altered" in section 413(a)(1) of the FDCA[56] "does not include the following physical modifications: minor loss of volatile components, dehydration, lyophilization, milling, tincture or solution in water, slurry, powder, or solid in suspension."[57] Such physical modifications in the production of cosmetics therefore should not preclude the use of "natural" claims.

From an FDA perspective, therefore, a cosmetic or cosmetic ingredient reasonably may bear a "natural" claim if it is derived from a plant, animal, or mineral source. If the product or ingredient of natural origin has been subjected to chemical or synthetic processes, however, it may be deemed to fall outside of FDA's "natural" policy.[58]

In the last few years, FDA has received citizen petitions requesting that it define the term "natural" by regulation. The agency has not yet made any substantive response to the pending petitions, nor otherwise indicated how it may proceed. In view of these developments, it is important to remain attentive to any future change in FDA "natural" policy.

5.4.1.3 Third-Party Standards Third-party standards may shape consumer perceptions of what "natural" means and therefore may be instructive in crafting "natural" claims that are truthful and nonmisleading. In late 2008, the Natural

[55] "Synthetic" is defined in the governing OFPA to mean "a substance that is formulated or manufactured by a chemical process or by a process that chemically changes a substance extracted from naturally occurring plant, animal, or mineral sources, except that such term shall not apply to substances created by naturally occurring biological processes." 7 U.S.C. § 6502(21).

[56] This term appears in the FDCA § 413 relating to new dietary ingredients, in which dietary ingredients are carved out of the new dietary ingredient requirements if they "have been present in the food supply as an article used for food in a form in which the food has not been chemically altered." 21 U.S.C. § 350b(a)(1).

[57] 140 Cong. Rec. 28961 (October 7, 1994).

[58] In concluding that high fructose corn syrup could bear a "natural" claim, FDA considered whether source material came into contact with a synthetic substance in the course of processing, which the agency suggested might disqualify the finished product from bearing a "natural" claim. See letter from Stephen F. Sundlof, Director, FDA Center for Food Safety and Applied Nutrition, to Gary Y. Yingling, Esq., June 24, 2008. The legal basis for such a consideration is questionable, particularly because substances that merely come in contact with food during the production process historically have not been considered to be included in or added to the finished food product.

Products Association (NPA) introduced its Standard for Personal Care Products.[59] To obtain certification under that standard, a product labeled as "natural" must be made with at least 95% all natural ingredients, for NPA asserts that 68% of people believe a personal care product labeled as "natural" should be at least 95% natural. The balance of the product may contain only synthetic ingredients that are specifically allowed under the standard. "Natural" under that standard means that the ingredients come from a renewable resource found in nature, with no petroleum compounds. Industrial processes including ethoxylation, sulfination, and polymerization are not permitted in "natural" products certified under this standard.

Cosmetics that meet the criteria outlined in the NPA standard reasonably may bear general "natural" claims even if they are not certified under the standard.[60] The NPA standard is at least as stringent as relevant U.S. government standards and is based on reported consumer perceptions of personal care products bearing "natural" claims. Going forward, it will be important to monitor whether the NPA standard affects consumer expectations for products labeled as "natural" more generally, in which case this standard effectively could set the bar for all "natural" claims for cosmetics.

5.4.2 Consideration of the Claim in Context

In considering whether a "natural" claim is truthful and nonmisleading, marketers must consider the overall message the consumer reasonably derives from the "natural" claim. That is, the product should meet the expectations a consumer reasonably would have in light of the net impression conveyed by the labeling or advertisement.

General "natural" claims for finished cosmetic products are likely to be perceived as claims that either the entire product or a vast majority of the product, such as the NPA 95% criterion, consists of naturally sourced ingredients. If this is not the case, then it may be necessary to call out and disclaim the presence of the nonnatural ingredients. For example, in an NAD challenge involving Tom's of Maine mouthwash, which was labeled and advertised as "natural" and being made from ingredients "in nature," the NAD concluded that these claims could be confusing because the product contained the synthesized substance poloxamer 335.[61] NAD emphasized that advertisers of "natural" products should be very specific when describing ingredients that may be inconsistent with consumers' expectations. It recommended that the advertiser avoid claiming (directly or by implication) that its product is entirely natural and more accurately communicate on its label and in its advertising that the product contains an ingredient (poloxamer 335) that is not inherently "natural" or "sourced from nature."

Where a product is not overwhelmingly natural, then other options exist to communicate the "natural" content of the product. The label or advertising may call out the natural ingredient or ingredients in the product or may identify the percentage

[59] See NPA, www.naturalproductsassoc.org.

[60] Such products must not refer to the NPA standard unless they are certified to the standard.

[61] Tom's of Maine (Tom's of Maine Natural Mouthwash), # 3470, NAD Case Reports (June 1, 1998).

of natural ingredients in the product. Even statements that are technically correct, however, could be misleading in context. For example, a claim on the PDP that a product is made with natural lavender oil may be misleading, even if truthful, if the product also contains synthetic preservatives. Their presence would likely be contrary to consumer expectations of a product bearing a prominent "natural" claim.

In sum, a "natural" claim in cosmetic labeling or advertising is likely to be deemed truthful and nonmisleading, and therefore lawful, if (1) the ingredient or product that is the subject of the claim is derived from plant, animal, or mineral sources; (2) the ingredient or product is processed through only physical methods or biological processes (i.e., enzymolysis or fermentation); and (3) the product meets the expectations a consumer reasonably would have in light of the "natural" claim in its context.

5.5 THE INFLUENCE OF CONSUMER ADVOCACY ON "ORGANIC" AND "NATURAL" COSMETIC CLAIMS

Cosmetics labeled as "organic" and "natural" have received significant attention from consumer advocacy groups. These groups play an indirect role in the development of certification standards for organic cosmetics and other organic personal care products. Their activities also can reveal information about what "organic" and "natural" claims for cosmetics mean to consumers, which in turn may shape the perceptions of regulators and courts as to what makes such claims truthful and nonmisleading.

5.5.1 The Organic Consumers Association

Organic Consumers Association (OCA) has organized a "Coming Clean Campaign" that is focused on developing standards for "natural" and "organic" claims used for personal care products. OCA supports organic cosmetic standards that mirror the NOP organic food standards. Specifically, the Campaign advocates that (1) certified organic agricultural raw materials are utilized exclusively as the key basic cleansing and conditioning ingredients versus petroleum or conventional vegetable raw materials; (2) manufacture of such ingredients is reasonably simple and ecological; (3) the toxicity of each ingredient is minimal; and (4) nonagricultural water is not counted in any shape or form as contributing to organic content.[62] The Campaign endorses the NOP standard that requires a mandatory 70% minimum weight of nonwater/nonsalt agricultural organic content in a product for a "made with organic..." label claim to be included on the front panel. On its web site, OCA provides a database of body care companies with one or more products that meet the NOP organic standards.[63]

[62] See OCA, Coming Clean Campaign, www.organicconsumers.org/bodycare/index.cfm.

[63] See OCA, Directory of USDA Certified Body Care Products, www.organicconsumers.org/bodycare/links.cfm.

5.5.2 The Environmental Working Group

Environmental Working Group (EWG) advocates for safe cosmetics and has compiled a database called "Skin Deep."[64] The Skin Deep database provides a list of chemicals in personal care products and pairs ingredients in approximately 42,000 products against 50 toxicity and regulatory databases.[65] EWG has an online petition to Congress to require premarket testing of the safety of cosmetic products and has produced a consumer guide entitled "Safety Guide to Children's Personal Care Products."

5.5.3 The Campaign for Safe Cosmetics

Campaign for Safe Cosmetics (CSC) urges industry to phase out the use of cosmetic ingredients linked to cancer, birth defects, and other health problems.[66] CSC has initiated an industry program linked to the EU Cosmetics Directive entitled Compact for Safe Cosmetics.[67] By signing the Compact, cosmetic companies pledge to meet or exceed the formulation standards and deadlines set by the EU Directive by banning ingredients in cosmetics that are known or strongly suspected of causing cancer, mutation, or birth defects.[68] The Compact encourages companies to implement substitution plans that replace hazardous materials with safer alternatives within 3 years and publicly report progress in meeting these goals. EWG's Skin Deep database of personal care products will eventually show a graphic representation of the level of signers' compliance with the Compact.

It may be important to consider some of these major consumer advocacy group initiatives when making "organic" or "natural" claims for cosmetics. For example, a product bearing a "natural" claim that also contains an ingredient that one of these groups (or the state of California) considers to be a carcinogen or reproductive toxicant could be vulnerable to a challenge that the "natural" claim is false or misleading in light of the complete product composition.

5.6 "GREEN" CLAIMS FOR COSMETIC PRODUCTS

Companies making "organic" and "natural" claims often are also interested in making "green" claims, including terms such as "recycled," "eco-friendly," "degradable," "biodegradable," "ozone safe," "refillable," "no CFCs," and other similar claims. This chapter addresses criteria to be considered in ensuring that such environmental claims are truthful and nonmisleading because federal and state antideception laws also govern these types of claims.

[64] See EWG, Skin Deep: EWG's Cosmetic Safety Database, www.cosmeticsdatabase.com.

[65] See ibid.

[66] See CSC, www.safecosmetics.org.

[67] See CSC, The Compact for Safe Cosmetics, www.safecosmetics.org/article.php?list=type&type=51.

[68] See ibid.

While FDA has not officially addressed "green" claims, the FTC has issued guidance called the "Green Guides" explaining how it reviews "green" marketing claims.[69] The Guides include specific examples of "green" claims and the substantiation the FTC requires to support such claims. For example, the Guides require "green" claims to be precise, specifying whether the claims refer to the product, the packaging, or both or just to a component of the product or its packaging.[70] In addition, the Guides recommend that marketers qualify environmental claims that are broad or vague or avoid them altogether. Broad claims such as "environmentally safe" or "environmentally friendly" should be qualified or avoided because they convey a wide range of meanings to consumers that may be difficult to substantiate.

The Guides require marketers to be able to substantiate all reasonable interpretations that consumers may draw from a "green" claim. In addition, the Guides give examples of how consumers may interpret various claims and identify the kinds of claims that should be explained or qualified to avoid deception. While the Guides provide direct guidance regarding the use of specific "green" claims, they do not scientifically define environmental terms or establish product performance standards.

An example from the Guides directed at the terms "degradable," "biodegradable," and "photodegradable" is instructive regarding the general approach the FTC takes to "green" claims. The Guides provide

> It is deceptive to misrepresent, directly or by implication, that a product or package is degradable, biodegradable, or photodegradable. An unqualified claim that a product or package is degradable, biodegradable, or photodegradable should be substantiated by competent and reliable scientific evidence that the entire product or package will completely break down and return to nature, i.e., decompose into elements found in nature within a reasonable short period of time after customary disposal.
>
> Claims of degradability, biodegradability, or photodegradability should be qualified to the extent necessary to avoid consumer deception about: (1) the product or package's ability to degrade in the environment where it is customarily disposed; and (2) the rate and extent of degradation.[71]

Although the Green Guides are administrative interpretations of the law and do not have the force and effect of law, the FTC can take action against a marketer who asserts claims that are inconsistent with the Guides. As a result, "green" claims on cosmetic products should conform to the recommendations set forth in the Guides. In October 2010, the FTC proposed revisions to its Green Guides and sought public comment.[72] Marketers interested in "green" claims should keep apprrised of FTC developments in this area.

[69] See 16 C.F.R. Part 260.

[70] See ibid. § 260.6(b); see also FTC, Complying with the Environmental Marketing Guides 19 (May 2000), available at www.ftc.gov/documents/bus42-complying-environmental-marketing-guides.

[71] See 16 C.F.R. § 260.7(b); see also FTC, Complying with the Environmental Marketing Guides (May 2000).

[72] 75 fed. Reg. 63552 (October 15, 2010).

5.7 CONCLUSION

In sum, claims in cosmetic labeling and advertising, including "organic," "natural," and other types of "green" claims, are all evaluated against the same touchstone—whether the claims are truthful, nonmisleading, and substantiated in light of reasonable consumer interpretations of the claims in their context. Aside from the California Organic Body Care Standards, no mandatory government standards have been established for the use of "natural," "organic," and other "green" claims for cosmetics. Cosmetic companies seeking assurance that their claims will be deemed nondeceptive and therefore lawful may wish to seek compliance with NOP standards for "organic" claims, FDA policy for "natural" claims, and the FTC "Green Guides" for other environmental claims. Given that interest in these types of claims is continuing to increase among consumers, advocacy groups, and regulators, it is important for marketers in this arena to remain aware with the latest developments, from both a legal and a public relations perspective.

CHAPTER 6

REGULATORY AND SAFETY ASPECTS OF NATURAL FRAGRANCE INGREDIENTS

Anne Marie Api[1] *and Matthias Vey*[2]

[1]*Research Institute for Fragrance Materials, Inc., Woodcliff Lake, NJ*
[2]*International Fragrance Association, Brussels, Belgium*

6.1 INTRODUCTION: ROLE OF NATURAL INGREDIENTS IN PERFUMERY

Natural materials like rose or myrrh belong to the materials that most likely marked the beginning of perfumery, independent of the question of who actually was the first one to intentionally change the odor of an object or a person by adding scented extracts. There is evidence that in the early days of perfumery, like in the ancient Egyptian culture, the use was often linked to religious ceremonies, but this does not seem to have been an exclusive application and over the centuries perfuming the home or body became the predominant application. However, to date incense sticks and frankincense are still playing a major role in religious ceremonies of Asian cultures and in Catholic ceremonies, respectively. Other natural materials supposed to have played an important role in ancient times like myrrh are still used today.

The creation of a new fragrance, for whatever application, despite all the modern techniques of analysis can still be regarded an art. Applications of fragrance are widespread and range from the so-called fine perfumery where new perfumes are created for the often glamorous brands or names, to technical perfumery, for use in dish washing liquids or toilet cleaners. Especially for technical perfumery, it is very important for the perfumer to have sufficient knowledge about the finished product properties, as these properties can impact the stability of the perfume compound and therefore exclude the use of certain fragrance ingredients. Somehow aggressive conditions such as caustic or acidic conditions in the finished product require a certain stability of the fragrance ingredient and therefore often prevent the use of natural ingredients. Publications by Jellinek [1], Storp [2], and Schreiber [3] describe in more detail the psychological and sociological aspects in the creation and widespread use of fragrances and the development of perfumes.

Formulating, Packaging, and Marketing of Natural Cosmetic Products, First Edition.
Edited by Nava Dayan and Lambros Kromidas.

Nowadays, when creating a perfume, the perfumer can choose from a huge palette of materials of natural origin (essential oils, concretes, absolutes, etc.), synthetic ingredients, including those that are synthesized as a duplicate of the material occurring in nature (the so-called "nature identical" ingredients) and accords, which are precompounded bases that, for example, provide a woody or musky note, itself consisting of natural and/or synthetic materials.

Based on a global volume of use survey carried out for the year 2008, the database maintained by the Research Institute for Fragrance Materials, Inc. (RIFM) has recognized the use of around 3000 odorous materials a perfumer can choose from. In reality, hardly any perfumer would use all the 3000 materials, but will have a specific subset that he or she is familiar with and likes to work with; however, even that includes several hundred materials.

What somehow makes the natural extracts so precious for perfumery is the richness and complexity of the odor. Besides the complexity, naturals may vary in composition due to various factors such as the growing region, agricultural conditions, the time of harvest, and/or the weather conditions during the growing season. Rose petals, for example, are usually harvested in the coolness of the morning, when the content of the essential oil is the highest—before environmental conditions had a chance to deplete them and/or change their nature.

This complexity and variation in the composition of the natural extract adds further to the complexity of a fragrance compound (see Figure 6.1) and from a safety assessment point of view does increase the challenge for the safety assessor.

The available spectrum of ingredients/odors, built on the basis of their natural ante type in more recent history, has further increased due to sophisticated headspace techniques [4]. This technique allows the trapping of odorous molecules, which via traditional techniques (such as distillation, pressing, or enfleurage), often due to

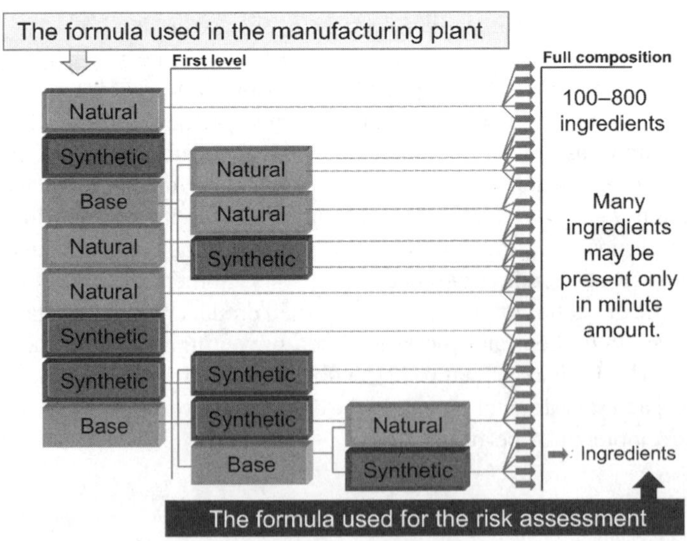

Figure 6.1 The complexity of a fragrance compound.

extremely low levels, high volatility, or delicate stability, previously escaped detection and use in perfumery. In addition, because of the headspace technique, the smell of plants delicate in both scent and structure, like lily of the valley, could be recreated. The method also allowed a better understanding of how the smell of a flower changes (when is it most powerful, when is it most typical) throughout the course of the day.

Both naturals and synthetic chemicals can be the driver for a certain fragrance family or give a certain fragrance its uniqueness. Nearly all olfactive families are built around one natural key material or an association (accord) of natural ingredients, as the following examples demonstrate.

Chypre is an example of a fragrance family in which without any doubt natural extracts play very important role. History tells us that its origin lies in the famous French perfumer Jean-Francois Coty, one of the key figures in the early history of modern perfumery. He developed an accord, which he used in a perfume in 1917, in which bergamot oil, labdanum, and oak moss extracts were central elements. The smell reminded him of the island of Cyprus, so he gave it the name Chypre. Over the years, other famous perfumes were developed around this accord, both for men and for women. Famous examples, to name a few, are Miss Dior (Christian Dior, 1947), Azzaro (Azzaro, 1973) for women or Aramis (Estée Lauder, 1965) and Fahrenheit (Christian Dior, 1988) for men. It goes without saying that all these are classics, which people still love today. The evolution of perfumery for various reasons, such as the availability of a certain raw material, in some cases led to the replacement of natural ingredients by synthetic ingredients (sometimes, following the example of Mother Nature by using nature identical materials). There is another kind of chypre-based fragrances using methyl-3-methylorsellinate, discovered as early as 1898, but used only in the second half of the twentieth century as a single molecule. Examples for its use include Polo (Ralph Lauren, 1978) and Narciso Rodriguez for Her (2003).

Citrus, as the name implies, is another family that also largely depends on the use of natural ingredients. Citrus-based fragrances have a very long history and two of them, created in the eighteenth century in the German city of Cologne, gave the whole family of such products its name. In the beginning of the eighteenth century, Giovanni Maria Farina (1685–1766) named his fragrance Eau de Cologne in the honor of his hometown and in 1792 Wilhelm Muelhens started to sell his 4711 Eau de Cologne. Citrus colognes never went out of fashion and experienced a remarkable revival with Calvin Klein's famous CK One in 1994.

The industrial revolution opened new opportunities for the fragrance industry by making important odorous materials more broadly available. Some famous examples might be coumarin, which was isolated from the tonka bean in 1820 and synthesized by William Perkin in 1868 [5, 6] or vanillin by Tiemann and Reimer in 1874 [7, 8]. Other examples are the synthesis of cinnamic aldehyde (Chiozza, 1856), Ionone (Krüger, Tiemann, 1893), or muscone (methyl cyclopentadecanone) by Weber and Ziegler in 1934 [9]. Today, the majority of these ingredients are still among the most widely used perfumery ingredients.

Like natural odorous material, synthetic odorous material also plays their role in the development of memorable products. A perfume that without any doubt gets some of its uniqueness from the presence of a synthetic material is the famous Chanel No. 5 created in 1920 by Ernest Beaux for Coco Chanel and launched a year later. There are countless myths around its creation and the selection of the name, but undoubtedly important protagonists in this product are aldehydic ingredients (2-methyl undecanal besides others) [10]. It became one of the first fragrances in the so-called floral aldehydic family, to which were added more famous names over the years, such as Madame Rochas by Rochas, in 1960, and True Love by Elizabeth Arden, in 1994.

An example where actually the "overdosing" of synthetic materials was a revolutionary step in creative perfumery is Cool Water, created by Pierre Bourdon for Davidoff in the late 1980s. It takes some of its ozonic freshness and marine aspects from high levels of allyl amyl glycolate, dihydromyrcenol, damascene, and ambroxan.

As a consequence of the important role natural and synthetic odorous material play in the development of perfumery products, it should be emphasized that the majority of today's perfumers acknowledge the benefits of natural extracts in combination with synthetic materials. When one talks to perfumers today, one hardly finds any who would prefer exclusively one class of ingredients (natural or synthetic). Experts in perfumery often regard the natural extracts as giving richness and complexity to a fragrance, while synthetic ingredients provide depth, provide originality, and make the perfume last. This is very nicely summarized in a citation of Luca Turin taken from his blog "Perfume Notes" [11], quoting the perfumer René Laruelle, "Synthetics are the bones of a perfume, naturals the flesh." He then continues, "Too much synthetics and you have bleached skeletons, too much naturals and you're stuck with invertebrates." Regardless, driven by the trend of turning back to "Mother Nature," there are some perfumers that prefer to work exclusively with natural materials [12].

6.2 CHARACTERIZATION OF NATURAL FRAGRANCE INGREDIENTS

Historically in the fragrance industry, natural fragrance ingredients were regarded as distinct entities. In 2002, the RIFM panel of experts (established 30 years earlier) was asked to review human health and environmental safety data on natural fragrance ingredients to determine whether they should be viewed as single entities or as mixtures of chemical components. After a careful review, the Expert Panel concluded that to date, there are no data to indicate that materials found in naturals, that is, the natural fragrance chemicals, are different in terms of (human health or environmental) toxicity from the same chemicals produced synthetically. Therefore, any International Fragrance Association (IFRA) Standards should address both synthetic and natural materials.

Of course, this does not exclude any safety testing on the natural fragrance ingredients. However, testing naturals can present unique challenges. When testing naturals, characterization of the test material with quantitative data on the components and possible variability in concentrations of the components is very important. Therefore, a full description of sourcing method of preparation and composition is needed. This includes a full description of the botanical species and inventory identification (e.g., Chemical Abstract (CAS) Number and European Inventory of Existing Commercial Chemical Substances (EINECS) Number). A detailed chemical analysis (quantitative and qualitative—including the ranges of percent composition and any unknown composition) is needed and must be representative of the natural. As discussed above, this can often present problems as the composition of a natural compound can change depending on the conditions when it was collected. The weather conditions can affect the composition of the natural ingredient. Furthermore, samples of the same plant may differ depending on the region where it is produced because of the impact of soil, temperature, and weather, and the time of harvest. A complete description of the type of natural product is also needed. For example, the composition of the natural compound can vary depending on how it was extracted—the composition of the steam distilled oil such as rose oil is different from when the same botanical is solvent extracted to produce the concrete and then alcohol extracted to arrive at the absolute. Rose absolute (Centifolia) is one of the most extensively used of all the floral absolutes.

As with any risk assessment, exposure information on the use of the natural compound is essential. For a natural concrete, an estimation of exposure to individual chemical components is also needed.

6.3 EVALUATION OF NATURAL FRAGRANCE INGREDIENTS

Safety testing of natural fragrance ingredients follows the same evaluation as a synthetic fragrance ingredient and is detailed in a paper by the RIFM Expert Panel [13]. The safety evaluation follows a systematic prioritization of data generation analysis, consideration of exposure, and critical analysis of the quality of the available information. The paper by Bickers et al. [13] provides the details of the approach used by the RIFM Expert Panel to examine the dermal effects, systemic toxicity, and environmental consequences of the use of and exposure to fragrance ingredients. The conclusions will allow a reliable determination of safe use under intended conditions. While the safety evaluation process remains the same for synthetic and natural fragrance ingredients, testing of natural compounds presents unique considerations.

When evaluating naturals for systemic toxicity, data on the components may preclude testing the natural. For example, sweet orange oil (*Citrus sinensis* (L.) Osbeck) is comprised of >90–95% limonene. As such, one may want to use the safety data available for limonene to support the use of orange oil. Alternatively, data on well-characterized naturals may preclude testing a major component. An example of this is the use of the reproduction toxicity data on coriander oil to support the safety of linalool. The sample of coriander oil tested in this particular study was well characterized and showed that the oil tested was 74% linalool and most of the remaining components were terpene alcohols. These data helped to support the safety evaluation of linalool [14].

Similarly, an environmental assessment of a natural can also be challenging. For example, a biodegradation or ecotoxicity study on a natural fragrance ingredient can be difficult to conduct because of solubility and volatility of the various components. When conducting environmental studies, it is important that the test concentration is maintained and measured. In addition, the test solution should be homogeneous. Most natural fragrance ingredients solutions in environmental studies may be extremely difficult to maintain, to measure test concentrations, and to attain homogeneous test solutions. Identification of components will also play an important role in environmental assessment. Analysis of components may provide useful information on environmental fate and effects. It may also be possible to avoid testing a natural if sufficient environmental data are available on the components concerned.

In preparation for the European legislation, Registration, Evaluation, Authorization and Restriction of Chemical or REACH [15], the fragrance industry has prepared a protocol for the environmental assessment of natural complex substances. The protocol will develop several case studies for evaluating environmental test methods and refining the protocol for natural complex substances' environmental assessments as needed.

The challenges in testing natural fragrance ingredients can be exemplified in more detail by reviewing a series of studies completed on some natural fragrance ingredients for dermal effects. Natural fragrance ingredients often contain dermal sensitizers. The purpose of the studies was to compare the dermal sensitization potential of the essential oils and compare it to the dermal sensitization of some of their components. The data obtained can provide information to (1) determine hazard, (2) estimate potency, and (3) determine any difference in sensitization for the major components arising from exposure to the natural fragrance. The studies provide an opportunity to investigate changes in dermal sensitization potential for individual components in the essential oil. The data may also help to elucidate the component that is the major cause of the dermal sensitization potential of the natural.

'Data from a series of murine local lymph node assays (LLNA) on seven different oils were presented by Lalko and Api [16] and carefully evaluated. Essential oils or any other mixture evaluated in an LLNA need to be very carefully studied. Knowledge of the constituents is essential in understanding the results and placing them in context. An expert group [17] summarized the general consensus on the current utility of the LLNA to evaluate mixtures. The panel concluded that the available animal testing methodologies are neither evaluated nor designed for the testing of preparations. They cautioned that further research would be needed in order to evaluate properly the utility of these methods for the safety assessment of preparations.

In general, this series of studies showed that the potency of each essential oil tested did not differ significantly from the potency observed from the main individual component if it was a dermal sensitizer. When the essential oil had no single major component as a dermal sensitizer, then no pattern was observed. Three examples will be discussed in more detail. Two oils, lemongrass and *Litsea cubeba,* have citral as a major component. Another oil, citronella, has no individual fragrance ingredient as a major component. The samples of the oils tested were analyzed and are detailed in Table 6.1.

The results from the LLNA tests for the three essential oils are presented in Table 6.2. Citral, the major component of both lemongrass oil and *L. cubeba* oil, is known to be a dermal sensitizer. Its dermal sensitization data are summarized by Lalko and Api [18]. Animal sensitization data showed that the material is a weak sensitizer. In the LLNA, the estimated concentration (EC3) required to elicit a positive

TABLE 6.1 Composition of Samples of Lemongrass, *L. cubeba*, and Citronella Oils Tested for Dermal Sensitization

Principal name	Analysis	%
Lemongrass oil	Citral	69.0
	Limonene	6.7
	Geraniol	6.1
	Geranyl acetate	2.2
	Caryophyllene	1.6
	trans-Isocitral	1.4
L. cubeba	Citral	86.0
	Limonene	2.9
	Linalool	1.7
	Citronellal	1.4
	Caryophellene	0.9
	Methyl heptanone	0.6
Citronella oil	Citronellal	36.7
	Geraniol	20.6
	Citronellol	11.2
	Limonene	4.1
	Geranyl acetate	3.7
	Citronellyl acetate	3.0
	Elemol	2.6
	Beta-bourbonene	2.2
	Delta-cadiene	1.9
	Isopugenol	1.6
	Germacrene D	1.4
	Eugenol	0.9
	Linalool	0.9

Source: This table is adapted from Ref. [16].

response was calculated and taken as a measure of relative potency. In this group of experiments, a pure sample of citral was retested with the natural fragrance ingredients. The data show that both lemongrass oil and *L. cubeba* oil have the same dermal sensitization potency as their major component, citral. The EC3 values for lemongrass oil and *L. cubeba* are very similar to that of citral (6.5, 8.4, and 6.3%, respectively). An analysis of the effective dose for each of these oils may help identify which component may be the ones contributing to the sensitization of the natural. For lemongrass oil and *L. cubeba* oil, citral is the component responsible for the dermal sensitization because the effective dose (ED) for citral is the one that is closest to the EC3 value for the components. Citronella oil showed no potential for dermal sensitization in this test when evaluated up to 50%. For citronella oil, this is somewhat unexpected given the level of geraniol in the oil. In fact, a review of the ED shows that geraniol may be responsible for the dermal sensitization because the ED for geraniol is the one that is closest to the EC3 value for the components. Citronella oil is a

TABLE 6.2 LLNA Results and Comparison of Components for Lemongrass Oil, *L. cubeba* Oil, and Citronella Oil

Natural fragrance ingredient	EC3[a] (μg/cm²)	Components		Component EC3[a] (μg/cm²)	ED[b] (μg/cm²)
Lemongrass oil	6.5% (1625)	Citral	69.0%	1,575[c]	1,100
		Limonene	6.7%	17,125	108
		Geraniol	6.1%	2,850	100
		Geranyl acetate	2.2%	NA	36
		Caryophylene	1.6%	NA	26
		trans-Isocitral	1.4%	NA	23
		6-Methyl-5-heptene-2-one	1.3%	NA	21
		Caryophylene oxide	1.2%	NA	20
		4-Nonanone	1.0%	NA	16
L. cubeba oil	8.4% (2100)	Citral	86.0%	1,575[c]	1,800
		Limonene	2.9%	17,125	61
		Linalool	1.7%	13,750	36
		Citronellal	1.4%	7,500	30
		Caryophylene	1.0%	NA	21
		Methyl heptanone	0.6%	NA	13
Citronella oil	>50% (<12,500)	Citronellal	36.7%	7,500	4,585
		Geraniol	20.6%	2,850	2,575
		Citronellol	11.2%	10,875	1,401
		Limonene	4.1%	17,125	520
		Geranyl acetate	3.7%	NA	473
		Citronellyl acetate	3.0%	NA	375
		Elemol	2.6%	NA	325
		Beta-bourbonene	2.2%	NA	286
		Delta-cadiene	1.9%	NA	238
		Isopugenol	1.6%	NA	203
		Germacrene D	1.4%	NA	182
		Eugenol	0.9%	1,325	122
		Linalool	0.9%	13,750	113

[a]EC3: estimated concentration required to elicit a positive; taken as a measure of relative potency.
[b]ED: effective dose (EC3 value of the material X percent of component).
[c]Citral (99.5%) was tested along with the essential oils; the EC3 value was 6.3% or 1575 μg/cm².

complex mixture and there are no LLNA data for the majority of the individual components. Additional data are needed in order to explain the mixed results.

In general, the dermal sensitization potency of each essential oil studied did not differ significantly from that observed for its main individual component when a natural comprised of more than approximately 70% of one individual component. Conversely, dermal sensitization potency for essential oils that lack a clear major component could not be easily determined [16]. The study demonstrated the importance of conducting assays on highly characterized samples of essential oils, and the extreme care that must be taken when interpreting the results of LLNAs on essential oils (or mixtures). At a minimum, information on the major components is required for a preliminary evaluation.

The conclusions made from this series of dermal sensitization studies can apply to systemic toxicity and environmental studies. The need to conduct assays on highly characterized samples of natural fragrance ingredients cannot be overemphasized. Knowledge about the toxicity of the components will enhance the interpretation of the results.

6.4 REGULATION OF NATURAL FRAGRANCE INGREDIENTS IN THE FRAMEWORK OF THE IFRA STANDARDS

6.4.1 The IFRA Standards in General

Over the past few decades, there has been a marketing tendency, driven by consumer demand, to use natural components. The driving factor is often the erroneous belief that natural products are safe per se. However, without any doubt a material is not generally safe because it originates from nature. In the end, making claims to the effect or implying that what is natural is also safe constitutes a disservice to this precious category of ingredients. Mother Nature has created some of the most toxic materials we know and the adverse effects of many naturals are well described in the literature [19–21]. Mitchell and Rook in their book *Botanical Dermatology* list more than 10,000 species that have been described as causing irritative or allergic contact dermatitis [22], and this does not include toxic plants that are estimated to occur globally at around 189,000 [23]. The list of adverse effects is long and ranges from contact dermatitis [19] to the Stevens–Johnson Syndrome [24], a severe form of erythema multiforme, marked by lesions of and sloughing of mucous membranes [25].

In conclusions, it is evident that the use of naturals, like any material from whatever origin, should be supported by an adequate set of safety data and a positive risk assessment. In addition, with regard to naturals the aspect of exploitation of resources (protection of endangered species) should also be considered.

Fragrance materials are used in a wide variety of consumer products ranging from perfumes to skin care products such as creams, lotions, cleansers, and various other personal care and household products. The potential for exposure to these materials therefore is very high, particularly for those products that come into direct contact with skin. Therefore, it is essential to minimize the number of potential skin disorders linked to the use of fragranced products including irritant dermatitis, allergic contact dermatitis, phototoxicity, and photoallergy associated with fragrances.

The fragrance industry has maintained a system of safety assurance since 1973. IFRA sets standards that are intended to ensure the safe use of fragrance ingredients for the consumer and the environment. These standards are based on scientific principles and therefore address both naturals and synthetic materials. IFRA promotes the self-regulatory practices of the industry in keeping with the idea that the adaptation of worldwide industry rules to new scientific findings can occur more quickly through self-regulation, as opposed to a change of legislation in different countries or different continents.

IFRA membership is open to all countries or regions in which a fragrance association exists and comprises national/regional associations of fragrance manufacturers from, for example, Australia/New Zealand, Europe, the Far East, and North and Latin America. IFRA has also provision for direct or supporting company membership.

The main thrust of IFRA's activities consists of developing, communicating, and implementing a code of practice for the international fragrance industry, based on broadly recognized scientific principles and utilized worldwide with the final objective to protect the consumer and the environment. The IFRA Code of Practice is comprehensive and applies to the manufacture and handling of all fragrance materials for all types of applications. The IFRA Standards are an integral part of the Code of Practice and of the reputation for safety of the global fragrance industry. Amendments and updates to the Code, if required, are issued on a regular basis, usually annually, taking into account new scientific developments. IFRA Standards are based on a scientific assessment of potential hazards (extensive set of toxicological data) and comprehensive information on the use of and exposure to fragrance materials by RIFM and subsequent evaluation by the RIFM Expert Panel, whose role over the years has been continuously strengthened.

RIFM is the scientific institute for the fragrance industry and is responsible for generating, evaluating, and distributing scientific data on the safety assessment of fragrance raw materials found in consumer products. The scientific program at RIFM is guided by the RIFM Expert Panel. This panel is an independent and international group of dermatologists, pathologists, environmental scientists, and toxicologists. The safety evaluation of a fragrance ingredient begins with research and testing being conducted by RIFM. All of RIFM's scientific information is published in peer-reviewed scientific literature. All aspects of the RIFM test program, from inception to planning a test, to interpretation of final results, is governed by the Expert Panel, which also reviews and approves all research and testing for safety evaluation.

The Expert Panel's safety conclusions include definition of the critical effects and a safety evaluation based on reported use. Once communicated to the industry (IFRA), the effects may trigger different actions or no actions at all depending on the type of effects. The following are some conclusion statements as examples:

- No recommended action or standards;
- No recommendation to change an existing standard (there is a general 5-year review of all standards);
- Determined "effect" to be the critical effect, such as sensitization, recommended maximum use level to be set in a standard;
- Critical effect of a material linked to its purity, recommended minimum purity criteria.

The process of the risk assessment strategy by the Expert Panel has been comprehensively described in the peer-reviewed scientific literature [13]. Today, the Expert

Panel's conclusions are the sole basis upon which the IFRA sets its standards. The consequence of the Panel's review, expressed in the form of the IFRA Standards, can be the restriction or ban of the use of an individual ingredient. IFRA in its activity as risk manager is responsible for the publication and enforcement of the standards.

On the basis of the safety evaluations by RIFM and the Expert Panel completed to date, IFRA has established 174 standards banning or restricting the use of fragrance ingredients (status 44th Amendment, July 2009).

The IFRA Code of Practice and its standards are distributed worldwide and are in the hands of all member associations and their member companies, the global customer industry and its main associations, and other interested stakeholders like government regulatory bodies, including the US FDA and the European Commission. All IFRA Standards are publicly available and can be accessed on the IFRA web site (http://www.ifraorg.org/).

Although in principle a voluntary approach for non-IFRA members, these standards are binding on all members of the association, all of whom understand that compliance with the Code of Practice and Standards is a condition of membership. The IFRA membership covers about 90% of the global market (in volume).

To effectively demonstrate the importance of the Code of Practice and the IFRA Standards to IFRA and its members, a compliance program was initiated in 2006. It focuses on the safety of consumer products by ensuring that the IFRA Code of Practice is fully applied and followed. It involves the analysis of a variety of randomly selected consumer products for the presence of fragrance ingredients regulated by IFRA Standards. The selection of the consumer products to be analyzed follows a specific protocol. Each year, IFRA selects a product/segment on a rotation basis. Ten countries where IFRA is linked to local fragrance associations are selected every year by a random computer program operated by a supporting external laboratory, which also operates as a third party administrator (TPA). In each country, the top leading 15 (20) products for each segment are identified on the basis of their ranking of the preceding year as follows:

- 15 in one segment of the fine fragrance category (e.g., Eau de Toilette, perfume)
- 15 in one segment of the cosmetics and personal care category (e.g., shampoo, deodorant)
- 20 in one segment of the household and detergents category (e.g., air freshener, fabric conditioner, washing powder).

The product selection and analysis is defined in specific procedures and is coordinated by the third party administrator in order to eliminate any conflict of interest. The third party administrator is well accredited and scientifically recognized and follows internal procedures to ensure confidentiality and impartiality in each step of the program. Additional protocols and procedures describe how potential cases of noncompliance are addressed. In case of a confirmed noncompliance, if the supplier agrees to correct the situation and there are no repeated violations, no further action is contemplated. Corrective actions are verified through the collection and analysis of additional sample products. If the supplier does not take corrective action, the name of that supplier is listed on the IFRA web site as not complying with the IFRA Code of

Practice and the national association to which that supplier belongs is asked to take appropriate steps regarding its membership.

Over the years, regulatory bodies around the world have bought into the system of the IFRA Standards by either having a direct reference to them as a basic requirement for compliance of products with regional or national regulation or, like in China or in the European Union, by adopting individual standards and making the respective requirements part of their regulation. A number of IFRA Standards (both restriction and prohibition of ingredients) are reflected by individual entries in Annex II and III, respectively, of the EU Cosmetics Directive (Council Directive 76/768/EEC of 27 July 1976 and its Amendments), which beginning 2010 was turned into a regulation (2009 Regulation (EC) No. 1223/2009). In China via the National Standard of the People's Republic of China GB/T 22731-2008, most of the IFRA Standards based on sensitization and the standards prohibiting the use of ingredients were supposed to come into effect by 2010.

6.4.2 Treatment of Naturals Under the Regime of IFRA Standards

If natural extracts show adverse effects, controlling them may include limiting the exposure of the whole mixture or limiting/reducing the natural component(s) responsible for the effects. The second course of action is only an option provided the effects are mainly linked to a limited number of known components and that there is an efficient mechanism to remove them without losing the olfactive properties that made the material important for perfumery purposes in the first place.

Natural materials can be affected by all three types of IFRA Standards: prohibition of the use of a material, restriction of the use of the material or purity requirement for its use, or a combination of standards. Furthermore, natural ingredients can be indirectly limited via the limitation of their constituents. Examples of all those cases are described in the following paragraphs.

One example of a natural material that has been prohibited by IFRA for use in fragrances both due to its phototoxic and due to its sensitizing properties [26] is fig leaf (*Ficus carica*) absolute (CAS 68916-52-9).

There are also several examples of purity requirements/specifications when it comes to natural ingredients. Birch wood pyrolysate (birch tar oil), a fragrance ingredient derived from the destructive distillation of the wood or bark of *Betula pubescens*, *B. pendula*, *B. lenta*, or *B. alba*, has a smoky burnt wood leathery note and may be used only if the oil has been rectified (purified) to reduce the content of polyaromatic hydrocarbons so that the concentration in the finished product does not exceed 1 ppb (parts per billion). The same limit applies whether the material is used alone or in combination with rectified cade oil.

Other examples of purity criteria for natural materials are the furocoumarin restrictions established for citrus oils such as bergamot or grapefruit oil (and other furocoumarin containing oils) where furocoumarins have been limited because of phototoxic properties. Another of efficient control measures addressing "impurities" are natural extracts with high contents of linalool or limonene like orange oil where the potential risk of the oxidation of these materials and the resulting creation of sensitizing hydroperoxides can be prevented by using antioxidants and by setting maximum limits on the peroxide value of the materials.

For some natural extracts, restrictions have been set (maximum use levels established). Since the 42nd Amendment to its Code of Practice, IFRA Standards for sensitizing ingredients are based on the quantitative risk assessment (QRA). For example, with the 43rd Amendment published in 2008, two standards were set for jasmine absolute. For jasmine, the composition and resulting (sensitizing) properties for the various types (different species) required IFRA to set two different standards, one for jasmine grandiflorum and one for jasmine sambac. Furthermore, for all individual ingredients for which use restrictions have been established because of their potential to induce sensitization above certain thresholds, contributions from other sources such as natural extracts have to be taken into consideration for the calculation of the total amount in the finished product. Using the example of cinnamic aldehyde, the Standard refers to Annex I to the IFRA Standards, which gives an indicative list of maximum levels of the restricted ingredients in various fragrance materials (including essential oils). Although the list is established on the best knowledge available to IFRA and intended to be as comprehensive as possible, yet in the end it remains illustrative only and cannot be regarded exhaustive. It is updated regularly as additional information becomes available, as part of an amendment to the IFRA Code of Practice. The indicative levels as provided in Table 6.3 for cinnamic aldehyde should be taken into account when determining the compliance of a fragrance compound under its conditions of use. If reliable analysis has shown that the level of the restricted ingredient in a specific fragrance material is not the same as the indicative level provided in the Annex, then the analyzed level can be used instead of the indicative level.

The following example should make the approach more transparent. The maximum use level for cinnamic aldehyde in an Eau de Toilette (level in category 3 of the QRA Standard, published as part of the 43rd Amendment) is 0.05%. If the respective fragrance compound does, for example, already lead to a concentration of 0.05% of Cassia oil in the finished product, there remains only 0.005% in the finished product that could result from the direct use of cinnamic aldehyde or be contributed via any of the other sources (see Table 6.3).

A very specific case is oak moss and tree moss extracts. Both materials have known dermal sensitizing properties and are often used in dermatology clinics as markers for fragrance allergy. IFRA very early started to establish policies and developed those over the years to prevent people from becoming sensitized. The first

TABLE 6.3 Natural Sources of Cinnamic Aldehyde

Cinnamic aldehyde	Cassia oil	<90
	Cinnamon bark oil	<75
	Cinnamon leaf oil	<3
	Hyacinth absolute	<3
	Tolu balsam resinoid	<2
	Styrax (*Liquidambar styraciflua*) oil	<1

43rd Amendment **IFRA STANDARD**

Oakmoss extracts

CAS N°: 90028-68-5 = Evernia prunastri extract
9000-50-4 = oils, Oakmoss resinoid
68917-10-2 =oils, Oakmoss

Empirical formula: N/A

Synonyms: Oakmoss absolute
Evernia absolute
Evernia prunastri, ext.
Mousse de Chêne absolute
Oakmoss absolute (*Evernia prunastri*)
Evernia prunastri (Oakmoss) extract

History:

Initial reviews:	April 1991, July 2001	
Current revision date:	2008	
Implementation date:	For new submissions*:	February 16, 2009
	For existing fragrance compounds*:	February 16, 2011
Next review date	2013	

* This date applies to the supply of fragrance compounds (formulas) only, not to the finished products in the marketplace.

RECOMMENDATION: RESTRICTED / SPECIFICATION

RESTRICTIONS:

Limits in the finished product:
For a description of the categories, refer to the QRA Information Booklet.

Category 1 See Note box (1)	0.02 %	Category 7	0.1 %
Category 2	0.03 %	Category 8	0.1 %
Category 3	0.1 %	Category 9	0.1 %
Category 4	0.1 %	Category 10	0.1 %
Category 5	0.1 %	Category 11	Not Restricted (2)
Category 6 See Note box (1)	0.5 %		

Note box:

For this material, for pragmatic reasons, restrictive levels allowed by the QRA for certain categories but actually being higher than those already in place before applying the QRA, will temporarily not be implemented until the end of a 5 year monitoring phase. At the end of the 5 years the position will be reevaluated again.

In the presence of tree moss extracts, the level of oak moss in the respective category has to be reduced accordingly such that the total amount of both extracts does not exceed the maximum permitted level in each category as listed in the table above.
If the same compound is intended for more than one IFRA QRA category, then the most restrictive limitation (based on foreseen use concentrations and maximum

IFRA Standard – Oakmoss extracts October 14, 2009 1/3

Figure 6.2 IFRA Standard on Oakmoss, 2008.

43rd Amendment **IFRA STANDARD**

Oakmoss extracts

permitted level) will apply.

(1) IFRA would recommend that any material used to impart perfume or flavour in products intended for human ingestion should consist of ingredients that are in compliance with appropriate regulations for foods and food flavourings in the countries of planned distribution and, where these are lacking, with the recommendations laid down in the Code of Practice of IOFI (International Organisation of the Flavor Industry) (http://www.iofiorg.org/).

(2) Category 11 includes all non-skin contact or incidental skin contact products. Due to the negligible skin contact from these types of products there is no justification for a restriction of the concentration of this fragrance ingredient in the finished product.

Fragrance material specifications: Oak moss extracts used in fragrance compounds must not contain added tree moss, which is a source of resin acids.

Traces of resin acids may be carried over to commercial qualities of oak moss in the manufacturing process. These traces must not exceed 0.1% (1000 ppm) dehydroabietic acid (DHA) in the extract.

The concentration of resin acids in oak moss can be measured with an HPLC Reverse Phase – spectrofluorometry method.

Further, levels of atranol and chloroatranol should each be below **100 ppm** in oak moss extracts.

Contribution from other sources: None to consider (see also the note on contributions from other sources in the **Introduction to the IFRA Standards**).

Critical effect: Sensitization

RIFM Summaries:

LLNA weighted mean EC3 values (μg/cm²)	Potency Classification Based on Animal Data[1]	Human Data			WoE NESIL[3] (μg/cm²)
		NOEL – HRIPT (induction) (μg/cm²)	NOEL – HMT (induction) (μg/cm²)	LOEL[2] (induction) (μg/cm²)	
970	Moderate	700[4]	1724[4]	1417	700

NOEL = No observed effect level; HRIPT = Human Repeat Insult Patch Test; HMT = Human Maximization Test; LOEL = lowest observed effect level; NA = Not Available

[1] Based on animal data using classification defined in ECETOC, Technical Report No. 87, 2003
[2] Data derived from HRIPT or HMT
[3] WoE NESIL limited to three significant figures

Figure 6.2 (*Continued*)

43rd Amendment **IFRA STANDARD**

Oakmoss extracts

[4] MT-NOEL = Maximum Tested No Effect Level. No sensitization was observed in human predictive studies. Doses reported reflect the highest concentration tested, not necessarily the highest achievable NOEL

REXPAN Rationale / Conclusion:

The RIFM Expert Panel reviewed the critical effect data for Oakmoss extracts and, based on the weight of evidence, established the No Expected Sensitization Induction Level (NESIL) as 700 ? g/cm². They recommend the limits for the 11 different product categories, which are the acceptable use levels of Oakmoss extracts in the various product categories. These were derived from the application of the exposure-based quantitative risk assessment approach for fragrance ingredients, which is detailed in the QRA Expert Group Technical Dossier of June 22, 2006.

References:

RIFM (Research Institute for Fragrance Materials, Inc.), 1973. Human Maximization Test. RIFM report number 1802, October 9a (RIFM, Woodcliff Lake, NJ, USA)

Research Institute for Fragrance materials, Inc. (1989). Human repeated insult patch test of oakmoss absolute. RIFM report number 12360, 31 October.

Research Institute for Fragrance Materials, Inc. (1989). Human repeated insult patch test of oakmoss absolute. RIFM report number 12361, 31 October.

Research Institute for Fragrance Materials, Inc. (1990). Human repeated insult patch test of oakmoss absolute. RIFM report number 12380, 1 March.

Research Institute for Fragrance Materials, Inc. (1990). Human repeated insult patch test on oakmoss absolute. RIFM report number 14118, 26 November.

RIFM (Research Institute for Fragrance Materials, Inc.), 2005. Local Lymph Node Assay. RIFM report number 50881, June 30 (RIFM, Woodcliff Lake, NJ, USA).

IFRA Standard – Oakmoss extracts October 14, 2009 3/3

Figure 6.2 (*Continued*)

standard was issued in 1988, limiting the level of oak moss absolute to 0.6% in finished product. The standard was revised in 1989 to also cover the concrete. In 1991, a standard on tree moss extracts (*Evernia and Usnea*) was added, the limit of 0.6% applying to both materials individually or in combination. The Standards were revised in 1997, reducing the level to 0.1%. In 2001, the standards were changed by introducing a limit of 0.1% for the maximum content of DHA (dehydroabietic acid) in the raw material, to control the formation of sensitizing oxidation products like 7-oxo-DHA. Despite all these measures, the number of people sensitized to the materials remains high.

The moss extracts have a very complex composition [27, 28] and some of their ingredients seem to contribute predominantly to their sensitizing properties. Research by Jean-Pierre Lepoittevin from the University of Strasbourg and others [29, 30] revealed that atranol and chloroatranol are two very potent sensitizers in the moss extracts. Atranol and chloroatranol are artifacts created during the production process of the moss absolutes. The industry developed a modified production process that results in moss products with low levels of both materials (below 100 ppm). IFRA notified new standards on oak and tree moss extracts in July 2008 (see Figure 6.2), setting new lower restrictions resulting from the QRA and introducing the purity requirement of 100 ppm maximum of each atranol and chloroatranol in the moss extracts.

It is hoped that these additional modifications will not only prevent new consumers from getting sensitized but also prevent already sensitized people from suffering skin reactions when using oak moss-containing products. Encouraging patch test data indicate that the oak moss with reduced atranol and chloroatranol levels has much lower potential to elicit reactions in already sensitized patients.

Oak moss is a particularly good example to adequately describe and summarize the complexity of trying to ensure the safe use of naturals. It is also an excellent example of a material that is unique and important for fragrance creation, as pointed out in the introduction to this chapter.

IFRA's and RIFM's focus will continue on ensuring the safe use of fragrance both from a consumer health and from environmental safety point of view. Profound knowledge about the composition of a material and the availability of a base set of data should be a basic requirement for materials used in consumer products of whatever origin. Ingredients used in perfumery should not be glorified or discriminated because of their origin or way of production but solely based on the evaluation of the available safety data including scientifically sound method of read-across and grouping, to prevent unnecessary testing.

REFERENCES

1. Jellinek JS. *Die Psychologischen Grundlagen der Parfümerie*, 4th edition. Heidelberg: Hüthig Verlag, 1999.
2. Storp F. PhD Thesis. Ludwig-Maximilians-University, Munich, 1997.
3. Schreiber WL. *Perfumes: Kirk-Othmer Encyclopedia of Chemical Technology*, 4th edition, vol. 18 Wiley-Interscience, 1996. pp. 172–201.
4. Kaiser R. *Meaningful Scents around the World: Olfactory, Chemical, Biological, and Cultural Considerations*, 1st edition. Weinheim: Wiley-VCH Verlag GmbH, 2006.
5. Vogel HA. Darstellung von Benzoesäure aus der Tonka-Bohne und aus den Meliloten - oder Steinklee Blumen. *Gilberts Ann. Phys.* 1820;64: 161.
6. Perkin WH. Perkin Reaction. *J. Chem. Soc.* 1868;21:53; 1877;31:388.
7. Reimer K. Ueber eine neue bildungsweise aromatischer aldehyde. *Berichte Deuts. Chem. Gesellsch.* 1876;9(1):423–424.
8. Tiemann F, Haarmann W. Ueber das coniferin und seine umwandlung in das aromatische Princip der Vanille. *Berichte Deuts. Chem. Gesellsch.* 1874;7(1):608–623.
9. Hall R, Klemme D, Nienhaus J. *H&R Lexikon der Duftbausteine*. Hamburg: Verlagsgesellschaft R. Glöss & Co., 1985.
10. Kraft P, Leard C, Goutell P. From Rallet No. 1 to Chanel No. 5 via Mademoiselle Chanel No. 1. *Perfum. Flavorist* 2007;10:36–48.

11. Perfume Notes, Blog of Luca Turin, http://lucaturin.typepad.com/perfume_notes/2006/01/blog_on_indefin.html, p. 58.
12. Aftel M. A comparison: aromatherapy, natural perfume and synthetic perfume. Fragrance Foundation Newsletter—Fragrance Forum, August 2008, p. 3.
13. Bickers D, Calow P, Greim H, Hanifin JM, Rogers AE, Saurat HJ, Sipes IG, Smith RL, Tagami H. The safety assessment of fragrance materials. *Regul. Toxicol. Pharmacol.* 2003;37:218–273.
14. Bickers D, Calow P, Greim H, Hanifin JM, Rogers AE, Saurat HJ, Sipes IG, Smith RL, Tagami H. A toxicologic and dermatologic assessment of linalool and related esters when used as fragrance ingredients. *Food Chem. Toxicol.* 2003;41:919–942.
15. European Union. Registration, Evaluation, Authorization and Restriction of Chemical (REACH), EC1907, 2006.
16. Lalko J, Api AM. Investigation of the dermal sensitization potential of various essential oils in the local lymph node assay. *Food Chem. Toxicol.* 2006;44:739–746.
17. Basketter DA, Andersen KE, Liden C, Van Loveren H, Boman A, Kimber I, Alanko K, Berggren E. Evaluation of the skin sensitizing potency of chemicals by using the existing methods and considerations of relevance for elicitation. *Contact Dermat.* 2005;52(1):39–43.
18. Lalko J, Api AM. Citral: identifying a threshold for induction of dermal sensitization. *Regul. Toxicol. Pharmacol.* 2008;52(1):62–73.
19. Aberer W. Contact allergy and medicinal herbs. *JDDG* 2008;6:15–24.
20. Schnuch A. Natur nicht ohne Nebenwirkungen (Nature not without side effects), Editorial. *JDDG* 2003;8(1):607–608.
21. Frosch PJ, Johansen JD, Menné T, Pirker C, Rastogi SC, Andersen KE, Bruze M, Goossens A, Lepoittevin JP, White IR. Further important sensitizers in patients sensitive to fragrances, II. Reactivity to essential oils. *Contact Dermat.* 2002;47:279–287.
22. Mitchell JC, Rook A. *Botanical Dermatology*. Vancouver: Greengrass, 1979.
23. Schempp CM, Schöpf E, Simon JC. Durch pflanzen ausgelöste toxische und allergische dermatitis (phytodermatitis). *Hautarzt* 2002;53:93–97.
24. Ernst E. Adverse effects of herbal drugs in dermatology. *Br. J. Dermatol.* 2000;143:923–929.
25. Lewis RA. *Lewis Dictionary of Toxicology*. Boca Raton, FL: Lewis Publishers, CRC Press LLC, 1998.
26. Opdyke DL, Letizia C. *Food Chem. Toxicol.* 1982;20:691.
27. Joulain D, Tabacchi R. Lichen extracts as raw materials in perfumery. Part 1. *Flavour Fragr. J.* 2009;24:49–61.
28. Joulain D, Tabacchi R. Lichen extracts as raw materials in perfumery. Part 2: Tree moss. *Flavour Fragr. J.* 2009;24:105–116.
29. Bernard G, Giménez-Arnau E, Rastogi SC, Heydorn S, Johansen JD, Menné T, Goossens A, Andersen KE, Lepoittevin J-P. Contact allergy to oak moss: search for sensitizing molecules using combined bioassay-guided chemical fractionation. GC-MS and structure–activity relationship analysis (part 1). *Arch. Dermatol. Res.* 2003;295;229–235.
30. Johansen JD, Andersen KE, Svedman C, Bruze M, Bernard G, Giménez-Arnau E, Rastogi SC, Lepoittevin JP, Menné T. Chloroatranol, an extremely potent allergen hidden in perfumes: a dose–response elicitation study. *Contact Dermat.* 2003;49(4):180–184.

CHAPTER 7

ADVERTISING SELF-REGULATION: A REVIEW OF COSMETIC CLAIMS AND NATURAL/ORGANIC CLAIMS

Annie M. Ugurlayan, Esq.

National Advertising Division, New York, NY

7.1 INTRODUCTION

Advertising self-regulation began nearly 40 years ago in response to growing public distrust of advertising and as a means to avoid excessive government regulation of advertising.[1] In 1971, the advertising industry's three major trade associations, the Association of National Advertisers, the American Association of Advertising Agencies, and the American Advertising Federation, together with the Council of Better Business Bureaus, formed an independent self-regulatory body called the National Advertising Review Council (NARC). NARC sets the policies for the divisions that make up the advertising self-regulatory system, including the National Advertising Division (NAD), which monitors and evaluates truth and accuracy in national advertising (in any medium, directed to consumers aged 12 and over, businesses, or other service professionals). Although NAD's self-monitoring efforts are limited due to funding constraints (most challenges are initiated by competitors), it routinely monitors claims of "natural" and claims implying health benefits to ensure that they are truthful.

The benefits of effective self-regulation are manifold, from speedily resolving complaints to creating high standards of truth and accuracy, to increasing public trust in the credibility of advertising, to enabling consumers to make better purchasing decisions, and to the promotion of fair competition. Government regulators, such as the Federal Trade Commission (FTC) and the states' attorneys general, can, as a result, devote their limited resources to investigate advertisers who engage in

[1] Cox, EF, et al. The Nader Report on the Federal Trade Commission, 1969.

Formulating, Packaging, and Marketing of Natural Cosmetic Products, First Edition.
Edited by Nava Dayan and Lambros Kromidas.

deliberate deception or endanger the health and safety of the public. The self-regulatory system is bolstered by the long-time support of the FTC.[2]

An NAD review of advertising may be initiated by a company (usually a competitor of the advertiser being challenged) consumers, advocacy groups, or by NAD itself through its ongoing monitoring program. There is no separate mechanism for reviewing the truth and accuracy of natural/organic claims. NAD has jurisdiction to assess the truth and accuracy of all claims in national advertising. To date, the issue of "natural" or "organic" cosmetics has not been the subject of competitor challenges. Companies who are trying to develop natural or organic cosmetics may themselves be unsure as to what constitutes a natural or organic cosmetic product. The terms are not defined by the US Food and Drug Administration (FDA), and industry groups have not yet provided coherent guidance on this issue. Companies also must consider how best to monitor the activities of their competitors and what remedies to pursue. Some manufacturers of cosmetic products are touting their products as "green" by virtue of their organic and natural ingredients that do not harm the skin with harsh chemicals and, by implication, that they are good for the environment, practices which are enabled by the plethora of seals that companies use to communicate the natural/organic/green benefits of their products.[3] In the absence of regulatory guidance as to what constitutes "natural" and "organic" in the context of cosmetic products, NAD, as the industry's self-regulatory program, plays an important role in reviewing advertising claims for cosmetic products to help alleviate consumer confusion and provide them truthful and accurate information while also leveling the playing field for advertisers of these products.

7.2 THE NAD PROCESS

The review process, though voluntary and informal, follows a detailed set of procedures available on NAD's web site.[4] Under the procedures, the parties to the proceeding provide letter submissions (including any relevant evidentiary exhibits) all of which remain confidential. An advertiser (or the company whose advertising is being called into question), but not a challenger (the company bringing the complaint before the NAD), may withhold from its competitor trade

[2] Pitofsky R, Chairman, FTC. Prepared Remarks at the D.C. Bar Association Symposium: Self-Regulation and Antitrust (Feb. 18, 1998). Available at http://www.ftc.gov/speeches/pitofsky/self4.htm.

[3] See, for example, http://www.cosmeticsdesign.com/Products-Markets/Good-Housekeeping-magazine-launches-green-seal-for-beauty-products; http://www.ecocert.com/Les-etapes-de-la-certification-en.html. The FTC's Guides for the Use of Environmental Marketing Claims ("Green Guides") provide guidance on "environmental claims included in labeling, advertising, promotional materials and all other forms of marketing, whether asserted directly or by implication, through words, symbols, emblems, logos, depictions, product brand names, or through any other means, including marketing through digital or economic means, such as the Internet or electronic mail. The guides apply to any claim about the environmental attributes of a product, package or service in connection with the sale, offering for sale, or marketing of such product, package or service for personal, family or household use, or for commercial, institutional or industrial use." See Section 260.2 at http://www.ftc.gov/bcp/grnrule/guides980427.htm.

[4] http://www.narcpartners.org/about/files/07_Procedures.pdf.

secrets and/or proprietary information or data (excluding any consumer perception data regarding the advertising in question). The challenging party may expedite the proceeding either by waiving its right to reply to the advertiser's submission or by requesting an expedited review at the outset of the proceeding. After the two rounds of submissions have been completed, parties to an NAD proceeding can meet *ex parte* with NAD to discuss their positions and allow the NAD the opportunity to ask any questions about the issues and evidence submitted. The meetings with the NAD and the informal nature of the self-regulatory process, which allows each party to contact the NAD attorney handling the matter with any questions at any time during the course of a proceeding, are among the reasons why NAD, unlike federal court or the FTC, is considered a user-friendly forum. Within approximately 60 business days[5] following the commencement of the proceeding, NAD releases a written decision in which it analyzes the supporting data, offering detailed findings, and determines whether or not the challenged claims are supported. If they are not, NAD will recommend that the advertising be modified or discontinued. Within five business days of receiving the decision, the advertiser provides NAD with an "advertiser's statement" in which it states whether or not it will comply with NAD's determination. The advertiser has the right to appeal NAD's determination.[6] If the advertiser refuses to comply with NAD's recommendations (or refuses to participate in the NAD process from the outset), NAD has the authority under its procedures to refer the advertiser to the appropriate regulatory agency. Unlike other dispute resolution forums where arbitration decisions are kept confidential, NAD's final decisions are published in the NAD/CARU Case Reports and made public through a press release.[7]

NAD's analysis of the advertising claims first involves a determination of what messages are communicated by the advertising at issue—not simply the express claims but also implied messages that may reasonably be conveyed to consumers. The parties may, but are not required to, submit consumer perception evidence of whether any implied messages are conveyed. In the course of its review, NAD assesses the reliability of any consumer perception evidence or other form of market research submitted to determine whether it is well conducted and methodologically sound. In the absence of reliable consumer perception evidence, the NAD, much like the FTC, steps into the shoes of consumers to determine the reasonable messages that are conveyed by the challenged advertisement. Second, NAD determines whether the advertiser has established a "reasonable basis" for its claims. Unlike the standard of proof for court proceedings under Section 43(a) of the Lanham Act, the advertiser has the initial burden to demonstrate that its claims are substantiated. NAD considers a variety of factors, including the six "*Pfizer* factors" set forth by the FTC: (1) the type of product, (2) the type of claim, (3) the benefits of a truthful claim, (4) the cost/feasibility of developing substantiation, (5) the consequences of a false claim, and

[5] Excluding extensions granted at the request of the parties and excluding the time involved in scheduling a meeting or teleconference with NAD.

[6] The challenger does not have an automatic right to appeal but may request permission to appeal from the Chair of the National Advertising Review Board, the appellate body of the self-regulatory forum.

[7] *Supra* note 4, § 2.9(C).

(6) the amount of substantiation that experts in the field believe is reasonable.[8] Where the advertiser is making claims that impact health and safety and establishment claims (i.e., explicit claims regarding the proof or support for other product claims), advertisers are typically required to produce "competent and reliable" scientific evidence to support such claims.[9] If a reasonable basis is established, the burden then shifts to the challenger to demonstrate that the evidence provided by the advertiser is fatally flawed or that the challenger possesses more reliable testing reaching a different result.[10]

In the case of "natural" and "organic" claims being made for cosmetics, NAD will, as always, look at the challenged claims in the context of the advertising at issue to determine the messages that are conveyed. In the absence of any defined standard of "natural" or "organic," the NAD considers industry usage, consumer expectation, and any testing and other scientific evidence that relate to the source and amount of each ingredient in the product. NAD appreciates that most cosmetic products contain synthetic ingredients and that consumers are not troubled by the inclusion of these ingredients, some of which are necessary to ensure the chemical stability of the product. Yet, there are many consumers who specifically seek out cosmetic products that contain "natural" and "organic" ingredients in whole or in substantial part and they are entitled to know whether the products that are marketed as such can support their claims.

Unlike the FTC, NAD does not determine that an advertiser has violated the law or engaged in deceptive or unfair advertising practices. Moreover, the self-regulatory process provides no award of monetary damages or other punitive measures, and no compulsory discovery is permitted. NAD's focus is on the truthfulness and accuracy of the advertising claims, and its authority is limited to making recommendations that claims not adequately substantiated be modified and/or discontinued.

NAD and competitors routinely follow up after a decision is issued to determine whether advertisers comply with NAD's recommendations. NAD may open a compliance proceeding where an advertiser has represented that it intends to follow NAD's recommendations, but a review of subsequent advertising shows otherwise. The advertiser is then asked to provide a status report detailing the steps it has taken to follow NAD's recommendations. If NAD determines the advertiser made a

[8] Pfizer, Inc., 81 F.T.C. 23 (1972); see also Basic Research, Report #4008, NAD/CARU Case Reports (Feb. 2003) (NAD applied the Pfizer factors in evaluating whether an advertiser provided a reasonable basis in support of performance claims); see also FTC Policy Statement Regarding Advertising Substantiation, appended to Thompson Medical Co., 104 F.T.C. 648, 839 (1984).

[9] FTC defines "competent and reliable scientific evidence" as "tests, analyses, research, studies, or other evidence based upon the expertise of professionals in the relevant area, that has been conducted and evaluated in an objective manner by persons qualified to do so, using procedures generally accepted in the profession to yield accurate and reliable results." See, for example, Consent Order, Schering Corp., 118 F.T.C. 1030 (1994) (No. 9232); Consent Order, The Quigley Corp., C-3926 (Feb. 10, 2000); Stipulated Final Order, Met-RX USA, Inc., Civ. No. SACV99-1407 DOC (ANX) (C.D.Cal. Nov. 24, 1999).

[10] Procter & Gamble Co. (Pampers Baby-Dry Diapers), Report #4599, NAD/CARU Case Reports (Nov. 2006); Colgate-Palmolive Co. (Palmolive Oxy Plus), Report #4252, NAD/CARU Case Reports (Oct. 2004); S.C. Johnson & Son (Shout Oxy Power Multi-Purpose Stain Remover), Report #4047, NAD/CARU Case Reports (May 2003); Mead Johnson & Co. (Enfamil with Iron), Report #4019, NAD/CARU Case Reports (Mar. 2003).

reasonable attempt to comply but issues remain, it will make additional recommendations in order for the advertiser to bring its advertising into full compliance with NAD's decision. If NAD determines that no effort was made to comply or that the advertiser has chosen not to make the recommended modifications, NAD will refer the matter to the appropriate regulatory agency for investigation, typically the FTC. The FTC has stated that it takes referrals from the NAD "very seriously,"[11] and some referrals have resulted in enforcement action by the FTC.[12] In 2008, NAD reviewed more than 150 cases and reported a compliance rate of approximately 95%.

7.3 NAD'S COSMETIC CASES

In the past several years, NAD has reviewed many advertisements for cosmetic products making very strong performance claims promising dramatic reductions in wrinkles and other unwanted signs of aging and, in many instances, posturing their products as equivalent to, or substitutes for, cosmetic procedures. NAD has opened a series of cases as part of its ongoing monitoring program to review the truth and accuracy of these types of performance claims.

One example where advertisers have taken NAD's recommendations to heart in a meaningful way is the case of The Procter & Gamble Company's Olay Regenerist line of skin care products. In 2004, NAD questioned print advertising claims made by the advertiser for its Regenerist Eye Lifting Serum such as the express claims "dramatically younger looking eyes" and "The Amino Peptide Complex hydrates to allow the natural regeneration of collagen and elastin" and the implied claim that Olay Regenerist works as good as (or similar to) eyelift surgery.

As support for its claims, the advertiser conducted many significant studies, including comprehensive, long-term (up to 4 months) double-blind clinical studies in which participants were asked to use facial moisturizers that contained varying doses of one amino peptide, palmitoyl pentapeptide-3, which is found in the amino peptide complex. These studies consisted of image (profilometry) and dermatological assessments. The advertiser also conducted danysl chloride testing,[13] which determined the product's exfoliating properties, visual attribute testing (expert and self-assessments), skin hydration testing using a corneometer (which measures the level of skin hydration), and transepidermal water loss (TEWL) studies, which measure the rate of moisture loss from the skin.

NAD determined that while the test results showed improvements in the skin's appearance (e.g., increased hydration) sufficient to support a general efficacy claim, they were not significant enough to support the strong "dramatically younger

[11] See, for example, "Self-Regulation in the Infomercial Industry," Remarks of FTC Chairman Deborah Platt Majoras (April 2006).

[12] United States District Court for the Northern District of Illinois Eastern Division) Civil Action No. 08 CV 3169, FTC File No. 062 3214 (Dietary Supplement Marketers Agree to Pay 2.5 Million to Settle FTC Charges for False Advertising Claims).

[13] Daily fluorescence readings were taken as the products (Regenerist Serum as sold and the Regenerist Serum without the amino-peptide complex) were applied twice daily with applications at least 4 h apart on days 2, 3, and 4.

looking eyes" performance claim—for example, while the dansyl chloride test revealed statistically significant results showing that the product makes skin smoother and softer, these results are typical of many moisturizers, including those containing exfoliating agents. Importantly, the test results did not directly speak to the "dramatically younger looking" claim and relevant aging parameters (e.g., wrinkle reduction). For example, in the 3 week single-product use test, many of the questions related to skin appearance ("makes eye area feel soft"; "makes eye area brighter/more radiant") and the "special instructions" given to each of the participants stressed that participants should "pay attention to the appearance and feel of your eye skin." The highest marks (excellent, very good, good) were reported for "makes eye area feel soft" and "makes eye area look smoother." However, none of the responses for the closed ended questions mirrored the "dramatically younger looking eyes" claim touted by the advertiser after using the products. NAD also determined that the test results from the instrument measurements and consumer assessments demonstrated significant improvements in the appearance of the skin (e.g., smoothness) due to increased hydration resulting from product use. However, the claim "regeneration of collagen and elastin" would need to be limited to the role scientists believe hydration plays in the regeneration of collagen and elastin because the test results did not address the impact, if any, of product use on collagen and elastin regeneration/production.

NAD looked to the advertisement as a whole and determined that it implied that Olay Regenerist works as well as (or similar to) eyelift surgery. The top of the advertisement contains the text "Three Zones. Two Eyes. One Total Eye Turnaround." Various parts of the woman's eye are highlighted and the text at the bottom of the ad discusses the serious nature of eyelifts and Regenerist's performance benefits. NAD determined that, taken together, the references to a "total eye turnaround" and "dramatically younger looking eyes" along with the reference to eyelifts being "serious business" and that women can use Regenerist without having to resort to "drastic measures" reasonably convey the message that the use of the Olay Regenerist Eye Lifting Serum can produce results comparable to eyelift surgery. Accordingly, NAD recommended that the implied claim be modified to avoid conveying this message.[14]

Of concern to NAD is the overstatement of performance benefits and, in this case, the comparisons to plastic surgery when the claims do not reflect the testing results. For example, in this case, the wording of the attribute questions in the consumer use testing that involved self-assessment of the product did not reflect the claims at issue—the participants were not asked whether their eyes looked "dramatically younger." As a general rule, performance claims based on self-assessments should reflect the wording of the attribute questions.

[14] See also, Poulter S, Edwards A. *Anti-ageing cream advert banned for misleading claims*, http://www.dailymail.co.uk/news/article-1209038/Anti-ageing-cream-advert-banned-misleading-claims.html (describing the ban by the United Kingdom's Advertising Standards Authority of a magazine advertisement for Nivea DNAge Cell Renewal Day Cream that claimed "An innovation in skincare, which boosts surface skin cell renewal leaving you with noticeably firmer looking skin" because it inferred that the product had a long-lasting or permanent effect on the skin).

NAD commenced another monitoring case a few years later for Olay Regenerist Eye DermaPod and the Regenerist Serum line. In the television commercial at issue, women are shown either with their face marked up with Xs as if they are about to undergo cosmetic surgery or are shown using the product on specific areas of the face in the manner a surgeon would use surgical instruments or a syringe when, in fact, it is the Olay products that are showcased as the "at home alternative." NAD identified express and implied messages in a television commercial as communicating that consumers can achieve results comparable to those achieved from injections and surgical procedures by using the Olay products. The advertiser agreed to voluntarily discontinue the television commercials. In the revised commercials, which show similar images of women about to undergo a procedure, the voiceover states that Olay Regenerist is not comparable to surgical procedures. The revised print advertisement states that products are "[n]ot for surgical results of course." NAD appreciates these improvements and hopes the willingness to embrace self-regulation and adhere to high standards of truth and accuracy will be emulated by other advertisers.

7.4 COSMECEUTICALS

In the past several years, the term "cosmeceuticals" has been widely used to describe a variety of cosmetic products. But what exactly are cosmeceuticals? As the name suggests, cosmeceuticals are generally defined as a cross between cosmetics and pharmaceuticals and are topically applied, as are cosmetics, but contain additional ingredients that are known to exhibit a biological function on the skin.[15] Cosmeceuticals are not defined or regulated by the FDA.[16] Cosmeceuticals are divided into categories based on their active ingredients: (1) antioxidants, which can reduce the harmful effects of free radicals that are molecules that injure the skin cells and cause inflammation; (2) peptides, which are smaller proteins that stimulate the production of collagen and thicken the skin; and (3) growth factors, namely, compounds that act as chemical messengers between cells and play a role in cell division, new cell and blood vessel growth, and in the production and distribution of collagen and elastin.[17]

A recent trend in cosmeceuticals is combination products, which include a variety of additional ingredients (e.g., vitamins A and C, which are known antioxidants), and takes an "inside out" product approach, claiming that the skin is "transformed" from within[18] (i.e., at a cellular level) based on the inclusion of certain ingredients. One such example is L'Oreal USA, Inc.'s Garnier Nutritioniste™ Ultra Lift™ and Garnier Nutritioniste™ Skin Renew. The print advertisement featured the

[15] Schwartz RM, MPH MD. *Cosmeceuticals*, http://emedicine.medscape.com/article/1067778-overview (May 21, 2008); http://www.aad.org/media/background/factsheets/fact_cosmeceuticals.html.

[16] Farris PK. *Cosmeceuticals*, American academy of dermatology (Nov. 2008); see also Heller, L. "Natural" will remain undefined, says FDA, http://www.foodnavigator-usa.com/content/view/print/5400 (Jan. 4, 2008).

[17] Ibid.

[18] See Montague-Jones G., *Beauty from within is going mainstream*, http://www.nutraingredients-usa.com/content/view/print/238016.

following claims: *Ultra Lift* —"Lifts wrinkles from the inside out"; "A clinical study proves: in 3 weeks wrinkles are visibly lifted and skin is noticeably firmer"; "Ultra-Lift with unique dermatological-nutrients refuels cells within skin's deepest surface layers: Omega 3 locks in moisture; vitamin A visibly reduces wrinkles; rice protein fortifies skin"; *Skin Renew*—"Lycopene is a powerful antioxidant, now it's in skincare proven to fight fine lines and dullness"; and "Only Garnier captures the power of our dermatological nutrients so Skin Renew transforms textures from the inside out for more flawless, visibly brighter skin."

As support for its claims, the advertiser presented numerous studies for each product. For Ultra Lift, the advertiser conducted a corneometer test with 24 participants that measured the level of their skin's hydration at 4, 8, and 24 h after the application of Ultra Lift on treated and untreated forearms. The results revealed a 15%–34% improvement in hydration on treated versus untreated sites after 4–24 h. There was also a 9 week clinical study involving expert assessments and self-assessments on 58 women subjects. The subjects were evaluated for four parameters (fine lines, shallow wrinkles, tactile roughness, and skin laxity) on a 10-point scale at baseline and at weeks 3, 6, and 9. The clinical grading results revealed a total improvement of 22% for fine lines, 29% for shallow wrinkles, and a similar improvement for skin firmness. The subjects also achieved improvements of over 60% in tactile roughness at weeks 3 and 9 and cutometer scores (measuring skin elasticity) showed significant improvements in skin elasticity, extensibility (which measures firmness), and resilience.[19] As for the self-assessment, 15.5% of the participants reported that Ultra Lift "visibly reduces the appearance of lines and wrinkles" on day 1, and over 65% so reported at weeks 3 and 9. Another study on 44 subjects aged 45–65 measured improvements in crow's feet wrinkles by silicone imprints of the skin taken 1 h after product application, and the results revealed a significant improvement in all four roughness parameters (average roughness, average dispersal, amplitude of the wrinkle, and surface smoothing). The advertiser also conducted a single-blind consumer use study in which 98 women used Ultra Lift once a day for 4 weeks. After 1 week's use, 72% of participants reported that their skin was firmer while 57% observed fewer lines and wrinkles, and after 4 weeks, 80% and 72% observed the same, respectively. The advertiser also referred to studies and scientific articles that explained that how each of the dermatological ingredients (vitamin A, rice protein, omega 3 and 6) helps reduce wrinkles by effecting an increase in collagen or helping skin retain its moisture.[20]

[19] Hydration—34% (4 h), 33% (8 h), and 15% (24 h); tactile roughness—62.2% (week 3) and 84% (week 9); cutometer—4.7% (week 3) and 13.5% (week 9); and "resiliency" of 21% (week 3) and 40.3% (week 9). The advertiser also conducted another study to assess skin firmness using a dermal torquemeter 4 h after application of Ultra-Lift on forearms and the results revealed that the treated sites were significantly firmer than the untreated sites.

[20] Vitamin A is known to increase collagen in the skin ("puffing" up skin from within and smoothing out wrinkles) and increase the thickness of the epidermis thereby reducing wrinkles, and retinyl palmitate, a form of vitamin A in Ultra-Lift, penetrates the outer layers of the skin and is released upon interaction with the skin's enzymes; *in vitro* studies on rice protein, which is rich in amino acids, revealed that it stimulates collagen production by increasing the proliferation of fibroblast cells in the skin; omega 3 and 6 are linoleic acids (fatty acids in the stratum corneum) which lock in skin's moisture.

As for testing on the Skin Renew product, the advertiser presented a 9 week controlled clinical study[21] consisting of expert and self-assessments. After 9 weeks, the expert assessments revealed a total improvement of 32.4% in fine lines, 20.7% more "radiant" skin, and a 92.6% reduction in "tactile roughness"; as for self-assessment results, for scores 6 to 9 (the most positive ratings, and the only ratings for which quantified improvements were reported), over 50% of the participants reported improvements in the "reduction in appearance of fine lines" and "looks radiant/glowing" attributes.[22] The advertiser also conducted a consumer use test in which participants reported similar results for more radiant skin and fewer lines attributes after the first application and after 4 weeks. Furthermore, the advertiser referred to a 3 week dansyl chloride test of 35 subjects that showed dansyl chloride stains from forearms treated with Skin Renew disappearing significantly faster than those on untreated forearms. The advertiser also maintained that Skin Renew contains lycopene that, it argued, had been proven to have a strong antioxidant effect by protecting skin cells from free radicals that work to destroy them.

NAD recommended that the claim "Now it's skincare that actually lifts wrinkles from the inside out" be discontinued or modified to reflect a reduction in the appearance of, and not elimination of, lines and wrinkles. NAD also recommended that the claim "A clinical study proves: in 3 weeks wrinkles are visibly lifted and skin is noticeably firmer" be modified to reflect the wording of the attribute questions in the studies (in which participants reported that their skin felt firmer not that it appeared "noticeably firmer"). NAD determined that there was a reasonable basis to support the "Ultra-Lift with unique dermatological-nutrients refuels cells within skin's deepest surface layers: Omega 3 locks in moisture; rice protein fortifies skin" portion of the ingredient claim but recommended that the "visibly reduces wrinkles" reference for "vitamin A" be modified to reflect the evidence in the record, that is, it helps reduce fine lines and wrinkles.

As for the Skin Renew claims, NAD determined that the advertiser had provided a reasonable basis for the claim that Skin Renew transforms skin from the "inside out" by accelerating the skin's cell regeneration process (namely, replacing surface skin cells with new cells created in the skin's deeper layers) but recommended that the claim "Lycopene is a powerful antioxidant, now it's in skincare proven to fight fine lines and dullness" be discontinued or modified to avoid conveying the unsupported message that lycopene can fight fine lines and dullness.

Performance claims require reliable product testing and the claims should reflect the testing submitted to support them. For example, testing at very short

[21] Clinical and self-assessments were conducted on 60 subjects who were evaluated at baseline and at weeks 3, 6, and 9 and evenly divided between those with normal, oily, and dry skin. Grading was done on a 10-point scale and three parameters were assessed: (1) fine wrinkles/lines; (2) tactile roughness; and (3) radiance. The self-assessment consisted of a questionnaire filled out at weeks 3, 6 and 9 where the participants rated the product's efficacy on several parameters (e.g., "looks radiant/glowing") on a 9-point scale.

[22] "Reduction in the appearance of fine lines," "looks radiant/glowing" attributes—53.3%, 70% and 68.3% reported an improvement at weeks 3, 6 and 9. Another study assessed the skin hydration levels of 24 subjects. Skin Renew was applied to one forearm while the other was untreated, and 4, 8 and 24 h after application, the treated forearms were 25%, 20.6% and 4.9% more hydrated than the untreated forearms, respectively.

intervals of time is necessary to promote claims that a product works immediately or after a single application. In Estée Lauder's Perfectionist Correcting Concentrate for Lip Lines,[23] among the claims at issue were "With a single stroke, this concentrated treatment visually improves the lip area in a way no injection can" and "Instantly, see up to 50% reduction in the look of deep vertical lines." The advertiser conducted a single-application study on 14 subjects aged 43–72. The lip area was digitally photographed immediately prior to and 10 min after product application. The photographs were analyzed with a powerful imaging device known as Optima 6.5. Based on this analysis, after 10 min, wrinkles were less visible by 50.24%. The advertiser also conducted another single-application study, also with 14 subjects (but not the same subjects as in the first single-application study), and the results showed that wrinkles were, on average, 54.84% less visible, and the advertiser decided to rely on the more conservative figure of 50%. NAD determined that this testing was sufficient to support the immediate benefit claims.

In some cases, advertisers do not conduct product testing and instead rely solely on studies that test individual ingredients (or a combination thereof) found in the product. While ingredient testing can be useful in determining the potential efficacy of a particular ingredient, it cannot typically serve as the sole basis of product performance claims.

For example, one case in which ingredient testing was submitted as support for product performance claims is IQ Cosmetics' IQ RestorEyes.[24] At issue were claims in Internet advertising such as "NO MORE DARK CIRCLES & CROW'S FEET," "Fight aging all day and night with 2 powerful treatments!," "Instantly decreases the appearance of dark circles," "Dramatically reduces the appearance of dark circles up to 45%," "Firms and Smoothes the appearance of wrinkles by 44–98%," and "Protects skin all day & all night with advanced anti-aging ingredients." The claims "No more dark circles and crow's feet" and "instantly decreases the appearance of dark circles" are juxtaposed to a visual of an eye with a dark circle and to the right the same image transforms and the dark circle disappears, a cycle that repeats continuously.

The advertiser's evidence consisted of materials from the ingredient suppliers that summarized the results of testing on various ingredients in the eye cream and concealer (in the same concentrations as they are found in the actual product). The advertiser argued that these ingredients, based on their properties, were designed to confer the claimed benefits. For example, dermaxyl was characterized as a "patented peptide formula" used to reduce the appearance of wrinkles and fine lines and combines a peptide fragment with skin reconstructive properties and a skin lipid related to skin moisture and renewal; the ingredient, haloxyl, was described as one that "eliminate[s] the two primary pigments that create dark circles under the eyes—bilirubin and iron" and is also an anti-inflammatory and aids in skin firmness and tone.

The advertiser also conducted a 6 week consumer use study on the eye cream only. The test, consisting of 32 subjects aged 32–69 who applied the product twice a

[23] Esteé Lauder, Inc. Perfectionist Correcting Concentrate for Lip Lines, Report #4368, NAD/CARU Case Reports (Aug. 2005).

[24] IQ Cosmetics/Intelligent Beauty, LLC (iQ RestorEyes), Report #4666, NAD/CARU Case Reports (May 2007).

day, was designed to "evaluate the consumer perception of various attributes of an eye cream." Among the results were as follows: 87% reported improvement in the appearance of deep wrinkles and lines around the eyes; 63% reported improvement in the appearance of lines and wrinkles within the first 5 minutes of application (which speaks to immediate results); 83% reported the skin around the eyes appeared firmer/better skin tone; and 77% reported improvement in the appearance of dark circles under the eyes.

NAD determined that the express and quantified performance claims were unsupported and recommended that they be discontinued. With respect to the ingredient testing, some documents provided a comprehensive summary of the testing conducted on certain ingredients, while other documents did not refer to the testing conducted or the methodology used, and none of the studies was provided to NAD. Moreover, no testing on the product in its full formulation (eye cream and concealer) was conducted to substantiate the impactful performance claims at issue.

The advertiser argued that the claims "the most effective system for controlling raccoon eyes, wrinkles and puffiness" and "a revolution in skin care" were puffery—that is, nonactionable claims. NAD did not agree. These claims appeared in connection with the visuals in the advertisement and other performance claims ("instantly decreases the appearance of dark circles") explaining to consumers what they can expect to achieve by using the product hence it cannot be puffery. As for the term "revolutionary" in the context of the advertisement, NAD determined that it communicates that the product is a breakthrough. NAD noted, however, that in other contexts, this term could be considered puffery. As for the consumer use study, while NAD was troubled by the small sample size, the subjects provided positive observations about the eye cream. Accordingly, NAD recommended that the advertiser limit its performance claims to the improvements observed by consumers in its consumer use study (e.g., reductions in wrinkling and dark circles under the eye and improvements in skin tone, texture, and firmness around the eye). Aside from the need for actual product testing to support performance claims, another important lesson from this case is that advertisers should beware of dramatic visuals that can convey product performance and transform a claim that, standing alone, is puffery into an actionable claim.

7.5 DIETARY SUPPLEMENTS

There is a great demand for dietary supplements that can confer cosmetic benefits either alone or in conjunction with a topical cosmetic product. In many instances, manufacturers of dietary supplements submit ingredient testing as support for product performance claims. Even when there is ingredient testing, only one ingredient or a combination of ingredients is tested that does not mirror the product as it is constituted. The amount of ingredient may also differ from what is found in the product. Moreover, the product may not be tested according to use instructions—for example, the participants took the supplement once a day instead of twice a day, or without a meal when they were supposed to take it with a meal. This can impact upon how the product is absorbed into the body and ultimately affects product performance.

NAD considered the submission of ingredient testing to support product performance claims in a case involving advertising for Biotech Corporation's DermaSilk AntiWrinkle Supplement. At issue was a television commercial making very strong establishment (i.e., establishing the level of proof to support the claims—for example, "clinically proven" claims) and performance claims: "Introducing DermaSilk AntiWrinkle. A revolutionary, age-defying anti-wrinkle supplement"; "It's like getting a face-lift without the invasive surgery"; "Helps repair damage caused by sun exposure"; "DermaSilk's natural skin formula helps increase firmness, hydration, elasticity and density"; and "Turn back the years with DermaSilk."

The advertiser offered no product testing or any clinical testing on a combination of ingredients similar to the product formulation. Instead, the advertiser referred to published human clinical and animal studies on individual ingredients: (1) carnosine which, in *in vitro* studies, has been shown to inhibit the aging process when it is exposed to connective tissue cells, rejuvenates cells that are far along in the aging process, and promotes collagen deposits[25]; (2) benfotiamine, a lipid-soluble form of thiamine, which inhibits the formation of advanced glycosylation end-products (AGE) which, in turn, stems the signs of aging (wrinkle formation, fine lines, and age spots)[26]; and (3) vitamin E and tocotrienols from vitamin E which help lower lipid oxidation and, hence, protect against the aging of skin cells.

The advertiser argued that the ingredients in DermaSilk are strong free radical scavengers that protect the skin from structural damage such as wrinkles and fine lines, oxidative stress, sun damage and lipid peroxidation. The advertiser also maintained that the claims "A revolutionary, age-defying anti-wrinkle supplement" and "It's like getting a face-lift without the invasive surgery" are puffery because they are not performance claims and they are ubiquitous in the marketplace; as such, consumers understand that they have no meaning. However, even if they were not puffery, the advertiser argued that the claims are substantiated because the product was designed to contain only those ingredients scientifically shown to reduce the signs of aging and that the product is revolutionary because of the unique blend of ingredients and the form in which it is administered (a capsule).

NAD did not agree and recommended that the advertiser discontinue the claims "DermaSilk is clinically proven to help reduce the appearance of wrinkles, fine lines, age spots and premature aging," "Introducing DermaSilk AntiWrinkle. A revolutionary, age-defying anti-wrinkle supplement," "It's like getting a face-lift without the invasive surgery," "Helps repair damage caused by sun exposure," and "Turn back the years with DermaSilk" because there was no product testing or testing of the formulation. As a general rule, ingredient testing is not sufficient to support product performance claims. As for the remaining claims ("DermaSilk's natural skin formula helps increase firmness, hydration, elasticity and density;" "For smooth, soft, younger looking skin, ask for DermaSilk Anti-Wrinkle"), NAD recommended that the claims be modified to reflect that there is emerging evidence that some of the

[25] Guiotto A, et al., *Carnosine and carnosine-related antioxidants*: a review. Curr. med. chem. 2005;12:2293.

[26] See, for example, Thomas MC, et al., *The role of AGEs and AGE inhibitors in diabetic cardiovascular disease*, Curr. drug targets 2005;6:453.

ingredients in DermaSilk have been shown to have potential benefits in improving the appearance of aging skin and in reducing oxidative stress on the skin.

The substantiation was far more comprehensive in NAD's review of advertising for Natrol, Inc's BioSil dietary supplement. Among the claims made in print and Internet advertising were as follows: "That's why BioSil is clinically proven to give you a more youthful look in just 20 short weeks"; "Reduces Fine Lines and Wrinkles by 19%"; "Creates Thicker, Stronger Hair"; and "Provides Stronger, More Break Resistant Nails."

In support of its claims, the advertiser presented two studies—both were published, double-blind, randomized placebo-controlled studies. The first, the Barel study, was a 20 week trial with 50 women, aged 40–65 years who had photodamaged skin. They were instructed to take two choline-stabilized orthosilicic acid capsules (the same form as it is found in the actual product—10 mg/day in accordance with product use instructions). Skin hydration was measured with a corneometer. Nail brittleness was a secondary endpoint in the study and was measured by a 4-point visual analog scale. After 20 weeks, a 19% statistically significant improvement was reported in the maximum roughness of the skin on the inner forearm based on skin visiometer measure of microrelief[27] and improved skin elasticity was also reported on the forehead as measured by a reviscometer.

In one study, the Wickett Study, 48 participants (ages 40–65) also took 10 mg/day of ch-OSA and the purpose of this study was to determine whether ch-OSA improves hair thickness and strength. A tensile strength analysis was conducted to quantify hair damage[28]—for example, hair with poor elasticity will stretch only to a limited extent and additional stress will cause permanent deformation or breaking. A cross-sectional area of the hair shaft was also measured to determine hair thickness.

NAD determined that the advertiser provided a reasonable basis for the claims "Creates Thicker, Stronger Hair" and "Provides Stronger, More Break Resistant Nails" because there was a statistically significant increase in the cross-sectional area of the hair (i.e., hair thickness) compared to placebo and in the objective and subjective measurements of hair thickness and strength. There were also statistically significant improvements in nail strength compared to placebo. However, NAD determined that the claims "That's why BioSil is clinically proven to give you a more youthful look in just 20 short weeks" and "Reduces Fine Lines and Wrinkles by 19%" were unsupported and had to be discontinued because they were based on an assessment of the roughness of the forearm skin that does not reflect whether the participants achieved a more "youthful look" (roughness on forearms does not necessarily correlate with fine lines and wrinkles on the face) and, in fact, there is no evidence that the improvements made the participants look more youthful. However, NAD noted that the advertiser may promote improved microrelief measurements (reducing maximum roughness of skin and improvements in skin viscoelasticity on the forehead).

[27] Skin microrelief is a measure of fine lines and wrinkles. Microrelief or microwrinkles develop over time into deeper, more visible wrinkles.

[28] A constant speed of extension is exerted on a single hair fiber to extend the fiber until it breaks from which measurements of break load and elastic gradient and break stress and elastic modulus are taken.

7.6 CURRENT TRENDS

In response to growing consumer interest in natural and organic products, consumers perusing the cosmetics aisle in virtually any drugstore in the United States will see many products characterized as "natural" and organic." While the US Department of Agriculture (USDA) has established standards for organic foods, and uses private and state agencies to inspect and certify food companies that market organic foods[29], no such standards apply to cosmetic products. The term "natural" remains undefined for both foods and cosmetic products. However, there has been an impetus from industry, as there has been for organic products, to define the term and develop a seal.[30] The potential for abuse and is great as consumers have no way of determining whether a product is, in fact, organic or natural or if one product is more "organic" or "natural" and, by extension, better for consumers and/or the environment.[31]

NAD has been increasingly called upon to review claims concerning the environmental (or "green") benefits products. One such case was a challenge brought by The Procter & Gamble against Church and Dwight Co., Inc's Arm & Hammer Essentials Liquid laundry detergent. Among the claims at issue were the following: (1) "Harnessing the Power of Nature™... to deliver cleaner, fresher laundry while being better for you and your family, more sensible for the environment"; (2) "100% Plant-Based Soaps"; (3) "No Petroleum-Based Cleaners"; and (4) "Naturally Cleaner & Fresher!" NAD determined that the net impression created by the challenged advertising was that this product provides a significant environmental benefit compared to the competition—that is, something unique to the Essentials product. The print advertisement features a picture of a mother and her daughter with the headline "1970: Mom's Secret: Use a More Sensible Detergent With No Compromises" under a box of Arm & Hammer detergent (in color) shown in a basket of laundry with "NON-POLLUTING" written on the top of the box and the next page showing a mother and son in 2007 with the headline "2007. Your Secret: The Totally New Cleaning Power of Baking Soda Plus 100% Plant-Based Soaps. Introducing Arm & Hammer® Essentials™." The text above this states: "Mom knew what was right for her family. So do you. New Essentials™ Now, all of your laundry will be whiter and brighter, every load cleaner and fresher than before, because only Essentials gives you the powerful combination of Baking Soda and 100% plant-based soaps. It's more sensible for you, your family, and the environment." At the bottom of the advertisement is the claim, in large typeface, "Harnessing The Power Of Nature."

[29] Organic foods, http://usda-fda.com/Articles/Organic.htm.

[30] See Duber-Smith, D. *A natural progression in personal care*, HAPPI (Jan. 2009), p. 58, citing the definition by the Natural Products Association (NPA), available at http://www.npainfo.org/index.php?submenu=NaturalStandard&src=gendocs&ref=NaturalStandard_new&category=NaturalStandard (defining a product as "natural" if it is "[m]ade up of only, or at least almost only, natural ingredients and be manufactured with appropriate processes to maintain ingredient purity," if animal testing was not conducted to develop it, there is no potential human health risk and the product uses "biodegradable ingredients and the most environmentally sensitive packaging." The NPA also has developed a Natural Standard and Certification Program.)

[31] Ibid.

In this advertisement, the 100% plant-based soaps is identified as the reason for Essentials being "more sensible" for the environment.

One source of guidance for environmental marketing claims is the FTC's Green Guides that require, among other things, that such claims be supported by "competent and reliable scientific evidence."[32] What makes one product better than another from an environmental standpoint is not clear at first blush. For instance, one product may produce less solid waste than another and while this may seemingly appear to be environmentally more beneficial, a cradle-to-grave life cycle analysis may show it is more harmful for the environment because it creates greater chemical emissions during production than the product that produces more solid waste. The parties submitted literature on life cycle analyses on detergent surfactants; however, none looked at any specific detergent but, rather, at the impact of various surfactants throughout their life cycle to determine their environmental impact which is not dispositive in determining the environmental impact of any one product.

NAD recommended that the claims "100% Naturally Derived Surfactants," "100% Percent Plant-Based Soaps," and "No Petroleum-Based Cleaners" be discontinued given that only about 35% of the surfactants in the actual product were plant-based. NAD recommended that the advertiser remove the reference to "naturally" in the claim "Naturally Cleaner & Fresher!" NAD determined that the advertiser had provided a reasonable basis for its "Harnessing the Power of Nature" claim because the surfactants are made from plants and the surfactants in the final product are nearly 35% plant-based. While the phrase "harnessing the power of nature" is unlikely to give rise to the impression that the product as a whole (or all of its surfactants) is 100% natural standing alone, NAD cautioned the advertiser to avoid juxtaposing the claim with elements (claims and/or visuals) that may give rise to an unsupported message that the product is natural, in whole or in substantial part. NAD further recommended that the advertiser discontinue the "more sensible for the environment" claim and avoid conveying the unsupported message that the Essentials product is good or better than competing products, for the environment or that it confers a unique environmental benefit.

NAD has also dealt with the issue of express and implied "natural" claims in food products. One such case was a challenge brought by the Center for Science in the Public Interest against The Procter & Gamble Company for claims made in television advertising for its Olean fat substitute.[33] The setting of the commercials is a farm and farm workers explain that Olean is a new kind of cooking oil "Starting from crops that grow here. Like soybeans. Though I don't like too much snacking, I do think that chips fried up in Olean are a better choice than the regular kind. It tastes real good, has a lot less fat and fewer calories."[34] NAD determined that the advertiser provided a reasonable basis for its claims that chips fried in Olean are "a little healthier" or

[32] See Section 260.5 at http://www.ftc.gov/bcp/grnrule/guides980427.htm, *supra* note 3.

[33] Procter & Gamble (Olean Fat Substitute), Report #3499, NAD/CARU Case Reports (Oct. 1998), quoted in Guidance for Food Advertising Self-Regulation: an historical overview of the investigations of food, nutrition and weight loss advertising of the Children's Advertising Review Unit (CARU) and the National Advertising Division (NAD), National Advertising Review Council, 2004, available at http://www.narcpartners.org/reports/NARC_White_Paper_6-1-04.pdf, pp. 56–57.

"a better choice" than regular chips (because they are lower in fat and calories), and that in the absence of any health or safety claims, it did not need to make the GI disclosures required by FDA on packaging.[35] However, NAD recommended that the advertisement be modified to avoid conveying a "natural" ingredient message (e.g., made from soybeans) because Olean is a synthetic product.[36]

In the absence of regulatory guidance defining "natural" and "organic" in cosmetic products, industry standards and usage as well as consumer expectations provide useful guidance. The efforts of trade associations and experts in the field play an important role in developing a consensus on definitions for these terms in order to alleviate consumer confusion and create a level playing field for all manufacturers. Although NAD is not authorized to independently define what "natural" and "organic" mean when used to describe a cosmetic product, or even in any other product, it can through its review of challenged advertising determine whether or not an advertiser's claims are truthful, accurate, and/or misleading based on any relevant test and industry usage as well as evidence of consumer understanding and expectation.

In conclusion, NAD's goal in its cosmetics cases is, as in any case, to ensure that advertising claims are truthful and accurate. This product area has evolved enormously over the years and NAD's decisions seek to ensure that claims about a product mirror the testing conducted on the product. In doing so, NAD upholds standards that enhance the credibility of brands, encourages product innovation, and improves consumer confidence in advertising.

[34] Ibid. p. 57.

[35] Ibid.

[36] Ibid.

PART III

SAFETY ASPECTS

CHAPTER 8

THE SAFETY ASSESSMENT OF PLANT-DERIVED INGREDIENTS IN COSMETICS

Viny Srinivasan[1] and Eric Antignac[2]

[1]L'Oreal USA Inc., New York, NY

[2]L'oreal R&D, Worldwide Safety Evaluation, Asnieres, France

8.1 INTRODUCTION

Today, there is a growing consumer demand for natural and organic cosmetic products. Accordingly, the beauty sector has seen a surge in interest and sales of these products from retail to high-end stores as well as from online purchases to salons. Thus, the distribution of natural and organic cosmetics is increasing in places such as supermarkets, drugstores, salons, and pharmacies around the world, especially in North America and Western Europe.

According to the statistics from Organic Monitor, a British-based business research and consulting company that specializes in global organic and related-product industries, natural and organic cosmetics were estimated to account for 3% of all cosmetic sales in Europe during 2008. Although the figure may seem a small percentage, the amount has been increasing by 20% in recent years, and in some countries, such as Germany, the sector's market share is expected to approach 10% by 2012. The European market for natural and organic cosmetics totaled €1.1 billion, or $1.5 billion, in revenue in 2006, while the global sales of natural and organic cosmetics continued to soar with revenues close to US$7 billion in 2007 [1].

The major driving force of this market is the public demand for both environment- and health-related products. Therefore, proliferation of organic and natural personal care products is making consumers scrutinize the ingredient composition and demand products with high levels of natural extracts as well as certified organic ingredients that are being used ethically. The certification of natural and organic cosmetic ingredients and products is mainly performed by a private owner's standards. However, the increase in number and complexity of ingredients and products, a lack of transparency associated with certification processes, and the nonexistence of

Formulating, Packaging, and Marketing of Natural Cosmetic Products, First Edition.
Edited by Nava Dayan and Lambros Kromidas.

internationally harmonized standards can lead to many products being launched in the global marketplace without meeting worldwide acceptable standards. Moreover, these standards deliver ethical and health claims associated with certified natural and organic cosmetic products but do not develop any scientifically sound approach aimed to assess safety in the use of these products. However, it is crucial to understand and remember that being natural does not mean the absence of unacceptable toxicity. During the past decade, there have been a significant number of cases of intoxication with plant-derived ingredients used as food supplements [2, 3]. These adverse effects have been shown to result from misuse, misidentification of the plant-derived ingredients, or from contamination with toxic plants [2, 4]. Adverse reactions of herbal drugs are also fairly well documented. Systemic toxicity, local adverse reactions, and allergic reactions resulting from the use of herbal drugs have been reported and reviewed [5, 6]. Together with the large variety and diversity of plants providing these ingredients, ranging from staple food sources such as cereals, fruits, and vegetables to herbal ingredients used in traditional medicines, as well as the multiplicity of extracts available, all these data raised safety concerns regarding plant-derived ingredient safety. Consequently, both food industry and regulatory authorities have extensively debated the safety assessment of plant-derived ingredients as food and food ingredients. These discussions clearly indicated that adequate specifications with regard to identity and composition were the key issues in the safety assessment of plant-derived ingredients. Accordingly, guidance tools including specific issues to consider in the safety assessment of plant-derived ingredients as food ingredients were published [7–10]. At the same time, both the European Medicines Agency (EMEA) and the Food and Drug Administration (FDA) published regulations relating to the safety assessment of herbal medicines placing particular emphasis on identification and characterization of these drugs [11–14]. Whatever be the regulatory status, the safety of the products must be established for their intended use. In this context, the European cosmetic regulation states that (1) a cosmetic product should not cause any damage to human health and (2) the manufacturer is responsible for placing the cosmetic products on the community market [15]. To achieve this goal, a safety assessment of the finished product should be performed taking into consideration the toxicological profile of the individual ingredients, their chemical structure, and their level of exposure. Accordingly, a guidance on the safety assessment of the cosmetic ingredients has been drafted by the Scientific Committee on Consumers Products [16, 17]. However, this guidance essentially dealt with the safety assessment of chemically well-defined ingredients but did not approach the safety assessment of natural substances, although the latter strongly appears as a priority. Therefore, one of the major challenges the cosmetic industry faces today is the lack of relevant approach aimed to specifically assess safety in the use of ingredients from natural or organic origin.

In this chapter, we propose a practical and pragmatic approach to assess safety in the use of natural and organic cosmetic ingredients and products. These approaches are mainly based on tools and methods that have already received both scientific and regulatory acceptance in the safety assessment of plant-derived ingredients as food supplements, novel foods, novel food ingredients, and herbal medicines.

The first step consists of a comprehensive botanical and chemical characterization of the ingredient under evaluation. Since most of the plant-derived ingredients can have a potential history of human use as foods and herbal medicines, the second step consists of comparing the ingredient under assessment with one or several comparators that have an established history of safe use in human. Obviously, the latter should be determined by exploiting all the available data. The two aforementioned steps will serve as a point of departure in order to determine if there is a need to generate additional safety data. This guidance tool should obviously be fully embedded in the risk assessment paradigm, that is, hazard identification, hazard characterization, intended use and exposure assessment, and risk characterization. A particular emphasis should be placed on the exposure assessment, since cosmetic products are topically applied.

8.2 DEFINITIONS AND SAFETY-RELATED ISSUES

Plant-derived ingredients used in cosmetics include a variety of plant extracts, tinctures, fatty or essential oils, expressed juices, and more purified components [18]. The variety of plants providing these natural ingredients is very large, ranging from staple food sources such as cereals, fruits, and vegetables to herbals used in traditional medicine. They may be obtained from cultivated or wildly growing plants.

A key issue in the safety assessment of plant-derived ingredients is a widely acknowledged misconception that natural means "safe." Although most of these ingredients are derived from plant organism with a history of traditional use as food and herbal medicine, it does not necessarily mean that they are safe under the proposed conditions of use.

It is well established that if food plants obtain an adequate dietary intake of essential and nutritional components such as proteins, amino acids, fats and fatty acids, complex carbohydrates, fibers, vitamins, and minerals, it can lead to human consumption of a variety of secondary plant metabolites formed as part of the plants own defense against herbivores or plant diseases. Some of these components such as carotenoids have a well-known and established nutritional value, whereas others such as isothiocyanates or indoles are potent toxicants. This is also true in the case of some glycoalkaloids, such as α-solanine and α-chaconine of Solanaceae plants, including potatoes [19, 20]. An extensive analysis of the available literature based on 12 separate reports of potato glycoalkaloid poisoning revealed more than 2000 cases resulting in about 30 deaths. Moreover, the review argues that most of the mild and even the more serious "solanine" poisonings go unrecorded and are solely diagnosed as "gastroenteritis" [21]. Based upon a risk assessment of the content of total glycoalkaloids in potatoes, the Joint Expert Committee on Food Additives (JECFA) of the World Health Organization (WHO) concluded in 1992 that normal levels of total glycoalkaloids in potatoes ranging from 20 to 100 mg/kg of unpeeled potatoes would not be of toxicological concern [22]. Another well-established example of plant food toxins are the cyanogenic glycosides contained in the roots of cassava and grass pea. These compounds may give rise to acute and chronic intoxications in human. The consumption of raw or inadequately processed roots may cause acute symptoms or chronic neurological disorders [23]. In spite of that, these plants are still used as food. The above-mentioned examples demonstrate that even edible plant foods can give rise to serious human health issues. However, many of these natural constituents in plants have excellent human health benefits at lower levels and therefore are widely used as food supplements. Typical examples of such products would be soy extracts for estrogenic activity of isoflavones and tomato extracts containing the antioxidant lycopene as well as concentrated polyphenols from grapeseed. Although the source materials (grapeseeds or green tea) have a "history of safe use" as food, the safety of these concentrated compounds should be evaluated with caution as the margin of safety between adverse physiological effects and the desired beneficial effect may be very narrow.

Other ingredients consist of, or may be derived from, plants that are traditionally used for health maintenance purposes such as herbal teas or used as traditional medicine products all around the world. These plants are used in specific cases to improve general health or in disease conditions. Their extensive use as food supplements has triggered significant concerns among scientific and regulatory communities. These products often fall into a gray area between food and medicines, which is typically undefined or underdefined from a regulatory perspective [24]. Furthermore, there have been a number of recorded cases of intoxication with plant-derived ingredients raising concerns regarding their safety [6, 25, 26]. For instance, the Food and Drug Administration has issued a final regulation declaring dietary supplements containing ephedrine alkaloids as adulterated under the Federal Food, Drug, and Cosmetic Act because they present an unreasonable risk of illness or injury [11, 12]. This action has been taken based upon the well-known pharmacological effects of ephedrine alkaloids, the peer-reviewed scientific literature, and the important number of adverse events reported to have occurred in individuals following consumption of dietary supplements containing ephedrine alkaloids [26]. Furthermore, several cases of intoxication reported with some plant-derived ingredient products have been shown to result from contamination with other plant species or through misidentification of species [2]. For instance, exposure to *Aristolochia* sp. in unlicensed herbal products has resulted in large numbers of patients having renal failure, which in some cases has proved fatal. Some patients have also subsequently developed urothelial cancers as a result of exposure to aristolochic acids. The latter have been shown to exert strong mutagenic effects in bacterial and mammalian cell systems and to be potent carcinogens in rats and mice at low dose levels and also in humans at micrograms per kilogram body weight (bw) levels [27, 28]. It has been evidenced that the origin of these adverse events was directly related with the contamination by *Aristolochia* sp. containing high levels of aristolochic acids or with the misidentification of plants [29].

It is important to point out that the potential adverse effects associated with plant-derived ingredients are not based on systemic toxicity alone. It is well known that certain plant-derived ingredients may induce local reactions when applied onto the skin, which is a critical issue for cosmetic ingredients. Thus, St. John's wort (*Hypericum perforatum*) extracts have been shown to be phototoxic in an applied dose and a UVA-dose dependent manner [30, 31]. Hypericin, one of the main constituents of *H. perforatum* extracts, has been found to induce phototoxic effects and was identified as the main phototoxic constituent of the extract. Since *H. perforatum* extracts are herbal medicines with a well-established use as dermatological drugs in Europe, health authorities have advised that during the treatment, intense UV exposure of the respective skin areas should be avoided. In addition, it is well established that essential oils, natural complex substances (NCS) (which have a wide range of traditional use as topical drugs), fragrances, and cosmetics can contain varying amounts of naturally occurring contact sensitizers [32].

The above examples clearly demonstrate that plant-derived ingredients with a human history of use as food and traditional medicines can present potentially significant health risks to the consumers. Therefore, they need to be reviewed based on their use and undergo a comprehensive safety assessment.

A second key issue in the safety assessment of plant-derived ingredients is that they are often highly complex mixtures that complicate their toxicological testing from a methodological point of view. It has long been recognized that whole foods cannot be tested in accordance with the well-established and standard safety evaluation principles [33] used for single substances such as pharmaceuticals, food additives, or food contaminants [34]. Thus, traditional toxicological testing procedures for whole plant extracts are not practical and have limited use because of their bulkiness and nutritional content, which can lead to nutritional imbalances in animals at high levels.

Finally, when available, the reliability of safety data on plant-derived ingredients should be carefully assessed. The safety assessor should verify that the natural variability of the plant extract tested has been considered before extrapolating these data to similar plant-derived ingredients. Given that the first prerequisite of a cosmetic ingredient is consumers' safety, there is a need to develop a pragmatic and reliable approach to ensure a globally acceptable assessment of safety of the plant-derived ingredients.

8.3 REGULATORY ASPECTS

8.3.1 Food

Plant-derived ingredients used as dietary supplements are very popular health care products in the United States. Therefore, in order to protect human health, the U.S. Congress passed the Dietary Supplement Health and Education Act (DSHEA) in 1994. Since then, the U.S. FDA promulgated a series of regulations to ensure both the quality and safety of dietary supplements, including plant-derived ingredients used in dietary supplements [35, 36]. Accordingly, the safety of these products has received a lot of attention. As a result, the U.S. NTP (National Toxicology Program) has performed toxicological studies, including carcinogenicity bioassays, on a series of plant-derived ingredients used as dietary supplements, such as *Ginkgo biloba* leaf extract, *Aloe vera*, *Panax ginseng*, and Kava [37]. In Europe, the use of plant-derived ingredients as food is regulated under the General Food Law [38], which attributes the primary legal responsibility for the safety of the products placed on the market to business operators. The regulation, however, does not provide any guidance on how the safety of these products should be assessed. However, in 2004, the EFSA expressed concerns about quality and safety issues of plant-derived ingredients widely available to consumers in the European Union. Therefore, the EFSA's Scientific Committee currently provides guidance on the scientific data needed to carry out a safety assessment of a plant-derived ingredient used as food supplement [10].

Plant-derived ingredients used as food are also regulated under novel food regulations in other countries such as New Zealand and Canada [39–41]. Generally speaking, novel foods are defined as food with no traditional history of use. The safety assessment of novel foods has been extensively debated. Therefore, guidelines have been developed by a number of regulatory authorities and others to assist those submitting products for consideration under legislation concerning novel foods [42–46]. Briefly, the safety assessment of novel foods and novel food ingredients is based upon the principle of substantial equivalence. This principle is designed to highlight the similarities as well as any differences between the new food and its traditional counterpart.

8.3.2 Drugs

Plants and plant-derived ingredients also have a long traditional history of use as herbal medicines among various ethnic groups. In the United States, when a herbal is claimed to cure, treat, or prevent human disease, it is classified as a drug and the manufacturer is required to provide scientific evidence showing that the product is safe and efficacious in humans according to the regulation promulgated by the FDA [47]. Considering the specificity of herbal medicines as drugs, the U.S. Food and Drug Administration also published a guidance document to assist academic and industry sponsors to develop such drugs [11, 12].

In Europe, traditional herbal drugs are regulated by the Directive 2001/83/EC as amended by Directive 2004/24/EC of the European parliament and the Council of 31 March 2004 with regard to traditional herbal medicinal products [48–50]. Chapter a of the Directive 2001/83/EC as amended establishes specific provisions for the simplified registration of traditional herbal medicinal products with a long-standing medicinal use of at least 30 years (including at least 15 years within the community). Accordingly, the EMEA issued guidance aimed to assess both the quality and safety of herbal medicines [13, 30, 31, 51, 52].

8.3.3 Cosmetics

In the United States, cosmetics are regulated by the U.S. Food, Drug, and Cosmetic Act that designated the U.S. Food and Drug Administration as the agency responsible for the safety of cosmetics. According to this regulation, the safety of cosmetic products and their ingredients is the responsibility of the manufacturer. There is no specific regulatory guidance on the safety assessment of cosmetics, including plant-derived cosmetic ingredients in the United States.

In the European community, a cosmetic regulation, namely, the European Directive 76/768/EC, was introduced in 1976 and is periodically updated. According to this regulation, cosmetic products do not require a premarketing clearance. However, the safety of cosmetic products and their ingredients is the responsibility of the manufacturer, with the exception of certain ingredients such as UV filters (Annex VII of the Directive), preservatives (Annex VI), colorants (Annex IV), and, most recently, hair dyes require approval prior to marketing. Similarly, banned ingredients are listed in Annex II of the Directive, whereas substances that are concentration limited are listed in the Annex III of the Directive. In the European Union, the approval process includes submission of a safety dossier to the EU Scientific Committee of Consumer Products/Safety (SCCP, recently renamed SCCS) that issues an *Opinion* on the safety of the ingredient in question. In 2006, the SCCP issued guidance on the safety assessment of cosmetic ingredients and products. However, this guidance is adapted to the safety assessment of chemically well-defined ingredients, but appears to be less suited for the safety assessment of complex plant-derived ingredients.

In summary, regulatory bodies have issued specific guidance for the safety assessment of plant-derived ingredients when they are used as food, food ingredients, and herbal drugs. However, similar approaches have not been put together for the safety assessment process of cosmetic products containing plant-derived ingredients. Therefore, there is an urgent need to propose a specific approach for the safety evaluation of cosmetic ingredients of natural or organic origin.

8.4 CHARACTERIZATION

The comprehensive characterization of plant-derived ingredients is crucial and is an important factor of an accurate and reliable safety risk assessment.

8.4.1 Plant-Derived Ingredient Identification

As previously mentioned, several cases of poisoning have been reported with products containing plant-derived ingredients as a result of intentional adulteration, accidental contamination with other plant species, or misidentification of species. This indicates that plant-derived ingredient identification is a critical factor in ensuring the quality and safety of plant-derived ingredients.

The taxonomic identity of the plant should be given according to the internationally accepted guidelines and the information provided should include the complete scientific name (family name, genus, species, author name, subspecies, and cultivar/breeding line), synonyms, and the vernacular name [53]. With regard to plant-derived ingredients that do not have a significant history of use within a region as traditional foods or herbal medicines, a complete scientific taxonomic identification will considerably speed up the identification of possible close relatives to the plant in question, thereby facilitating a prediction of chemical composition of the plant food and a better assessment of its safety profile.

The geographical origin of the plant used as a source material is another key parameter in the characterization of the plant-derived ingredient since it is a potential source of variability in the composition. Additional information to aid the risk assessment evaluation is whether the plant is collected from wild population or is bred in a specific way. If bred in the wild, proper botanical characterization should be done to minimize the risk of adulteration and misidentification. Thus, growth and harvesting conditions as well as plant health treatment should be correctly detailed.

Then, the part of the plant used as a source material for the product has to be clearly defined. This is of particular importance if toxic constituents are present in higher concentrations in the part of the plant used as source material for the ingredient of choice. The primary processing of the starting material, including washing, cutting, fumigation, freezing, distillation, extraction, and so on, as well as packaging, storage conditions, and transportation should be adequately defined. Processing should also be carried out after harvesting as soon as possible to ensure there is no contamination from a chemical or microbiological origin.

All together, the aforementioned process aims to ensure the traceability, quality, and, therefore, safety of plant-derived ingredients. Their efficacy will obviously be supported by an appropriate and consistent quality assurance system. Accordingly, the development and the implementation of standard operating procedures based on existing guidelines on good agricultural and harvesting practices should be highly encouraged. These documents provide standards regarding the personal identification, cultivation, collection, harvest, processing, and so on of the plant-derived ingredient [51, 54]. These guidelines may also need to be adjusted according to each country's conditions, with their main overall objectives being not only to improve the quality and safety of plant-derived ingredients but also to support the sustainable cultivation and collection of plants and thereby help the development of community trade. Therefore, producers, suppliers, traders, and processors of plant-derived ingredients should be encouraged to comply with standardized guidelines and procedures. Likewise, standard operating procedures aimed at controlling both biological and chemical contaminations that can occur during the whole production chain, from harvesting to distribution, should be established and integrated in the quality assurance system. Approach such as Hazard Analysis of Critical Control Point [55] should be adopted and applied with the necessary flexibility and adapted to each plant-derived ingredient preparation.

8.4.2 Chemical Characterization

The safety assessment of plant preparations is complicated by important sources of variability in composition, such as the natural variances due to the geographical origin of the plant, conditions of growth, and the impact of production processes. It is now well established that processes such as extraction may significantly modify the composition of the plant-derived ingredients and consequently affect its safety. Therefore, the phytochemical characterization is a critical step in assessing safety of the plant-derived ingredients. The extent of information necessary for a consistent characterization of the plant should be decided on a case-by-case basis and by considering all available relevant data. In the case of a plant-derived ingredient consisting of or derived from a plant with a history of use as food could be characterized by its content of macronutrients, such as proteins, fats, carbohydrates, and fibers, and micronutrients, such as vitamins and minerals. This basic characterization can further be completed by data on the content of specific proteins, individual amino acids, fraction of oils, fatty acids, individual sugars and fibers, and simple and complex carbohydrates. In the case of a plant-derived ingredient or if derived from a plant traditionally used as herbal medicines, analytical data on constituents, including those with known therapeutic activity, as well as components suitable as active markers or even as analytical markers can be very useful [30, 31]. The characterization of a given plant-derived ingredient (fingerprinting) might also be based on a fixed plant extract ratio. Fingerprinting approaches should be encouraged particularly in cases where appropriate markers are not known. It should also be mentioned that fingerprinting may be helpful not only in defining the variability of a plant-derived ingredient but also in detecting adulteration and therefore in confirming authenticity. Obviously, analytical data on the level of naturally occurring known toxicants and of constituents known to be responsible for adverse effects should be included in the set of phytochemical markers. Whatever be the marker, the appropriateness of its choice should always be justified [10].

8.4.3 Specifications

The characterization of a plant-derived ingredient preparation, which includes a detailed evaluation of its ingredients and phytochemical aspects of the plant and processing of the preparation, is therefore essential to allow specifications to be established, which are both comprehensive and relevant. Specifications are primarily intended to define the quality of plant-derived ingredients rather than to establish full characterization, and should focus on those characteristics found to be useful in ensuring safety of the herbal substances/preparations. Specifications consist in a set of criteria to which a herbal substance/preparation product should conform to so as to be considered acceptable for its intended use. Specifications are one important component of a total quality control strategy for the herbal substance/preparation and herbal medicinal product designed to ensure product quality and consistency. The setting of specifications for a herbal substance/preparation and herbal medicinal product is part of an overall control strategy that includes control of raw materials and excipients, in-process testing, process evaluation/validation, stability testing, and testing for consistency of batches. When all these elements combine together, they provide assurance that an appropriate and acceptable quality of the product will be maintained. If present and relevant, internationally accepted methods and specifications such as those mentioned in pharmacopoeias should definitively be applied [52].

8.5 THE COMPARATIVE APPROACH

8.5.1 History, Definition, and Regulatory Applications in Food Safety Assessment

Once the plant-derived ingredient under evaluation has been comprehensively characterized, we recommend initiating its safety assessment using the comparative approach currently accepted globally for novel foods and food ingredients.

Historically, the comparative approach is based on the concept of "substantial equivalence." The latter was introduced through a joint FAO/WHO consultation on biotechnology and food safety in 1990 and was further developed in 1992 by the Food Safety Working Group of the OECD Group of National Experts in Safety in Biotechnology [56]. This concept was primarily intended to provide a practical approach to the safety evaluation of foods that are genetically modified (GM) organisms or are produced from them. It describes the safety assessment of GM plants and GM-derived food and feed using appropriate methods and approaches as well as comparing them with non-GM comparators. Since many GM foods are obtained from plants that are conventional food plants, the comparators chosen to facilitate the safety evaluation are usually the parent (host) or other edible varieties of the same species or edible products derived from them.

During the above period, a number of exotic food plants with no traditional history of use in Europe were introduced to the food market. These foods, such as kiwi fruit, were placed on the market based on the assumption that their safety could be ensured by their traditional use as food in other parts of the world. However, their introduction was associated with health problems such as allergic reactions and raised safety concerns among the EU food safety authorities. Accordingly, a Novel Food legislation was introduced in Europe in 1997 [39] in order to regulate the placing on the market of foods or food ingredients that have not hitherto been used for human consumption to a significant degree within the European Union. Therefore, the principles employed in the safety assessment of novel foods have been extensively discussed and guidelines have been developed by a number of regulatory bodies and the food industry [43, 57–59]. However, the safety assessment of novel foods and food ingredients raised several technical issues. As foods are usually highly complex and contain many different substances, it has long been recognized that whole foods cannot be tested using the standard safety evaluation principles [33] employed for single substances such as pharmaceuticals, food additives, or food contaminants [34]. Traditional toxicological testing procedures are of limited use and not very practical for whole foods because when fed to animals in high levels, it can lead to nutritional imbalances. On the other hand, although many traditional foods are considered to have a "history of safe use," no food can be considered to be absolutely safe under all circumstances—individuals may tolerate the same food differently. Moreover, traditional food is considered safe only within the context of its typical use by certain population groups with prevailing dietary, preparation, processing regimes, and cultural practices, that is, some foods that have a "history of safe use" in one part of the world may be deemed to be novel if introduced into another part of the world. This means that it is necessary to develop alternative approaches to properly assess safety of the novel foods and novel food ingredients.

Based on the assumption that novel foods should be at least as safe as traditional counterparts, that is, using comparators where they exist, or at least they should not add to risks, both the regulatory bodies and the food industry concluded that the comparative approach developed for the safety evaluation of GMOs can be the logical starting point in the safety assessment of novel foods and novel food ingredients. Briefly, to assess safety of a novel food, the first step should be to determine what (if any) existing foods should be used as comparators. If there is no comparator, the novel food is not necessarily unsafe; rather, it indicates that an extensive safety assessment program is required. If the comparators exist, deemed to be traditional within the context of its use, the maximum safety information should be gathered to compare it with the novel food. In some instances, the use of more than one comparator may be needed to address different safety issues. Once the comparators have been identified, the comparison needs to be performed on the basis of plant-derived ingredient characterization, chemical composition, methods of processing, previous human experience and exposure, intended intake, and target groups. This approach is designed to highlight equivalence, similarity, or differences between the new food and its traditional counterpart, that is, the comparator. This approach is a very practical tool that allows reasonable safety assessment to be conducted on a "case-by-case" basis without extensive additional toxicological testing. Furthermore, the approach of targeting equivalence and similarity can highlight dissimilarities and allows us to focus on their significance. Obviously, the final objective of this approach is to derive a human safe level for the novel foods or food ingredients under evaluation [60]. The framework of the concept must be scientifically valid, the results of the safety assessment must be reproducible and acceptable to the responsible health authorities, and, finally, the outcome must satisfy and convince the consumer. In addition to the regulatory safety assessment of novel foods and food ingredients, the comparative approach has also been recognized recently as a reliable method to assess safety of the plant-derived ingredients used as food supplements in Europe [7, 10].

8.5.2 Application in the Safety Assessment of Plant-Derived Cosmetic Ingredients

According to the aforementioned principles, the first step of a plant-derived cosmetic ingredient safety assessment should therefore be to determine what, if any, available existing foods and herbal drugs can be used as comparators.

8.5.2.1 Choice of the Comparator In many cases, the plant sources from which a cosmetic ingredient is obtained are plants that are traditional foods or food sources. In these cases, the plant source becomes the comparator. Various databases can be used to help establish whether a particular product has a "history of safe use" as a food or food source. These include national food survey reports and global, regional, and national surveys of plants with food uses [61, 62]. Overall, the choice of the comparators is a crucial issue since the latter should provide a benchmark against which to assess compositional differences, differences in processing, and exposure for the plant-derived cosmetic ingredient under evaluation.

8.5.2.2 Plant-Derived Ingredient and Phytochemical Characterization Once identified, both the comparators and the cosmetic plant-derived ingredients should be accurately characterized according to the principles previously described. Briefly, this includes a precise plant-derived ingredient identification (e.g., taxonomy, phenotype, and genotype) using appropriate methodologies. In addition, the origin, geographical distribution, and genetic diversity of the plant source should be described. Then, the chemical composition of the comparators should be determined. Obviously, the variation in the natural composition of the comparators should be given and should serve as a basis to set specifications.

Once the comparators have been fully characterized as previously described, the safety assessor may then proceed to the compositional comparison with the plant-derived ingredient under evaluation. If no significant differences are evidenced, the next step of the comparative approach, that is, the comparison of the previous exposure versus the intended cosmetic exposure can be initiated.

8.5.2.3 Exposure Assessment

Previous Human Exposure The determination of the previous human exposure to the comparators is a crucial issue in the comparative approach. Exposure data should include serving size, daily intake, frequency of consumption, and period of use (number of years of consumption of the product). In cases where the comparator is a traditional herbal, current allowed posologies are easily available on medicinal databases and herbal medicine monographs [63–65]. In this case, special attention should be given to any sensitive groups in the population exposed to the material of reference (e.g., young, old, pregnant, immunocompromised persons, and those receiving medication) as well as to know cautions and undesirable effects.

Intended Cosmetic Exposure It should be mentioned that exposure data by the dermal route, that is, the route of application of a cosmetic product, will certainly be limited. Therefore, in order to proceed to the comparison of previous exposure and intended exposure by the dermal route, it will be necessary to measure or at least estimate the percutaneous absorption of the constituents of both the comparators and the cosmetic plant-derived ingredients. The differences between oral and dermal uptake of chemicals have been extensively discussed recently by Kroes et al. [66]. It is well known that topical application and oral ingestion often result in different proportions of the applied dose entering the body. These differences in bioavailability result mainly from

- a more extensive metabolism in the intestine and liver, compared to the skin, prior to reaching the general circulation, and
- a slower and incomplete transfer across the skin compared to the intestinal wall, which is essentially linked to the physicochemical properties of the compound.

It is important to appreciate that absorption will also depend upon the manner in which the chemical is presented to the skin (i.e., the "formulation" in which it is applied), the area of skin exposed, and the duration of skin contact/exposure. In addition, exposure would also need to consider the duration and frequency of human exposure to the respective cosmetic end product and whether it is a rinse-off or leave-on cosmetic product.

The percutaneous absorption of chemicals, including cosmetic ingredients, may be estimated using *in vitro* studies. Numerous guidelines exist for the conduct of *in vitro* skin absorption studies, such as the U.S. EPA proposed guidelines [67],

European Chemicals Bureau Technical Guidance Document [68], OECD Guidelines [69, 70], recommendations by a CEFIC expert panel [71], the EC Dermal Exposure Network [72], Guidelines issued by the EC and the EU Scientific Committee on Plants [50, 73], and the Guidelines of the EU Scientific Committee of Consumer Products [16, 17]. It should be mentioned that due to variability in methodology and study design, there can be a lack of consistency of data obtained from different *in vitro* percutaneous absorption studies [74]. Also, if such data are not available, it is proposed that conservative default adjustment factors derived from physicochemical properties of the compound under evaluation be assigned to the different levels of permeability as defined by Kroes et al. [66].

Determination of Safe Human Use It is well recognized that whole foods cannot be tested in their own right according to the standard safety evaluation principles [33] used for single substances such as pharmaceuticals, food additives, or food contaminants and that traditional toxicological testing procedures are of limited use for whole foods. Moreover, although foods that are commonly eaten are presumed to be safe, the lack of any side effects or adverse reactions, however, is not an evidence of the absence of toxicity under new conditions of usage. The history of use (food or medical application) provides valuable information concerning safety [7, 8]. However, for some plant-derived ingredient products, relevant epidemiological and clinical human studies addressing both efficacy and safety are available and should be considered as part of the risk assessment [3]. Moreover, it has recently been demonstrated that all available data, including human data derived from history of traditional use of plants, should be fully exploited in the safety assessment of plant-derived ingredients. The weight of evidence-based approach is the so-called concept of "history of safe use." Recently, this concept has been accurately described by Constable et al. [60] and its application in the safety assessment of plants and plant-derived ingredients have been comprehensively discussed by Knudsen et al. [75].

In summary, all evidence of safety and any health concerns should be provided by exploitation of any available data in order to determine a human safe dose for the comparators and by extension for the novel plant-derived cosmetic ingredients.

8.6 CHEMICAL GROUPING

According to OECD [76], the term grouping or chemical grouping describes the general approach to assessing more than one chemical at the same time. It can include formation of a chemical category or identification of a chemical analogue for which "read-across" of existing data on known chemicals can be applied. The latter is defined as a technique used to predict endpoint information for a chemical using the same endpoint information from another chemical that is considered to be similar on the basis of structure, properties, and activities. A chemical category is a group of chemicals whose physicochemical, human health, and environmental toxicological properties as well as environmental fate properties are likely to be similar or follow a regular pattern as a result of structural similarity (or other similarity characteristics). The similarities can be based on the following:

- Common functional groups (e.g., aldehyde, epoxide, ester, and specific metal ion).
- Common constituents or chemical classes or similar carbon range numbers. This is frequently the case with complex substances often known as substances of "Unknown or Variable composition, complex reaction products or Biological" material (UVCB substances).
- An incremental and constant change across the category (e.g., a chain-length category), often observed in physicochemical properties, for example, boiling point range.
- The likelihood of common precursors and breakdown products, via physical or biological processes, which result in structurally similar chemicals (e.g., the metabolic pathway approach).

In principle, more than one member is generally present in a chemical category, enabling the detection of trends across endpoints. Within a chemical category, data gaps can be filled by using the read-across technique. In summary, chemical grouping is a logical and pragmatic *in silico* approach, alternative to animal testing, which allows evaluation of chemicals, not on a one-by-one basis but in groups.

Chemical grouping and read-across techniques have recently received regulatory acceptance in the new European regulation on chemicals, that is, REACH. Annex XI of this regulation contains the following wording for the use of grouping methods (read-across and chemical categories):

"Substances whose physicochemical, toxicological and ecotoxicological properties are likely to be similar or follow a regular pattern as a result of structural similarity may be considered as a group, or 'category' of substances. Application of the group concept requires that physicochemical properties, human health effects and environmental effects or environmental fate may be predicted from data for reference substances within the group by interpolation to other substances in the group (read-across approach). This avoids the need to test every substance for every endpoint."

Interestingly, complex substances often known as substances of Unknown or Variable composition (UVCB substances), including natural complex substances such as plant extracts, are mentioned as typical example of substances for which chemical grouping should apply. The application of chemical grouping to the safety assessment of UCVB substances of natural origin is definitively justified by the fact that they are complex mixtures and therefore cannot be separated into their constituent chemical species [77]. There are many different types of complex substances, although generally they all have common characteristics:

- They contain numerous chemicals (typically closely related isomers or chemical classes with defined carbon number or distillation ranges) and cannot be represented by a simple chemical structure or defined by a specific molecular formula.
- They are not intentional mixtures of chemicals.

- Many are of natural origin (e.g., crude oil, coal, and plant extracts) and cannot be separated into their constituent chemical species.
- The concept of *impurities* typically does not apply to complex substances.
- They are produced according to a performance specification related to their physicochemical properties.

8.6.1 Essential Oils

Historically, the chemical grouping approach has been widely used for the safety assessment of flavoring agents by the JECFA since 1998 and more recently by EFSA [78]. In 1996, the JECFA began a review of the safety of flavoring agents used in food. The procedure used by JECFA for the safety evaluation of flavoring substances integrates data on intake, metabolism, and structure–activity relationships with toxicity data [79]. To perform these evaluations, flavoring substances are first compiled into groups of structurally related materials, the so-called congeneric groups. The members of each of these groups would be expected to have common routes of metabolism and similar toxicity. According to the JECFA approach, substances with simple structures, which are known or presumed to be readily metabolized to innocuous products, can be evaluated without toxicity data. Substances for which their metabolism is not well defined or are expected to be metabolized to reactive products require toxicity data in order to perform a safety evaluation in accordance with the procedure. Since metabolism or toxicity data are rarely available for all members of a group, the safety evaluation of substances lacking data depends upon the availability of data for other group members. Using this approach, the safety of more than 1400 flavoring substances was evaluated between 1996 and 2004 [80, 81]. Flavoring agents are secondary metabolites of plants and can be the major constituents of essential oils [82]. All the chemical constituents present in essential oils are formed by one of five or six major biosynthetic pathways. Since all these pathways operate to a varying extent in plants, many of the same chemical constituents are present in a wide variety of essential oils. A consequence of having a limited number of plant biosynthetic pathways is that the structural variation of chemical constituents in an essential oil is also limited. Therefore, essential oils usually contain 5–10 distinct chemical classes or the so-called congeneric groups (e.g., terpene aliphatic and aromatic hydrocarbons). Overall, the chemical grouping approach is a practical tool for assessing safety of essential oils that are widely used as cosmetic ingredients. However, safety of essential oils can also be assessed using the comparative approach or the TTC approach (discussed below), provided that the essential oil under evaluation is accurately characterized.

8.7 THE THRESHOLD OF TOXICOLOGICAL CONCERN

The threshold of toxicological concern (TTC) is a principle that refers to the establishment of a human exposure threshold value for all chemicals, below which there would be no appreciable risk to human health. The concept that safe exposure thresholds or levels can be identified for individual chemicals is already widely used for chemicals with known toxicological profiles. However, the TTC concept goes further in proposing that a *de minimis* value can be identified for many new chemicals, including those of unknown toxicity, based on their chemical structural similarity to known chemicals. The TTC concept can be used to assess and set a safe level of exposure for a chemical in the absence of chemical-specific toxicity data, based on the available toxicity data from a wide range of structurally similar chemicals; in other words, existing knowledge from the "world of chemicals" can be used to justify acceptable low exposures of a chemical under evaluation. The concept of a TTC evolved from the review by Munro [83] of the threshold of regulation in the United States and from further developments [84] based on analysis of the chronic toxicity data of chemicals in three structural classes identified according to the Cramer et al. [85] decision tree. The latter consists of a series of 33 questions regarding chemical structure that can be used to classify a chemical as class I (a low order of oral toxicity), class II (indeterminate toxicity; no presumption of safety or significant harm), or class III (no strong presumptions of safety; structure may even suggest significant toxicity). Human exposure thresholds (in mg/day) for the three Cramer classes were based on the fifth percentile NOEL (no observable effects limit), an assumed human body weight of 60 kg, and a safety factor of 100 [84]. The 100-fold safety factor was considered to provide an adequate margin-of-safety considering the size of the database and the nature of the supporting toxicological data. These human exposure thresholds were recommended as TTC exposure limits for the three Cramer classes (1.8 mg/day for class I; 0.54 mg/day for class II, and 0.09 mg/day for class III).

The TTC concept has been accepted by various international and national regulatory agencies for consideration of safety of substances in the human diet and human pharmaceuticals, and was also proposed for other chemicals that produce minimal human systemic exposure. Regulatory TTC values have been defined for migrant substances from packaging material in food [86], flavoring substances in food [87, 88], and genotoxic impurities in pharmaceuticals [89–91]. The approach was also used by the former EC Scientific Committee on Food and is now used by the European Food Safety Authority to evaluate flavoring substances [78]. The TTC approach has also been endorsed by the WHO International Program on Chemical Safety for the risk assessment of chemicals [92] and by the EU Scientific Committee on Toxicology, Ecotoxicology and the Environment [93]. Recently, the approach has also been suggested for application to aquatic environmental exposure [94].

Subsequent work by an ILSI Europe Expert Group [95, 96] extended the approach adopted by the JECFA [87, 97–99] and incorporated the analyses of Cheeseman et al. [100] into an additional TTC value for compounds with certain structural alerts for possible activity as genotoxic carcinogens. In the TTC decision tree developed by Kroes et al. [96], proteins, heavy metals, and polyhalogenated dibenzo-*p*-dioxins and related compounds were excluded because the databases used to derive the TTC values did not include toxicity data on proteins, heavy metals such as cadmium, lead, and mercury and because the uncertainty factors used to derive the TTC values would not allow the extreme species differences in elimination seen with polyhalogenated-dibenzo-*p*-dioxins, polyhalogenated dibenzofurans, and polyhalogenated biphenyls. In addition, a well-established risk assessment method exists for the evaluation of dioxin-like compounds and, therefore, the TTC approach is not necessarily applied for this type of chemical. This decision tree also excludes substances with structural alerts for high potency genotoxicity (aflatoxin-like compounds, *N*-nitroso compounds, and azoxy compounds) from consideration and applies a generic highly conservative threshold for all other structural alerts of 0.15 μg/(person day) (0.0025 μg/kg(bw day)). Analysis of neurotoxicity data showed that the cumulative distributions of no observed adverse effect levels (NOAELs) for organophosphates were lower than that for other neurotoxins. Therefore, a TTC of 18 μg/(person day) was proposed for organic phosphates.

8.7.1 Case Study with Calendula

Based on the above case studies, the TTC concept has gained an extensive acceptance for the safety assessment of numerous compounds following an oral exposure. However, in 2007, Kroes et al. [66] expanded the TTC concept to chemicals administered by the topical route. They provided guidance for the application of TTC approach to the safety assessment of cosmetic ingredients, placing particular emphasis on both skin metabolism and percutaneous absorption. The application of TTC for dermal application of chemicals and specifically for cosmetics was fully supported by Munro et al. [101]. Finally, using Calendula extracts as an example, we have recently demonstrated that the TTC concept can be effectively used to evaluate safety of the complex plant-derived ingredients if a comprehensive plant-derived ingredient and phytochemical characterization of the plant-derived ingredients are available [102]. The authors' main point was to ensure safety of the complex materials, such as plant-derived ingredients, which have low-level human systemic exposures, without relying on animal testing or high level of resources. In the case of Calendula, more than 150 distinct chemicals were identified from various literature sources. In order to make the number of individual evaluations workable, the authors excluded mineral matter, carbohydrates, fatty acid esters, amino acids, resins, components with MW > 1000 (negligible skin penetration), and inert material such as cellulose as well as components present at < 0.5% from analysis. The justification for eliminating components present at < 0.5% is that it would result in a systemic exposure below 90 μg/day (assuming 100% absorption). This is based on the use of up to 0.1% of a plant material in a personal care product that is applied to the body at 18 g/day (i.e., maximum amount used in a body lotion product as defined by EU SCCP).

For the remaining substances, that is, approximately 30 components, chemical structures were identified and classified according to the Cramer decision tree [85]. The molecular weight and $\log P_{o/w}$ (oil and water partition coefficient) were calculated for each component to estimate their skin penetration potential. It was determined that the potential human systemic exposure from each of the components was below its respective TTC value. The mixture was therefore deemed to be safe for use for any cosmetic application of up to 18 g/day. Therefore, for low-level exposures to chemicals without an established toxicology profile, the TTC approach can be very useful, provided one knows the chemical structure of the substance, its $\log P_{o/w,}$ the concentration in the formulation, and application rate to the skin.

Overall, by using the approach described by Kroes et al. [66], we demonstrated that the TTC concept is a practical and efficient tool to reliably assess safety of the plant-derived cosmetic ingredients after they have been well characterized and when their intended exposure is low by the dermal route.

8.8 CONCLUSIONS

The buzz regarding health-related products has led to a huge public demand for natural and organic cosmetic products. While this presents opportunities for companies to enter this growing market and although standards exist for certification of organic and natural cosmetics, there are no well-established, globally acceptable standards for the safety assessment of these types of products. Since they are natural in origin, the assumption is that they are inherently safe. However, this is not true at all. Many botanical ingredients have intrinsic activity and their toxicity profiles and adverse reactions have been well documented in the literature. Therefore, a safety assessment process is needed to ensure safety of these ingredients. Although there are established, globally acceptable guidelines for the safety assessment of foods, food supplements, and botanical drugs, there is no clear or prescribed list of studies to evaluate safety of the botanical ingredients used in cosmetic products. They incorporate a certain level of acceptable risk based on daily consumption during the lifetime of an individual.

The goal of this chapter is to propose a practical approach to conducting safety assessments of botanical ingredients and natural extracts. Plant-derived ingredients typically consist of a highly complex mixture of chemicals. While the ingredients or the major components themselves may be well known and have biological activity, they vary widely in terms of identity and composition. The phytochemical composition of plant-derived ingredients depends on a number of factors such as part of plant used, growing and harvesting conditions, geographical and natural variability of the plant source, and so on. The poor characterization along with limited or lack of adequate safety data is a major challenge in the safety evaluation process for these ingredients. However, if these ingredients have been historically used as traditional food, it is helpful in establishing an acceptable level of use as drugs, food supplements, or cosmetics.

It is important to note that the traditional standard safety testing and assessment is not possible with foods and natural ingredients due to the fact that they are mostly mixtures of substances. In addition, classical toxicology testing requires high doses that are not practical in case of foods as they would cause nutritional imbalances. Comparative approach for foods and food ingredients can be used based on the concept of substantial equivalence. The safety assessment of plant-derived ingredients can be done using methodologies that have already gained scientific and regulatory acceptance. Therefore, the comparative approach, the chemical grouping, and read-across approaches as well as the TTC concept offer practical, logical, and reliable tools for the safety assessment of plant-derived cosmetic ingredients. One or more of these *in silico* approaches can be used on a case-by-case basis by exploiting all existing data. While these tools are relevant for systemic toxicity endpoints, they do not necessarily help in supporting local toxicity endpoints, which should be evaluated according to more conventional methods. Finally, it should be emphasized that in addition to their practical character, scientific relevance, and regulatory acceptance, these alternative approaches also fully comply with the animal testing ban and its accompanying progressive marketing ban required by the European regulation on cosmetics. Furthermore, these alternative approaches are also in agreement with the consumer concerns as well as natural and organic standards requirements on animal testing.

REFERENCES

1. The Organic, Monitor. The North American market for natural & organic personal care products, Report: 3001–60, September 17, 2007.
2. Nortier JL, Muniz Martinez M-C, Schmeiser HH, Arlt VM, Bieler CA, Petein M, Depierreux MF, De Paw L, Abramowicz D, Vereerstraeten P, Vanherwegheim J-L. Urothelial carcinoma associated with the use of a Chinese herb (*Aristolochia fanghi*). *New Engl. J. Med.* 2000;342(23):1686–1692.
3. Ernst E. The risk-benefit profile of commonly used herbal therapies: *Ginkgo,* St. John's wort, Ginseng, Echinacea, Saw Palmetto, and Kava. *Ann. Inter. Med.* 2002;136:42–53.
4. Fu PP, Chiang HM, Xia Q, Chen T, Chen BH, Yin JJ, Wen KC, Lin G, Lin G, Yu H. Quality assurance and safety of herbal dietary supplements. *J. Environ. Sci. Health C* 2009;27:91–119.
5. Ernst E. Adverse effects of herbal drugs in dermatology. *Br. J. Dermatol.* 2000;143(5):923–929.
6. Ernst E. Risks of herbal medicinal products. *Pharmacoepidemiol. Drug Saf.* 2004;13(11):767–771.
7. Schilter B, Andersson C, Anton R, Constable A, Kleiner J, O'Brien J, Renwick AG, Korver O, Smit F, Walker R. Guidance for the safety assessment of botanicals and botanical preparations for use in food and food supplements. *Food Chem. Toxicol.* 2003;41:1625–1649.
8. Kroes R, Walker R. Safety issues of botanicals and botanical preparations in functional foods. *Toxicology* 2004;198:213–220.
9. Rousseaux CG. and Schachter H. Regulatory issues concerning the safety, efficacy and quality of herbal remedies. *Birth Defects Res. B* 2003;68:505–510.
10. EFSA, Scientific, Committee., Guidance on safety assessment of botanicals and botanical preparations intended for use as ingredients in food supplements. *EFSA J.* 2009;7(9):1249.
11. Food and Drug Administration 21, CFR, Part, 119., Final rule declaring dietary supplements containing ephedrine alkaloids adulterated because they present an unreasonable risk. Final rule. *Fed. Regist.* 2004;69(28):6787–6854.
12. Food and Drug, Administration. *Guidance for Industry: Botanical Drug Products,* Center for Drug Evaluation and Research, 2004;
13. European Medicines Agency, Committee on Herbal Medicinal, Products. Guideline on non-clinical documentation for herbal medicinal products in application for marketing authorisation (bibliographical and mixed applications) and in applications for simplified registration, EMEA/HMPC/32116/2005.
14. Wu KM, Ghantous H, Birnkrant DB. Current regulatory toxicology perspectives on the development of herbal medicines to prescription drug products in the United States. *Food Chem. Toxicol.* 2008;46:2606–2610.
15. EC., Council Directive 76/768/EEC of July 27, 1976 on the approximation of the laws of the member states relating to cosmetic products, 1976.
16. SCCP. The SCCP's notes of guidance for the testing of cosmetic ingredients and their safety evaluation, 6th revision, December 19, 2006.
17. Scientific Committee on Consumer, Products. Basic criteria for the *in vitro* assessment of dermal absorption of cosmetic ingredients, 2006 (updated March 2006, adopted March 28, 2006).
18. Allemann I. B, Baumann L. Botanicals in skin care products. *Int. J. Dermatol.* 2009;48:923–934.
19. British Industrial Biological Research, Association, Toxicity profile of potato glycoalkaloids, Issue 19, 1996.

20. Andersson C. *Glycoalkaloids in Tomatoes, Eggplants, Pepper and Two Solanum Species Growing Wild in the Nordic Countries.* TemaNord, No. 599, 1999.
21. Morris SC, Lee TH. The toxicity and teratogenicity of Solanaceae glycoalkaloids, particularly those of the potato (*Solanum tuberosum*): a review. *Food Technol. Australia* 1984;36:118–124.
22. JECFA. Evaluation of certain food additives and naturally occurring toxicants: thirty-ninth report of the Joint FAOIWHO Expert Committee on Food Additives. WHO technical report series 828, 1992.
23. Mlingi N, Poulter NH, Posling H. An outbreak of acute intoxication from consumption of insufficiently processed cassava in Tanzania. *Nutr. Res.* 1992;12:677–678.
24. McNamara SH. Dietary supplements of botanicals and other substances: a new era of regulation. *Food Drug Law J.* 1995;50(3):341–348.
25. Chan TY, Critchley JA. Usage and adverse effects of Chinese herbal medicines. *Hum. Exp. Toxicol.* 1996;15(1):5–12.
26. Woolf AD, Watson WA, Smolinske S, Litovitz T. The severity of toxic reactions to ephedra: comparisons to other botanical products and national trends from 1993–2002. *Clin. Toxicol.* 2005;43(5):347–55.
27. Cosyns JP. Aristolochic acid and "Chinese herbs nephropathy": a review to the evidence to date. *Drug Safety* 2003;26(1):33–48.
28. National Toxicology, Program. Aristolochic acids. Reports on carcinogens, 2008.
29. European medicines Agency, (EMEA). Public statement on the risks associated with the use of herbal products containing *Aristolochia* species. EMEA/HMPC/138381/2005.
30. European Medicines Agency, (EMEA). Reflection paper on markers used for quantitative and qualitative analysis of herbal products and traditional herbal medicinal products. EMEA/HMPC/253629/2007.
31. European Medicines Agency, (EMEA). Draft assessment report on *Hypericum perforatum* L., herba. EMEA/HMPC/101303/2008 Corr.
32. Lalko J, Api AM. Investigation of the dermal sensitization potential of various essential oils in the local lymph node assay. *Food Chem. Toxicol.* 2006;44:739–746.
33. WHO., Principles for the safety assessment of food additives and contaminants in food. No. 70. WHO, Geneva, 1987.
34. Barlow SM, Greig JB, Bridges JW, Carere A, Carpy AJM, Galli CL, Kleiner J, Knudsen I, Koëter HBVM, Levy LS, Madsen C, Mayer S, Narbonne JF, Pfannkuch F, Prodanchuk MG, Smith MR, Steinberg P. Hazard identification by methods of animal-based toxicology. *Food Chem. Toxicol.* 2002;40:145–191.
35. Food and Drug, Administration., Proposal for dietary supplements: current good manufacturing practices (CGMPs), 2003.
36. Federal, Register., Final rule and interim final rule, June 25, 2007.
37. National Toxicology, Program. Fact sheet: herbal medicines. ntp.niehs.nih.gov/ntp/htdocs/Liason/factsheets/HerbMedFacts.pdf.
38. EC, Regulation. Directive No. 178/2002 of the European Parliament and of the Council of January 28, 2002 laying down the general principles and requirements of food law, establishing the European Food Safety Authority and laying down procedures in matters of food safety. Official Journal of the European Communities, L 31/1, 1.2.2002.
39. EC, Regulation. Directive No. 258/97 of the European Parliament and of the Council of January 27, 1997 concerning novel foods and novel food ingredients. Official Journal of the European Communities, L 43/1–7, February 14, 1997.
40. FSANZ. Standard 1.5.1 (Novel Foods) of the Food Standards Code, Australian New Zealand Food Standards Code, 1999.
41. Health, Canada. Amendment (Schedule No. 948) to Division 28 of the Food and Drug Regulations, Sections B.28.001–003, 2003.
42. Food and Drug, Administration. *Fed. Regist.* 1992;57(104):2284.
43. Scientific Committee for, Food. Opinion on the assessment of novel foods: food science and techniques. Reports of the Scientific Committee for Food (40th series). Directorate-General Industry, European Commission, 1997.
44. Health, Canada. Guidelines for the safety assessment of novel foods, 2006.

45. FSANZ. Standard 1.5.1 novel foods including Amendment 70 (FSC 12), Australian New Zealand Food Standards Code, April 29, 2004.
46. FSANZ. General information to assist in applying to amend the Australian New Zealand Food Standards Code: novel foods updated April 2004.
47. Wu KM, DeGeorge J, Atrakchi A, Barry E, Bigger A, Chen C, Du T, Freed L, Geyer H, Goheer A, Jacobs A, Jean D, Rhee H, Osterberg R, Schmidt W, Farrelly J. A special update from the United States Food and Drug Administration: preclinical issues and status of investigation of botanical drug products in the United States. *Toxicol. Lett.* 2000;111:199–202.
48. EC, Regulation. Directive No. 2001/83/EC of the European Parliament and of the Council of November 6, 2001 on the Community code relating to medicinal products for human use. Official Journal of the European Communities, L 311/67, 28.11.2001.
49. EC, Regulation. Directive No. 2004/24/EC of the European Parliament and of the Council of March 31, 2004 amending, as regards traditional herbal medicinal products, Directive 2001/83/EC on the Community code relating to medicinal products for human use. Official Journal of the European Union, L 136/85, 30.4.2004.
50. EC, Regulation. Guidance document on dermal absorption. European Commission, Health and Consumer Protection Directorate General, Directorate E, Food Safety, Plant Health, Animal Health and Welfare, March 19, 2004.
51. European Medicines Agency, (EMEA). Guideline on good agricultural and collection practice (GACP) for starting materials of herbal origin. EMEA/HMPC/246816/2005.
52. European Medicines Agency, (EMEA). Guideline on specifications on test procedures and acceptance criteria for herbal substances, herbal preparations and herbal medicinal products/traditional herbal medicinal products. CPMP/QWP/2820/00 Rev 1 & EMEA/CVMP/815/00 Rev 1, 2006.
53. König A, Cockburn A, Crevel RWR, Debruyne E, Grafstroem R, Hammerling U, Kimber I, Knudsen I, Kuiper HA, Peijnenburg AACM, Penninks AH, Poulsen M, Schauzu M, Wal JM. Assessment of the safety of food derived from genetically modified (GM) crops. *Food Chem. Toxicol.* 2004;42:1047–1088.
54. WHO. Guidelines on good agricultural and collection practices (GACP) for medicinal plants, 2003.
55. Codex, Alimentarius. Recommended International Code of Practice: General Principles of Food Hygiene Including Annex on Hazard Analysis and Critical Control Point (HACCP) System and Guidelines for Its Application, 1997.
56. WHO. Application of the principles of substantial equivalence to the safety evaluation of foods or food components from plants derived by modern biotechnology. Report of a WHO Workshop. World Health Organisation, Geneva, Switzerland, 1995.
57. Jonas DA, Antignac E, Antoine JM, Classen HG, Huggett A, Knudsen I, Mahler J, Ockhuizen T, Smith M, Teuber M, Walker R, De Vogel P. The safety assessment of novel foods. *Food Chem. Toxicol.* 1996;34:931–940.
58. ILSI. The safety assessment of novel foods and concepts to determine their safety in use. ILSI Europe Report Series, 2003.
59. Edwards G. Safety assessment of novel foods and strategies to determine their safety in use. *Toxicol. Appl. Pharmacol.* 2005;207:S623–S627.
60. Constable A, Jonas D, Cockburn A, Davi A, Edwards G, Hepburn P, Herouet-Guicheney C, Knowles M, Moseley B, Oberdorfer R, Samuels F. History of safe use as applied to the safety assessment of novel foods and foods derived from genetically modified organisms. *Food Chem. Toxicol.* 2007;45:2513–2525.
61. NETTOX. Nettox list of food plants: information on inherent food plant toxicants, Report 2. Danish Veterinary and Food Administration, Søborg, Denmark.
62. IPGRI (International Plant Genetic Resources Institute). Conserving and increasing the use of neglected and underutilized crop species, 2004.
63. The American Botanical, Council. The complete German Commission E monographs: therapeutic guide to herbal medicines, M. Blumenthal (Ed.), 1998.
64. European Scientific Cooperative on Phytotherapy, (ESCOP). Monographs, 2nd edition, 2003.
65. European Medicines Agency, (EMEA). Overview of HMPC assessment work: status September 2009. EMEA/HMPC/278067/2006.

66. Kroes R, Renwick AG, Feron V, Galli CL, Gibney M, Greim H, Guy RH, Lhuguenot JC, van de Sandt JJ. Application of the threshold of toxicological concern (TTC) to the safety evaluation of cosmetic ingredients. *Food Chem. Toxicol.* 2007;45(12):2533–2562.
67. USEPA. Proposed test rule for *in vitro* dermal absorption rate testing of certain chemicals of interest to Occupational Safety and Health Administration, 2006.
68. ECB. Technical guidance document (TGD) on risk assessment in support of Commission Directive 93/67/EEC on risk assessment for new notified substances. European Commission, Part 1, 2003, pp. 1–303.
69. OECD. OECD guidelines for the testing of chemicals: Guideline 428—skin absorption: *in vitro* method, Paris, April 13, 2004.
70. OECD. OECD Environmental Health and Safety Publications: series on testing and assessment No. 28, guidance document for the conduct of skin absorption studies, Paris, March 5, 2004.
71. Jones AD, Dick JP, Cherrie JW, Cronin MTD, Van de Sandt JJM, Esdaile DJ, Iyengar S, Ten Berge W, Wilkinson SC, Roper CS, Semple S, De Heer C, Williams FM.CEFIC workshop on methods to determine dermal permeation for human risk assessment. Research Report TM/04/07, December 2004.
72. Sartorelli P, Andersen HR, Angerer J, Corish J, Drexler H, Göen T, Griffin P, Hotchkiss SAM, Larese F, Montomoli L, Perkins J, Schmelz M, van de Sandt J, Williams F. Percutaneous penetration studies for risk assessment. *Environ. Toxicol. Pharmacol.* 2000;8:133–152.
73. SCP. Opinion on the draft guidance on dermal absorption, European Commission, Health and Consumer Protection Directorate General, Directorate C—Scientific Opinions C2—Management of Scientific Committees, Scientific Co-Operation And Networks, April 30, 2002.
74. Lademann J, Richter H, Jacobi U, Patzelt A, Hueber-Becker F, Ribaud C, Benech-Kieffer F, Dufour E, Sterry W, Schaefer H, Leclaire J, Toutain H, Nohynek, GJ. Human percutaneous absorption of a direct hair dye comparing *in vitro* and *in vivo* results: implications for safety assessment and animal testing. *Food Chem. Toxicol.* 2008;46:2214–2223.
75. Knudsen I, Søborg I, Eriksen F, Pilegaard K, Pedersen J. Risk management and risk assessment of novel plant foods: concepts and principles. *Food Chem. Toxicol.* 2008;46:1681–1705.
76. OECD. Guidance on grouping of chemicals (No. 80). ENV/JM/MONO(2007)28, 2007.
77. EU. Regulation (EC) 1907/2006 of the European Parliament and the European Council of December 18, 2006 concerning the Registration, Evaluation, Authorisation and Restriction of Chemicals (REACH), establishing a European Chemicals Agency, amending Regulation 1999/45/EC and repealing Council Regulation (EEC) No. 93/793 and Commission Regulation (EC) No. 1488/94 as well as Council Directive 76/769/EEC and Commission Directives 91/155/EEC, 93/677/EEC, 93/105/EEC, and 2000/21/EC. Official Journal of European Commission, L 369, December 30, 2006, pp. 1–849.
78. EFSA (Panel on Food Additives, Flavourings, Processing Aids and Materials in Contact with Food). Opinion on Flavouring Group FGE.03: acetals of branched- and straight-chain aliphatic saturated primary alcohols and branched- and straight-chain saturated aldehydes, and an orthoester of formic acid, from chemical groups 1 and 2. Opinion expressed on October 7, 2004.
79. JECFA. Safety evaluation of certain food additives and contaminants. In: *49th Meeting of the Joint FAO/WHO Expert Committee on Food Additives,* World Health Organization (WHO), International Programme on Chemical Safety (IPCS), Geneva, Switzerland, WHO Food Additives Series No. 40, 1998.
80. Munro IC, Kennepohl E. Comparison of estimated daily per capita intakes of flavouring substances with no-observed-effect levels from animal studies. *Food Chem. Toxicol.* 2001;39:331–354.
81. Munro IC, Danielewska-Nikiel B. Comparison of estimated daily intakes of flavouring substances with no-observed-effect levels. *Food Chem. Toxicol.* 2006;44:758–809.
82. Smith RL, Adams TB, Cohen SM, Doull J, Feron VJ, Goodman JI, Hall RL, Marnett LJ, Portoghese PS, Waddell WJ, Wagner BM. Safety evaluation of natural flavour complexes. *Toxicol Lett.* 2004;149 (1–3):197–207.
83. Munro IC. Safety assessment procedures for indirect food additives: an overview. *Regul. Toxicol. Pharmacol.* 1990;12:2–12.
84. Munro IC, Ford RA, Kennepohl E, Sprenger JG. Correlation of structural class with no-observed-effect levels: a proposal for establishing a threshold of concern. *Food Chem. Toxicol.* 1996; 34(9):829–867.

85. Cramer GM, Ford RA, Hall RL. Estimation of toxic hazard: decision tree approach. *Food Cosmet. Toxicol.* 1978;16(3):255–276.
86. Food and Drug Administration. Food additives: threshold of regulation for substances used in food-contact articles. *Fed. Regist.* 1993;58(195):52719–52727.
87. JECFA. Summary of evaluations performed by the Joint FAO/WHO Expert Committee on Food Additives (JECFA 1956–2003). In: *First Through Sixty-First Meetings,* 2003.
88. Renwick AG. Toxicology databases and the concept of thresholds of toxicological concern as used by the JECFA for the safety evaluation of flavouring agents. *Toxicol. Lett.* 2004;149:223–234.
89. European Medicines Agency, (EMEA)., Guideline on the limits of genotoxic impurities, The European Medical Control Agency, Committee for Medicinal Products for Human Use. London, June 23, 2003.
90. European Medicines Agency, (EMEA)., Guideline on the limits of genotoxic impurities, CPMP/SWP/5199/02, European Medicines Evaluation Agency. Committee for Medicinal Products for Human Use (CHMP), London, June 23, 2004.
91. Müller L, Mauthe RJ, Riley CM, Andino MM, De Antonis D, Beels C, DeGeorge J, De Knaep AGM, Ellison D, Fagerland JA, Frank R, Fritschel B, Galloway S, Harpur E, Humfrey CDN, Jacks AS, Jagota N, Mackinnon J, Mohan G, Ness DK, O'Donovan MR, Smith MD, Vudathala G, Yotti L. A rationale for determining, testing, and controlling specific impurities in pharmaceuticals that possess potential for genotoxicity. *Regul. Toxicol. Pharmacol.* 2006;44:198–211.
92. IPCS. World Health Organisation, WHO Food Additives Series 35, WHO, Geneva, 1998.
93. Bridges J.Strategy for a future chemicals policy. The view of the Scientific Committee on Toxicology, Ecotoxicology and the Environment (CSTEE), University of Surrey, UK, 2003.
94. Wolf de W, Sieber-Sauer A, Lecloux A, Koch A, Holt M, Feytel T, Comber M, Boeije G. Mode of action and aquatic exposure thresholds of no concern. *Environ. Toxicity Chem.* 2005;24:479–485.
95. Kroes R, Galli CL, Munro I, Schilter B, Tran L-A, Walker R, Würtzen G. Threshold of toxicological concern for chemical substances present in the diet: a practical tool for assessing the need for toxicity testing. *Food Chem. Toxicol.* 2000;38:255–312.
96. Kroes R, Renwick AG, Cheeseman M, Kleiner J, Mangelsdorf I, Piersma A, Schilter B, Schlatter J, van Schothorst F, Vos JG, Würtzen G. Structure-based thresholds of toxicological concern (TTC): guidance for application to substances present at low levels in the diet. *Food Chem. Toxicol.* 2004;42:65–83.
97. JECFA. Evaluation of certain food additives and contaminants: safety evaluation of flavouring agents. Forty-first Report of the Joint FAO/WHO Expert Committee on Food Additives, WHO Technical Report Series 837, World Health Organization, Geneva, 1993.
98. JECFA. Evaluation of certain food additives and contaminants: safety evaluation of flavouring agents. Forty-Fourth Report of the Joint FAO/WHO Expert Committee on Food Additives, WHO Technical Report Series 859, World Health Organization, Geneva, 1995, pp. 2–3.
99. JECFA. Evaluation of certain food additives and contaminants: procedure for the safety evaluation of flavouring agents. Forty-Ninth Report of the Joint FAO/WHO Expert Committee on Food Additives, WHO Technical Report Series 884, World Health Organization, Geneva, 1999, pp. 3–6.
100. Cheeseman MA, Machuga EJ, Bailey AB. A tiered approach to threshold of regulation. *Food Chem. Toxicol.* 1999;37:387–412.
101. Munro IC, Renwick AG, Danielewska-Nikiela B. The threshold of toxicological concern (TTC) in risk assessment. *Toxicol. Lett.* 2008;180:151–156.
102. Re TA, Mooney D, Antignac E, Dufour E, Bark I, Srinivasan V, Nohynek G. Application of the threshold of toxicological concern approach for the safety evaluation of calendula flower (*Calendula officinalis*) petals and extracts used in cosmetic and personal care products. *Food Chem. Toxicol.* 2009;47:1246–1254.

CHAPTER 9

APPROACH TO ASSESSING CONSUMER SAFETY OF BOTANICAL INGREDIENTS WITH EMPHASIS TO TYPE I ALLERGY

Esperanza Troyano,[1] Donna Mcmillan,[2] Katherine Sarlo,[3] Lijuan Li,[2] and Rohan Wimalasena[2]

[1]The Procter & Gamble Company, Sharon Woods Innovation Center, Cincinnati, OH
[2]The Procter & Gamble Company, Mason Business Center, Mason, OH
[3]The Procter & Gamble Company, Miami Valley Innovation Center, Cincinnati, OH

9.1 INTRODUCTION

Botanicals and natural extracts include a very diverse number of substances including leaves, roots, flowers, fruits, seeds, berries, or parts or mixtures thereof, as well as exudates or natural substances such as silk, pearl, or propolis. They can be the crude plant and natural exudates or preparations obtained by traditional extraction methods or by more complex processes such as fermentation.

The consumer demand for products containing these natural-derived extracts has grown significantly over the past years, partly because of the common belief that they are the healthiest option. However, this public perception of "natural" equaling safe does not always hold true as evidenced by a significant number of recorded cases of adverse health effects with botanical products [1–4]. Botanicals tend to be poorly characterized substances with great variability in composition, for which limited safety information may be available. To ensure safety, botanical and natural substances should be evaluated as thoroughly as any other cosmetic ingredient ahead of market introduction.

Because of the chemical complexity associated with botanical and natural extracts, it is often difficult to have a good understanding of what is being used. However, this knowledge is critical to assess the safety of these materials. Failure to

Formulating, Packaging, and Marketing of Natural Cosmetic Products, First Edition.
Edited by Nava Dayan and Lambros Kromidas.

understand the composition and origin of the extract can result in adverse events in the marketplace as illustrated by reported cases of adverse health effects from botanicals as a result of contamination or confusion with species different than the one intended to be used [5]. Therefore, the first step in the assessment of a botanical or natural material should be to ensure that accurate information from the species and the part of the plant that is being used is available. This is particularly important if toxic constituents are known to be present at high concentration in any part of the plant (e.g., alkaloids in comfrey plant). One must also be aware of residuals from processing of naturals, such as residual organic solvents from solvent extraction.

Correct identification of the species and the part of the plant used to prepare the material is not sufficient to characterize the material. It is important to also understand all the other factors that contribute to the variability of the composition of the extract such as the developmental stage of the plant, period of harvest, growth conditions, use of pesticides, as well as details on the processing and preparation of the exact material to be used.

The quality and chemical constituents of a plant can vary widely based on geographical and even seasonal differences in soil, rainfall, fertilization, and so on. The appropriate level of information required to correctly identify the starting material needs to be evaluated on a case-by-case basis, taking into account everything that is known about the botanical. For example, it is well known that some plant extracts, such as those derived from several aloes or rhubarb, can contain anthraquinones. Anthraquinones are gastrointestinal irritants and also phototoxic. For their phototoxicity potential, their levels should be kept low in cosmetic products [6]. Thus, the characterization of botanical substances from these plants should include a step to measure levels of anthraquinone. Processing the material is important because it can lead to significant changes in its composition. For example, an aqueous extract will have a different composition from a solvent extract; heat/bleaching or ultra filtration processes can help eliminate and/or destroy proteins in the material, which will help decrease the risk for Type I allergy as will be discussed below. Finally, the chemical and biological properties of the preparations need to be reproducible between lots so it is important to set clear specifications that the material must meet to be used in a cosmetic preparation.

In summary, there are three key aspects to characterization that are widely recognized as critical for the production of a consistent botanical material:

1. Identification of the starting material.
2. Processing or other treatment applied.
3. Setting clear specifications for the composition of the material for batch-to-batch reproducibility.

A good approach to understanding the raw material and setting clear specifications for reproducibility has been proposed by Schilter et al. [7]. This approach contains an extensive list of details that need to be investigated in order to characterize the botanical substance and includes aspects such as botanical source, growth conditions, process applied, botanical preparation, and methodology used for analytical confirmation.

9.2 TYPE I ALLERGY ASSESSMENT

For cosmetic products, the botanical assessment should include evaluation of four relevant routes of exposure: inhalation, ocular, dermal, and systemic. However, one toxicological endpoint that can be easily overlooked when assessing safety of botanical and natural extracts is Type I allergy or immediate hypersensitivity. Immediate hypersensitivity is classically defined as the development of IgE allergic antibody to an allergen (usually a protein) with potential development of symptoms ranging from rhinitis, conjunctivitis, hives, asthma, and anaphylaxis. The most severe response is anaphylactic shock where the allergic response leads to cardiopulmonary collapse and possible death. Botanical and natural extracts can have a wide range of levels of protein. Fortunately, there have been few reports of immediate allergic responses to proteins used in personal care products but these reports, plus a shift in the view of the role of skin exposure and development of respiratory allergy, prompts an approach to the assessment of immediate hypersensitivity to botanicals in personal care products. The following are some examples of where protein in personal care products has led to the development of IgE antibody and/or the development of allergic symptoms. Incorporation of a quaternized hydrolyzed bovine collagen preparation (Crotein Q) into hair care products led to development of allergic symptoms in hairdressers and in consumers [8–10]. Diagnostic follow-up testing showed that these individuals had IgE antibody to the derivatized collagen. One cosmetic manufacturer (P&G) had received several reports of sudden onset urticaria and one report of shock in consumers using shampoo or conditioner containing this material with confirmation of IgE antibody in these subjects (unpublished data; [11]). A Finnish study of 2199 clinic subjects revealed that 12 individuals (0.5%) had IgE antibody to Crotein Q and 2 of these 12 individuals also reacted to another protein hydrolysate, Nutrilan H [12]. Four of the 12 reported some symptoms (urticaria, conjunctivitis) upon exposure to products containing these ingredients. The Crotein Q material was widely used for over 10 years in a variety of personal care products so it was not clear how these individuals developed their IgE antibody but certainly a small percent of consumers were able to experience allergy symptoms.

Prospective clinical studies designed to assess the safety of prototype beauty products such as a bar soap or a body lotion containing a protease enzyme revealed that these exposures were sufficient to induce IgE antibody in some portion of the study populations [13, 14]. In the bar soap study, 4 of the 62 test subjects developed IgE antibody to the enzyme after 4–6 months of use of the bar soap [13]. Laboratory measures showed that the soap aerosolized in the shower with an average exposure to enzyme of 10.9 ng/m^3. In the body lotion study, 3 of the 864 test subjects developed enzyme-specific IgE antibody after 12–14 months of intermittent use of the lotion [14]. Laboratory measurements showed that the enzyme was still present on skin 8–12 h after application and this low amount could aerosolize in the shower. Other exposures such as direct mucosal contact or inhalation of enzyme on shed skin cells [15, 16] likely also played a role in the sensitization.

There have been reports that hydrolyzed wheat protein in cosmetic creams elicited allergic reactions in some users [17, 18]. Similarly, incorporation of milk protein into creams has been associated with allergy in milk-allergic children [19]. A longitudinal study of parent and children showed a significant relationship between use of skin products containing peanut oil and peanut allergies [20]. Finally, investigative work in the food allergy area has shown it to be possible to induce allergen-specific IgE antibody in mice via dermal exposure [21–23]. This has prompted several clinical groups to seriously consider that dermal exposure, especially to skin with some compromised barrier function, can be a relevant route of exposure to induce immediate hypersensitivity as well as elicit symptoms in a subset of people [24].

Public awareness that use of consumer products containing botanical or natural substances could elicit an allergic reaction in individuals with a preexisting allergic condition has grown over the last few years. The Food Allergy and Anaphylaxis Network (FAAN) advise patients with food allergies such as soy, wheat, or peanut to read labels on cosmetic products to avoid exposure. More recently, the European Commission issued a request for scientific data on peanut oil and wheat protein extracts due to concerns related to the potential for consumers to become sensitized to these materials as a result of their use in cosmetic products.

9.2.1 Parameters Needed for the Protein Risk Assessment

Based on the market experience with enzymes and proteins described above, we developed a pragmatic approach to assessing the risk of immediate hypersensitivity on exposure to proteins in botanical extracts. The backbone of the approach is avoiding known sources of allergens, using botanicals with low protein content (if the protein is not a key component of the extract) and minimizing exposure.

The first step is to identify the source of the extract and the amount of protein in the extract. The total protein content is needed to estimate potential exposure upon use of product containing the extract. A tiered approach can be followed to determine the level of protein in the botanical extract. The complexity and cost of the method used increases with the tier, but so does the accuracy of the measurements. Therefore, lower tier methods are usually only sufficient for botanical or natural substances used at very low levels in the cosmetic product. For those botanical or natural extracts for which there is a relatively high prevalence of sensitization (i.e., food allergens), highest tier methods are recommended.

9.2.1.1 Analytical Characterization of Protein Content A tiered approach as shown in Figure 9.1 was developed and implemented at P&G for characterization of protein content in botanical/natural substances. In this scheme, determination of total nitrogen content by Kjeldahl nitrogen method [25] is the first tier test for majority of botanical/natural substances. The total nitrogen method is a versatile method that can handle very complex matrices such as botanical and natural substances for easy determination of protein content. Even though the total nitrogen method is very nonspecific for total protein determination, it is performed as the first tier screen, since a negative test result or a protein level estimated to be below the threshold of toxicological concern (TTC) would qualify the substance without further testing. The TTC is described in Section 9.2.1.3.

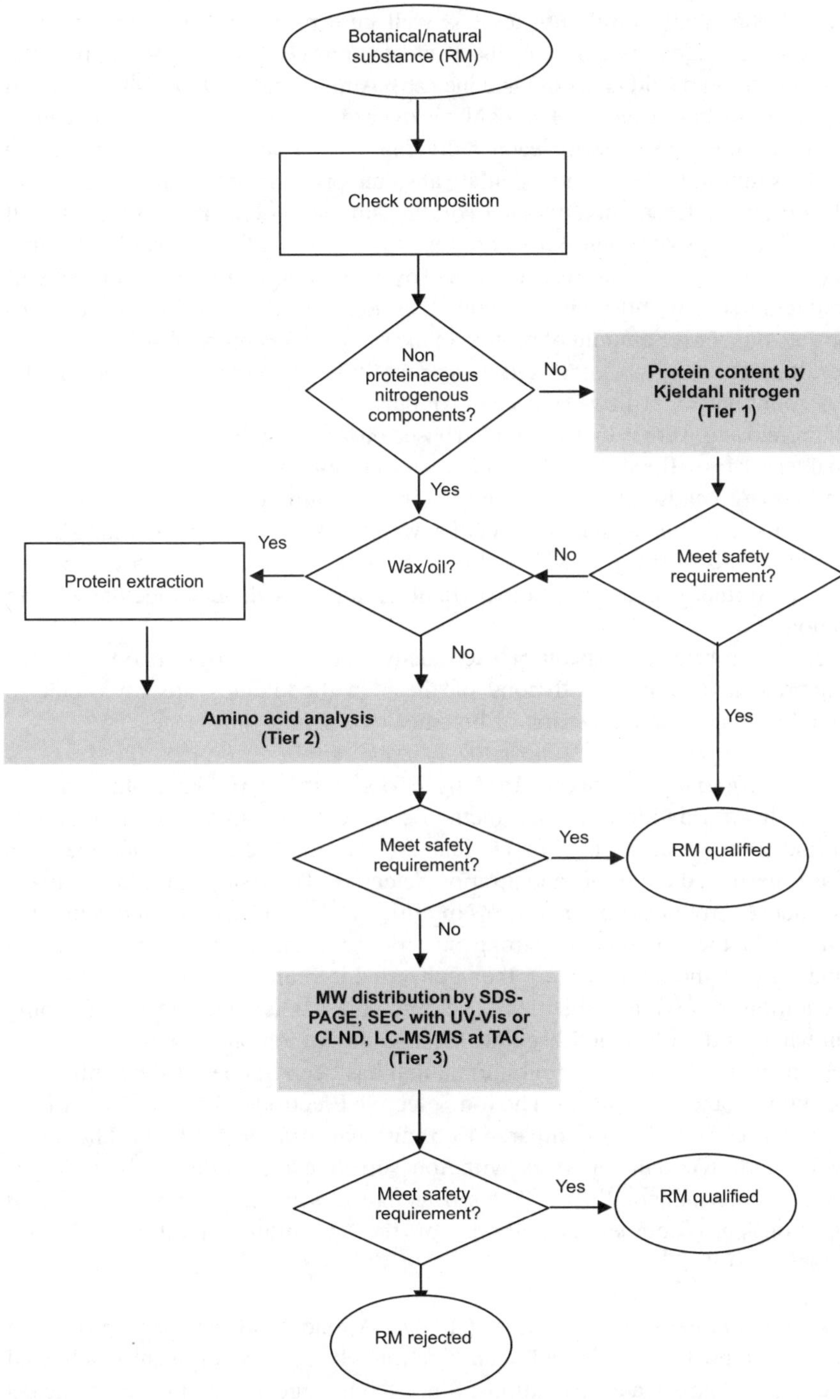

Figure 9.1 Flow chart: tiered approach for protein characterization in botanical/natural substances.

If the botanical/natural substance is well known to contain protein or non-proteinaceous nitrogenous components, Kjeldahl nitrogen analysis is skipped and Tier 2 testing (amino acid analysis) is done [26]. Amino acid analysis is also done if protein exposure based on the Kjeldahl nitrogen data exceeds TTC. Amino acid analysis determines the level of each individual amino acid from proteins in the sample. This in turn will allow to calculate absolute protein content in the botanical/natural substance. Even though some of the simpler colorimetric methods are available for total protein determination, they are not suitable for botanical/natural substance due to potential interference caused by other nonproteinaceous components (e.g., reducing sugars and plant pigments.) In addition, the colorimetric methods provide data on relative amount of protein in the sample and not the absolute amount. As a result safety margins calculated using protein levels obtained by colorimetric methods could have a positive or a negative bias.

If protein exposure values calculated based on total protein level from the amino acid analysis data still exceeds the TTC then one must focus only on the protein fraction known to cause allergic responses for calculation of exposure levels. This protein fraction in general has a molecular weight (MW) greater than 2.5 kDa or 20–30 amino acids in size [27–31]. The level of protein with MW > 2.5 kDa can be obtained by further characterization of total protein based on molecular weight distribution.

Below is a brief discussion of each analytical methodology involved in the tiered approach. The pros and cons of these methodologies are important to understand protein characterization of botanicals or natural substances.

Tier 1—Kjeldahl Nitrogen Analysis (Total Nitrogen) Kjeldahl nitrogen analysis is the worldwide reference method for the quantitative determination of protein content in a wide variety of materials (26). The method consists of three steps including digestion, distillation, and titration/detection. These steps have to be carried out in sequence. Proteins and other forms of nitrogen are broken down and converted to ammonia in the presence of strong sulfuric acid and catalysts under heated condition. The ammonium ions are then converted into ammonia gas under strong basic condition, heated, and distilled. The ammonia gas is collected in a trapping solution where it dissolves and becomes an ammonium ion once again.

Historically, the amount of ammonia that has been trapped is determined by titration with a standard solution. The Ion Selective Electrode Detection system has greatly increased sensitivity compared to traditional titration methods. The quantitation limit of Kjeldahl analysis with ion selective electrode is about 5 ppm nitrogen under normal conditions. The nitrogen results are multiplied by the factor 6.25 based on AOAC compendia methods from the Agricultural literature to obtain a total protein result [32].

Tier 2—Amino Acid Analysis (AAA) Amino acid analysis procedure includes two steps: hydrolysis and quantification. Hydrolysis is typically achieved under strong acid and heated conditions. Since the botanical or natural substance is often a complex matrix, it sometimes is necessary to extract the protein content from the matrices prior to hydrolysis. For example, protein extraction needs to be done for wax/oil type of samples and a fraction of the extract will be dried down under vacuum before undergoing hydrolysis.

Hydrolyzed samples are then subjected to chromatographic analysis. Ion-exchange chromatographic separation with postcolumn ninhydrin derivatization provides very reliable data for amino acid analysis. The postcolumn reaction between ninhydrin and amino acid eluted from column is monitored at 440 nm for most amino acids and 570 nm for imino acid such as proline. The accurate limit of detection for this method is about 10 pmol for most amino acids, but 50 pmol for proline. As little as 5 pmol is detectable for some amino acids.

To obtain good composition data, samples larger than 1 μg before hydrolysis are best suited for amino acid analysis of protein/peptide. The estimated protein concentration of an unknown botanical/natural substance can be calculated by the sum of the masses of the recovered amino acids divided by the total mass of the sample. The results are normally reported as total protein concentration expressed in % weight (ppm).

Since amino acid analysis measures the total protein content in a given sample, in some cases it may greatly overestimate or underestimate the amount of proteins in a sample. If the sample contains significant amount of free amino acids, the total protein content calculated based on the amino acid analysis will be exaggerated. Filtration of such samples with molecular weight cut off (MWCO) membrane filter at 2.5 kDa prior to hydrolysis may provide more accurate estimation on the protein content. Protein content can also be underestimated in proteins that are rich in sensitive amino acids because these amino acid residues are usually destroyed during hydrolysis and cannot be measured.

Tier 3—Molecular Weight Distribution SDS-PAGE (sodium dodecyl sulfate polyacrylamide gel electrophoresis) or size-exclusion chromatography (SEC) provides information on molecular weight distribution of protein composition and on an estimated concentration of proteins >2.5 kDa in botanical/natural substance. Ideally the lower limit detection of the method should be 1 ppm or less based on standard proteins. This approach may be done concurrently with AAA if the protein content of a botanical/natural substance is well known to be high.

As in amino acid analysis, liquid and solid (powder) samples can be directly prepared in SDS sample buffer for SDS-PAGE analysis. Wax/oil samples need to be extracted first prior to SDS-PAGE. Recovery of proteins from wax and oil samples varies from sample to sample and based on limited spiked and recovery data, these values could vary from 7 to 40%.

Using the SDS-PAGE method, the samples are preferably analyzed at three different concentrations, the targeted use concentration, 10 times greater than the use concentration, and a 10-fold dilution of the use concentration. For example, when the use concentration of a botanical/natural substance is 0.1% (1:1000) in formulation, the substance will be preferably tested at three concentration levels of 0.1% (1:1000), 1% (1:100), and 0.01% (1:10,000) using SDS-PAGE and Coomassie Blue stain. At each concentration level, the protein level will be estimated for the protein bands >2.5 kDa using 1 ppm as detection limit. If a protein band >2.5 kDa is visible at 1% concentration and invisible at 0.1 and 0.01% concentration, the conclusion can be made that the protein concentration for the particular band is between 1 and 10 ppm at 1% concentration and less than 1 ppm at the use concentration (0.1%). Streaking or broad smear band on the gel can make the estimation of protein level impossible. Carbohydrate in the matrix can have this effect on the gel.

Size-exclusion chromatography is an alternative method for determining protein molecular weight distribution. Although the resolution is less than SDS-PAGE, SEC offers several advantages over SDS-PAGE, including instrument automation and quantitation based on UV absorbance of protein standards. SEC with UV detection can overestimate the total protein level in some matrices. It is very common to see interference from polysaccharides for protein concentration determination using this method. Therefore, it is important to verify the UV-Vis spectrum pattern of the high molecular weight species acquired by online photodiode array (PDA) detection matches the spectrum pattern of proteins and not polysaccharides before using the absorbance values for calculating the protein concentrations.

The presence of polysaccharides may hinder the ability to use SDS-PAGE and SEC to evaluate the MW distribution of the protein content. Chemiluminescence nitrogen detector (CLND) coupled with SEC separation may overcome the problem and provide more accurate estimation of proteins >2.5 kDa in the botanical/natural substance. If needed, liquid chromatography-tandem mass spectrometry (LC-MS/MS) or matrix-assisted laser desorption ionization time-of-flight mass spectrometry (MALDI-TOF/MS) with sensitivity 1–20 ppm can be used to confirm the absence of certain proteins of specific safety concerns.

Characterization of proteins in botanical/natural substance is very important to facilitate reliable safety assessment of such substance and allow manufacturer to include the substance as a value adding ingredient in consumer products. The tiered approach implemented for this characterization work provides a roadmap for manufacturer to facilitate the safety assessment by providing the necessary protein data needed for exposure calculations.

9.2.1.2 Source of Protein Knowing the source of the extract is important since one principal of the approach is to avoid the major food allergens. These food allergens are responsible for 90% of the food allergies in the population so avoiding their use in personal care products will eliminate the risk of eliciting symptoms in already allergic individuals [33, 34]. The food allergens are egg, milk/milk products, soy, wheat, peanut, tree nuts, fish, and shellfish. The tree nuts only include those that have been documented to cause allergy in humans. These include walnuts, almonds, macadamia nuts, pecans, hazelnuts, cashews, pistachios, pine nuts, and brazil nuts. The European Food Safety Authority [35] also considers celery, mustard, and sesame seed (36) as food allergens. If a food allergen extract is used then there are ways to greatly reduce the risk of allergy to the food extract. The protein can be removed by the extraction and processing method or the protein can be hydrolyzed to very small peptides and amino acids. In general, the immune system will recognize proteins of a certain size (around 10 kDa and larger) in order to initiate an immune response. Once initiated, allergens need to be of a certain size in order to effectively interact with IgE antibody to elicit allergic responses. Work done with hydrolysates and immunotherapeutic peptides showed that allergens need to be around 2.5 kDa or about 20 amino acids [27–31] to elicit a response. Peptides smaller than this are too small to cross-link IgE antibodies and elicit allergic responses. Heat denaturation of the protein can disrupt the three-dimensional nature of some IgE binding site but it is not always optimal since some allergens, such as peanut, become more effective after heat treatment.

9.2.1.3 Exposure Assessment If the nonfood allergen or botanical extract contains protein and the protein is not hydrolyzed to small peptides or amino acids, then it is necessary to conduct an exposure assessment. While no thresholds of exposure has been identified for inducing and/or eliciting Type I allergy to proteins, benchmark clinical experiences can be used to gauge whether a particular type of exposure is higher risk for allergy. In the case of personal care products, we use the information from several clinical studies of enzyme containing products to provide guidance. Analytical measurements done on enzyme containing personal care products have demonstrated that aerosols generated in the shower are one source of exposure to the enzyme [13, 14]. Aerosol generation was apparent for products such as bar soaps, body washes, shampoos, and conditioners. However, other products, such as lotions, are not typically used in the shower but can aerosolize off the skin once the user goes into the shower and generate significant exposure to the enzyme. Other exposures via use of personal care products such as direct contact with mucosal tissue or inhalation/mucosal exposure to enzyme associated with shed skin flakes also occurred but were harder to measure reproducibly. Therefore, our approach to assessing exposure to protein or enzymes in products was based on the use of shower exposures, as surrogate to account for all exposures. This methodology uses the aerosol exposure as the standard to compare across clinical studies and to build exposure models. The prospective clinical studies showed that exposure to enzyme from use of a personal care product that led to enzyme aerosol levels in the shower greater than 0.1 ng/m^3 were sufficient to sensitize test subjects [13, 14].

Aerosols in the shower are calculated using models built from laboratory measures of product use. For bar soaps, body washes, shampoo, and conditioner the amount of protein aerosolized is proportional to the amount of protein in product and amount of product used in the shower. There is also a model for aerosols of protein off of skin in the shower 8–12 h following use of a lotion. In this case the amount of protein aerosolized is proportional to the amount of protein on skin; the amount of protein on skin is proportional to the amount of protein in lotion and the amount of lotion applied. Once the level of protein in the botanical or natural extract, the source of the material, and the exposure to the consumer from the product have been defined, the protein risk assessment can be done using the tiered approach described below.

9.3 TIERED TOXICOLOGICAL/SAFETY APPROACH TO ASSESSING BOTANICALS

Botanical substances may be used at variable levels in personal care products. This, together with the type of product to which the extract is being added will define the level of exposure to the consumer. Therefore, the safety assessment approach can be flexible based on the level of the botanical or natural substance in the product. The approach to botanical and natural extracts evaluation is based on tiers, where first tier would be based on very conservative assumptions using worst-case scenarios and last tier would require an in-depth knowledge of the exact composition of the botanical extract to be used and specific toxicological data to address all relevant endpoints. This approach has three tiers and each tier addresses Type I, dermal and systemic endpoints:

- Tier 1 for botanical and natural substances used at very low levels (typically <50–100 ppm) in the consumer product.
- Tier 2 for botanical and natural substances used at higher levels (>50–100 ppm) in the consumer product and have a history of safe human use.
- Tier 3 for botanical and natural substances used at higher levels (>50–100 ppm) in the consumer product *without* a history of safe human use.

9.3.1 Tier 1—Assessing Ingredients Present at Very Low Levels Using the TTC Concept

When specific toxicity data on the botanical or natural extract are not available, exposure to a botanical substance used at low levels in personal care products can be assessed using the TTC approach.

The TTC approach was initially developed to assess the safety of indirect food additives. The approach is based on the principle of establishing a threshold exposure below which there is a very low probability of risk to human health [37]. The World Health Organization (WHO) has adopted the use of this approach to evaluate low level, direct food additives [38]. Use of TTC has expanded to include natural flavor complexes [39, 40], cosmetic ingredients [41, 42], and recently, a botanical used as a cosmetic ingredient [43].

The TTC methodology is used in Tier 1 for the botanical or natural substance itself. It recognizes that botanicals are often complex mixtures and that the identity and/or concentration of each natural chemical constituent in the substance may not be readily available or even technically feasible to be obtained. The exposure to the entire amount of the botanical extract (i.e., by assuming it is a single chemical) added to the product is then compared to the TTC.

9.3.1.1 Type I Allergy Assessment As described in the previous section, the protein source and content of the botanical or natural substance must be determined in order to assess the potential for a botanical or natural extract to induce hypersensitivity. Exposure to the protein (>2.5 kDa) from the botanical or natural substance via the inhalation route is then calculated using consumer habits and practices data or exposure guidance from regulatory agencies, for example, RIVM ConsExpo model, SCCP exposure models [44, 45]. The consumer exposure is compared to the Type I allergy benchmark of 0.1 ng/m^3. The amount of protein in the botanical or natural substance is either acceptable, that is, below the benchmark of 0.1 ng/m^3, or not acceptable and needs to be reduced. Reduction can be done by lowering the level of extract in the product or by elimination, reduction, or hydrolysis of the protein present in the raw material.

Using a botanical with a level of protein that results in an exposure above the allergy benchmark set for botanicals requires a higher level of assessment as described in Tier 2.

9.3.1.2 Dermal Assessment The potential for skin irritation and sensitization and phototoxicity/photoallergy should always be considered for the ingredients in dermally applied products, including botanical and natural substances.

A conservative benchmark dose for dermal sensitization of 1 μg/cm^2 can be used when conducting a skin sensitization risk assessment for consumer exposure to botanical or natural extracts (46). The conservatism of the approach is based on the assumption that the botanical or natural extract is a single chemical and on the use of a conservative value as the threshold for skin sensitization. The benchmark dose was selected based on data from potent sensitizers and thus represents a conservative, worst-case scenario. This benchmark dose can be used for all botanical/natural extracts that are not concentrated, that is, contain the chemical constituents at approximately the same levels as found in nature. In this instance, an uncertainty factor is not applied to the benchmark dose because it is highly likely that a natural chemical constituent of concern represents only a small fraction of the entire botanical substance component. In other words, using 1 μg/cm^2 as the benchmark dose or the limit for consumer exposure provides a high degree of assurance that the likelihood for induction of contact allergy from exposure to the botanical extract through intended use will be negligible. This same conservative benchmark dose can also be used for the assessment of dermal irritation, phototoxicity, and photoallergy potential of the substance under evaluation.

Exposures to botanical extracts exceeding the benchmark dose and exposures to concentrated extracts (i.e., essential oils) require further evaluation and thus are outside what can be supported using the Tier 1 approach. For these cases as well as cases in which exposure to the botanical exceeds the TTC of 1 μg/cm^2 a more refined assessment is required as described in Tier 2.

9.3.1.3 Systemic Assessment A conservative TTC value to use for the Tier 1 systemic assessment is 0.15 μg/day, commonly used for chemicals with structural alerts for genotoxicity [41]. In the absence of specific data on dermal penetration of the botanical extracts and its components, dermal absorption for the material should be assumed to be 100%.

9.3.2 Tier 2—Assessing Botanicals with a History of Safe Human Use

9.3.2.1 Type I Allergy Assessment If the calculated exposure to the botanical as indicated for the Tier 1 is greater than 0.1 ng/m^3 then advise is given to lower the level of extract in the formula to lower exposure, consider a substitute extract, or determine if the protein can be removed or hydrolyzed. If this cannot be done and the extract plays a significant role in the product performance profile then additional work will be needed before the extract can be used.

Since there are no valid *in vitro* models that can be used to test the allergenic potential of the extract, then additional work that can include clinical work is the only option to move forward. Not all proteins are allergens but there are no methods to identify which proteins are or are not allergens. If the source of the extract has not been associated or rarely associated with allergy in humans even though there has been considerable exposure via other uses then the likelihood of the protein in the extract to be allergenic in a cosmetic product is low. An extensive literature search along with use of public allergen databases such as Allergome.org should be done to confirm lack of allergy reports. In addition, the Tier 2 assessment requires understanding the uses and exposures to the botanical or natural substance. An example is guar extract. The ground endosperm of the guar bean has been used as a source of galactomannan for many applications since it is a good thickener, emulsifier, and stabilizer [47]. The textile, paper, machine, and mining industries use guar extracts. Guar extracts are also used in foods (e.g., ice cream, puddings) and drugs and personal care products such as toothpastes and shampoo. Varied exposures to guar extracts have occurred for decades but reports of allergy or asthma related to guar are rare [48]. This weight of evidence informs one that exposure greater than the benchmark can likely be tolerated, but effort should be made to keep the exposure as low as possible and confirmation via a clinical program is recommended ahead of full market introduction. While this approach can work for a source material that has a wide distribution of use and exposure, the assessor must use caution. A lack of reports of allergy or asthma may be related to lack of exposure. For example, allergy to kiwi was uncommon in North America until the fruit became popular and more common to the diet [49].

9.3.2.2 Dermal Assessment When exposure to the botanical following the Tier 1 approach is above the benchmark dose of 1 $\mu g/cm^2$, skin sensitization, irritation, phototoxicity, and photosensitization data on the specific extract or each of the extract components will be required. Occasionally, to conduct this assessment, a history of safe dermal use of the botanical can be documented based on traditional medicinal uses of the substance. One example is aloe extracts. Extracts from aloe have been widely used in topical applications for many years because of their skin conditioning properties [6, 50, 51], thus the use of published data available on the safe uses of these extracts could be used to assess the dermal local effects of the use of the specific aloe extract under evaluation in the cosmetic product.

9.3.2.3 Systemic Assessment When systemic exposure to the botanical extract exceeds the Tier 1 TTC, the safety assessment becomes more complex. Tier 2 safety assessment involves an in-depth search for relevant information, establishing a history of safe human exposure typically from use as a food (e.g., fruit, vegetable), food component (e.g., fruit/vegetable oils, fiber), food additive (e.g., spice, flavoring, edible colors/dyes), or dietary supplement/herbal medicine (in appropriate cases) that can be used to address potential adverse systemic effects. Occasionally a history of safe dermal use can be documented.

Kroes and Walker [52] described a decision tree from an International Life Science Institute (ILSI) workshop that can help determine the information required for the safety assessment of botanicals as foods. This decision tree can also be helpful for botanicals to be used in personal care products. Information should be gathered on the typical form, preparation and exposure to the botanical. Active principles and adverse event profiles can also be useful information if the plant is used as an herbal medicine.

It is important to distinguish a history of safe human use from a history of traditional use because the latter typically involves limited information to address safety. Traditional use is based on knowledge and experience in the relevant population but may have very limited scientific documentation. Traditional use may provide information on acute toxicity but is unlikely to provide information on chronic toxicity or delayed effects such as cancer and developmental or reproductive toxicity. Traditional use will be influenced by the general health of the population and the available health care systems for that population. With a history of safe use, there is evidence for safety for multiple generations in a large, genetically diverse population. Detailed usage information should include defined geographical area with information on exposure, usage habits and practices, years of use, preparation methods and impact on human health, both efficacy and adverse events. There should also be a scientific evaluation of the data with conclusions on safe use [53–55].

Several authoritative bodies have produced evidence-based monographs for a selection of herbal medicines. These include the EMEA European Medicines Agency Community Monographs from the Committee on Herbal Medicinal Products (HPMC), the European Scientific Cooperative on Phytotherapy (ESCOP) Monographs, The Scientific Foundation for Herbal Medicinal Products, 2nd Edition along with the Supplement 2009, the German Commission E Monographs as described in Herbal Drugs and Phytopharmaceuticals [50, 51]. Herbal Medicines and WHO [56]. These monographs typically provide information on the correct identification and preparation of the herbal medicine, as well as indications, contraindications, and efficacy as well as adverse event profiles.

For personal care products, comparisons should be based on an initial assumption of equal oral and dermal absorption. If systemic exposure associated with the intended new use in a personal care product does not significantly increase daily exposure beyond already existing exposure occurring via a history of safe use (typically in food), the safety assessment can be completed without entering Tier 3. If the exposure to the botanical exceeds exposure based on a history of safe human use or if a history of safe human use could not be documented, a Tier 3 safety assessment should be done.

9.3.3 Tier 3—In-Depth Evaluation of a Botanical Extract

The Tier 3 approach is recommended for the safety support of a botanical to be used at higher than TTC levels with an inadequate history of safe human use or where serious concerns exist regarding endpoints that are difficult to detect through human use (e.g., reproductive/developmental toxicity and carcinogenicity). Identification of a botanical with consistent quality and characterization/standardization are prerequisite to assessing at this level. The safety assessment approach could be based on the assessment of the botanical itself, if applicable safety data are available. If no safety data are available on the botanical then the safety assessment will be based on the extract major constituents using safety assessment methods similar to those used for single chemicals.

9.3.3.1 Analytical Characterization Appropriate characterization and quantification of the chemical constituents in the botanical or natural substance is prerequisite prior to conducting the safety evaluation of the botanical or natural substance. No single analytical method is capable of comprehensively identifying all constituents due to the complexity of botanical materials. Characterization should strategically employ quantitative analytical techniques that are specific for key known components, and also qualitative techniques that are more universal and capable of defining a fingerprint that is reproducible and characteristic for the material. This qualitative approach is a critical element for comparison of new batches and sources of substances in order to establish whether the original safety evaluation is applicable, or whether there are new and significant unknown constituents that may warrant further analytical characterization and safety assessment.

9.3.3.2 Type I Allergy Assessment If the protein component of a botanical/natural material is a key component to its performance profile and safety cannot be established in Tier 2 then additional work that can include clinical evaluation will be needed. As with a thorough analytical assessment of the chemical components, a similar assessment should be done with the protein. Isolation and identification of the protein is needed. If the primary amino acid sequence is not available in a database then the sequence should be established. The sequence can be compared to the sequences of known allergens using databases such as allergenonline.com [57]. Similarities can be checked by testing antibody to the known allergen(s) against the protein of interest. The assessor can use this information to develop a risk profile for a clinical study; the more similar the protein is to a known allergen the greater the chance that the clinical study will fail. There is no hard and fast definition for "similar" but the assessor is referred to the work done in the genetically modified food arena to obtain guidance [58, 59]. If a clinical study is done it should be of sufficient size (up to 1000 test subjects) and conducted over sufficient time (at least 6 months to 1 year) to ensure the power of the study is sufficient to predict what would happen in the marketplace.

9.3.3.3 Dermal Assessment When exposure to the botanical following the Tier 1 and 2 approaches cannot be supported, skin sensitization, irritation, phototoxicity, and photosensitization data on the specific extract or each of the extract components will be required. Then the extract can be assessed using a more traditional approach, such as the quantitative risk assessment method [60].

9.3.3.4 Systemic Assessment In addition to assessing each component in the botanical as a single chemical, botanicals typically used as herbal medicines may have pharmacological effects themselves and also have the potential to affect the potency and efficacy of pharmacological products the consumer may also be using. This potential for interaction must not be ignored. If systemic exposure is high enough to affect metabolic processes, consideration must be given as to the potential of the material and/or the natural chemical constituents to have an effect on the cytochrome P450 enzymes (i.e., grapefruit juice is a known cyp P450 inhibitor).

The safety assessment for the botanical's identified chemical constituents involves a comprehensive evaluation of available chemical and toxicological data on each of the individual chemical constituents. A chemically based safety evaluation approach has recently been well accepted by external scientific organizations such as the Flavor and Extract Manufacturers Association (FEMA) Expert Panel and the WHO Joint Expert Committee on Food Additives (JECFA) for the evaluation of flavoring agents [40, 61, 62]. The same approach has also been recommended by both organizations for the safety evaluation of natural flavoring complexes [40, 63].

Some of the identifiable chemical constituents of an extract may have available toxicity data and can be assessed by the usual safety assessment methods. Other constituents of the same botanical may also be found in foods so that a Tier 2, history of safe human use assessment can be done if the levels in foods are quantified. For those chemical constituents for which there is no relevant toxicity data, a structure—activity assessment can be done to search for relevant toxicity data on structurally similar chemicals. Finally, the TTC method described in the Tier 1 can be used if the identified chemical constituents are present at very low levels in the botanical.

Exposure assessments for each of the identified and unidentified chemical constituents are then conducted using analytical quantification data or the most conservative default assumptions for exposure. The safety of unidentified chemical constituents of the botanical can only be assessed by the conservative TTC method described in Tier 1.

If the Tier 3 approach is not able to provide sufficient assurance to support the safety of the botanical, additional safety testing may be needed. Testing that requires the use of animals may not be possible if the personal care product will be sold in the European Union because of the 7th Amendment of the EU Cosmetics Directive. The type of safety testing to be done will depend on the gaps in the data set for the botanical and the regulatory requirements of the region in which it will be sold.

In summary, the Tier 3 safety assessment is used when a history of safe human use cannot be demonstrated for the botanical in the personal care product. The Tier 3 assessment requires extensive chemical characterization of the material. Usually the assessment in Tier 3 is a mixture of all the above tiers but addresses specific constituents within the botanical instead of considering the botanical as a single chemical. It requires an in-depth knowledge of the composition of the botanical and an assessment tailored to each of its constituents using TTC, toxicity data from the literature, toxicity data from structurally related compounds, and the other tools typically used for safety assessment of single chemicals.

9.4 SUMMARY

Botanicals and natural substances can safely be used in cosmetic products provided appropriate care is taken to ensure that the assessor has a good understanding of the composition of the material and sufficient toxicological data to assess the expected exposure. Characterization of the material comprises multiple aspects and requires detailed information on the sourcing, processing, harvesting, and many other parameters. Defining acceptable specifications for the substance is critical for substance compositional reproducibility and ensuring safety. In addition, the safety assessment of the botanical or natural substance should be thorough and address all relevant routes of exposure. This assessment can be done using conservative assumptions, such as TTC, when no other data are available and the consumer exposure to the substance is very low. For materials leading to higher exposures, safe human history or toxicological data on the botanical or natural material will be required. Given the complexity of botanical and natural substances, exposures above TTC may require a significant analytical and safety program if no previous relevant history of safe use exists. The high tier assessment often requires individual assessments of each constituent present in the material. The amount of effort, time, and cost for such assessment should not be underestimated.

REFERENCES

1. Fowler JF. Contact Urticaria with systemic symptoms caused by bovine collagen in a hair conditioner. *Am. J. Contact Dermatitis* 1996;7:56–58.
2. Burkhard PR, Burkhard K, Haenggeli CA, Landis T. Plant-induced seizures: reappearance of an old problem. *J. Neurol.* 1999;246:667–670.
3. Haller CA, Benowitz NL. Adverse cardiovascular and central nervous system events associated with dietary supplements containing ephedra alkaloids. *N. Engl. J. Med.* 2000;343:1833–1838.
4. Willett KL, Roth RA, Walker L. Workshop overview: hepatotoxicity assessment for botanical dietary supplements. *Toxicol. Sci.* 2004;79:4–9.
5. Nortier JL, Muniz Martinez MC, Schmeiser HH, Arlt VM, Bieler CA, Petein M, Depierreux MF, De Paw L, Abramowicz D, Vereerstraeten P, Vanherwegheim JL. Urothelial carcinoma associated with the use of a Chinese herb (*Aristolochia fanghi*). *N. Engl. J. Med.* 2000;342:1686–1692.
6. Cosmetic Ingredient Review. Final report on the safety assessment of *Aloe andongensis* extract, *Aloe andongensis* leaf juice, *Aloe arborescens* leaf extract, *Aloe arborescens* leaf juice, *Aloe arborescens* leaf protoplast, *Aloe barbadensis* flower extract, *Aloe barbadensis* leaf, *Aloe barbadensis* leaf extract, *Aloe barbadensis* leaf juice, *Aloe barbadensis* leaf polysaccharides, *Aloe barbadensis* leaf water, *Aloe ferox* leaf extract, *Aloe ferox* leaf juice and *Aloe ferox* leaf juice extract. *Int. J. Toxicol.* 2007;26:1–50.
7. Schilter B, Andersson C, Anton R, Constable A., Kleiner, J, O'Brien J, Renwick AG, Korver O, Smit F, Walker R. Guidance for the safety assessment of botanicals and botanical preparations for use in food and food supplements. *Food Chem. Toxicol.* 2003;41:1625–1649.
8. Pfeiff B, Kalveram CM. Airborne contact urticaria as a reaction to collagen hydrolysate. *Akt. Dermatol.* 1989;15:21–23.
9. Pasche-Koo F, Claeys M, Hauser C. Contact urticaria with systemic symptoms caused by bovine collagen in a hair conditioner. *Am. J. Contact. Dermatitis* 1996;7:56–58.
10. Kousa M, Strand R, Makinen-Kiljunen S, Hannuskela M. Contact urticaris from a hair conditioner. *Contact Dermatitis* 1990;23:279.
11. Freeman S, Lee MS. Contact urticaria to hair conditioner. *Contact Dermatitis* 1996;35:195–196.
12. Niinimaki A, Niinimaki M, Makinen-Kiljunen S, Hannuksela M. Contact urticaria from protein hydrolysates in hair conditioners. *Allergy* 1998;53:1078–1082.
13. Kelling CK, Bartolo RG, Ertel K, Smith LA, Watson DD, Sarlo K. Safety assessments of enzyme-containing personal cleansing products: exposure characterization and development of IgE antibody to enzymes after a 6-month use test. *J. Aller. Clin. Immunol.* 1998;101:179–187.

14. Sarlo K, Adamson GM, Hollis VL, Innis JD, Babcock LS, Kirchner DB. Development of allergic antibody to an enzyme in a body lotion: results of an 18 month clinical study. *J. Immunotoxicol.* 2004;1:71–77.
15. Blaikie L, Richold M, Whittle E, Lawrence RS, Keech S, Basketter DA. Airborne exposure from topically applied protein (proteolytic enzyme). *Hum. Exp. Toxicol.* 1999;18: *528* (abstract).
16. Pocalyko DJ, Chander P, Harding CR, Blaikie L, Watkinson A, Rawlings AV. The efficacy, stability and safety of topically applied protease in treating xerotic skin. In: Leyden JJ, Rawlings AV, editors. *Skin Moisturization*. New York: Marcel Dekker, 2002. pp. 365–384.
17. Varjonen E, Petman L, Makinen-Kiljunen S. Immediate contact allergy from hydrolyzed wheat in a cosmetic cream. *Allergy* 2000;55:294–296.
18. Hann S, Hughes M, Stone N. Allergic contact dermatitis to a hydrolyzed wheat protein in cosmetic cream. *Contact Dermatitis* 2007;56:119–120.
19. Jarmoc LM, Primack WA. Anaphylaxis to cutaneous exposure to protein in a diaper rash ointment. *Clin. Pediatr.* 1987;26:154–155.
20. Lack G, Fox D, Northstone K, Golding J. Factors associated with the development of peanut allergy in childhood. *N. Engl. J. Med.* 2003;348:977–985.
21. Wang LF, et al. Epicutaneous exposure of protein antigen induces a predominant Th2-like response with high IgE production in mice. *J. Immunol.* 1996;156:4077–4082.
22. Herrick CA, MacLeod H, Glusac E, Tigelaar RE, Bottomly K. TH2 responses induced by epicutaneous or inhalational protein exposure are differentially dependent on IL-4. *J. Clin. Invest.* 2000;105:765–775.
23. Hsieh KY, Tsai CC, Wu CH, Lin RH. Epicutaneous exposure to protein antigen and food allergy. *Clin. Exp. Allergy* 2003;33:1067–1075.
24. Strid J, Strobel S. Skin barrier dysfunction and systemic sensitization to allergens through the skin. *Curr. Drug Targets—Inflamm. Allergy* 2005;4:531–541.
25. Kjeldahl J. A new method for the determination of nitrogen in organic matter. *Z. Anal. Chem.* 1883;22:366.
26. Stone K, Williams KR. High-performance liquid chromatographic peptide mapping and amino acid analysis in the sub-nanomole range. *J. Chromatogr.* 1986;359:203–212.
27. Huby RD, Dearman RJ, Kimber I. Why are some proteins allergens? *Toxicol. Sci.* 2000;55:235–246.
28. Akiyama H, Sakata K, Yoshioka Y, Murata Y, Ishihara Y, Teshima R, Sawada J, Maitani T. Profile analysis and IgE reactivity of wheat protein hydrolysates. *Int. Arch. Allergy Immunol.* 2006;140:36–42.
29. Haseldon BM, Kay AB, Larche M. Peptide-mediated immune responses in specific immunotherapy. *Int. Arch. Allergy Immunol.* 2000;122:229–237.
30. Kane PM, Holowka D, Baird B. Cross-linking IgE receptor complexes by rigid bivalent antigens greater than 20a in length triggers cell degranulation. *J. Cell Biol.* 1988;107:969–980.
31. van Berestjein EC, Peeters RA, Kaper J. Molecular mass distribution, immunological properties and nutritive value of whey protein hydrolysates. *J. Food Prot.* 1994;57:619–625.
32. AOAC Official Method. Kjeldahl nitrogen method: combustion analysis (LECO). AOAC official method, 990.03, 2006; Kjeldahl; AOAC official method 984.13 (A–D), 2006; standard report of crude protein utilizes the calculation 6.25 × nitrogen value.
33. FAO. Report of the FAO Technical Consultation on Food Allergies, Rome, 1995.
34. Hefle SL, Nordlee JA, Taylor SL. Allergenic foods. *Crit. Rev. Food Sci. Nutr.* 1996;36:S69–S89.
35. European Food Safety Authority, 2004. http://www.efsa.europa.eu/EFSA/efsa_locale-1178620753812_1178620761196.htm.
36. Gangur V, Kelly C, Navuluri L. Sesame allergy: a growing food allergy of global proportions? *Ann. Allergy Asthma Immunol.* 2005;95:4–11.
37. Kroes R, Renwick AG, Cheeseman M, Kleiner J, Mangelsdorf I, Piersma A, Schiliter B, Schlatter J, van Schothorst F, Vos JG, Wurtzen G. Structure-based thresholds of toxicological concern (TTC): guidance for application to substances present at low levels in the diet. *Food Chem. Toxicol.* 2004;42:65–83.
38. WHO (World Health Organization). Safety Evaluation of Certain Food Additives and Contaminants. WHO Food Additive Series, 48, 2002.

39. Smith RL, Adams, TB, Cohen SM, Doull J, Fero, VJ, Goodman JI, Hall RL, Marnett LJ, Portoghese PS, Wadell WJ, Wagner BM. Safety evaluation of natural flavor complexes. *Toxicol. Lett.* 2004;149:197–207.
40. Smith RL, Cohen SM, Doull J, Feron VJ, Goodman JI, Marnett LJ, Portoghese PS, Wadell WJ, Wagner BM, Hall RI, Higley NA, Lucas-Gavin C, Adams, TB. A procedure for safety evaluation of natural flavor complexes used as ingredients in foods: essential oils. *Food Chem. Toxicol.* 2005;43:345–363.
41. Blackburn K, Stickney JA, Carlson-Lynch HL, McGinnis PM, Chappell L, Felter SP. Application of the threshold of toxicologic concern approach to ingredients in personal and household care products. *Regul. Toxicol. Pharmacol.* 2005;43:249–259.
42. Kroes R, Renwick AG, Feron V, Galli CL, Gibney M, Greim H, Guy RH, Lhuguenot JC, van de Sandt JJM. Application of the threshold of toxicological concern (TTC) to the safety evaluation of cosmetic ingredients. *Food Chem. Toxicol.* 2007;45:2533–2562.
43. Re TA, Mooney D, Antignac E, Dufur E, Bark I, Srinivasan V, Nohynek G. Application of the threshold of toxicological concern approach for the safety evaluation of calendula flower (*Calendula officinalis*) petals and extracts used in cosmetic and personal care products. *Food Chem. Toxicol.* 2009;47:1246–1254.
44. RIVM ConsExpo. http://www.rivm.nl/en/healthanddisease/productsafety/ConsExpo.jsp.
45. Hall B, Tozer S, Safford B, Coroama M, Steiling W, Leneveu-Duchemin MC, McNamara C, Gibney M. European consumer exposure to cosmetic products, a framework for conducting population exposure assessments. *Food Chem. Toxicol.* 2007;45:2097–2108.
46. Felter SP, Ryan CA, Basketter DA, Gilmour NJ, Gerberick GF. Application of the risk assessment paradigm to the induction of allergic contact dermatitis. *Regul. Toxicol. Pharmacol.* 2003;37: 1–10.
47. Undersander DJ, Putnam DH, Kaminski AR, Kelling KA, Doll JD, Oplinger ES, Gunsolus JL.Guar in the Alternatives Field Crops Manual. University of Wisconsin-Extension Cooperative Services and University of Minnesota, Center for Alternative Plant and Animal Products and the Minnesota Extension Services, 1991. http://www.hort.purdue.edu/newcrop/afcm/guar.html.
48. Lagier F, Cartier A, Somer J, Dolovicg J, Malo JL. Occupational asthma caused by guar gum. *J. Aller. Clin. Immunol.* 1990;85:785–790.
49. Lucas JS, Lewis SA, Hourihane JO. Kiwi fruit allergy: a review. *Pediatr. Allergy Immunol.* 2003;14:420–428.
50. Witchl M, editor. *Herbal Drugs and Phytopharmaceuticals*, 4th ed. Stuttgart: Medpharm Scientific Publishers, translated from the German by JA Brinckmann and MP Lindenmaier, CRC Press, 2002.
51. Barnes J, Anderson LA, Phillipson JD. *Herbal Medicines*, 3rd ed. Pharmaceutical Press, 2007.
52. Kroes R, Walker R. Safety issues of botanicals and botanical preparations in functional foods. *Toxicology* 2004;198:213–220.
53. Bast A, Chandler RF, Choy PC, Delmulle LM, Gruenwald J, Halkes SBA, Keller K, Koeman JH, Peters P, Przyrembel H, Ree EM, Remwick AG, Vermeer ITM. Botanical health products, positioning and requirements for effective and safe use. *Environ. Toxicol. Pharmacol.* 2002;12:195–211.
54. Constable A, Jonas D, Cockburn A, Davi A, Edwards G, Hepburn P, Herouet-Guicheney C, Knowles M, Moseley B, Oberdorfer R, Samuels F. History of safe use as applied to the safety assessment of novel foods and foods derived from genetically modified organisms. *Food Chem. Toxicol.* 2007;45:2513–2525.
55. Knudsen I, Soborg, I, Eriksen F, Pilegaard K, Pedersen J. Risk management and risk assessment of novel plant foods: concepts and principles. *Food Chem. Toxicol.* 2008;46:1681–1705.
56. WHO Monographs on Selected Medicinal Plants, Vol. 1–3. http://whqlibdoc.who.int/publications/1999/9241545178.pdf; http://whqlibdoc.who.int/publications/2002/9241545372.pdf; http://whqlibdoc.who.int/publications/2002/9241545372.pdf.
57. FARRP (Food Allergy Research and Resource Program Database). University of Nebraska, Lincoln, NE, 2007. Available at www.allergenonline.com.
58. Cressman RF, Ladics G. Further evaluation of the utility of "sliding window" FASTA in predicting cross-reactivity with allergenic proteins. *Regul. Toxicol. Pharmacol.* 2009;54:S20–S25.
59. Taylor SL, Hefle SL. Will genetically modified foods be allergenic? *Allergy Clin. Immunol.* 2001;107: 765–771.

60. Api AM, Basketter DA, Cadby PA, Cano M-F, Ellis G, Gerberick GF, Griem P, McNamee PM, Ryan CA, Safford R. Dermal sensitization quantitative risk assessment (QRA) for fragrance materials. *Regul. Toxicol. Pharmacol.* 2008;52:3–23.
61. Munro IC. A procedure for the safety evaluation of flavoring substance. WHO Food Additive Series, No. 35, Annex 5, 1996.
62. Renwick AG. Toxicology databases and the concept of thresholds of toxicological concern as used by the JECFA for the safety evaluation of flavoring agents. *Toxicol. Lett.* 2004;49:223–234.
63. Munro IC, Mattia A.The safety evaluation of natural flavoring complexes. WHO Food Additive Series, No. 52, 2004, pp. 435–458.

CHAPTER **10**

PRESERVING COSMETICS WITH NATURAL PRESERVATIVES AND PRESERVING NATURAL COSMETICS

James Flanagan

Coty Inc., New York, NY

10.1 INTRODUCTION

"Natural" has been the main buzzword in the cosmetic industry for much of this young twenty-first century. It has been the darling of cosmetic marketing departments, ensuring almost immediate consumer attention along with the willingness to pay premium prices for natural cosmetics. According to a Natural Marketing Institute (NMI) survey, 59% of women indicate that 100% natural ingredients are very/somewhat important when purchasing personal care products [1]. In addition, countless web sites have sprung up, claiming the safety of natural cosmetic preservatives and ingredients because they have always existed in nature and will most likely not be recognized as invaders to our immune system. Originators of these web sites may be self-proclaimed experts consisting of housewives, former beauty queens, or home remedy hucksters relying mostly on anecdotal evidence [2–4].

Due to these sites and the major media, any ingredient with a long chemical sounding name is suspected of being both a threat to one's health and to the environment. "Natural" has erroneously come to be associated with all that is pure and good.

From the point of view of an experienced microbiologist, I have come to envision the word "natural" with both skepticism and dread. To me, and to most of my colleagues, a naturally derived product or ingredient is a material that, because of its source and nature, is likely to contain a high microbial load, making it a potential hazard in the manufacturing and production environment.

Formulating, Packaging, and Marketing of Natural Cosmetic Products, First Edition.
Edited by Nava Dayan and Lambros Kromidas.

This chapter will discuss the shortcomings of using ingredients that are perceived as natural in the cosmetic industry and the fact that many of these so-called "natural preservatives" are neither natural nor effective preservatives.

10.2 UNIQUE MICROBIOLOGICAL ISSUES TO GOING GREEN

The natural personal care market is one of the fastest growing specialty sectors within personal care. The public's desire for more naturally derived ingredients in cosmetics has prompted marketers to "go green" with many of their product lines. Specialty chemical companies are following the money trail into green cosmetics by expanding their portfolios with organics and natural ingredients. In the past several years, these vendors have bombarded formulators with countless ingredients with claims of natural antimicrobial effectiveness.

The issue of parabens has further fueled the consumer desire for natural preservation. Parabens are the most common, effective, and inexpensive group of cosmetic preservatives. In 2003 and 2004, parabens came under attack in several scientific articles because of concern over their potential estrogenicity or endocrine disruption capabilities and possible link to breast cancer. Although the attacks have not been proven to be accurate (authors of the original articles have refuted their findings) [5], the media and other groups continued to fuel paranoia of these preservatives to the public. Even though a 2006 report from the Cosmetic Ingredient Review (CIR) stated no issue with parabens, many companies experienced public pressure to replace them. Specialty chemical companies, sensing potential profits, began marketing their natural/organic preservative systems with a "paraben-free" label. This increased the demise of paraben use despite further safety evidence and support from reputable organizations such as the EU's (European Union) Scientific Committee on Consumer Products (SCCP) [6].

Marketers, being unaware of the complications of this initiative, have encouraged formulators to develop "green" and paraben-free formulations. Unfortunately, the current list of natural preservatives [7] contain materials that have several drawbacks for a major cosmetic manufacturer. In general, they are very poor preservatives. Natural substances that display antimicrobial activity are inadequate for broad-spectrum protection that is a requirement for large-scale manufacturing. Most are not active against *Pseudomonas* spp. and other Gram-negative bacteria that are the major source of water-borne contamination in the cosmetic production environment. Others, such as natural essential oils, require very high concentrations to be effective. Some have offensive odors (neem seed oil) or colors (usnic acid) that would pose an aesthetic issue in skin care products. Many require special handling and become inactivated by scaled up manufacturing procedures. Formulators are challenged with a group of preservatives that do not mix well in emulsions, are not generally compatible with many cosmetic ingredients, and do present a safety hazard due to irritation (limonene) or allergenic potential (cinnamon oil). Finally, an important drawback of the so-called natural preservatives is that they are generally too expensive to be used for a mass market cosmetic product.

Replacement of parabens presented a triple dilemma to cosmetic microbiologists and formulators. First, parabens are very effective at low concentrations; this cannot be accomplished by natural preservatives (e.g., fragrance oils such as sage oil require 100 times the concentration of methyl paraben to inhibit Gram-positive bacteria). Second, parabens are very stable and can be used in a variety of cosmetic formulations; even providing improved texture in some systems [8]. I have already mentioned the incapability of most natural preservatives to mass market cosmetic systems. Third, unlike naturals, parabens are very inexpensive at use levels. For example, methyl paraben at a cost of $4.66/kg adds about a penny in a formulation at 0.15%. Companies were finally made to face the harsh reality that parabens could not be replaced without an increase in formula/manufacturing costs and a decrease in product quality.

10.3 NATURAL COSMETICS

10.3.1 "Green Certifiers"

In the ongoing effort to become a part of the growing natural and organic personal care market, specialty chemical companies have looked to expand their portfolios with organic and natural ingredients.

To attain legitimacy for these ingredients, companies have sought the certification of various agencies that would place their seal of approval on ingredient labels. This has resulted in the proliferation of seals from various groups, each describing them as the standard for natural and/or organic certification. In reality, these groups can be permissive and/or misleading and often have low requirements for an ingredient to be approved as natural or organic.

The next section will review some of the major certifiers for they consider as acceptable preservatives [9].

10.3.1.1 ECOCERT The current leading European standard and certification, ECOCERT certifies outright "organic" product claims on products with as low as 10% organic content. Stating that a product must consist of, at least, 95% naturally occurring materials, ECOCERT does not tell you that much of this natural/organic content can be just plain water (up to 80%). They also allow various petrochemicals such as cocamidopropyl betaine. Touting their "great principles to Ecocert's Natural and Organic Cosmetics certified by Ecocert" [10], it can be both the most permissive and misleading standard available.

In the case of preservatives (Table 10.1), inquiries revealed that phenoxyethanol and parabens are still allowed in preserving ingredients (bought). This illustrates the dilemma faced by cosmetic companies when dealing with suppliers who choose to preserve ingredients with the most economic preservative available. The formulation of a "natural" cosmetic product is complicated by ingredients containing synthetic preservation. A preservative added to preserve cosmetic ingredients must appear on the ingredient label since they are intentionally added.

10.3.1.2 BDIH Primarily a German/European standard, Bundesverband Deutscher Industrie- und Handelsunternehmen (BDIH) grants only natural

TABLE 10.1 Preservatives Permitted by ECOCERT

Preservative
Parabens (for bought ingredients only)
Phenoxyethanol (for bought ingredients only)
Benzylic alcohol
Sorbic acid and its salts
Benzoic acid and its salts
Salicylic acid and its salts
Dehydroacetic acid (DHA)

certifications. BDIH does not have organic standards. BDIH makes great use of the term "nature-identical" when only allowing certain synthetic preservatives such as organic acids, their salts, and their ethyl esters (benzoic acid, salicylic acid, sorbic acid), and benzyl alcohol.

10.3.1.3 NaTrue Based in Belgium, NaTrue was created by a group of European natural cosmetic companies who decided to create their own standards. In this way, a member achieves instant credibility by joining up. Their 1 to 3 star ratings make NaTrue one of the most permissive standards for "organic" personal care. Like ECOCERT, NaTrue has a low organic percent content requirement and allows sulfation, hydrogenation, and synthetically preserved ingredients utilizing the nature-identical term (see Table 10.2).

10.3.1.4 OASIS [11] The current standard of Organic And Sustainable Industry Standards (OASIS) contains no definition of "natural" and does not contain a definitive list of allowed preservatives.

TABLE 10.2 Nature-Identical Preservatives Approved by NaTrue

Preservative	Examples of occurrence in nature
Benzoic acid, its salts, and its ethyl ester	In benzoin (Styrax benzoin) and in the defense secret of water beetles *Dytiscus* sp.
Benzyl alcohol	Up to 6% in jasmine flower oil, in its free form or as esters in many other essential oils
Formic acid	Occurrence in insects known since 1670, used for protection by beetles and other articulates. Also occurs in stinging nettles and fir needles
Propionic acid and its salts	Formed during propionic acid fermentation. Carbohydrates are converted into propionic acid by *Lactobacillus casei*, *Bacillus subtilis*, or *Propionbacterium pentosaceum*
Salicylic acid and its salts	Free acid, for example, in meadowsweet (*Filipendula ulmaria*), chamomile blossom (*Chamomilla recutita*), and willow bark
Sorbic acid and its salts	In rowanberry seeds (mountain ash, *Sorbus aucuparia*)

Like its European counterpart, NaTrue, the US-based OASIS was formed by trade groups and cosmetic companies (L'Oreal and Este Lauder are members) in order to create their own standards and, thus, legitimize their so-called naturals and organics. The Organic Consumers Association has pointed out that OASIS allows questionable ingredients to be incorporated in their products, making it the most permissive US standard as far as synthetic preservative allowances are concerned. OASIS will certify products that include ethylhexylglycerin and petrochemical preservatives such as phenoxyethanol, yet will assign "organic" claims to products with only 85% organic content. There is no requirement that main cleansing ingredients, such as sulfated surfactants, be made from organic material. Allowing sulfation, hydrogenation, and synthetic preservation of ingredients, OASIS can be highly misleading to consumers looking for wholly organic personal care.

10.3.1.5 USDA All of the above groups, with their different standards and labeling, serve to create consumer confusion surrounding the words "natural" and "organic." Yet, they are sought out by cosmetic companies who seek to capture a piece of the growing natural/organic market.

In reality if a consumer seeks 100% organic certified products, they should look to the US Department of Agriculture (USDA) organic certification that has a 95% minimum organic content for outright "organic" claims with no synthetic preservatives or petrochemicals allowed [9]. However, the USDA's rigorously enforced compliance cannot be met by the major cosmetic companies without them radically reworking their manufacturing/production processes, resulting in major investments. As a consequence, "natural" became the favored term in cosmetic labeling and claims. Since the Food and Drug Administration has not established a definition of "natural," you might find different interpretations of the word between competing cosmetic companies. Given the nebulous meaning of "natural," the various "green" certifiers were only too happy to provide product legitimacy and to give their standards and labels to cosmetic companies, for the proper fee.

10.4 EFFECTIVENESS OF NATURAL PRESERVATIVES

For years, cosmetic chemists and microbiologists have fought hard against the idea that natural ingredients are more effective and inherently safer than their synthetic counterparts. Consumers have been made to feel that if it occurs in nature, it must be good and safe. Companies wishing to achieve a "natural" claim, sometimes look to unapproved natural alternatives that are less effective and may be more toxic than approved synthetic ingredients.

10.4.1 Fact versus Fantasy

Many bloggers, suppliers, and self-styled experts have long proclaimed that there are natural food preservatives available (e.g., rosemary leaf oil) and, therefore, it is possible to use these in cosmetics. The fact is that (1) fresh food preservatives such as rosemary leaf oil are sprayed on fresh produce and are not part of a formulation. (2) When they are an ingredient in a processed food formula, they are at best a weak preservative added after a manufacturing process that includes a high heat or baking phase. There is no better "lovin" (preservation) than baking in the oven!

I have met with many suppliers touting the advantages of their "natural" preservatives and how it can solve one's formulation problems. The fact is that some preservative suppliers may fail to detail the exact formulation of their agents. An example is an herb extract (*Lonicera japonica* or Japanese honeysuckle) that was peddled as completely natural with broad-spectrum antimicrobial activity. Chemical analysis revealed that this agent contained parabens. When confronted with these results, the company representative declared that these are "natural' parabens. To my knowledge, to date, there are no commercially available natural parabens. As a result, this rep and their company were banned from making sales calls to my company.

Another example is the great hoax of grapefruit seed extract (GSE) (INCI: *Citrus grandis* (grapefruit) seed extract). Originating in the 1970s, this agent has resurfaced every few years as the great remedy to marketing demands for natural preservatives. The fact is that GSE a solution of triclosan, benzethonium chloride, or parabens in glycol.

These are but two out of many examples of the "natural preservative" fantasies that some manufacturers and distributors pitch to formulators, wasting their valuable time using unapproved and bogus ingredients. It is therefore recommended to involve a research microbiologist to help you to screen out the hucksters and the misrepresentations.

10.4.2 Methods of Determining Effectiveness

As with synthetic preservatives, proposed natural preservatives should be tested to determine their antimicrobial effectiveness.

One of the supplier-preferred methods to determine effectiveness is the *minimum inhibitory concentration(MIC)/minimum bactericidal concentration (MBC) method* [12] as a classic broth dilution susceptibility test. It is mostly used in vendor presentations of their individual materials. A general procedure will be described here.

To perform the classic broth dilution susceptibility test, a standard inoculum of the test microorganism is added to an equal volume (usually 1 mL) of each concentration of the test preservative and to a tube of the growth medium (broth) without the preservative, which serves as a growth control. An uninoculated tube of the medium is incubated to serve as a negative growth control. After sufficient incubation (usually 24 h at 35°C), the tubes are examined for turbidity, indicating growth of the microorganism. The organism will grow in the control tube and in any other tube that does not contain enough antimicrobial to inhibit growth. The lowest concentration of the test preservative that inhibits growth of the organism, as detected by lack of visual turbidity (matching the negative control), is designated the minimum inhibitory concentration.

After the MIC has been determined, a known quantity (usually 0.1 mL) from each broth tube that showed no visible turbidity after incubation is subcultured to solid agar plates. After incubation, the colony forming units (CFUs) on each plate are counted and compared with the CFU/mL in the original inoculum. Within the tubes that showed no turbidity, microorganisms can still be viable or they are completely killed by the test preservative. The lowest concentration of the preservative that allowed less than 0.1% of the original inoculum to survive is designated as the minimum bactericidal concentration.

The MICs and MBCs of a given preservative can be determined for any bacterium or a fungus that is able to grow in liquid media. It is common practice for a vendor to list MIC/MBCs of an agent against several common microbial contaminants.

While the submitted MIC/MBC of a preservative provides initial information on its activity, the cosmetic chemist is more concerned with a preservative's activity within a formulation. That is because ingredients or compounding conditions may inactivate a preservative. It is here that the *preservative efficacy test(PET)* or preservative challenge test is employed to determine if a contaminant would survive against a particular preservative system. In this way, the formulator can determine the shelf life of a product by providing a worst-case scenario to mimic a contamination that might occur in the field.

Most companies develop their own test methods with one or more of the following organizations used as guidelines [13]:

- Personal Care Products Council (PCPC)
- American Society for Testing and Materials (ASTM)
- US Pharmacopeia (USP)

Basically, individual representatives of bacteria and fungi are grown and harvested into high concentrated inocula (10^4–10^8 CFU/mL or g). A known amount of test formulation is inoculated with the challenge inoculum. At established intervals, an aliquot of the test sample is plate counted to determine the number of surviving organisms.

Criteria for determining preservative adequacy vary. Surviving organisms are evaluated either as $\log_{10}$ unit reductions or as percent survivors (i.e., no more than 0.1% of inoculum surviving). While some companies advocate complete eradication of the inoculum, others have adopted a guideline of a 3 log decrease at the end of the study.

10.5 "FREE FROM" DILEMMA

Since a true broad-spectrum "natural" antimicrobial has not yet been identified, cosmetic marketers have elected to employ the "free from" or "preservative free" claim. Given the bad press that parabens and other synthetic preservatives have received, both marketers and vendors have found that using a "free from parabens" or "preservative free" label enhances a products attractiveness to customers. However, this could be seen as misleading the consumer about the actual formulation of the product.

10.5.1 Drawbacks of "Preservative Free" Claim

While some eye area products can have a shelf life of 1 or even 2 years, most cosmetics have more than a 3 year shelf life. A product that is truly free of preservatives would have a much shorter shelf life, could be easily contaminated (during production and by the consumer), would require special packaging or would need to be refrigerated. Such marketing drawbacks would greatly increase the cost of the product.

Instead of using preservatives, some of the "preservative free" products have a high wax, oil, or glycerin content that makes these formulations antagonistic to microbial growth. Unfortunately, such products are more greasy, sticky (tacky), and, therefore, less appealing to consumers.

In recent years, companies have gravitated toward the use of other ingredients to replace the standard preservatives. The majority of these materials belong to the class of diols, primarily 1,2-octanediol also known as caprylyl glycol. These ingredients are classified as humectants and wetting agents in cosmetic and skin care formulations in addition to providing viscosity modification in some formulations. Since diols have been found to possess antimicrobial activity but are not classified as preservatives, some companies have felt confident in using the "preservative free" claim. This practice came under attack during 2008 and 2009.

10.5.2 Recent Statements from PCPC, CTPA, and DGCCRF

During the March 2009 meeting of the Personal Care Products Council (http://www.personalcarecouncil.org/) Microbiological Committee, there was some discussion on the hazards of using the claim "preservative free" and "free from preservatives." Committee members stated that the European Union may be looking more closely at formulations using these types of claims, citing that, to avoid misleading the consumers, ingredients such as diols may soon need regulatory approval and be classified as preservatives.

The United Kingdom cosmetics trade association (CTPA, http://ctpa.org.uk) has stated that "free from" claims can trade on false safety concerns and has advised manufacturers to avoid using this claim in marketing their ingredients. CTPA Head, Chris Flowers, stated that the labeling of a product as "free from" certain chemicals such as parabens can imply that there are safety concerns over the ingredient in question and that replacement ingredients are safer and better. He advised against using safety as a route to commercial benefit [14].

Also in 2009, the DGCCRF, the French body for competition and fraud control (http://ww.dgccrf.bercy.gouv.fr/anglais.htm) conducted an investigation in conjunction with France's authority on the safety of health products (AFSSAPS, http//www.afssaps.fr/). This study compared the actual chemical makeup of natural/organic products with their ingredient labels. Of the 28 products tested, 12 claimed to be preservative free and the other 16 as free from parabens and/or phenoxyethanol (preservatives used in the latter group were dehydroacetic acid and salts of benzoic acid or sorbic acids). Chemical analysis showed many formulations with traces of preserving compounds that were not marked on the label. As stated above, this was mainly due to chemicals used to preserve ingredients—ingredients that should be listed on the label since they are not considered incidental per European Union's Cosmetic Directive. Regardless, as the DGCCRF stated, emphasizing the absence of certain preservatives when the product contains others, is a way of intentionally giving incomplete information and misleading the consumer [15]. This is especially true for those consumers who expect their natural/organic products to be free from preservatives or preserved via nature's means.

10.6 SUMMATION

The issues concerning the marketing, labeling, and certification of natural, organic, and "free from" cosmetic systems are constantly evolving. As recently as January 2010, the French advertising regulatory body ARPP (Autorite' de Regulation Professionelle de la Publicite) has recommended tighter labeling standards for natural products (a minimum of 95% natural ingredients is required), more positive use of the term "free from," and closer monitoring of the use of the word organic [16]. There have been more reports on misleading claims concerning natural and organic cosmetics [17]. In addition, new data on the skin absorption and the distribution of parabens in the body suggest that these preservatives are not as harmful as originally believed [18]. Time will tell if the furor over parabens will subside to a point when "free from parabens" will fall into disuse.

The major problem has always been that the public is not properly educated about what natural or organic actually means since there is no federally mandated reference point due to an absence of regulations. More federal guidance will be needed to provide the cosmetic industry with baseline definitions of naturals and organics so that hucksters and false claims can be properly discovered.

REFERENCES

1. HAPPI, June 2007, p. 75.
2. http:/www.organicdivas.com
3. http:/gmy.news.yahoo.com/v/12390836/belo_austin_kvue/20090308/av_localcat_dontlaun/_goodmorningyahoo_belo090308_339617.
4. http:/www.fromnaturewithlove.com.
5. CTFA. CTFA Response Statement Concerning Parabens, November 2003.
6. European, Commission. The Scientific Committee on Consumer Products (SCCP) Extended Opinion on Parabens, Underarm Cosmetics, and Breast Cancer, SCCP/0874/05, 2005.
7. Steinberg D. *Preservatives for Cosmetics*, 2nd edition. Allured Publishing Corp., 2006. pp. 94–99.
8. Bayliss J. Mineral cosmetics and their ingredients: part two. Available at http://www.articlesbase.com.
9. David B. A 5 Star Comparison and Ranking of US and European 'Organic', 'made with Organic' and 'Natural' Personal Care Standards, November 10, 2008.
10. Ecocert.com.
11. Organic and Sustainable Industry Standards, Inc., Health and Beauty Products, Beta Version #4, March 15, 2009.
12. Baron E, Finegold S. In: *Bailey & Scotts Diagnostic Microbiology*. St. Louis, MO: The CV Mosby Co. Chapter 13, 1990, pp. 175–178.
13. Geis PA (ed.), *Cosmetic Microbiology: A Practical Approach*, 2nd edition. Informa Healthcare, 2006. pp. 111–130.

14. Katie B. "Free from" claims can trade on false safety concerns says trade body. *Cosmetics Design Europe,* March 13, 2009.
15. http://www.mulondon.com/blog/preservatives_cosmetics.
16. Katie B. France changes rules for natural, organic and 'free from' advertising claims. *Cosmetics Design Europe,* January 6, 2010.
17. Nichol K. German Consumer Association highlights problem of greenwashing in cosmetics. *Cosmetics Design Europe,* December 15, 2009.
18. Katie B. New data on parabens suggests no adverse hormonal effect on the body. *Cosmetics Design Europe,* November 18, 2009.

CHAPTER 11

MICROBIAL RISKS AND ECO-FRIENDLY PACKAGING

John I. Yablonski and Sharon E. Mancuso

Bio-Control Consultants, Inc., Westfield, NJ

11.1 INTRODUCTION

The evolution of cosmetic and personal care product packaging from simplistic utilitarian containers to commercial works of art that entice the consumer to purchase the contents has now given way to an entirely new and even more dynamic concept referred to as "eco-friendly" or "green" packaging. Not only does a package have to be visually and sensually appealing but it must also encompass the very essence of naturalness, environmental compatibility, and healthfulness. Over the past number of years, consumers have become significantly more environmentally conscious. In so doing, they have also become more aware of the potential negative impact of ill-conceived and/or superfluous packaging and packaging materials on the environment. The excessive use of large, overblown package formats composed of complex mixtures of synthetically derived materials and nondegradable plastics often derived from unsustainable resources appears to be a major factor in driving this phenomenon. Whether valid or not, there is a growing common perception that packaging and packaging concepts of the past are no longer in tune with the sound and fiscally responsible environmental attitudes of the future. Needless to say, savvy and foresighted manufacturers and suppliers are beginning to take advantage of this new trend and focus considerable effort and resources on the development of more eco-friendly packaging and green packaging materials. As a result of this evolution, a new lexicon has emerged to describe the various attributes of this relatively new technology. Terms such as biodegradable, sustainable, recyclable, compostable, natural, organic, reusable, refillable, bulk, concentrates, bioplastics, dematerialization, energy efficiency, green washing, and carbon footprint have been introduced and are becoming a part of the common parlance when referring to eco-friendly and green issues [1–6].

On the surface, most of these concepts appear to be more than appropriate to satisfy the desire to become more ecologically conscious. Without proper evaluation and determination of long-term consequences, however, there may be some serious

Formulating, Packaging, and Marketing of Natural Cosmetic Products, First Edition.
Edited by Nava Dayan and Lambros Kromidas.

TABLE 11.1 Green Packaging Concerns

Preservative package compatibility and stability
Preservative package interaction and inactivation
Package permeability and product weight loss
Package shelf life and biodegradation
Decreased package integrity
Refill contamination and biofilm development
Higher risk potential for microsensitive products
Lack of adequate disposal and recycling infrastructure
Negative impact on other resource sectors, for example, agriculture

negative ramifications resulting from attempting to apply these "green" principles. Some of the potentially negative implications that may be associated with blindly pursuing the development, manufacture, and implementation of fundamental "green" packaging ideas involve pitfalls and areas of concern such as increased product microbial contamination risk, product/package compatibility and stability issues, product preservative inactivation, decreased package integrity, reduced product shelf life, increased product weight loss, lack of an adequate disposal/recycling infrastructure [7–10], increased energy consumption, and possible negative impact on other resource application sectors, for example, agriculture and food supply (Table 11.1). Any design or material modification that contributes to increasing the probability of microbial contamination can, in turn, negatively affect consumer risk and product safety. Although the challenges related to contamination potential are the most critical, they are often the least likely to be fully appreciated and the last to be considered [11]. The bottom line is that these and other related concerns and considerations must be satisfactorily evaluated and the associated issues resolved early on in the package development process in order to be successful in implementing comprehensive eco-friendly reforms in the basic fundamental concepts of package engineering and design [12].

11.2 MAJOR MICROBIAL RISK FACTORS

Cosmetic packaging comes in a multitude of forms, sizes, and configurations with a wide variety of components and parts designed to enclose and protect the product, facilitate convenience, improve application, and, most importantly, enhance consumer appeal. The right package accompanied by an innovative and ecologically appealing message or presentation will give a brand a considerable competitive marketing advantage. On the technical side, however, these may or may not be compatible with or contribute to the goal of minimizing the potential for microbial contamination. Probably the most significant risk factor associated with cosmetic packaging involves product exposure and excessive consumer access. Whether designing an eco-friendly primary container or a conventional package, the basic preventative concept of minimizing product exposure as much as possible is paramount to a successful product presentation. The more a container allows the product to be exposed, the greater the likelihood that product may become contaminated and the greater the potential risk to the consumer. For instance, a large mouth jar or container for creams, by design, allows excessive product exposure and consumer contact during use. As the consumer uses and applies the cream, the product can receive multiple insults from the normal and abnormal microbial flora of the hands, body, and air. Products presented in this fashion generally require a relatively robust preservative system to be able to compensate for frequent consumer exposure and resist this form of normal usage of microbial contamination. When evaluating preservative robustness in this and similar high exposure packaging formats, it is highly recommended that *in vivo* testing of the final package form be conducted in addition to the conventional recommended protocols for product preservative efficacy testing (PET). In terms of the conventional methodologies, it is strongly suggested that the preservative challenge protocols be modified so as to include at least two or more microbial inoculations spaced relatively close together. The spacing can be critical if one is seriously attempting to demonstrate or predict robustness based on simulated multiple insults. Yablonski and Mancuso presented and published an accelerated double challenge protocol that appears to be capable of predicting and measuring both rate of kill and the product's preservative capacity to withstand multiple microbial insults [13]. The Personal Care Products Council (PCPC, formally CTFA) Microbiology Guidelines on preservative challenge include the option for additional inoculations where appropriate [14, 67]. Many cosmetic and personal care companies have also developed their own in-house multiple challenge protocols in an attempt to measure the impact of multiple product insult on various product forms [15]. The bottom line is that the determination of preservative system robustness is a critical factor in evaluating package safety and integrity and should be an integral part of the product package development process [16, 17].

Consumer product abuse, on the other hand, is a microbiological risk factor that is generally not as easily controlled as is product accessibility reduction. The approach and applicable design principles for minimizing abuse, however, may be very similar. Consumer abuse can come in a variety of forms and often bypasses the safeguards built into a well-designed primary container. That being said, the effort should still be made to construct a package in such a manner as to deter or, at least make more difficult, the attempt to compromise the product system. Some of the more common abuse practices include such activities as diluting product with tap water, mixing different products, adding natural food ingredients, leaving the primary container open to the environment, refilling containers, and using product for other than intended purposes. Some of these cannot be easily avoided with many existing easy access packaging configurations, for example, adding foreign material and inappropriate product use. In addition, a relentless consumer can and often does go to almost any length to access the container and achieve the goal of modifying or compromising the product. The more attractive the original container, the more likely consumers will find some means of reusing it for their own purposes. Many green marketing advocates support this practice that, on the surface, appears to be appropriate. The potential microbial contamination risks that may be involved, however, may far outweigh the benefits and should be thoroughly evaluated before supporting this meaningful effort. Some of the more successful approaches to deterring consumer abuse have included the use of restrictive closures, sealed/molded containers, single-use packages, and airless pump systems. Alerting the consumer to the risks involved in these questionable practices has also met with some success. In the age of blogs, twitters, and other forms of social communication, it is somewhat easier to get the word out that certain common practices are risky and may not be in the best interest of the consumer. Conversely, any perceived defect or health risk, whether valid or not, associated with cosmetics and cosmetic packaging can also be completely blown out of perspective and spread like wildfire via the media and social communication networks throughout a consumer population eager for news of potentially questionable behavior on the part of big business. Accordingly, it behooves responsible manufacturers to encourage the development and design of products and packaging concepts that are not only ecologically friendly but also performance effective in protecting the health and safety of the consumers [18–22].

11.3 PACKAGE CHARACTERISTICS THAT INFLUENCE RISK ASSESSMENT

The basic forms and characteristics of the primary cosmetic package significantly influence the potential for microbial risk. These characteristics include not only the physical configuration of the package but also the nature and composition of the materials used to create it. Composition obviously becomes much more important as more and more packaging components are being fabricated from eco-friendly and sustainable raw materials. The trend to use more of these naturally derived "green" materials in any number of unique package configurations and presentations can lead to a significantly greater likelihood of product package interaction and potential product preservative system compromise. The more common primary packaging configurations can be found in Table 11.2. Certain of these container types, for example, jars and bottles, are more likely to develop microbial contamination issues than are the less accessible configurations of aerosol cans, airless pumps, flex packages, and other sealed systems. Tub and canister wipe containers, if not properly managed, can present a serious microbial risk by both the nature and openness of the package and the mode of consumer access [23, 24]. The use of creative resealable closures and similar restrictive wipe delivery mechanisms, however, can significantly

TABLE 11.2 Primary Packaging Configurations

Jars and bottles
Tubes and delivery systems
Flexible soft packages
Tubs and canisters
Aerosol cans and airless pump systems
Pans and trays
Bulk containers—buckets, pails, jugs, and bottles

reduce the microbial risk potential. Pressed powders and other similar makeup products presented in a pan configuration are characteristically exposed and easily accessible to the consumer. This could present an even more significant microbial risk were it not for the low moisture content, for example, water activity, of many of these products that renders them somewhat hostile to microbial growth and proliferation. Again, the addition of water or other materials to the product, which for some reason is frequently endorsed in many "how to beauty columns," can seriously compromise product integrity and affect contamination potential. With the exception of uncontrollable product abuse, clean manufacturing practices coupled with clean raw materials, exposure prevention package design, and, where appropriate, effective preservation systems can assist in mitigating some of the risk issues associated with many packaged cosmetic products.

Both the size of the package and the dimensions of the delivery orifice can also have an effect on exposure and microbial risk. Small packaging is often favored by consumers as being convenient, portable, and easy to use. From a microbiological safety point of view, smaller containers are used for a shorter period that, in turn, can reduce the amount of repeat insult to the product and thus lower the risk to the consumer. The problem from an eco-friendly perspective is that small packaging generally uses considerably more packaging material and energy to produce and fill than does larger package configurations. In order to be more "green," some cosmetic manufacturers are beginning to turn to larger package forms, products in bulk form, and the questionable practice of promoting refills. Others, including many in the household products industry, are moving in the opposite direction of using product concentrates in small containers. The use of concentrates allows one to reduce the package size, while gaining the advantage of lowering energy costs associated with manufacture, storage, and transportation. Since the number of uses from concentrated products would be relatively the same as from larger sized nonconcentrates, there should not, theoretically, be a significant increase in the number of packages sold or the amount of packaging material required. In some cases, empty containers are sold or provided with instructions to dilute the concentrate with tap water and refill the container with new concentrate and water as needed. While the reduction in package size, container reusability, and the effort to minimize energy costs is to be commended, in the case of cosmetics, the potential increase in the likelihood of microbial contamination and corresponding negative impact on the consumer can be a cause of serious concern. On the basis of experience, any scenario in which the consumer is allowed to intimately interact with the integrity of a product or package system dramatically increases the likelihood of the product becoming contaminated over time [3–6].

One of the most critical primary packaging characteristics influencing microbial risk has to do with the mechanism of application or delivery of the product. As with other packaging characteristics, the overriding concern is the degree of product exposure and contamination transfer to and from the consumer and/or the environment. Skin, hair, eyelashes, and the body in general are primary sources of microorganisms and cellular debris that can contribute significantly to the biological load of the product during use. Repeated biological insult of the product in the primary container via applicators or nonrestrictive closures can eventually lead to a degradation of product preservative capacity. A weakened preservative system would not be able to withstand additional incremental microbial insult and the product would eventually become irreversibly contaminated. In cases where the packaging components may be constructed of materials derived from organic fiber, plant-based polymers, or similar natural materials, the opportunity for enhanced package/product interaction and possible preservative inactivation may be even greater and could contribute to a much more rapid breakdown of the preservative system [25–30].

Although significant progress has been made, many questions still remain to be answered regarding the compatibility and stability of many eco-friendly packaging materials when used in conjunction with complex natural ingredient formulations. To avoid some of the pitfalls associated with these natural packaging concepts, it is mandatory that all new or compositionally modified packages, components, and materials be subjected to rigorous product stability and compatibility studies conducted over an extended range of temperatures and environmental conditions. Typical accelerated stability studies often include at least 12 weeks of incubation time at temperatures of 4, 25, 40 (or 45), and 50°C. Higher temperatures such as 55 or 60°C have been used, however, for much shorter periods, for example, 2–4 weeks. Cadwallader [31] published an excellent paper in 1989 regarding the role of stability testing in the development and formulation of cosmetic products. Table 11.3, from this paper, presents a suggested "Accelerated Stability Sampling Program" that proposes evaluation test points as related to time and temperature of incubation. Other incubation times and temperatures can be chosen based on the type of product and package involved and the anticipated vulnerabilities of the product. The length of time of stability studies is generally determined by the expected product shelf life, the package configuration and composition, product use characteristics, and the formulation makeup of the particular product. Ambient temperature real-time studies for more extended periods should always be included for verification purposes. Normally, ambient temperature studies are conducted between 24 and 36 months although some products may have an even longer projected use life. Table 11.4, also derived from Cadwallader, suggests a long-term storage and testing guideline for 4 and 25°C stored samples with test points up to 60 months. Studies such as these, coupled with in-use studies, can often be useful in developing appropriate period after opening (PAO) estimations or expiration dating in markets where they may be required by regulation.

TABLE 11.3 Typical Accelerated Stability Sampling Program

	Sampling time (weeks)						
Temperature (°C)	0	1	2	3	4	8	12
4		×			×	×	×
25	×	×			×	×	×
40		×	×	×	×	×	×
50		×	×		×		
60		×	×				
25°C, 50% relative humidity				40°C, 75% relative humidity			

TABLE 11.4 Long-Term Stability Storage and Testing

Temperature (°C)	Sampling time (months)									
4	0	3	6	9	12	18	24	36	48	60
25	0	3	6	9	12	18	24	36	48	60
37/40	0	3	6	9	12					
37/40°C, 75–80% relative humidity										

The specific product and package chemical, physical, and compositional characteristics to be evaluated at each of the test points can vary considerably depending on the general nature of the product system being evaluated. Typical suggested stability observations and evaluations commonly employed in the industry are presented in Table 11.5 [32–34].

11.4 REFILLABLE PACKAGING CONCEPTS

Refillable packaging is cited by some experts as "one of the most promising ideas in sustainable beauty packaging" [35]. Others have referenced the now outdated glass milk bottle as being one of the most established and successful examples of refillable packaging where the consumers had their milk delivered and left out their empty bottles to be refilled. Using this basic approach, a number of beauty and personal care products companies have explored the opportunity of applying the refillability concept to their product lines. In one case, a glass perfume container was designed specifically to be refilled in order to encourage consumers to reduce their package use. In another marketing experiment during the 1990s by a different company, it was demonstrated that returning cleaned bottles to a store to have them refilled was an

TABLE 11.5 Typical Stability Observations

Product attributes
pH and viscosity
Color, odor, and appearance
Separation, bleeding, and ingredient precipitation
Weight loss
Preservative/active analytical verification
Preservative efficacy verification
Other critical product-specific chemical/physical properties
Package attributes
Size, color, finish, labeling
Cap, closure, and primary container integrity
Seal and disk distortions
Delivery system function
Leakage

inconvenience that the average consumer did not want to endure. Some refillable concepts incorporated the use of concentrates for shampoos and shower gels. The consumer could purchase the concentrate and a reusable bottle and then add water at home to make the final product. As previously mentioned, this same approach has been successfully introduced into the household products industry in the form of various cleaning products. The major downside to most of these ideas related to cosmetics has to do with convenience, aesthetic perception, and the potential risk of microbial contamination. Consumers may be favorable to diluting concentrates for cleaning products since they have been already conditioned to dilute many of them prior to use, but in the realm of personal care items, consumers have to readjust their thinking to accept this less than convenient format. For acceptance, both the concentrate package and the refill bottle must be attractive to the consumer and the process of mixing the concentrate with water relatively easy. Although this may be utilitarian, the concept of elegance, beauty, and pampering associated with the allure of cosmetics can often be lost in translation. On the technical side, there is absolutely no way to control the quality of the water used to dilute the product. Water in different regions can vary significantly in terms of pH, mineral content, color, odor, chlorine concentration, suspended solids, hardness, and possible microbial contamination. Any or all of these variables can seriously affect the functionality of the product. Overdiluting the concentrate is another common problem that can negatively affect preservative system capability by reducing the preservative concentration to sublethal levels that will not function effectively to prevent microbial proliferation. In a global distribution market, adapting a cosmetic concentrate to all possible water types and consumer use habits and conditions can be a major challenge [3, 11, 35–37].

11.5 REUSABLE/REFILLABLE PACKAGING RISKS

One of the more compelling examples of the reusable refill concept going astray occurred during the late 1960s and 1970s. During this period, many major cosmetic companies marketed mascara tube refills without the brush applicator. Consumers were encouraged to continue to use their old "favorite" brush by merely transferring it to the new mascara refill container. Owing to their physical configuration and the manner in which they are used, mascara brushes can be extremely vulnerable to becoming microbial contaminated. During normal use, dirt, microorganisms, and cellular debris can become attached to the stiff bristles and transferred from the lashes of the user to the mascara tube. With continued use, the enclosed product receives frequent and multiple inoculations of contamination and debris from the eyelashes. Over time, this repeated microbial insult has been shown to be able to exhaust the product preservative system and overwhelm the product. As the preservative becomes exhausted, the rapid proliferation of contaminating microorganisms becomes inevitable. Once established, the resulting contamination can then be transferred back to the consumer via the brush, creating a vicious cycle that sets the stage for a potential major health and safety risk. The mascara brush, being stiff bristled, has the potential to inadvertently damage the cornea if scraped across or poked into the eye. Women applying mascara while in a moving vehicle or performing other functions have a significantly higher probability of serious mascara brush eye injury. Couple the injury with contaminated mascara, there is a significant possibility of eye infection and, as was documented by Wilson and Ahearn in 1977 [38], corneal ulcers, and loss of vision. At the time, manufacturers were totally unaware of the potential safety risk impact of this seemingly promising, to before its time, green concept that would be both profitable and reduce the amount of component material delivered to the consumer. The reason for this lack of awareness can be mostly attributed to the lack of sophistication and guidance regarding product safety and microbiological testing during this period. Cosmetic microbiology in the 1960s and 1970s was in its infancy and many of the protocols and methodologies that currently exist were not available. The concept of "consumer in-use testing," which would have most likely identified this potential risk, was not yet developed and the relationship between consumer use and product contamination poorly understood. During an early study published in 1971 [39], Wilson et al. initiated a series of studies to evaluate a variety of used and unused mascara, eyeliner, and eye shadow products for microbial contamination. They found that approximately 43% of the used eye products contained significant microbial contamination, whereas all of the unused eye products evaluated were found to be essentially free of microorganisms. These findings were, at the time, controversial at best. Companies honestly believed the eye products produced in their facilities to be safe and free of microbial contamination as confirmed by microbiological test data developed in their own or at outside testing labs. In addition to the production line quality control data, the contention that their products were free of microbial contamination was further supported by the availability of acceptable in-house preservative effectiveness testing data derived using evaluation protocols based on the USP, the CTFA technical guidelines, and similar standardized methodologies. Later in 1977 [38], however, Wilson and Ahearn published the results of additional investigations on used mascaras that reinforced their original findings and further emphasized the microbial contamination risks associated with used mascaras. These studies clearly documented the association of seven cases of severe *Pseudomonas* eye infection resulting from contaminated mascara [40, 41]. The resulting uproar jolted both the industry and the FDA prompting the formation of a joint FDA, industry, academia task force that was charged with the responsibility of developing appropriate guidelines for the safe production and use of mascara products. In view of these devastating findings and in support of this effort, a number of cosmetic companies initiated "in-use" panel testing programs to determine the potential microbial risks associated with mascara usage in normal populations. The objectives were to verify the potential risks involved and to define the time usage limitations for eye products of this nature. The data developed contributed significantly to a better understanding of the dynamics of microbial contamination as related to product, package, applicator, and consumer use. Final recommendations of the task force included, among others, banning the use of mascara refills, suggested maximum product usage time after opening, discouraging the reuse of mascara brushes, and providing instructions to the consumer on the proper use and care of mascara brushes and eye area products in general.

In recent years, a number of companies are once again attempting to tap into the potential of eye and lip product refillability under the now environment-friendly banner of sustainability. On the basis of a report by the marketing research company Pira International, entitled "The Future of Personal Care Packaging," refill packaging formats are emerging as a leader in the efforts at sustainability. This is largely due to such influencing factors as being lighter in weight, allow the use of more alternative material choices, use less inner and outer packaging materials, take up less space, and are generally cost effective while using less energy during the manufacturing process. One company is currently marketing a click pen for lip gloss or concealer application that is coupled with refill cartridges that allows the primary case to be reused multiple times by simply replacing the cartridge when the product runs out. Another company is also using the replaceable cartridge concept for both lip and eyeliners so as to cut down on packaging waste. Similarly, other product refill approaches embrace the use of two-part packaging systems in the form of a dispenser and a refill pack. This latter concept is projected by some to result in potential package material savings of between 50% and 75% over a 3–4 year period. While this may all be true, a major safety concern regarding these applications has to do with product contamination potential and the seemingly increased probable risk of cross-contamination from applicator to refill as well as from applicator and refill to the consumer. Some have even suggested that many of these refill sustainability approaches could be leading us down a path similar to that taken by the unaware mascara manufacturers of the 1960s and 1970s. This is especially true for eye products and other sensitive area product forms that under the best of conditions may present higher potential risk. Obviously, some product forms by their basic nature are less sensitive to microbial contamination than are others. All, however, do have the potential to present additional risk to the consumer if they have not been properly evaluated and the potential risks involved thoroughly assessed and minimized [18, 19, 42].

11.6 BULK PRODUCT CONSIDERATIONS

The marketing and sale of consumer products in bulk configurations has been around long before the concept was ever associated with eco-friendly packaging, package size considerations, or green movement. The idea was to economically produce and deliver a product to the consumer in a larger volume container from which the consumer could fill or refill smaller containers and dispensing units. Although the primary bulk package can be considerably larger and can utilize substantially more packaging material than does the comparable smaller format, it is believed that the total numbers of individual packaging units would essentially be reduced that, in turn, would significantly reduce the overall package material mass. An additional frequently claimed eco value benefit spin-off would be that producing fewer units using less material could contribute to a reduction in greenhouse gas emissions, water consumption, waste, and energy used in the package production process. At this time, however, many of these projected environmental savings are purely theoretical having been largely developed using models designed specifically to demonstrate the potential benefits. As these progressive concepts take hold and real-time numbers begin to accumulate, however, a better and more accurate picture as to the true eco impact of this and similar eco-friendly approaches will evolve as will a better understanding of the balance between being eco appropriate and the reality of introducing additional microbial risk to both the products and the consumers.

Generally, the most common bulk product formats are buckets, pails, jugs, and bottles of various larger shapes and sizes. Other more creative decorative, natural composition bulk container offerings are currently in the development and design stages and have been or soon will be presented to the market in the near future. Hand and body soaps, lotions, gels, and liquids are common personal care items currently sold in bulk form. It should be noted that household and commercial cleaning and maintenance products have been provided in bulk for many years with various degrees of success. From a microbiological vulnerability point of view, many of the household and commercial cleaning products are high in pH and are often composed of ingredients that are relatively hostile to microbial contaminants. This will soon be changing, however, as more and more household product manufacturers begin taking up the eco-friendly marketing banner. In doing so, they are beginning to move toward the development of greener, more natural product forms and/or the use of greener, naturally sourced raw ingredients and packaging materials. With these changes, products previously protected from incidental microbial contamination by high pH and/or hostile raw materials are now becoming more vulnerable to microbial contamination and are requiring the addition of preservative materials that were previously unnecessary. To complicate the matter, for those products that previously did require preservation, many of the established and proven safe and effective preservatives are no longer considered to be green enough or natural enough to be allowed in the new green household product [43, 44]. In many cases, to meet the new green product criteria and associated marketing challenges, the manufacturer is forced to formulate with less effective "natural" preservatives at significantly greater cost and higher concentrations than the levels used with the "unacceptable" preservatives they are replacing. This whole issue of increased microbial vulnerability has become one of the potentially major repercussions of the change in direction to green marketing. The bottom line is that there is a greater probability of a significant increase in microbial vulnerability with the newer green formulation concepts. In a business where it is customary for the product pH and chemical composition of a formulation to protect the product from microbial contamination, many household products companies may not fully appreciate the potential negative ramifications that these changes can create. As a result, many may not have the internal infrastructure or be proactive enough to be able to conduct the types of susceptibility testing that would be necessary to determine the degree of product vulnerability. In addition, increased microbial vulnerability can have a significant impact on the manufacture and production of these green reformulated, and now microbiologically sensitive, household products. Converting to more susceptible natural formulations with less effective preservative systems would result in the need for the development and implementation of expanded preventative operational and quality control requirements. These additional new requirements would minimally include considerations such as product susceptibility testing, preservative stability evaluation, PET, routine production QC microbiological testing, development of effective cleaning and sanitization procedures, possible aseptic filling, initiation of equipment and environmental testing, and the development and implementation of applicable SOP protocols. The issue becomes even more complicated when the potential effects of using alternative packaging materials such as fiber-based composites, organic fibers, wood flour, bamboo or bioplastics are factored in [45, 46]. Any or all of these very promising material concepts can have the potential of becoming major liabilities if they are not evaluated and managed appropriately. Issues relating to long-term stability, product package compatibility, and preservative system interaction must all be considered and properly evaluated prior to going to market. On the cosmetic side, many of these same concerns regarding increased formulation vulnerabilities and packaging material impacts are also faced by the cosmetic developer and manufacturer. Since most of the responsible cosmetic manufacturers have already put the appropriate systems and protocols in place for properly evaluating these new and eco-friendly formulations and progressive packaging concepts, they are generally in a better position. It does not, however, make the job any easier or the products involved less susceptible. With the minimal number of acceptable natural and organic preservative materials available for this application as well as the various regulatory requirements and restrictions constantly being introduced across the global market, it is becoming extremely difficult to balance the need for contamination-free cosmetics and personal care products with some of the newer concepts currently being cultivated as being eco-friendly.

11.7 BULK PRODUCT AND REFILL DISPENSERS

In the current green and eco-friendly business environment and in spite of the legitimate and seemingly obvious microbial susceptibility concerns, many cosmetic and personal care manufacturers are now seriously considering the potential economic and marketing value of entering into the world of bulk product configurations. Associated packaging concepts would include refillable bulk dispensers, refill replacement packaging units, reusable delivery systems, and multiproduct consumer packaging. The range of products that are now being considered for these applications is being expanded far beyond those commonly presented in the past. Once the domain of soaps, shampoos, and bath products, the industry is now entertaining the application of bulk technology to such product concepts as refillable decorative lotion and fragrance dispensers, bulk makeup and foundation presentations, refillable replacement eye products, and multiuse refillable lip and skin products. With any bulk or refillable product form, the major risk issues are generally the same and are most often related to inadvertent and repeated exposure of the product to microbial insult during use. Potential contamination risks most commonly occur through direct person to product contact when dispensing, during the bulk transfer or refilling process, and from inadequate precleaning of the dispensing unit prior to refilling. Any of these risks can be further compounded by improper handling and storage of the remaining bulk product and primary container [47–53]. In terms of dispensing units, any dispenser that allows direct consumer contact with the delivery orifice, nozzle, or applicator significantly increases the potential for product contamination and significant risk to the consumer. Wall mounted and undercounter mounted bulk dispensers appear to have a somewhat greater potential for developing a growing and active microbial population. This is due to a variety of reasons including their general inaccessibility for proper cleaning and maintenance and the overall difficulty in refilling these units without excessive handling and spillage. Repeated direct contact with the dispensing orifice during use has also been found to be a major contributor to microbial infestation of these dispenser units. There is some evidence that appears to suggest that the greater the frequency of use, the greater is the probability of significant microbial contamination and the more rapid is the development. Good examples of high-risk product dispensers include kitchen counter soap and lotion dispensers, decorative cream and lotion refill containers, wall mounted home bathroom fixtures, and public restroom soap dispensers. Of all the product dispenser systems, public facility soap dispensers appear to be especially vulnerable to microbial contamination and may have the potential to pose a major health risk concern. Bulk dispenser units have been the subject of a number of significant research studies and presentations over the past few years [54–58, 59, 60, 62]. The findings appear to clearly demonstrate that nearly 25% of the soap samples taken from refillable bulk soap dispensers in public restrooms were heavily contaminated with a wide range of bacteria, many of them potentially pathogenic (Table 11.6). Soap in bulk restroom dispensers can be prone to microbial contamination due to any number of factors including constant exposure to bacteria from the environment, handling during the refilling process, spray of toilet/sink water, hands and body of the persons using the soap, stagnation, reusable nozzles, dust in the air, and the general poor sanitary quality and maintenance of many public restrooms. Bulk dispensers are usually refilled by pouring soap from a large container into an open reservoir. Typically, the nozzle that dispenses the soap is not replaced. In contrast, sealed dispensing systems utilize sealed bags or cartridges that contain soap along with a new nozzle. In a primary study by Gerba et al. [54], liquid hand soap from more than 500 dispensers across the United States was sampled and tested to determine the prevalence of contaminated soap in public restrooms. The majority of samples were obtained from public restrooms in restaurants, health clubs, office buildings, and retail stores while others were collected from employee break rooms, airplane restrooms, and kitchen areas. The liquid soap samples collected were from refillable dispensers that are also referred to as "open systems" or "bulk soap" systems. The study reported that of the 541 sampled, 133 or nearly 25% had bacterial counts greater than 500 cfu/mL and that 87% or 16% of these contaminated samples contained coliform bacteria. All the contaminants isolated were found to be Gram-negative bacteria. This is not an unusual finding in liquid soap systems where the surfactant composition of such formulations generally inhibits yeast, mold, and

TABLE 11.6 Opportunistic Pathogens Isolated from Liquid Soaps

Klebsiella oxytoca and pneumoniae
Enterobacter aerogenes
Serratia marcescens
Pseudomonas aeruginosa
Burkholderia cepacia
Enterobacter sakazakii

Gram-positive bacteria from proliferating. The number of bacteria detected in the contaminated soaps ranged from approximately 590 cfu/mL to $>5.0 \times 10^7$ cfu/mL, whereas average coliform counts when present were in the order of 4.0×10^6 cfu/mL. In a subsequent study conducted by Zapka et al. [57], the prevalence of contamination in both regular or "bland" liquid hand soaps and "antimicrobial" hand soaps was investigated. In terms of rate of contamination, the type of soap did not appear to make any difference. Both types of soap were found to be equally contaminated. Interestingly enough, no contamination was detected in any of the dispensers that used a sealed disposable replacement system as contrasted to the refillable open bulk system. The data strongly suggest that replaceable sealed dispensers may be far superior to bulk refill dispensers in preventing or at least minimizing the proliferation of microbial contamination in liquid soap dispensing systems. This same principle of incorporating an easily replaceable sealed product unit in the package design could also be applied to other cosmetic and consumer product forms resulting in a major reduction in the potential for microbial cross-contamination. Extrapolating these findings to the current eco-friendly marketing trends of moving toward bulk product or multiple use product packaging designs, it would appear that there may be some major microbial contamination concerns that could significantly magnify the health risk potential of refillable bulk packaging and presentation formats.

In a related study reported by Zapka et al. [57], an attempt was made to assess and possibly define some of the dispenser/product factors that may contribute to or influence the potential for microbial contamination. In this study, 155 samples were obtained from public restroom areas in Ohio. The samples represented a wide range of soap types and a variety of dispenser models, brands, and configurations. As with the Gerba study, bacterial contamination was found in approximately 21% of the samples and averaged approximately 6.3×10^6 cfu/mL. Again, coliforms and related opportunistic Gram-negative bacteria were detected with *E. coli* and species of *Pseudomonas*, *Klebsiella*, *Serratia*, *Burkoholderia*, and *Enterobacter* most predominant. In terms of the refill bulk soaps, there was no apparent difference in contamination incidence attributable to type or brand. The type, brand, and model of bulk dispenser also did not appear to make a significant difference in the frequency of contamination. Both plastic and metal wall and undercounter dispensers were found to be equally susceptible. A small number, 17 of 155, of pump bottle dispensers included in this study were also found to have a contamination rate of 24%. The finding of pump bottle dispenser contamination could be of serious concern since this is a typical packaging configuration that has been widely promoted for any number of applications in the cosmetic and personal care industries. When coupled with many of the green concepts such as eco-friendly preservatives, organic chemistries, and sustainable packaging, it would appear that as these new packaging and product concepts are introduced to the market in bulk form, the possibility of major microbial contamination issues arising is very likely. The study concluded that the fundamental and apparently widespread problem of contamination in bulk liquid hand soaps is not limited to any particular soap, dispenser type, or dispenser composition and that it appears to be primarily associated with the open refillable design of the unit itself.

On the basis of these findings, additional studies were conducted to determine the potential for contamination transfer from using contaminated bulk soap. This

TABLE 11.7 Contamination Transfer

OH Elementary School Hand Wash Field Study [56] (Glove Juice Sampling Method)	
Before washing	21% contamination detected
After control soap wash	10% contamination detected
After contaminated soap wash	63% contamination detected

study was conducted with volunteer students in an Ohio elementary school [56, 61]. The results are presented in Table 11.7. After accounting for control levels of background microflora, the hands of the subjects were exposed to both contaminated and uncontaminated liquid bulk soap. Hands were found to harbor 10 times more Gram-negative bacteria after washing with the contaminated soap than before washing. In addition, it was also demonstrated that this same contamination could be easily transferred from the hands to other inanimate surfaces, suggesting that contaminated bulk soap may play a potentially significant role in the transmission of bacteria in both schools and other public venues. In terms of cosmetic bulk and refill packaging, a similar possibility exists. As previously discussed regarding mascara refills, cross-contamination and transfer of contamination between the product and the consumer via the applicator was determined to contribute to multiple incidents of eye infections. As with bulk soap, the repeated insult of a product in a relatively open refill system can allow the establishment of a growing population of microoganisms, which can be a potential threat to both the product and the consumer.

In another study conducted at Montana State University, Center for Biofilm Engineering that could have far-reaching implications for the future of refillable packaging, the possible implication and impact of biofilm on bulk soap refill dispensers was evaluated [58]. Plastic and metal wall and counter mounted liquid hand soap dispensers were collected from various public access locations and analyzed for the presence of suspended and biofilm heterotrophic and coliform bacteria. Methods included bacterial viable plate counts and community analysis using PCR and gene sequencing of DNA recovered from the dispensers. In addition to verifying the presence of biofilm in bulk dispensers, three cleaning procedures were also evaluated for effectiveness in preventing recontamination (Table 11.8). The washing study results showed that bacterial counts in the bulk soap returned to

TABLE 11.8 Biofilm Removal Studies

Plastic and metal bulk soap dispensers	
Removal techniques	Effectiveness
Water rinsing	Rapid regrowth of biofilm
Rinse and scrub	Rapid regrowth of biofilm
Rinse, scrub, 5000 mg/L bleach	Regrowth within 2 weeks

Residual biofilm persisted and recontaminated soap in dispensers after all three treatments.

prewash levels within 2 weeks of cleaning a dispenser and then rinsing it with 5000 mg/L bleach. The studies also demonstrated that dispensers contaminated with bacteria suspended in the bulk soap also had high levels of attached biofilm bacteria that would be available to recontaminate a dispenser even if the old soap is emptied and the dispenser washed and treated with bleach. Considering the implications of this study, it becomes obvious that there appears to be no easy approach to controlling or minimizing the potential for product contamination with multiple use refill bulk dispensers. Regardless of the product involved, the composition of the dispenser package, or the cleaning method suggested, based on the data, the refill product/package configuration appears to be extremely vulnerable to microbial contamination during normal use. It is advised that extreme care be taken when considering expanding the use of the refill concept to other cosmetic and personal care product forms. The only approach that appears to be viable would be the one in which the product is contained in a replaceable, access restricted package that protects the product and the consumer from overexposure and potential microbial cross-contamination. Examples of low to medium risk bulk dispensers would include dispenser systems that minimize direct contact with the nozzle, utilize automatic rather than manual dispensing, and restrict easy access to the bulk compartment. By far, the safest and least potentially problematic in terms of microbiological contamination are dispensers that use replaceable sanitary sealed bags or cartridges that include a new nozzle or delivery device.

11.8 COSMETIC APPLICATORS AND MODE OF APPLICATION

Contamination retention and contamination transfer are two of the most significant microbiological risk factors associated with cosmetic product applicators. The function of an applicator is to facilitate the transfer of product from the primary cosmetic container to the targeted area of application in a relatively controlled manner. With makeup products, these targeted areas most often include the eyelids, eyelashes, eyebrows, lips, face, and neck. With other products, areas of the body such as the arms, legs, torso, nails, and hair may also be involved. The products themselves come in a wide range of compositions and configurations that are often customized for specific applications. The product types range from mascaras, eyeliners, eye shadows, lip balms, and lip liners to foundations, moisturizing creams, powders, suntan products, hair products, and other facial and body treatment products. Product formulation and physical/chemical composition include a tremendous variety of creams, lotions, gels, liquids, foams, aerosols, and powders composed of a multitude of synthetic, natural, organic, inorganic, and extracted raw materials. The applicators and packages can be equally complex in nature, especially considering the current trends toward the use of green, sustainable, and more eco-friendly materials. Further complicating the situation in terms of potential microbial risk is the basic nature, design, and composition of cosmetic applicators. Those applicators designed specifically for use in applying product to the sensitive areas around the eyes and mouth are of most concern since they may have the greatest potential vulnerability to microbial contamination during use [42, 66]. Much of the concern is related to the fact that many applicator formats and mechanisms can provide the ideal environment for the capture, retention, and proliferation of microorganisms. Depending on the applicator composition, structure, and configuration, the risk factors associated with intended use can be considerably influenced. Applicators that can physically trap and retain moisture, sebum, skin cells, dirt, and microorganisms create the greatest problems and have the greatest probability of contributing to significant, high-risk product/consumer contamination transfer. During normal applicator use, the product is often subjected to frequent and multiple insults from the debris and contamination carried from the user via the applicator to the primary product container [42, 63, 65]. Over time, this repeated insult can compromise the integrity of the product and its ability to resist microbial contamination, resulting in a situation that essentially feeds on itself. Microorganisms and debris from the skin are transferred to the product, where they may eventually be capable of reproducing. Once established, the organisms are then transferred back to the user via the applicator. The risk potential of various applicator concepts and configurations are presented in Table 11.9. Examples of commonly used high-risk applicator concepts include sponge tip eye product applicators and sponge or puff-type makeup applicators. Mascara brushes are generally considered to have a medium high risk potential due to their ability to trap debris and microorganisms, as well as having the potential to damage the cornea during use. Other makeup brushes also have a significant degree of risk associated largely due to the increased potential for contamination transfer and the often poor handling and maintenance habits of the consumer. The increased trend toward the use of natural bristled brushes and natural sponge material may also contribute to the issue since they can provide additional organic nutrient to the product applicator system, which in turn can contribute significantly to the potential for microbial growth and proliferation.

Tube applicator units combine the compactness and security of a squeezable plastic tube with the convenience of a built-in delivery device. The delivery or applicator portion of these tubed products can consist of mechanisms such as roller balls, domed perforated caps, sponges, brushes, wands, and stylized units for

TABLE 11.9 Applicator Risk Potential

High risk potential
Sponge tip applicators
Sponges and puff applicators
Mascara brushes
Medium high risk potential
Tube roller balls
Domed and flat perforated caps
Eye makeup brushes
Low risk potential
Dry powder brushes, puffs
Single-use swabs, sponges, puffs

specialized product delivery applications [63, 66]. Typical tubed products that use built-in applicators include product types such as lip balms and moisturizers, eye and face makeup products, and a wide variety of skin and body treatments. Generally speaking, products that are by their nature and composition susceptible to microbial contamination and contain preservative materials to protect them are of greater risk when combined with high-risk packaging and applicator concepts. The plastic tube configuration is generally considered to be a more protective packaging concept than many of the more open and accessible package presentations that can allow excessive product exposure. Combining the tube with certain applicator devices, however, can significantly alter the degree of vulnerability of the package and increase the overall potential for microbial contamination. The risk potential generally increases proportionally to the degree of direct product contact and/or the degree of contamination and debris transfer. During normal use, any attached applicator device that allows exposed product to be returned to the primary container has the potential for compromising the microbiological integrity of the product system. Roller balls, sponges, and perforated caps all have the capability of enhancing the overall vulnerability of the tubed product. During typical use, applicators such as roller balls make direct contact with the skin surface while transferring product and skin debris to and from the primary container. Perforated heads contact the skin in a similar manner but, in this case, allow the debris contaminated product to remain exposed on the outer surface of the applicator and be sucked back into the primary container. Sponges, on the other hand, trap cellular debris contaminated product while providing the perfect environment for microbial survival and growth. Natural sponges may contribute additional organic nutrient to the mix that could serve to further exacerbate the problem. Frequency of product application also plays a major role in the risk potential associated with these applicator concepts. The greater the number of product applications and the shorter the time period between applications, the greater is the probability of product contamination. A constant and continuous repeat insult to the product via the applicator can eventually have a negative impact on the product by slowly degrading the robustness of the product's preservative system, as well as the overall safety and integrity of the product package system. Being aware of some of these concerns, some companies are investigating the possibilities associated with alternative low-risk applicator concepts. One example is a single-use swab applicator that contains the product in the hollow barrel of the swab stick. From a contamination risk perspective, this approach significantly minimizes the potential for microbial cross-contamination; however, from a sustainability point of view, single-use applicators significantly increase the amount of component package material and the associated carbon footprint required to deliver the same dose of product from a multiple use package form. Certain product types such as loose powders and foundations, however, are more easily adaptable to low-risk package applications. For example, loose powders can be delivered using single unit powder dispensers that allow the powder to be dispensed directly from the primary container through the brush. This reduces the amount of component material necessary for product delivery since it eliminates the need for a separate brush applicator. Owing to their relatively low water activity, loose powders are generally less susceptible to microbial growth and proliferation than are typical creams, lotions, and other water containing formulations. Combining the flow through brush dispenser and the primary container into a single compact design that both protects the product and safely delivers the powder results in the perfect balance between convenience, safety, and economy of packaging material. This same flow through concept, however, modified to include a sponge applicator rather than a brush or change the product to a liquid rather than a powder could have a dramatic negative effect on product risk potential. Either modification would significantly alter the vulnerability of the product and increase the overall probability of potential microbial contamination. Utilizing organic or natural fiber applicator materials could even further compound the concern of increased susceptibility if the impact of the modified component or packaging material on product integrity were not thoroughly and properly evaluated. To do this, both in use and conventional product package compatibility studies would have to be conducted.

The issue of microbial risk as applied to applicators continues to be a major concern for most cosmetic and personal care manufacturers. Applicators, by their very nature and composition, are highly susceptible to contamination due to their continuous exposure to microbial insult during the course of normal usage. There is no silver bullet or single strategy that will reduce the risk involved with all applicator forms. There are, however, approaches that can be taken to minimize the negative impact and reduce the probability of microbial risk. The first and most important consideration is to determine exactly what is the degree of potential risk associated with a particular applicator or applicator package combination. Knowing and understanding the vulnerability of the product, package, and applicator allows one to select the most probable and appropriate approaches or courses of action to reduce the potential risks. Obtaining this information may require a combination of in-use clinical panel evaluations, literature reviews, in-house experience with similar concepts, and, in some cases, common sense. The next step in the process is to reevaluate the parameters, characteristics, and mode of use of the applicator and primary package. Modifications in design, materials, configuration, composition, and manner of usage can have a very positive effect on mitigating potential contamination pitfalls. The difference between using a brush and a sponge material or a dry application as opposed to a wet application could be the difference between marketing a high-risk product and marketing a low-risk product. The goal is to creatively develop an applicator concept that combines market appeal, functionality, and product safety with an aesthetically pleasing and convenient packaging format. A number of suggested approaches to reducing contamination risk potential are presented in Table 11.10. Products to be used in higher risk package applications should be formulated specifically to withstand the challenges of multiple insults during the course of normal usage. This can be accomplished by formulating with

TABLE 11.10 Applicator Risk Reduction Approaches

Rapid kill and highly robust preservative system
Verification of enhanced capacity (conventional PET and in-use studies)
Presterilization of applicators (ETO, gamma, electron beam)
Incorporation of antimicrobials into applicator material
Proper cleaning and use instructions

microbiologically resistant raw ingredients, judiciously choosing effective and compatible preservatives or a combination of both. In terms of preservation, the preservative system should be designed so as to be extremely robust and capable of a rapid rate of kill in order to compensate for the repeated and aggressive microbial insult it is expected to receive in the hands of the consumer. This may be considerably more difficult to achieve than it is to recommend since many of the most effective preservatives and preservative combinations have been restricted in their use and are not natural enough to qualify for use in organic or eco-friendly products. The methods for evaluating both capacity and rate of kill, as previously indicated, must also be of appropriate design and capability to be able to demonstrate and predict both robustness and overall effectiveness during normal and expected product exposure. Using a single challenge PET procedure, such as the USP <51> Antimicrobial Effectiveness Testing (AET) protocol [67], as primary evaluation tool will not provide the data necessary to make a proper and meaningful decision about the preservative robustness of a vulnerable high-risk applicator product system. Although the USP AET and similarly constructed single challenge preservative efficacy testing methodologies play a valuable role in evaluating relatively low-risk OTC product forms and aseptically filled products, they fall far short in providing the assurances that are necessary to demonstrate that a highly susceptible and microbiologically vulnerable applicator dispensed product will be adequately protected from normal in-use contamination. For evaluating products in this highly susceptible category, both the evaluation methodology and the criteria of acceptability must be far more rigorous than those conventionally suggested for less susceptible OTC, cosmetic, and personal care products. A proper approach for determining product vulnerability and overall preservative effectiveness should minimally include an aggressive double or triple challenge PET evaluation coupled with a real-time consumer in-use panel test. If at all possible, product package microbial susceptibility and preservative efficacy evaluations should always be conducted with the product in the primary container and with all product contact components in place. This is especially critical for products whose packaging and components are constructed from materials of natural, plant-based, or recycled origin. This would include such things as bioplastics, oxo-biodegradable plastics, corn-based polylactic acid (PLA), postconsumer recycled (PCR) paperboard, and other similar eco-friendly nonpetroleum-based component materials.[12–14, 16, 17, 19–21, 27, 45, 68, 69]

An additional approach that has been taken by some manufacturers attempting to minimize the future risk potential of applicators and associated components involves the decontamination of these items prior to assembly. Typical packaging materials that are treated include both natural and synthetic bristled brushes, foam applicators, sponges, wands, seals, cap liner discs, and a variety of caps and closures. The primary treatment processes include ethylene oxide gas, gamma irradiation, and electron beam irradiation. Dry heat, wet heat (steam), and cold sterilization or decontamination are less commonly used mainly due to issues associated with the treatments often being destructive to or incompatible with the component materials. Ultraviolet (UV) has also been used with some degree of success in selected situations. One of the most interesting and successful examples of the application of UV for decontamination of components involves the installation of UV tunnels on makeup assembly lines. In the case of pressed powders, for instance, all components, including the pressed cake(s) and applicator(s), are assembled appropriately in a suitable single or multiple celled compact or holder. Prior to adding the final protective glassine slip sheet, the open compacts are passed through the UV tunnel for decontamination. Post-UV, the slip sheet is added and the compact closed protecting the decontaminated product(s) and components from further microbial insult until it is opened and used by the consumer. The major limiting factor with UV is that it is a surface treatment only and when used for solid materials, such as packaging components, it cannot penetrate more than a few millimeters or so depending on the density of the article being treated. Anything that is not directly exposed to the UV light source will not be treated. When contemplating the use of any bioload reduction method or sterilization procedure for applicators, components, or packaging materials, there are also certain issues and areas of concern that should be addressed. Probably the two most critical factors to consider and evaluate are the effectiveness of the decontamination or treatment process and the overall effect of the sterilization process on the physical, chemical, and aesthetic integrity of the treated material. Prior to proceeding with full-scale batch or lot sterilization, it is first recommended that baseline microbial bioburden assessments be established for each of the components to be treated. The determinations are normally conducted on a minimum of three typical production lots of a particular component. Once established, trial runs can be conducted to determine and establish the appropriate dosage parameters and exposure times necessary to achieve the desired level of efficacy and confidence. A final posttreatment evaluation of the components is required to ensure that the selected treatment process has not in any way adversely affected the structure and functional composition of the treated items. For many applicators constructed from conventional plastics and synthetic materials, there is a considerable amount of information available regarding potential negative effects of the various treatment approaches. For the newer more eco-friendly packaging material concepts, however, there is considerably less real-time experience resulting in a current lack of reliable data and information regarding potential long-term material stability and compatibility. These issues and the potential problems associated with them are essentially unknown at this time although this will obviously change in the future as the newer green materials become more commonplace and the issues associated with compatibility and structural integrity are more thoroughly researched. In terms of available technical expertise, companies that offer sterilization services will generally assist in setting trial parameters, establishing criteria of acceptability, and conducting the preliminary evaluations. For guidance in terms of irradiation, the Association for the Advancement of Medical Instrumentation (AAMI) [70] has developed specific guidelines for both gamma radiation and electron beam radiation. Although originally designed to provide guidance for the process control of the irradiation sterilization of medical devices, the guidelines have long become the standard references for the irradiation decontamination of many materials and products. In all cases, a predetermination of compatibility of the treatment process with the components is mandatory. This is especially important when considering eco-friendly materials that may or may not have the same degree of durability and resistance to structural change as do conventional noneco-friendly packaging materials.

An area that is beginning to become more popular regarding protecting applicators from excessive microbial contamination has to do with the incorporation of antimicrobial agents directly into components. Significant early success has been achieved with sponge-type applicators where antimicrobial is introduced into the sponge during the manufacturing process. The nature and sequence of the sponge formation process allows a relatively easy means of incorporating antimicrobials or other additives into the sponge matrix. Other efforts in this area have been directed at the addition of antimicrobials to various plastics and polymeric materials during production. Suggested applications for this technology range from cosmetic components and packaging materials to toys, consumer products, medical devices, and construction materials. A standard test method for evaluating the antimicrobial efficacy of agents incorporated into or bound to plastics and related polymeric surfaces has been established by the ASTM in the form of ASTM E2180-07 Standard Test Method for Determining the Activity of Incorporated Antimicrobial Agent(s) in Polymeric or Hydrophobic Materials [71]. The method primarily focuses on the quantitative assessment of antibacterial activity; however, it can easily be adapted to evaluate activity against both yeast and mold species. In regard to research into green and organic-based natural packaging materials and the application of antimicrobial incorporation technology, the literature has been relatively sparse. Theoretically, it would appear that organic-based packaging chemistries such as bioplastics, corn-based PLA, polyhydroxyalkanoate (PHA), and other similar biopolymers and biomaterials would adapt well to this technology since a number of antimicrobial agents are naturally substantive to organic materials [16, 24–27, 41–44]. This would allow the antimicrobial properties of the chemical agent to be retained by or bound to the packaging biomaterial and still maintain its biological activity. Considering the possible increase in potential microbiological susceptibility associated with natural and organic packaging materials, this approach would appear to be one worth pursuing. The most obvious drawback with this technology revolves around the mechanism of action of component impregnated antimicrobials. The three main functional options for incorporated biocides include surface treatment, total impregnation, and substantive binding. In all these systems, the critical functional factor for effectiveness is based on the ability of the biocide to make direct physical contact with the microbial contaminants. Since the antimicrobials are in a fixed or static state and are bound to the component, they may not be able to easily migrate into the product system and make contact with the invading microorganisms. Leachable biocides that can be more mobile have been considered; however, the impact of these materials on the safety of the product must be firmly established. The issue of product compatibility would also have to be addressed since the presence of additional biocide and the rate of leaching could have a significant impact on the physicochemical integrity of the product package system.

11.9 CLOSURES AND DELIVERY SYSTEMS

Aside from applicators, probably the most critical packaging attribute that can significantly influence product exposure and play a major role in microbiological susceptibility is the design and functional nature of the package closure and delivery system. A closure that allows excessive product exposure to inadvertent microbial contamination whether from the environment or the consumer has the potential of being one of the major contributors to microbial risk and a significant threat to the safety and integrity of the total product system. Package configuration and closure design are the first line of defense in the protection of the product from microbial insult during normal conditions of storage and consumer use. Closures that are open in design and allow unrestricted access to the product optimize exposure conditions and set the stage for potential contamination incidents. An ideal package closure system would allow the product to be easily accessed, minimize overexposure to the environment, reduce or eliminate backflow and dripping during delivery, restrict or discourage direct contact with the delivery mechanism, be composed of recyclable or sustainable materials, and be presented in a stylish and user-friendly design (Table 11.11). Standard bottle closures essentially come in four basic designs, although newer, more progressive concepts are constantly coming on the scene. The most common conventional designs are the screw cap, the dispenser cap, the pump top, and the atomizer or spray top. Dispenser caps, which are normally used for both bottles and tubes, come in a fairly wide range of specialty configurations with orifices of various shapes and sizes. For cosmetic and personal care products such as shampoos, lotions, and creams, the most frequently used dispenser closures include the flip or snap cap with a circular orifice, the flip cap with a slit or slotted orifice, the flip spout dispenser cap, the disk top toggle-type dispenser cap, and the dropper tip cap. Most are constructed of plastics such as low-density polyethylene (LDPE), polypropylene, and other suitable petroleum-based plastic matrices. With the recent explosion of the green phenomenon, however, considerable interest is now being shown toward the development and possible use of more eco-friendly materials for standard closures and dispenser caps. In the case of bottles and other primary containers, some success has already been achieved using bioresins and similar natural or organically sourced materials. The adaptation of these same materials for closures, however, may not be as easily attainable considering some of the intrinsic material qualification attributes and functional design specifications required for closure composition. As advances are made, however, there is no question that the successful use of bioplastics and similar eco-friendly materials will eventually make its way into typical closure and delivery system applications.

In terms of potential microbial risk related to closure design, probably the most informative study and discussion on the subject to date was published by Brannan and

TABLE 11.11 Ideal Closure

Easily accessible
Minimizes product exposure
Reduces or eliminates backflow and dripping during delivery
Restricts or minimizes direct contact with delivery mechanism
Compatible with product and preservative system
Composed of recyclable or sustainable materials
Aesthetically appealing

Dille in 1990 [72]. In their study, the investigators attempted to determine the potential role of closure design in protecting cosmetics from normal in-use microbial contamination. The study involved the use of unpreserved shampoo and a skin lotion product that were verified using a standard PET as not being hostile to microbial contamination. The two products were packaged into identical bottles using commonly used closure types that theoretically provided different degrees of perceived product protection and potential microbial vulnerability. The shampoo bottle closures included a screw cap closure considered to be the most vulnerable during use, a flip or snap cap with a circular orifice considered to offer the greatest degree of protection, and a flip cap with a slit orifice whose protective abilities were unknown at the time. For the skin lotion, both the screw cap and the flip cap with a circular orifice were also used. Since the pump top dispenser is one of the most popular closure systems for typical lotion products, a pump top was substituted for the slit dispenser flip cap. The aseptically packaged test units of both products were then distributed to selected consumer use panels. Panelists were instructed to use the products as they normally would over the term of the study and to return the test units on completion. The shampoo home use study was conducted for a 3-week period, whereas the lotion study was of 2-week duration. Upon return, the samples were evaluated for microbial content immediately and after 4–7 days. Unused and unopened control units were also reevaluated for contamination at this time. Products with a microbial content of >100 cfu/g or contained Gram-negative bacteria at any level at both testing times were considered to be contaminated. Final results of the 3-week shampoo consumer use test confirmed that the flip or snap cap with a circular orifice was the most effective in protecting the shampoo from contamination, with 0% of the samples being contaminated. The slit dispenser flip cap and the screw cap did not fare as well, with 21% and 29% of the test units, respectively, found to be contaminated. The skin lotion, which apparently was much more susceptible to contamination during use than the shampoo, provided results that were quite interesting. The pump top dispenser clearly outperformed the other two closure systems with only 10% of the returns being positive for contamination. The flip or snap cap with a circular orifice that did so well with the shampoo performed poorly with 39% of the returned units being positive. This may be attributable in part to the manner in which the lotion dispenser is used in that normal usage usually involves direct consumer hand contact with the exposed orifice of the dispenser. The worst closure system of the three was, as expected, the screw cap with 71% of the returned samples being heavily contaminated (Table 11.12). On the basis of data, a number of conclusions can be drawn. The most obvious is that package closures and dispensers vary considerably in their abilities to protect the product from inadvertent microbial contamination during use. The pump dispenser was the most contamination-resistant delivery system of the closures evaluated for lotions, whereas the flip cap dispenser appeared to be the most effective protector for shampoos. It also appears that consumer product evaluation studies can be an effective means for measuring the differences in closure contamination potential and, in this case, be used to rank closures by degree of protectiveness. Less obvious is that both the susceptibility of the product and the manner in which the consumer dispenses the product can have a significant influence on the microbial susceptibility of the total package format. This is evidenced by comparing the data

TABLE 11.12 In-Use Closure Study [72]

Unpreserved shampoo (home use conditions for 3 weeks)	
Flip cap	0% contamination
Slit cap	21% contamination
Screw cap	29% contamination
Unpreserved lotion (home use conditions for 2 weeks)	
Pump top	10% contamination
Flip cap	39% contamination
Screw cap	71% contamination

derived on the flip cap with a circular orifice for the shampoo with the same closure for the skin lotion. In one case, 0% contamination was observed, whereas in the other case, 39% contamination was detected. Since the packaging and dispensing closures were equivalent, the only differences between the two were the microbial susceptibility of the product and the degree of direct product exposure and contact associated with the manner in which the consumer dispenses the product. Accordingly, when choosing a closure and or dispensing delivery system for a particular application, it is necessary to consider not only the protective design characteristics of the unit but also both the degree of product susceptibility and the manner in which the product will be dispensed and used by the consumer. As newer closure and dispenser designs are developed using both conventional and more eco-friendly packaging materials [73], the need for additional studies of this type becomes even more apparent. Standard preservative efficacy testing alone cannot provide the kind of information necessary to make the critical decisions regarding microbial susceptibility that are so vital for the proper evaluation of new and old package and closure concepts.

11.10 RATING PACKAGING MICROBIAL RISK

Package development requires the blending of both scientific and creative disciplines. A critical part of this process is the ability to identify and assess the various potential risk factors associated with a particular design or material concept. This is especially true with regard to package, closure, and applicator attributes that can significantly influence product microbial vulnerability. Some of these attributes are purely design and configuration features, while others can be directly related to the composition of the package system or the manner in which the consumer accesses and uses the enclosed product. Assigning a microbial risk rating for a package or component involves a detailed evaluation of every aspect and property of the item with the intent of identifying any feature that may contribute to potential product susceptibility. These susceptibility factors can include product overexposure to environmental contamination, excessive consumer contact, cross-contamination between the user and the primary container, the package or component containing materials that may negatively affect the stability or compatibility of the product, and any other major or minor characteristic of the package, applicator, dispenser, or closure that has the potential to be detrimental to the overall microbial integrity of the product/package system. The various configurations and packaging formats of wet wipe products provide a good example of the range of potential risk factors that can be found within the same product category [23, 24]. The open lidded wet wipe tub design presents a significant or high contamination risk by allowing overexposure of the product and excessive direct contact by the consumer during normal use. The canister configuration, on the other hand, reduces excessive exposure and is generally considered a low to medium risk provided the dispenser cap is replaced between uses. Unfortunately, in most cases during normal usage, this is not done, so the risk factor associated with this package increases from low to medium. The development of the resealable closure or flap for soft or flexible package wipes was a major step forward for this product category and has contributed significantly to reducing the overall microbial vulnerability of the product by providing an easy to use mechanism to reseal the package between uses. Some of the earlier versions did, however, have seal contamination issues that were later corrected by using more tenacious adhesive materials. The seal failures experienced were caused by product liquid residue contaminating the outer surface of the package and subsequently interfering with the flap closure adhesive. A spin-off positive effect of the resealable closure is the enhancement of the stability of the product by protecting the wipes from excessive dehydration and weight loss. As the wet wipe industry moves toward a greater use of natural fibers, bioplastics, and other eco-friendly materials that may be more porous and interactive, improving stability and reducing wipe exposure during use may become an even bigger challenge. A more recent wipe flexible packaging closure design involves the use of a hard snap style lid similar to that used on tub configurations attached to a flex package. This closure type is capable of minimizing product exposure provided that, as with other lidded packages, the consumer actually closes the lid. In terms of potential microbial risk, this package system could be rated as low to medium. From a green sustainability point of view, however, the snap lid closure does add bulk and mass to the package and, unless the lid is made from bioresins, may also not qualify as being truly eco-friendly. Another concept in wipe packaging that is worth commenting involves using a zip lock closure system on a bulk style flexible bag. This format is most commonly used in high wipe count bulk units and provides a sturdy tight seal and adequate protection to the product. Extraneous microbial contamination is generally minimized; however, when the product is repeatedly used directly from the primary container rather than as a bulk refill, it is somewhat clumsy to handle and can lead not only to overexposure of the product but also to excessive consumer contact with the wipes. This packaging style can be rated as being low risk; however, the rating could change dramatically depending upon the manner in which the product is used and/or abused. As with any of these wipe packaging concepts, the only way to truly evaluate the potential microbial risk associated with their use is to submit them to appropriately designed consumer use studies. Panel testing coupled with specialized wet wipe preservative efficacy testing as recommended in the PCPC Technical Guidelines [74, 75] would provide the kind of information and data necessary to predict in-use product robustness while gaining a better understanding of the possible vulnerabilities of these packaging formats.

There are other very good examples of more recently created closure and packaging concepts that can be considered to be rated as having low or, in some cases, medium microbial risk. Among these are a number of pump dispenser and tube closure designs that appear to have been specifically developed to reduce product exposure and minimize or eliminate potential microbial contamination during use. Pump dispensers are generally regarded as being considerably safer in terms of minimizing microbial contamination compared to most other bottle or tube dispenser designs. Although this may be true in many cases, as revealed in the Brannan study [16, 72], the conventional pump can be compromised based on the microbiological susceptibility of the product being dispensed. Using a hand lotion as an example, direct hand contact with the nozzle is quite common during normal usage as is the presence of residual product around the dispenser orifice. In addition, most pump designs allow some degree of flow back into the dip tube. Under the right circumstances, the combination of these variables could have the potential of increasing the degree of microbial vulnerability. Of the newer pump concepts, the airless pump [76] appears to be gaining popularity and a number of variations of the design have recently come on the market. The attraction is apparently related to this pump type being able to effectively dispense higher viscosity products while minimizing exposure to potential microbial contamination. Among the more interesting variations is a unit that includes a rubber spout valve seal preventing product flow back into the primary container while protecting the product from drying out between uses. It is claimed that the combination of airless pump and valve seal ensures that the product continues to be protected until it is dispensed by the consumer. Another interesting low-risk pump is designed so that the pump spring is located outside the dosing chamber and uses a glass ball to control flow characteristics. This approach not only protects the products from exposure but also significantly reduces the potential for product/package incompatibility. The airless pump approach has also been adapted for use on tube packaging formats. Other approaches taken to minimizing product exposure and possible microbial contamination in tubed products include designs such as the no-drip tube dispenser and the "airbackless" system. Both are claimed to reduce or eliminate the possibility of product either dripping out of the tube or being sucked back into the tube during use. In addition, another tube closure claims to keep the contents of the tube "sterile" by utilizing a four-piece closure system that contains a membrane that moves when the tube is squeezed, thus minimizing contact with the other dispenser components. Refill packs and combination refill closure dispenser concepts have also entered into the market with promises of eco savings and environmental friendliness. One design provides a two-part system composed of a cream refill jar or pot that attaches to a reusable piston actuated dispenser top. The dispenser expels a predetermined dose of cream through an orifice to a collecting surface for easy access by the consumer. The design, however, falls into one of the many traps that most refill concepts do—namely, it allows any excess product to be sucked back into the primary container and it suggests the reuse of a dispenser that can serve as a potential reservoir of microbial contamination that can be cross-transferred between the product and the consumer. As interesting as this design is, from a microbiological point of view, the potential risk would be rated as being medium to high. As previously discussed in relation to bulk product refills, there is a very high probability of potential microbial risk associated with nearly every refill scenario unless both the product and the dispensing system can be so designed as to minimize product exposure and prevent dispenser-related microbial cross-contamination. As with many design approaches, there are many variables that can significantly influence the degree of risk associated with a particular packaging concept. A very good example of an eco-friendly packaging approach that has received considerable attention is based on the relatively old style packaging concept of using paperboard as the primary container material. One company has developed an all-paperboard package ideally suited for packaging a wide range of gel- and oil-based solid product forms. The package is a completely biodegradable and recyclable container composed of up to 95% recycled material and available with more than 80% postconsumer content. For nonmicrosensitive atypical anhydrous products, the potential microbial risk rating is very low. As the microsensitivity of the product increases, the potential microbial risk rating also increases proportionally.

11.11 SUMMARY

Cosmetic and personal care product packaging has evolved from simplistic utilitarian containers to commercial works of art and now has branched into an entirely new and even more dynamic concept referred to as "eco-friendly" or "green" packaging. A new lexicon for this technology has emerged and includes terms such as biodegradable, sustainable, recyclable, compostable, natural, organic, reusable, refillable, bulk, concentrates, bioplastics, dematerialization, energy efficiency, green washing, and carbon footprint. The major microbial risk factors associated with eco-friendly cosmetic and personal care packaging are essentially the same as those challenging conventional packaging and are directly related to the degree of product exposure, the frequency of direct consumer product contact, normal and unexpected consumer use behaviors, and the availability of opportunities for consumer abuse. Depending on the green approach taken, however, many of the new eco-friendly concepts have the potential to further exacerbate potential microbial risk if not properly evaluated and controlled. This is largely due to the lack of significant history and experience with many of the new materials and associated technologies.

The physical and compositional characteristics of the primary cosmetic package can significantly influence microbial risk potential. Material composition is more important when using packaging components fabricated from eco-friendly and sustainable natural raw materials. The more use of naturally derived "green" materials can lead to a greater likelihood of product package or preservative package incompatibility that could lead to preservative system failure.

The determination of preservative system robustness is a critical factor in evaluating product and package microbial risk potential and should be made an integral part of the product package development process. Preservative efficacy protocols for measuring robustness should minimally include at least two or more microbial challenges spaced relatively close together. In addition to the conventional recommended protocols for product PET, it is also highly recommended that *in vivo* consumer use testing of the product in the final package form be conducted.

Many questions still remain regarding the compatibility and stability of many eco-friendly packaging materials when used in conjunction with preservatives or complex natural ingredient formulations. Accordingly, it is mandatory that all new or compositionally modified packages, components, and materials be subjected to rigorous product stability and compatibility studies conducted over an extended range of temperatures and environmental conditions. Part of the stability criteria should include the verification that the accelerated aging process has not compromised effectiveness of the preservative system.

The product applicator or delivery mechanism is one of the most critical factors influencing consumer use contamination and microbial risk. The primary concerns are the degree of product exposure and the transfer of contamination to and from the consumer and/or the environment. Vivid examples exist as to the negative ramifications associated with certain reusable applicator concepts. Applicators designed specifically for use in applying product to sensitive areas are of most concern since they have the greatest vulnerability to microbial contamination. The concern is that many applicator formats and mechanisms can provide the ideal environment for the capture, retention, and proliferation of microorganisms. Repeated microbiological insult of the product in the primary container via applicators or nonrestrictive closures can lead to degradation of preservative system capacity. A weakened preservative system cannot withstand additional incremental microbial insults, resulting in the product becoming irreversibly contaminated. Packaging components constructed of materials derived from organic fiber, plant-based polymers, or similar natural materials may provide the opportunity for enhanced package/product interaction and possible preservative inactivation, both of which could contribute to a more rapid breakdown of the preservative system. High-risk applicator concepts include sponge tip eye and lip product applicators, sponge or puff-type makeup pads, and roller ball tube dispensers.

Many cosmetic and personal care manufacturers are now considering the potential economic and marketing value of marketing bulk product configurations. Associated packaging concepts would include refillable bulk dispensers, refill replacement packaging units, reusable delivery systems, and multiproduct consumer packaging. The major risk associated with bulk or refillable product is related to the inadvertent and repeated exposure of the product to microbial insult during use. Contamination risks most commonly occur through direct person to product contact when dispensing, during the bulk transfer or refilling process, and from inadequate precleaning of the dispensing unit prior to refilling. Any dispenser that allows direct contact with the delivery orifice, nozzle, or applicator significantly increases the potential for product contamination. Major studies have been conducted that clearly demonstrate that there are significant microbiological hazards associated with bulk soap dispensers in schools and public restrooms. Of equal significance is the fact that microorganisms from contaminated dispensers not only can be easily transferred during hand washing but can also be retransferred from the hands to inanimate objects. An additional serious concern regarding refillable bulk dispensers is related to the development of biofilms [58, 77] that cannot be removed by conventional cleaning and sanitizing practices.

Minimizing bioburden contamination in packaging materials can be accomplished by utilizing decontamination treatments prior to assembly. Typical packaging materials that are treated include both natural and synthetic bristled brushes, foam applicators, sponges, wands, seals, cap liner disks, and a variety of caps and closures. Primary treatment processes include ethylene oxide gas, gamma irradiation, and electron beam irradiation. Dry heat, wet heat (steam), and cold sterilization or decontamination processes are often incompatible with packaging materials and are less commonly used, whereas UV is used for selective applications.

Package configuration and closure design are the first line of defense in the protection of the product from microbial insult. Closures that are open in design and allow unrestricted access to the product set the stage for potential contamination incidents. The ideal package closure system allows easy product access, minimizes overexposure, reduces or eliminates backflow and dripping during delivery, restricts or discourages direct contact with the delivery mechanism, and can be composed of recyclable and sustainable materials and presented in a stylish and user-friendly design. Consumer product use studies are an effective means for measuring closure contamination potential and degree of protectiveness. Both product susceptibility and the manner in which the consumer uses the product have a significant influence on the microbial susceptibility of the total package format. Studies have demonstrated that pump top closures are more protective in preventing contamination than are flip snap cap, slit cap, and screw cap bottle dispenser designs.

A critical part of the package development process is the ability to identify and assess potential risk factors associated with design or material concepts. Package, closure, and applicator attributes can significantly influence product microbial vulnerability. Some of these attributes are purely design and configuration features, while others can be directly related to the composition of the package system or the manner in which the consumer accesses and uses the enclosed product. Assigning a microbial risk rating for a package or component involves a detailed evaluation of every aspect and property of the item with the intent of identifying any feature that may contribute to potential product susceptibility.

REFERENCES

1. Armstrong L. Simplicity, functionality and personality: key trends in cosmetic packaging. CosmeticsDesign.com, *Decision News Media,* May 20, 2010.
2. Conroy J. More consumers, manufacturers and suppliers are making moves toward sustainable packaging, Special Section: Green Growth. *CPC Packaging Mag.,* March/April, 2009.
3. Pitman S. Sustainability trend pushes packaging companies to rethink. CosmeticsDesign.com, *Decision News Media,* February 2, 2009.
4. Pitman S. Recycling and all things green define packaging in 2009. CosmeticsDesign.com, *Decision News Media,* December 8, 2009.
5. Packaging News, Chicago Paper Tube awarded for Ecopak, *HAPPI,* July 2009, p. 129.
6. Hill A. Beauty packaging identity trends. *GCI magazine, Packaging Design,* May 2009, pp. 52–56.
7. Harrington R. Israel plans packaging recycling revolution. CosmeticsDesign.com, *Decision News Media,* February 25, 2010.
8. Nichol K. Survey reveals low consumer awareness over cosmetics packaging recycling. CosmeticsDesign.com, *Decision News Media,* March 18, 2010.
9. Bird K. Green packaging is more about improving recycling than new materials. CosmeticsDesign-Europe.com, *Decision News Media,* September 29, 2009.

10. Sweeney C. Now, the cosmetics jar matters, too, Skin Deep: Friendly Packaging for Cosmetics. *New York Times,* March 24, 2010.
11. Harrington R. Sustainable packaging—threats and opportunities. CosmeticsDesign.com, *Decision News Media,* September 24, 2010.
12. Cosmetic Packaging quality control. *Household Pers. Prod. Ind.* 1965, 36.
13. Yablonski JI, Mancuso, SE. Preservative efficacy testing: accelerating the process. *Cosmet. Toilet.* 2007;122(10):51–62.
14. Determination of preservation adequacy in cosmetic formulations, CTFA Microbiology Guidelines. The Cosmetic, Toiletries and Fragrances Association, Washington, DC, 2007.
15. Shaw AH. Preserving the future. *Soap Cosmet. Chem. Spec.* 1992;68:26–28.
16. Brannan DK. The role of packaging in product preservation. In: Kabara JJ, Orth DS, editors. *Preservative-Free and Self-Preserving Cosmetics and Drugs*, Chapter 10. New York: Marcel Dekker, 1977.
17. Brannan DK, Dille JC, Kaufman DJ. Correlation of *in vitro* challenge testing with consumer use testing for cosmetic products. *Appl. Environ. Microbiol.* 1987;53:1827–1832.
18. Microbial risk factor assessment off atypical cosmetic products, CTFA Microbiology Guidelines, Supplement. The Cosmetic, Toiletries and Fragrances Association, Washington, DC, 2003.
19. Yablonski JI. Risk factor assessment of anhydrous/atypical cosmetic products. In: Orth DS, et al., editors. *Cosmetic and Drug Microbiology*, Vol. 31, Chapter 8. Informa Healthcare, 2006, pp. 133–141.
20. Farrington JK, et al. Ability of laboratory methods to predict in-use efficacy of antimicrobial preservatives in an experimental cosmetic. *Appl. Environ. Microbiol.* 1994;60:4553–4558.
21. Lindstrom SM, Hawthorne JD. Validating the microbial integrity of cosmetic products through consumer-use testing. *J. Soc. Cosmet. Chem.* 1986;37:481–488.
22. Spielmaker RJ. Report of a cosmetic industry survey concerning correlation of preservative challenge and consumer use test results. *J. Soc. Cosmet. Chem.* 1985;36:376.
23. Yablonski JI. The microbiology of wet wipes. *NJ SIM Meeting,* Newark, NJ, September 18, 2002.
24. Yablonski JI, Mancuso SE. Personal care wipes: manufacturing practices and microbiological control. *Cosmet. Toilet.* 2004;119(8):53–56.
25. McCarthy TJ. Interaction between aqueous preservative solutions and their plastic containers. *Pharm. Weekbl.* 1970;105:557–563, 1139–1146.
26. McCarthy TJ. Storage studies of preservative solutions in commonly used plastic containers. *Cosmet. Perfum.* 1973;88:41–42.
27. Gilbert R. Bottles for natural shampoos fade away: containers biodegrade quickly. *Food Drug Pac* 1992;56:4.
28. Dempsey G. The effect of container materials and multiple-phase formulation components on the activity of antimicrobial agents. In: Bloomfield SF, et al., editors. *Microbial Quality Assurance in Pharmaceuticals, Cosmetics and Toiletries*, Chapter 7. Chichester: Ellis Horwood Ltd., 1988.
29. Bahal SM, Romansky JM. Sorption of benzoic acid, sorbic acid, benzyl alcohol and benzalkonium chloride by flexible tubing. *Pharm. Dev. Technol.* 2002;7:49–58.
30. Song YS, Al-Taher F, Sadler G. Migration of volatile degradation products into ozonated water from plastic packaging materials. *Food Addit. Contam.* 2003;20:985–994.
31. Cadwallader DE. Stability testing: its role in pre-formulation and formulation of cosmetic products. *Cosmet. Toilet.* 1989;104:87–102.
32. Dukes GR. Building a stability program. *Pharm. Tech.* 1982;6:80–86.
33. Reich RR, Sharpe DC, Anderson HD. Accelerated aging of packaging: considerations, suggestions and use in expiration date verification. *Med. Dev. Diag. Indus.* 1988;10:34–39.
34. Zographi G. Physical stability assessment of emulsions and related disperse systems: a critical review. *J. Soc. Cosmet. Chem.* 1982;33:345–358.
35. Montague-Jones G. Thinking outside the box: refillable packaging opportunities. CosmeticsDesign.com, *Decision News Media,* June 24, 2009.
36. Bird K. Refill packaging promises eco savings while retaining luxury feel. CosmeticsDesign.com, *Decision News Media,* March 26, 2010.
37. Nichol K. Refill packaging is winning formula for sustainability: Pira International. CosmeticsDesign-Europe.com, *Decision News Media,* April 29, 2010.

38. Wilson LA., Ahearn DG. *Pseudomonas*-induced corneal ulcers associated with contaminated eye mascaras. *Am. J. Ophthalmol.* 1977;84:112–119.
39. Wilson LA, et al., Microbial contamination in ocular cosmetics. *Am. J. Ophthalmol.* 1971;71:1298–1302.
40. Wilson LA, Jillian AJ, Ahern DG. The survival and growth of microorganisms in mascara during use. *Am. J. Ophthalmol.* 1975;79:596–601.
41. Ahearn DG, et al. Mascara contamination: in-use laboratory studies. *J. Soc. Cosmet. Chem.* 1978;29:127–131.
42. Dawson DL, Reinhardt DJ. Microbial flora of in-use, display eye shadow testers and bacterial challenges of unused eye shadows. *Appl. Environ. Microbiol.* 1981;42:297–302.
43. Anastas P, Warner J. Twelve principles of green chemistry. *Green Chemistry: Theory and Practice.* Oxford University Press, New York, 1998.
44. Pitman S. Formulating with natural and organic ingredients still throws up challenges. CosmeticsDesign.com, *Decision News Media,* March 10, 2010.
45. Harrington R. Oxo-bio additives come under attack from UK report. CosmeticsDesign.com, *Decision News Media,* March 19, 2010.
46. Bird K. Metabolix' field investigate bioplastic-producing crops. CosmeticsDesign.com, *Decision News Media,* October 26, 2009.
47. Graf WD, Kersch D, Scherzer G. Microbial contamination of liquid soap wall dispensers with one way bottles. *Zentralbl. Bakteriol. Mikrobiol. Hyg. B* 1988;186:166–179.
48. Buffet-Batallon S, et al. Outbreak of *Serratia marcescens* in a neonatal intensive care unit: contaminated unmedicated liquid soap and risk factors. *J. Hosp. Infect.* 2009;72:17–22.
49. Rabier V, et al. Hand washing soap as a source of neonatal *Serratia marcescens* outbreak. *Acta Paediatr.* 2008;97:1381–1385.
50. Becks VE, Lorenzoni NM. *Pseudomonas aeruginosa* outbreak in a neonatal intensive care unit: a possible link to contaminated hand lotion. *Am. J. Infect. Control* 1995;23:396–398.
51. Amemiya K, Taguchi F. Survey of the bacterial contamination of hand washing liquids. *J. Antibac. Antifung. Agents* 1992;20:459–463.
52. Sartor C, et al. Nosocomial *Serratia marcescens* infections associated with intrinsic contamination of nonmedicated soap. *Infect. Control Hosp. Epidemiol.* 2000;21:196–199.
53. Weber D, Rutala W, Sickert-Bennett E. Minireview: outbreaks associated with contaminated antiseptics and disinfectants. *Antibac. Agents Chem.* 2007;51:4217–4224.
54. Gerba CP, Maxwell S. Bacterial contamination of liquid soaps used in public restrooms. *71st Annual Educational Conference*, National Environment Health Association, Atlantic City, NJ, June 18–21, 2007.
55. Zapka C, et al. Handwashing with contaminated soap results in hand contamination and transfer of bacteria. *73rd Educational Conference*, National Environment Health Association, Atlanta, GA, June 21–24, 2009.
56. Zapka C., et al. Opportunistic pathogens from contaminated bulk soap on the hands of students and staff in an elementary school. *41st Annual Conference*, National Association of School Nurses, Boston, MA, June 23–26, 2009.
57. Zapka C., et al. Open refillable bulk soap dispensers in public restrooms: a public health risk? *137th Annual Meeting*, The American Public Health Association, Philadelphia, PA, November 7–11, 2009.
58. Lorenz L, et al. Evaluation of contaminated bulk soap dispensers for biofilm bacteria: comparison of two methods of analysis and effectiveness of dispenser washing procedures. *5th ASM Conference on Biofilms*, Cancun, Mexico, November 15–19, 2009.
59. Ashkin S. Green soap and hand sanitizer options. *Contracting Magazine*, Trade Press Media Group, cleanlink.com, October 19, 2009.
60. Chattman M, Maxwell S, Gerba CP. Bacterial contamination of liquid hand soaps. *ASM 107th General Meeting*, Toronto, ON, Canada, May 21–25, 2007.
61. Cleanlink News, Study: bulk soap in schools contaminated. Trade Press Media Group, cleanlink.com, February 16, 2010.
62. Weber DJ, Rutala WA, Sickbert-Bennett EE. Minireview: outbreaks associated with contaminated antiseptics and disinfectants. *Antimicrob. Agents Chemother.* 2007;51:4217–4224.

63. Tran TT, Hitchins AD. Microbial survey of shared-use cosmetic test kits available to the public. *J. Ind. Microbiol.* 1994;13:389–391.
64. Campana R, et al. Microbiological study of cosmetic products during their use by consumers: health risk and efficacy of preservative systems. *Lett. Appl. Microbiol.* 2006;43:301–306.
65. Behravan J, Bazzaz F, Malaekeh P. Survey of bacteriological contamination of cosmetic creams in Iran. *Int. J. Dermatol.* 2000;44:482–485.
66. Bird K. Alcan creates rolling sponge applicator for L'Oreal. CosmeticsDesign.com, *Decision News Media,* September 22, 2009.
67. United States Pharmacopeia Convention, USP 30-NF25, Rockville, MD, 2007.
68. Yablonski JI, Mancuso SE. Preservation of "atypical" cosmetic product systems. *Cosmet. Toilet.* 2002;117(4):31–40.
69. Preservation Subcommittee of the, CTFA. A study of the use of rechallenge in preservation testing of cosmetics. *CTFA Cosmet. J.* 1981;13:19–22.
70. ANSI/AAMI/ISO 11137-1, 11137-2, 11137-3, 2006.
71. Standard test method for determining the activity of incorporated antimicrobial agent(s) in polymeric or hydrophobic materials, ASTM E2180-07, American Society for Testing and Materials, West Conshohocken, PA.
72. Brannan DK, Dille JC. Type of closure prevents microbial contamination of cosmetic during use. *Appl. Environ. Microbiol.* 1990;56:1476–1479.
73. Bird K. New packaging closure cuts out need for preservatives. CosmeticsDesign.com, *Decision News Media,* February 2, 2010.
74. Determination of Preservation Efficacy in Nonwoven Substrate Products, CTFA Microbiology Guidelines. The Cosmetic, Toiletries and Fragrances Association, Washington, DC, 2007.
75. M-5 Methods for Preservation Testing of Nonwoven Substrate Personal Care Products, CTFA Microbiology Guidelines. The Cosmetic, Toiletries and Fragrances Association, Washington, DC, 2007.
76. Bird K. MWV expands its airless range as trend gathers pace. CosmeticsDesign.com, *Decision News Media,* May 27, 2010.
77. Khardori N, Yassien M. Biofilms in device-related infections. *J. Ind. Microbial.* 1995;15(3):141–147.

PART IV

USE OF NATURAL INGREDIENTS

CHAPTER 12

FORMULATING COSMETICS WITH NATURAL OILS, FATS, BUTTERS, AND WAXES

Mark Garrison[1] *and Nava Dayan*[2]

[1]*Avon Products, Suffern, NY*
[2]*Lipo Chemicals Inc., Paterson, NJ*

12.1 INTRODUCTION

Formulating with natural ingredients presents challenges that may differ when compared to the use of synthetic ingredients. In many cases, synthetic ingredients are designed to eliminate certain limitations associated with natural ingredients. For example, when formulating with natural ingredients, the formulator may experience ingredient instabilities and aesthetic limitations, with the individual ingredients and in the finished product. These may include crystallization such as with natural triglycerides, oxidation with unsaturated compounds, development of undesired colors or odors in raw ingredients and in formulations, and incompatibilities between natural and synthetic ingredients. This chapter focuses on the use of fats, oils, and waxes in cosmetic and personal care products. These are widely used in almost every formulation. The aim of this chapter is to provide the formulator with tools to overcome (and ideally prevent) these challenges by trying to analyze and understand the causes resulting from the chemistry involved. Ideally one will be able to observe the structure of a natural ingredient, or one that is naturally derived, and therefore know what will be the best approach in formulating with it. A thorough understanding of the chemistry of natural ingredients, and the chemistry involved in their instabilities, should lead to better formulations while minimizing development obstacles.

12.2 NATURAL VERSUS NATURALLY DERIVED

The definition of a natural ingredient versus an ingredient that is naturally derived is not always clear. Bench chemists will probably view the question differently from a marketing professional or a regulatory expert. Professionals in the cosmetic/

Formulating, Packaging, and Marketing of Natural Cosmetic Products, First Edition.
Edited by Nava Dayan and Lambros Kromidas.

personal care industry may have a perception different from consumers. Further confusion results from the existence of terms such as green, organic, certified organic, and renewable that may not be fully defined or recognized by regulatory authorities. A compound may be considered "green," for example, if it is derived from a renewable source and is biodegradable, yet it may still not be "natural." From a purely scientific standpoint, an argument can be made that all ingredients are natural or naturally derived. Even the "synthetic" monomers that form polymers may be considered as naturally derived because they are built of molecules that are part of nature, such as H_2, CO_2, CO, and water.

For purposes of describing the chemistry and properties of natural ingredients, if a compound is found in nature, it is obviously natural. If this substance is extracted from its source, but not altered chemically, it can still be defined as natural. Employing this criterion, oils and extracts that are distilled, pressed, or extracted by any physical means (even with solvents) are considered natural. A natural butter that is fractionated into a lower melting fraction or a higher melting fraction is considered natural. This is because it did not undergo chemical modification but only a physical separation. Any composition extracted out of a fruit, seed, or plant part and even a compound derived from an animal (although most of the cosmetic and personal care industry does not advocate that) can be considered as natural.

The criteria for "naturally derived" are somewhat fuzzy and we leave it to the regulators to clarify. In our case, where a molecule was modified from its natural state, but keeps a very similar chemistry to its original form, it can be considered naturally derived. For example, when a double bond is hydrogenated or a product undergoes short organic synthesis, where at least one of the starting materials is natural, its chemistry is not significantly modified. Of the latter, examples would be preparing esters that do not exist in nature by starting with fatty acids and alcohols that are naturally derived. Some ingredients that are naturally derived are clearly closer to the natural state than others, and therefore the number of steps in the synthetic modification should be kept to a minimum. Also, bear in mind as instrumentation and detection limits improve, chemicals we think of as unnatural may be found to exist naturally in small quantities. The broad scope of this chapter is to assist formulators in using oils, fats, butters, and waxes that are considered natural and naturally derived when composing natural cosmetic products.

12.3 NATURAL LIPIDS

Lipids are hydrophobic substances, soluble in organic solvents but not in water, that are principal components of living cells. Fats, waxes, and oils, fatty acids, sterols, and triglycerides are all lipids. Lipids can be pure substances or complex mixtures [1].

12.3.1 Oils, Fats, and Butters

Natural fats and oils are primarily composed of triglycerides. The usual distinction between oil and a fat is its physical state at room temperature (RT). While an oil is often a liquid, a fat is a solid. Although fats are sometimes called waxes, more accurately, waxes are lipids that are not triglycerides, consisting primarily of fatty esters, sterols, and even hydrocarbons [2]. The difference in composition stems from fats and oils being primarily energy reserves in nature, while waxes tend to serve other purposes in nature, such as functioning as barriers against water loss on leaves. The terms may be confusing because in everyday use cosmetic chemists usually refer to the solids as waxes and the liquids as oils. Butters are substances with a consistency in between, soft or low melting point solids. The common butters are triglycerides [3].

12.3.1.1 Triglycerides Natural oils and fats are primarily triglycerides. Triglycerides are triesters of glycerin (also known as glycerol) with three fatty acids. Below is a typical structure of a triglyceride. Triglycerides are also called triacylglycerols. The three fatty acids of the triglyceride can be similar or different (Figure 12.1). In cases where the four groups bonded to the central carbon are all different, the carbon is chiral and the triglyceride can exist as optical isomers (mirror images).

Since the esterification reaction is reversible, water may be added to cleave the ester bond back into its glycerin and fatty acid components. This is called hydrolysis. In general, the hydrolysis of triglycerides is not actually a common problem when these are formulated into aqueous systems (such as emulsions). There are a few reasons for this. First, the solubility of the triglycerides in water (and also of the water in the triglycerides) is extremely low. Reactions occur fastest when compounds are totally miscible. The second reason is that the hydrolysis is accelerated by acid or base catalysis, so the reaction is fastest at low or high pH, and most cosmetic products are not formulated at extreme pH. Hydrolysis of triglycerides is also called saponification.

Fatty Acid Composition of Natural Oils/Butters/Fats In natural triglycerides, the fatty acids may be the same or different, saturated or

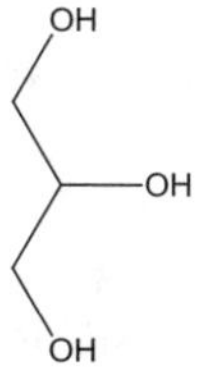

Glycerin

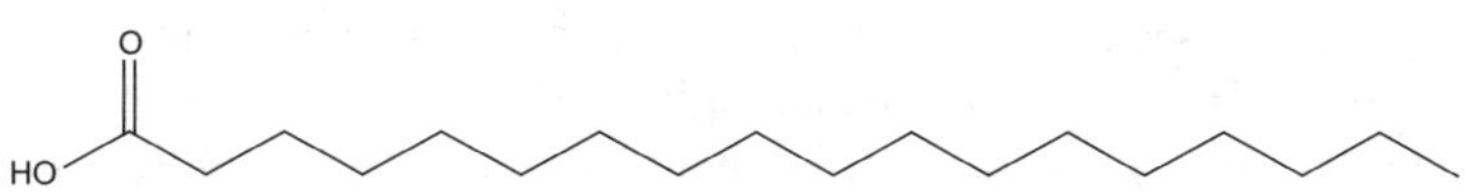

Typical fatty acid (in this case stearic acid)

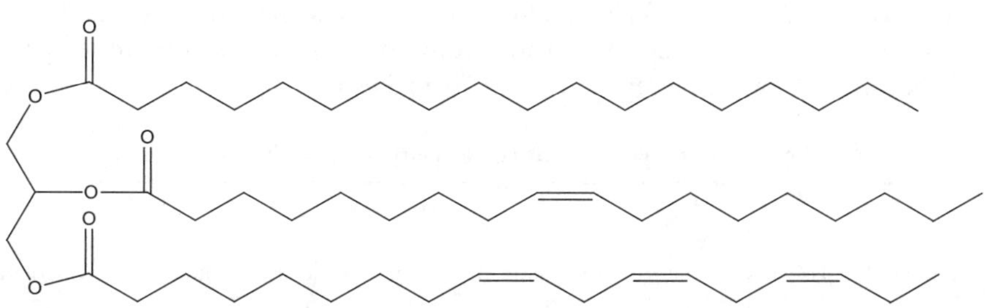

Typical triglyceride, and in this case composed of esters of stearic, oleic, and linolenic acids.

Figure 12.1 Triglyceride structure.

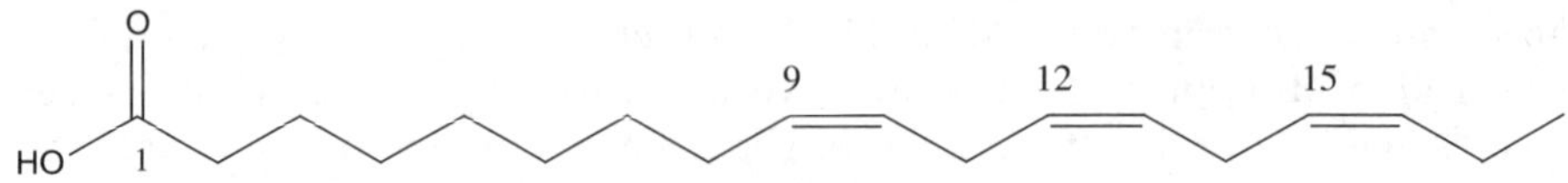

Figure 12.2 Fatty acid nomenclature and numbering in terms of its structure. It is named all-*cis*-9,12,15-octadecatrienoic acid (common name linolenic acid).

unsaturated. To understand the properties of an oil or fat, we first need to learn the properties of the fatty acids. There are several systems of nomenclature for fatty acids and the numbering conventions vary in the various nomenclatures [4]. For this discussion, we will use either common names or the numbering system illustrated with linolenic acid (Figure 12.2). Fatty acids in oils can be analyzed by first converting them to their methyl esters and then by different chromatographic techniques such as high-pressure liquid chromatography (HPLC) or gas chromatography (GC) [5, 6].

Carbon Number Since natural oils are usually characterized by their fatty acid composition, a useful parameter to know is the "carbon number" of the fatty acid composition [7]. This is a weighted "average fatty acid" length. For example, stearic acid, having 18 carbons, has a carbon number of 18. Myristic acid, having 14 carbons, has a carbon number of 14. In a 50:50 mixture of these (by weight), the carbon number would be 16, that is, $(18 \times 0.5) + (14 \times 0.5) = 16$. It is important to realize that many mixtures of different fatty acids will provide the same carbon number; for example, 100% palmitic acid would also have a carbon number of 16 as would a mixture of 25% stearic acid, 50% palmitic acid, and 25% myristic acid. Also, the carbon number does not consider variables such as unsaturation. For example, oleic acid has the same carbon number as stearic acid, but has very different properties. Thus, the information derived from determining the carbon number is limited, but when combined with other information, it can be useful, helping to explain how the fatty acid composition influences the properties of the triglycerides and ultimately the oil (or butter or fat) as a whole.

Table 12.1 lists triglycerides consisting of saturated fatty acids and illustrates that as the fatty acids of the triglycerides increase in chain length, the melting point of the triglycerides increases [8–10]. This mirrors the melting points of the individual fatty acids as seen in Table 12.2 [8–10]. The melting point values have been rounded for these examples. Also note that triglycerides can exhibit complex melting behavior and different crystal forms. (The melting points given here are for the more stable crystal form.) The reason for this is that the total intermolecular forces of attraction along the chain are greater as chain length increases. Thus, the fatty acids, and also their corresponding triglycerides, transform from liquids to solids at room temperature as the carbon number increases. Chart 12.1 makes it easy to observe this trend.

TABLE 12.1 Melting Points of Triglycerides of Linear Fatty Acids—Effect of Chain Length

Fatty acid	C8	C10	C12	C14	C16	C18	C20	C22
Name	Tricaprylin	Tricaprin	Trilaurin	Trimyristin	Tripalmitin	Tristearin	Triarachidin	Tribehenin
Melting point (°C)	10	33	45	56–57	66	72	75–78	83

TABLE 12.2 Melting Points of Simple Saturated Fatty Acids

Fatty acid	C8	C10	C12	C14	C16	C18	C20	C22
Saturated	Caprylic	Capric	Lauric	Myristic	Palmitic	Stearic	Arachidic	Behenic
Melting point (°C)	16	32	44	54	63	70	75	80

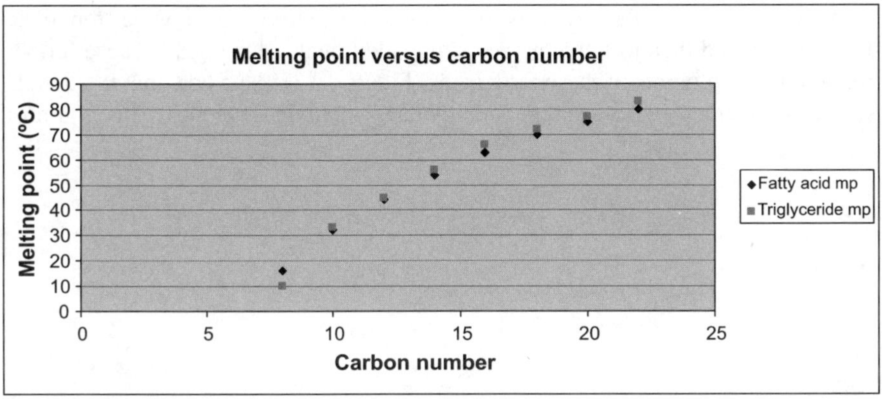

Chart 12.1 Fatty acid melting points versus carbon number.

TABLE 12.3 Melting Points of C18 Triglycerides—Effect of *cis* Double Bond Position

Double bond position	Name	Melting point
None	Tristearin	72
At C9	Triolein	−4
At C6	Tripetroselinin	17

Unsaturation Table 12.3 shows the effect of unsaturation on the melting point of triglycerides [8–11]. Generally, as unsaturation is introduced into the chain, there is a decrease in melting point. Comparing stearic acid to oleic acid, the corresponding triglyceride shows a marked reduction in melting point, from about 72°C to about −5°C. This mirrors the change in the individual fatty acid melting points shown in Table 12.4 [8–11].

TABLE 12.4 Melting Points of C18 Linear Fatty Acids—Effects of Unsaturation

Number of double bonds	Position	Name	Melting point
None	NA	Stearic	70
1	C9	Oleic	13
2	C9, C12	Linoleic	−5
3	C9, C12, C15	Linolenic	−11
1 (*trans*)	C9	Elaidic	44
1	C6	Petroselinic	30

The explanation for this resides in the packing of the molecules into a lattice-like structure. Natural fatty acids, when they are unsaturated, usually have *cis* double bonds. These bonds introduce a "kink" into the molecule's shape. The kink is rigid, so that particular bond can no longer rotate, and it turns the rest of the molecule off in a different direction. As a result, the packing between molecules is less efficient, so the intermolecular forces of attraction are weaker. It will therefore require less energy to disrupt these weaker forces, which manifests itself in a lower melting point.

As we consider the fatty acids with unsaturation, in the series moving from oleic acid to linoleic, and then to linolenic acids, the additional double bonds cause further decrease in melting points of these fatty acids (Figure 12.3). Also note that the double bonds are not conjugated but are separated by CH_2 (methylene) groups. This is almost

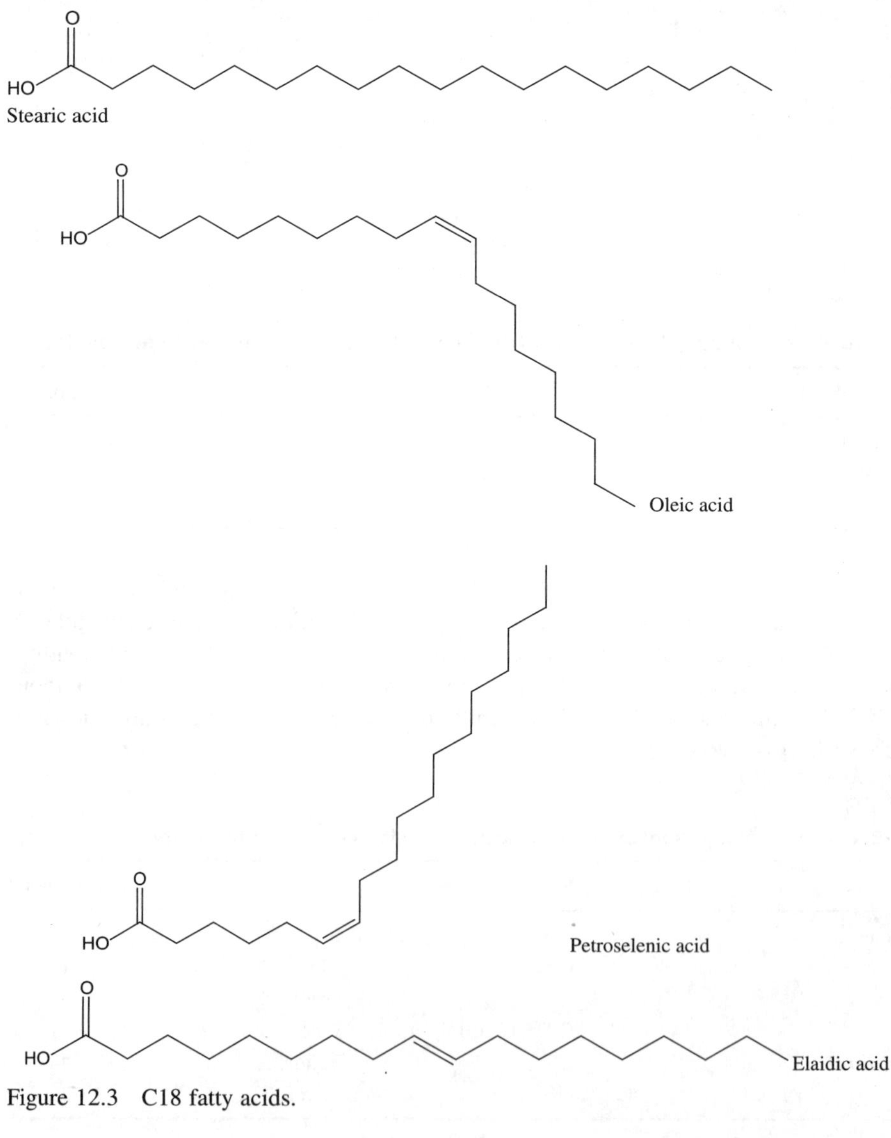

Figure 12.3 C18 fatty acids.

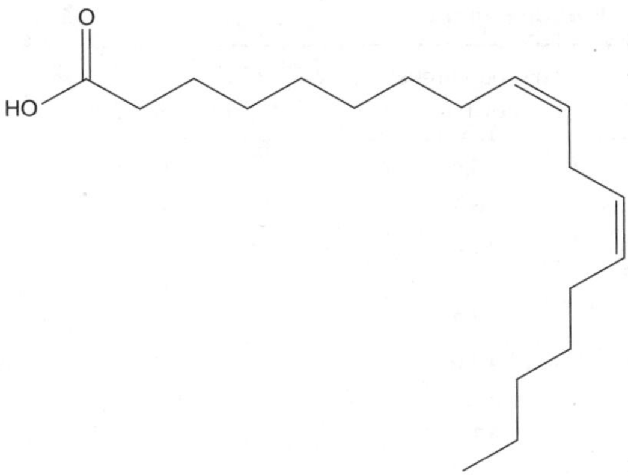

Linoleic acid

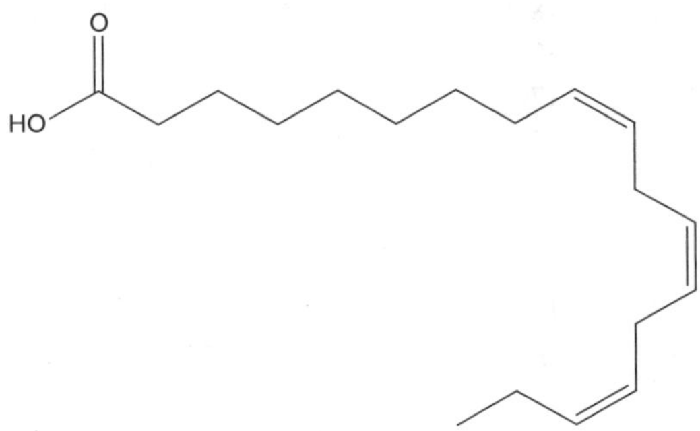

Linolenic acid

Figure 12.3 (*Continued*)

always the case with natural fatty acids (a couple of interesting exceptions to this are discussed later) [12].

The position of the double bond influences the melting point as well. Comparing oleic acid, where the double bond is at C9, to petroselenic acid, where it is at C6, it is found that the melting point rises from 13 to 30°C. This is because the "kink' has now moved to a more favorable spot, and the molecules are packed tighter. Finally, when a *trans* double bond is present at C9, in elaidic acid, versus the C9 *cis* bond in oleic acid, the melting point rises substantially, from 13 to 44°C. This is because the *trans* double bond, although still rigid, introduces less of a kink than a *cis* bond and does not interfere as much with optimum packing. The melting point however is still not as high as the saturated fatty acid, stearic acid, which having free rotation about all its carbon–carbon bonds can assume the best packing of all of these.

Iodine Value A common measurement of the unsaturation in a triglyceride is known as the iodine value. Iodine (I_2) reacts with the double bonds, and the amount of iodine taken up by a particular triglyceride is an indication of the degree of unsaturation. The iodine value will vary from lot to lot of oil as the composition of the extracted oil varies because of factors such as season, location, and climate [13]. Thus, typical values reported are often ranges. The result is reported in grams of I_2 per 100 g of the substance tested. Theoretical iodine values can also be calculated. For example, measuring pure tristearin and trilaurin that have no double bonds, the iodine value would be zero. For pure triolein, the calculated iodine value would be 86. Table 12.5

TABLE 12.5 Iodine Value Versus Carbon Number

	Carbon number (of major fatty acid)	Iodine value (of total mixture)
Bayberry wax	15.6	3
Japan wax	16.3	6
Coconut oil	12.8	8
Babassu oil	13.4	15
Palm kernel oil	13.3	19
Kokum butter	18.0	31
Illipe butter	17.7	35
Cocoa butter	17.5	37
Milk fat (also known as butter)	15.5	39
Cupuacu butter	18.2	40
Mango butter	17.5	43
Palm oil	17.1	50
Palm butter	17.0	52
Shea butter	17.9	59
Papaya seed oil	17.7	70
Avocado oil	17.6	75
Marula oil	17.6	75
Moringa oil	18.0	75
Olive oil	17.8	84
Castor oil	18.0	85
Hazelnut oil	17.9	86
Meadowfoam seed oil	20.5	95
Peanut oil	18.0	98
Rapeseed oil (high erucic)	20.6	100
Apricot kernel oil	17.9	102
Sweet almond oil	17.9	102
Rice bran oil	17.6	105
Cottonseed oil	17.6	108
Sesame oil	17.8	110
Pumpkin seed oil	17.7	113
Corn oil	17.8	123
Soybean oil	17.9	130
Sunflower seed oil	17.9	130
Tall oil	18.0	130
Wheat germ oil	17.7	130
Grape seed oil	17.8	135
Borage	17.8	147
Walnut oil	17.9	150
Evening primrose	17.9	152
Raspberry seed oil	18.0	161
Cranberry seed oil	18.0	164
Tung oil	17.9	170
Linseed oil	17.8	190

describes a few triglycerides and their iodine values [14]. Measured iodine values may differ from calculated values and from each other, especially if different methods are used. The most common method is the Wijs method. The iodine value is most useful as a quality control measure for natural oils, and reported values will vary. Like the carbon number, it provides useful information, and in this case on the degree of unsaturation, but here as well there will be many different triglyceride combinations that would give the same iodine value. Also, the method does not accurately measure conjugated double bonds, giving lower values than expected [15].

Triglyceride Structure and Physical Properties Data are available for the composition of triglycerides, their fatty acid composition, and carbon number, as well as iodine value for many oils and fats [16, 17]. The examples given so far illustrate the effects of chain length (expressed by carbon number) of the fatty acids that compose the triglycerides and the effects of unsaturation on fatty acids and simple triglycerides. Since natural oils/butters/fats are mixtures of different triglycerides, and triglycerides themselves are usually mixed (i.e., composed of different fatty acids, rather than of the same fatty acids), the analysis of the influence on melting point and consistency of a natural oil, butter, or fat is extremely complex. Plotting the iodine value versus the carbon number of a variety of fats and oils can be helpful and can illustrate what nature has to offer.

Iodine Value of Natural Oils/Butters/Fats Versus Carbon Number The carbon number of fatty acid for most of the natural oils falls around 18. With two competing factors, increasing carbon number tending to raise the melting point and unsaturation (as indicated by iodine value) tending to lower it, it is interesting to see where each effect dominates. Starting at the lower left side of the Chart 12.2, in yellow are coconut, babassu, and palm kernel "oils." These will most likely not be in a liquid form at room temperature. The melting point of coconut oil is low due to the low carbon number, but not very broad because of the high proportion of one

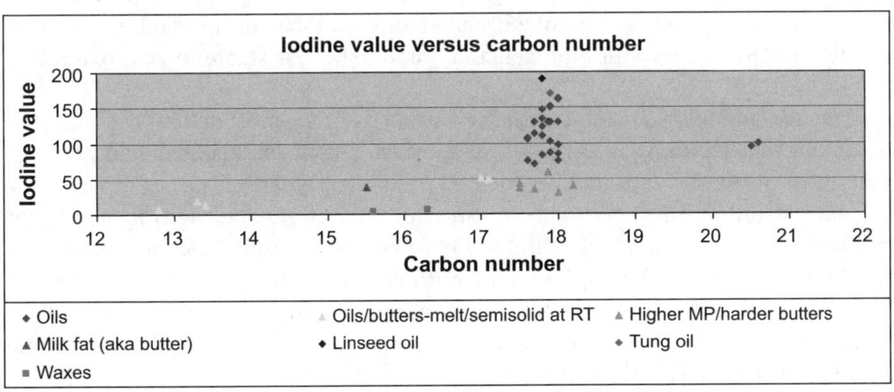

Chart 12.2 Iodine value versus carbon number of natural oils, butters, and fats. (*See the color version of this chart in Color Plate section.*)

acid, lauric acid, comprising its triglycerides. So although the iodine values of these three are low, and their carbon numbers are also low, these will melt at around room temperature (~20–25°C). As a point of comparison milk fat (butter) is included on the plot as the red triangle, a compound we are all familiar with in terms of softening point and consistency. Palm oil and palm butter are positioned around carbon number 17. Here, again represented by the yellow triangles, increasing carbon number has been offset by increased unsaturation, as evidenced by an iodine value of about 50. These again are compounds that melt or soften at around room temperature. To the right of these, with a small increase in carbon number, more solid, harder butters are situated. Examples are shea butter and cocoa butter. These are clustered at iodine value of about 60 and below and carbon numbers of about 18. At room temperature, these are hard to semisolid, but soften or melt at around body temperature. When the iodine value is higher, it becomes the dominant factor, and compounds on the graph are oils (marked in blue). There are, in fact, many combinations of fatty acids that can give the same iodine values, and the same carbon number, but will behave differently. This is because there are additional factors influencing the melting point of the compound, such as the presence of sterols and terpenes (although usually at low levels). The chart, therefore, provides only the general trend. It is also interesting to note that a few compounds are exceptions to the general trends. For example, rapeseed and meadowfoam seed oils, with higher carbon numbers and iodine values around 100, are liquids at room temperature. Finally, along the bottom of the chart, with carbon numbers around 16 and extremely low iodine values, there are bayberry and Japan waxes (technically fats as they are triglycerides, so it is legitimate to include them on this chart) and definitely solids at room temperature. Linseed oil and tung oils are "drying oils" that are high in unsaturated fats and may oxidize and polymerize into a film. The iodine value for tung oil illustrates a limitation of the iodine value method, as it is lower than expected, since tung oil contains mostly conjugated double bonds that the method does not detect well [18].

The choice of fat to use in a formulation will depend on many factors and is usually done in coordination with marketing requirements. Generally, since saturated and also longer chain (higher carbon number) oils will tend to feel heavier when applied to skin, aesthetics and feel should be considered too. Oils that have high degrees of unsaturation generally tend to be less stable toward oxidation.

Simple and Mixed Triglycerides As stated above, most natural triglycerides are mixed, that is, composed of two or more different fatty acids. The position where each particular fatty acid is located on the triglyceride (i.e., if a particular fatty acid is on an end versus in the middle) will affect the melting point and consistency. Most often, if unsaturated fatty acids are present, they are located in the middle position of the triglyceride. With no other changes, the melting point of triglycerides will follow this order, with S representing saturation and U being unsaturation and the position of the abbreviation indicating the ends and middle positions of the triglyceride.

$$\text{SSS} > \text{SUS, SSU} > \text{SUU, USU} > \text{UUU}$$

Triglycerides with high levels of disaturates (SSU and SUS) tend to melt sharply near body temperature. An example is cocoa butter. Monosaturates (SUU and USU) are semisolids [19].

Polymorphism The triglyceride structure itself is very flexible, with a central core and three long arms extending out. There are multitudes of low-energy conformations available to such a molecule, having free rotation around every single bond, while the arms are free to stretch out in different directions. This enables the molecules to pack in several different ways. As a result of this, triglycerides exhibit complex melting behavior and usually can exist in several crystal forms, each with its own melting point. This is called polymorphism and is one of the key concerns when formulating with natural triglycerides, which are solids or semisolids (the butters and fats). An enormous amount of work has been conducted on this phenomenon, and melting points of the different crystal forms of many triglycerides, both simple and mixed, have been determined [20, 21].

Triglycerides with three identical fatty acids pack into three crystal forms, α, β, and β' with β being the most stable. The β packing is triclinic, the β' suggests orthorhombic packing, and the α packs hexagonally. Rapid cooling leads to the α form that then slowly reorganizes to the more stable forms. Mixed triglycerides pack with defects in the crystal structures. When unsaturated fatty acids are part of the triglyceride, they must pack to accommodate the bend in the alkyl chain from the *cis* double bond(s). The crystallization, remelting, and recrystallization to a more stable form can be observed using differential scanning calorimetry (DSC) [22].

Much is known about this type of behavior, especially with cocoa butter because of its role in chocolate. Cocoa butter may exhibit up to seven different polymorphic forms. Problems arise in the chocolate if cocoa butter is blended with another fat having different crystallization tendencies. The "foreign" fat can crystallize separately and a whitish coating appears on the surface, a process called blooming. Much work has been done to understand and minimize/prevent blooming in chocolate, as well as to compose triglyceride blends that will minimize blooming [22].

Issues related to polymorphism are not limited to cocoa butter or chocolate. Polymorphism can cause problems in any formula or type of product that uses natural butters, fats, or waxes. These can be anhydrous products or aqueous emulsions, sticks (such as in deodorants and antiperspirants), creams, or lotions. In sticks one can get blooming, similar to what happens in chocolate. In an emulsion, problems such as viscosity differences over time or a visible change in the look and texture of the emulsion may occur (i.e., the emulsion may no longer look smooth). The processing temperatures, cooling rates, and product temperature drop can all influence the type of crystal formed, and over time the less stable forms can and will convert to the more stable crystal forms. Manufacturing conditions should be tightly controlled, and storage and transportation conditions should avoid extreme temperatures. With a caveat of numerous exceptions, it is suggested in that β' stable fats such as hydrogenated coconut oil, palm kernel wax, and soybean wax are easier to handle and will not require special tempering regimes to become stable when compared to the β stable fats such as shea butter and illipe butter. However, the former may be more sensitive to temperature fluctuations during storage and also more sensitive to changes in the composition of the formulation [23]. Rapid cooling of an emulsion will produce more crystals and slower cooling will lead to fewer larger crystals (i.e., allows crystal growth). With β stable fats, rapid cooling followed by a tempering step (i.e., held at 25°C for a few hours) will help achieve a transformation to the most stable crystal form [24]. Of course, after subjecting a formula to a month at 40 or 50°C, where any natural butters have remelted, the "stable" crystal form is now long gone and may not quickly return when the stability samples cool back down. It is particularly important during development of the formulation to prepare and store different versions under different temperature conditions in order to see how prone the formula is to this type of change, and then conduct adequate stability testing for these different variations. The minimum testing should be 1 month of storage at room temperature, 4, 40, and 50°C and three freeze/thaw cycles. Conditions such as 37°C and also alternating cycles between 4 and 40°C can be helpful to better understand behavior in storage conditions. At times, even after passing the above described stability tests, changes will still occur in a finished product being sold. These seem to occur only slowly in real time or because of other factors. Sometimes the changes are brought about only by the unique temperature profile a product experiences in a particular warehouse in a particular country at a particular time of year. When these issues are discovered and resolved, they may lead to additional stability testing for future products. If time allows, it is always a good idea to keep stability samples long term and recheck them periodically.

To minimize polymorphism issues, natural butters can be processed or fractionated, a practice that can also minimize or eliminate blooming. Many of the natural butter suppliers have expertise in this area and can advise on how to minimize this problem. Crystal modifiers can also be added to the formulation. With a structure close to that of triglycerides, the naturally derived ingredient sorbitan tristearate is used as a crystal modifier in fat products and a bloom inhibitor in chocolate [25]. When one understands the structures and interactions of the ingredients in the formulation, one may be able to predict polymorphism problems and therefore control and prevent their occurrence (Figure 12.4).

12.3.1.2 Other Constituents of Natural Oils/Butters/Fats While natural oils, butters, and fats are primarily composed of triglycerides, they typically contain several other types of compounds. As these tend to be present at low levels, they will usually have minor impact on the formulation.

Unsaponifiables This is a general term for compounds that will not saponify, that is, hydrolyze with base. These are not triglycerides or even esters. They include plant-derived sterols, terpenes, tocopherols (vitamin E derivatives), carotenoids (vitamin A and related compounds), and others [26].

Terpenes Terpenes are compounds derived from 5-carbon branched units known as isoprene. Classic monoterpenes are composed of two such units, that is, 10 carbons. Diterpenes have 20 carbons, while sesquiterpenes have 15 carbons. Being branched and relatively low in molecular weight, these compounds tend to be present

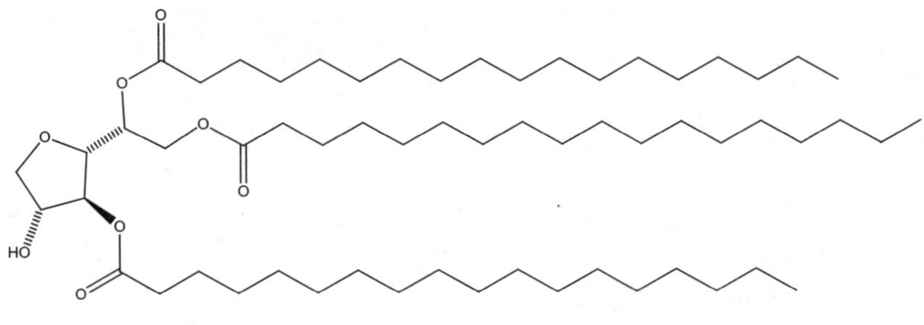

Sorbitan tristearate

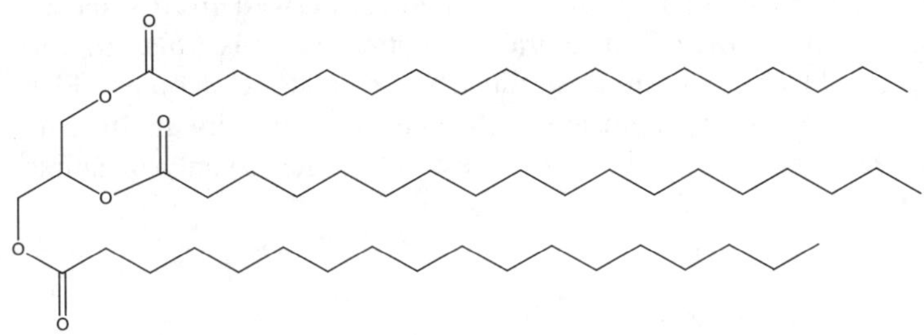

Glyceryl tristearate

Figure 12.4 Chemical structures of crystal modifiers.

as liquids. When containing ring structures, they move more toward being solids at room temperature. Squalene is an example of a triterpene, which is a component of many natural oils at low levels (0.05% or less) but present in olive oil at about 0.15–0.7% [27]. Because squalene is susceptible to oxidation, hydrogenated squalene (squalane, which can be considered "naturally derived") is frequently used in cosmetics (Figure 12.5).

Terpenes that undergo oxidation or other biochemical modification are called terpenoids. Classic examples are compounds such as menthol and camphor. Terpenes/terpenoids frequently contain a level of unsaturation, so they can be susceptible to oxidation, and being usually low molecular weight may be volatile and have some odor. Terpenes and terpenoids are usually found in high levels in essential oils rather than in seed oils [28]. They are common components in fragrances (Figure 12.6).

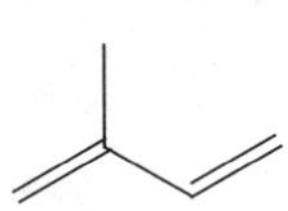

Isoprene

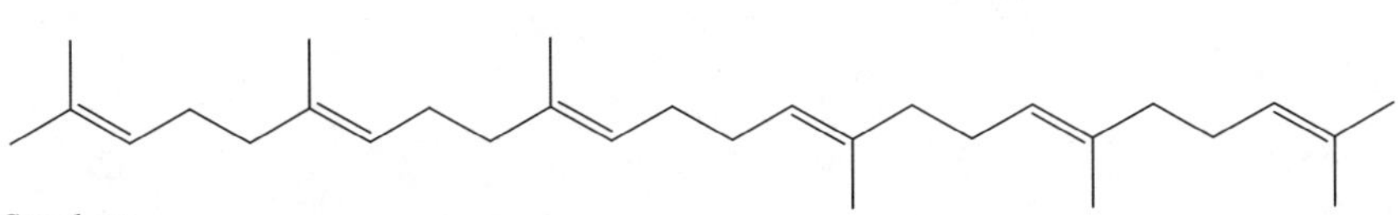

Squalene

Figure 12.5 Isoprene and squalene.

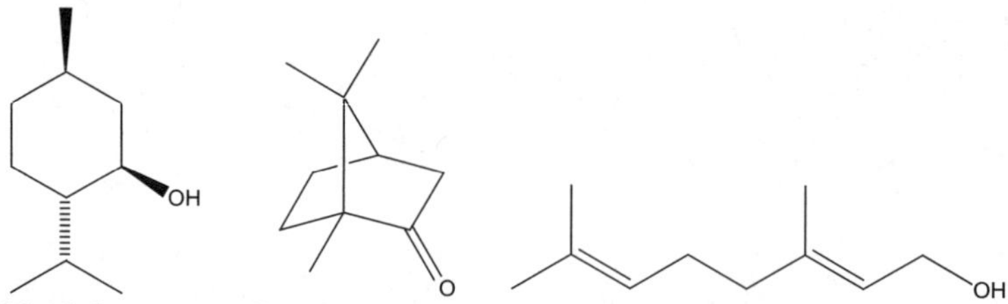

Menthol Camphor Geraniol

Figure 12.6 Chemical structure of terpenoids.

Phytosterols The plant steroids are triterpenoids, derived from squalene. Plant steroids are common constituents of natural fats, butters, and oils, usually at very low levels. Frequently, they are present as sterols (steroids that are alcohols). Plant sterols are commonly known as phytosterols. Notice the similarity in structure between the plant sterol beta sitosterol and cholesterol, which is only of animal origin [29] (Figure 12.7).

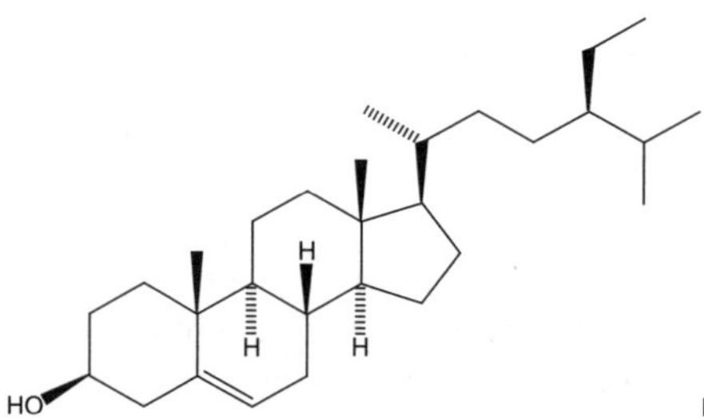

Beta sitosterol

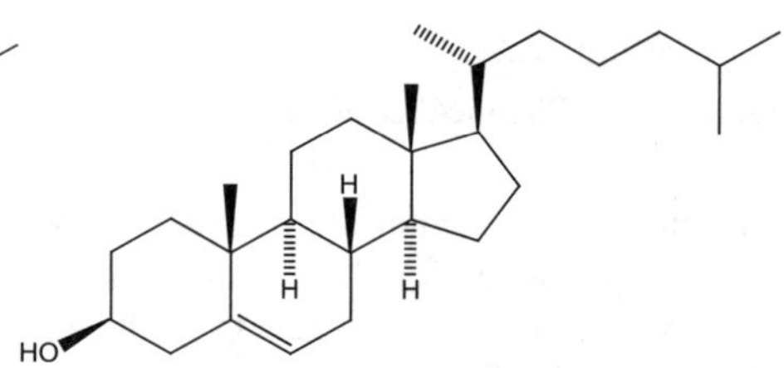

Cholesterol

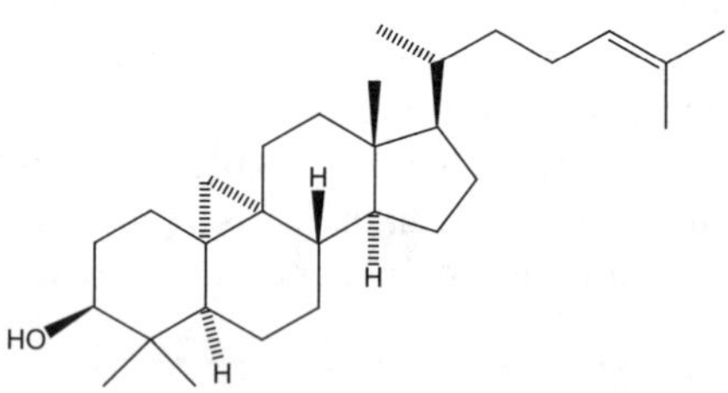

Cycloartenol

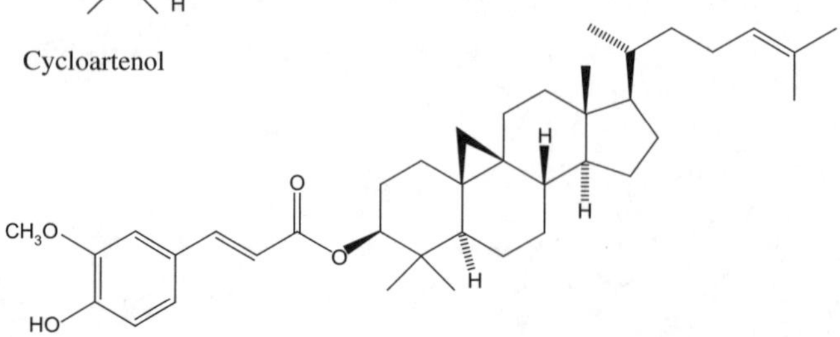

Gamma oryzanol (cycloartenol esterified with ferulic acid)

Figure 12.7 Chemical structures of sterols.

Phytosterols can be present in their free form, and as such as part of the unsaponifiables, or can also be present as esterified with fatty acids, in which case they can hydrolyze. In this form, they would not be classified as unsaponifiable. Structurally, they have a core of four fused rings in the order shown. Fused rings cannot change conformation the way linear structures can. Because of the rigidity of the structure, the melting points of sterols are higher compared to linear compounds with similar molecular weights. Typical oils contain only small percentages of phytosterols [30]. Because of their high melting points, when present in greater proportion, they can contribute to the texture and higher melting points of butters and waxes. Shea butter is an example of a butter that is rich in sterols, containing about 6% sterol esters [31]. Soy sterols and their derivatives such as polyethylene glycol (PEG) sterols are used in cosmetic products.

Phospholipids Phospholipids are glycerides esterified by fatty acids at the 1 and 2 positions and by phosphoric acid at the 3 position. As such they are saponifiable. These are commonly referred to as phosphatides or lecithins. The most common is phosphatidyl choline or lecithin. Lauric oils, such as coconut and palm kernel, contain very low levels of phospholipids. Most oils contain 0.1–0.5% of phospholipids. Corn and cottonseed oils contain almost 1% phospholipids and soybean oil up to 3% [32].

Miscellaneous Components Compounds such as free fatty acids, vitamins (vitamin E as tocopherol and tocotrienes that are related compounds), natural colors, or antioxidants may also be present in the natural oils [33]. Many of the vitamins are prone to oxidation. These will be covered later.

12.3.1.3 Unusual Oils

Castor Oil Castor oil is a triglyceride oil, but unusual for its high content of ricinoleic acid, approximately 87–90% [34]. The structure of this acid is analogous to oleic acid but with the addition of a hydroxyl group on C12. The presence of the hydroxyl group results in a marked difference in ricinoleic acid compared to oleic acid. Therefore, the viscosity of castor oil is approximately 650–800 centipoise, compared to other oils such as olive oil in which the main component is oleic acid with viscosity scale of about 63 centipoise [35]. This is explained by the hydrogen bonding of the OH groups between triglyceride molecules. The solubility of the oil also differs from other triglycerides, and it is more soluble in polar organic media and less soluble in aliphatic hydrocarbons (Figure 12.8) [36].

Castor oil is heavily used in lipsticks and is an ideal dispersant for pigments. Because of the additional functional group on the molecule, there are numerous cosmetic ingredients derived from castor oil. Waxes, emulsifiers, emollients, solubilizers, esters, and even polymers have all been derived from castor oil [37].

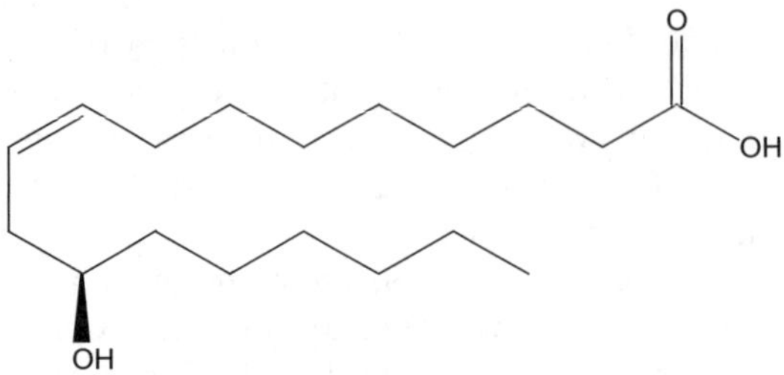

Figure 12.8 Ricinoleic acid.

12.3.1.4 Lipid Peroxidation Unsaturation also has important ramifications in terms of oxidative stability. Oxidation of oils results in rancidity. The degree of unsaturation of a molecule will largely determine its susceptibility to oxidation. While the iodine value provides information about the degree of oxidation, it does not describe the arrangement of double bonds, the presence of conjugation, and the presence of natural antioxidants, all of which may influence susceptibility to oxidation.

Lipid oxidation (also referred to as lipid peroxidation) can be triggered by three different mechanisms. These are a radical-mediated mechanism (i.e., autoxidation), a nonradical, nonenzymatic mechanism (i.e., photooxidation), and an enzyme-induced mechanism. Of these the most relevant and understood is autoxidation [38]. The oxidation reaction of oils is complex and proceeds via three steps. In this process, O_2 interacts with the oil to form peroxides that subsequently degrade into secondary oxidation products [39]. Oxidation is driven by free radicals, and during the process new radicals are formed. Radicals are molecules with an unpaired electron and are usually highly reactive and unstable. When a molecule is "attacked" by a radical, it may become a radical itself, attacking another molecule to create a chain reaction. The degree of susceptibility to oxidation will depend on its ability to form radicals and on their stability.

In natural oils, the first step of oxidation, the initiation, involves the homolytic cleavage of the C–H bond alpha to the double bond, so that the radical formed is allylic [40]. The allylic radical forms more easily because it benefits from resonance stabilization, where the unpaired electron charge is delocalized over more than one atom. The radical is still highly reactive, but forms easiest at this position. Understanding lipid oxidation and the relative reaction rates of natural oils and other natural compounds is largely about understanding this radical. Looking at linoleic acid, with two double bonds separated by a methylene unit, the radical formed could be of a few different types, and one possibility is doubly allylic, where the radical forms on the carbon between the two double bonds. This has three resonance structures giving extra stability compared to radicals formed at the other allylic carbons. Thus, in polyunsaturated fatty acids, peroxy radicals attack the bisallylic hydrogens selectively (Figure 12.9) [39].

The oxidation process can lead to the cleavage of carbon–carbon bonds, generating volitile short chain aldehydes and ketones. These oxidation products can have unpleasant "rancid" odors. When radicals combine with neighboring

Oleic acid hydrogen atom abstraction, the radical formed is allylic.
Resonance structures

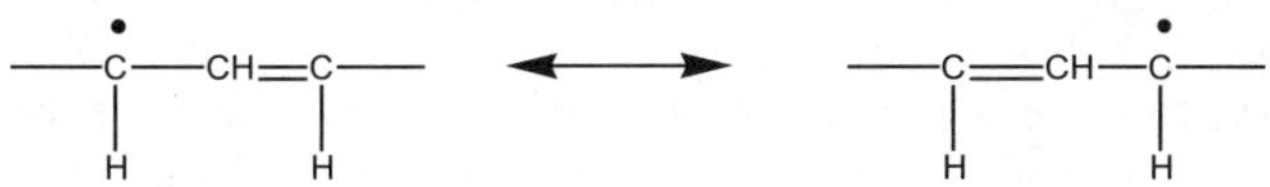

Linoleic acid hydrogen atom abstraction, the radical formed is double allylic.
Resonance structures

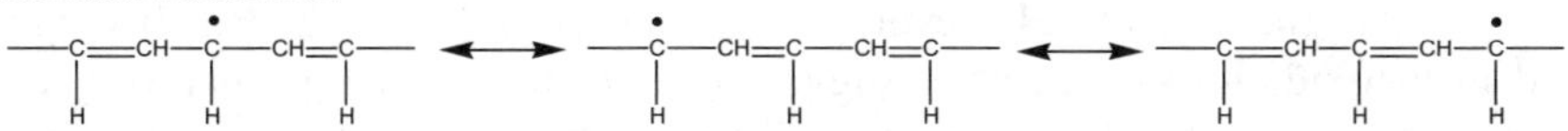

Figure 12.9 Allylic radicals resonance structures.

chains, the result is polymerization. In cases where the oil contains a high degree of unsaturation, it can form a polymer film that may be referred to as drying. Linseed oil and tung oil are two oils commonly referred to as drying oils. They form tough films and have been used for paints and coatings. The main fatty acid in linseed is linoleic and that in tung is eleostearic acid. In general, oils with an iodine value of 140 or greater are considered drying oils, values below 125 are considered nondrying, and values in between are "semidrying" (Figure 12.10) [41]. Note that in eleostearic acid, the radical formed is allylic but also conjugated with three double bonds, not just two as in linolenic acid. Also, methods for determining the iodine value may not give an accurate result for conjugated systems [18].

Relative Oxidation Rates With one double bond, oleic acid is still relatively resistant to oxidation, but linolenic acid autoxidizes about 200 times faster than stearic acid, and arachidonic about 400 times faster. The acid α-eleostearic is the fastest to get oxidized, about 1000 times faster compared to stearic acid [42]. From the data, there seems to be a *cis/trans* influence as well. In general, the more double bonds in a natural fatty acid/oil, the faster the autoxidation rate. This is of concern especially when the bonds are conjugated. Molecules with more than one isolated double bond, where the separation is larger that a methylene unit, would not be expected to show as great an effect. Data on the oxidative stability rates of specific natural oils are sometimes available from the suppliers of these oils. Oxidation reactions are not limited to fatty acids and triglycerides. Any other compounds

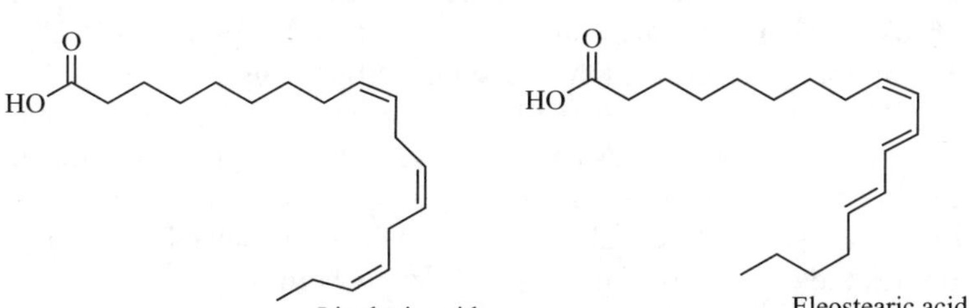

Figure 12.10 Chemical structures of linolenic and eleostearic acid.

present in the natural oils/butters/fats/waxes that are unsaturated, such as terpenes or vitamins, are also susceptible to oxidation and degradation, leading to unfavorable colors and odors.

In photoxidation, oxygen is elevated to an excited singlet state in the presence of UV light and usually a photosensitizer molecule. The singlet oxygen formed can react directly with double bonds. The reaction with singlet oxygen is not catalytic, but it is stoichiometric; that is, one molecule of oxygen reacts with one molecule of substrate via this mechanism. However, this oxidation product can then become involved in autoxidation. It has been suggested that photooxidation mainly generates hydroperoxides that break down into free radicals initiating autoxidation reactions [40].

Factors that Accelerate Oxidation Oxidation reactions are intensified with the presence of energy in the form of heat or light, the presence of peroxides, and of course by exposure to the oxygen in the air. Transition metals can catalyze these oxidation reactions. Transition metals are metals having multiple oxidation states, which means they can lose or gain an electron relatively easily, thus facilitating free radical reactions. Copper, for example, can show a significant effect at as little as 0.05 ppm [43]. The best way, therefore, to protect compounds susceptible to oxidation is to refrain from exposure to air, heat, and light. Sensitive ingredients should be kept tightly closed, in dark containers, away from sunlight, and in cool areas or refrigerated. Chelators such as citric acid can help sequester transition metals and antioxidants can be added.

Antioxidants Antioxidants can stop or slow down the free radical chain reaction by forming stable oxidized entities. Antioxidants counteract oxidation in two ways, by blocking the initiation phase and by stalling the propagation phase. In the first case, they hinder the formation of reactive oxygen species responsible for initiation. In the second case, they intercept radical oxidation propagators or indirectly participate in stopping the radical chain propagation. Antioxidants often act via mixed mechanisms that combine different types of antioxidation [40]. There are both natural and synthetic antioxidants. Since the oxidation reaction will often result in unfavorable colors or odors, the antioxidant, as it is being consumed, should not produce any color or odor. These undesired changes are frequently the case with the use of natural antioxidants, so in this regard synthetics are usually superior. Antioxidants are effective only in preventing oxidation and do not reverse its effects, so they must be added before the oxidation occurs. Below are the structures of several common antioxidants, some are vitamins and some are synthetic (Figure 12.11). The structures reflect their ability to form resonance-stabilized radicals.

Tocopherol, BHA (actually a mixture of isomers), and BHT are oil-soluble antioxidants with a phenolic OH group that can split to give H• and a phenolic radical. The phenolic radical is stabilized by resonance in the aromatic ring and also benefits from steric stabilization via the adjacent alkyl groups on the ring. When R1, R2, and R3 are methyl groups, the tocopherol shown is alpha tocopherol.

Tocopherol should not be confused with tocopheryl acetate, which does not possess antioxidant activity. The tocopherol can form a resonance-stabilized radical,

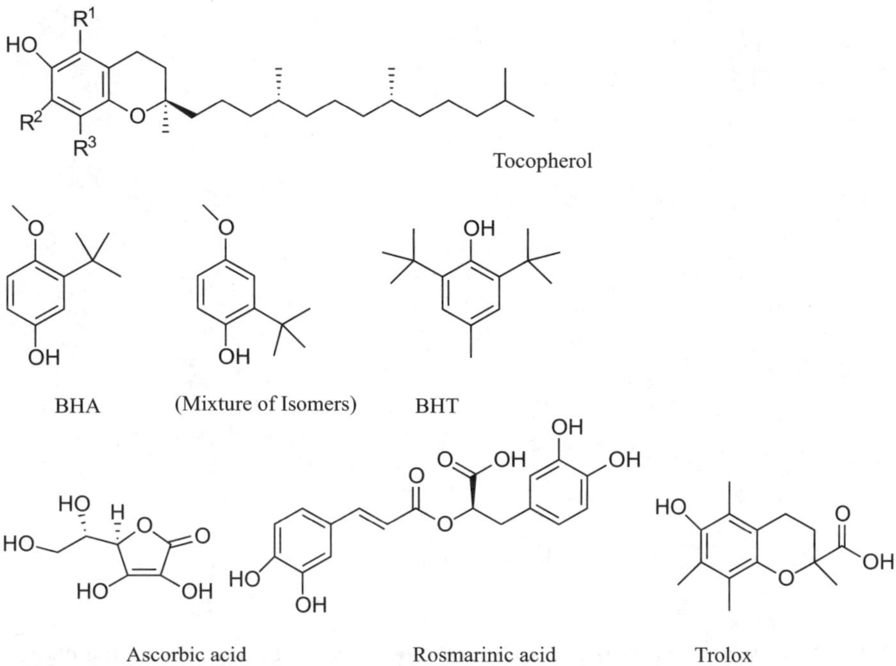

Figure 12.11 Structures of common antioxidants.

while in vitamin E acetate the business end of the molecule is tied up, and the reaction cannot occur (Figure 12.12). Oxidation of tocopherol leads to opening of the ring and formation of tocoquinone, which is red in color [44]. Also important in preventing oxidation are chelators that sequester metal ions. Copper and iron are particularly detrimental to lipids [43].

Rosmarinic acid and Trolox are water-soluble antioxidants. They can form resonance-stabilized radicals in the same manner as tocopherol and BHT. In fact, Trolox can be considered as a water-soluble version of vitamin E. Ascorbic acid (vitamin C) is a unique molecule. It has an enediol structure, which is stabilized by conjugation with the carbonyl group in the ring. When it loses a proton from the OH of carbon 3, the resulting anion is resonance stabilized, hence the ease of deprotonation and its acidity. As an antioxidant, it can form a resonance-stabilized radical, but one that is not aromatic. In addition to being an antioxidant, ascorbic acid is also a reducing agent, meaning a source of H^-, and easily oxidized. Ascorbic acid in its oxidized form is known as deydroascorbate.

Choosing the appropriate antioxidant for a system may be complex. Some considerations can be type of exposure of the system (e.g. autooxidation, photooxidation) or the type of formulation (e.g., emulsion, anhydrous, etc). Another consideration is the need to include a natural or synthetic compound. The above-mentioned natural antioxidants have been used in numerous commercial products. BHA and BHT are very effective synthetic antioxidants.

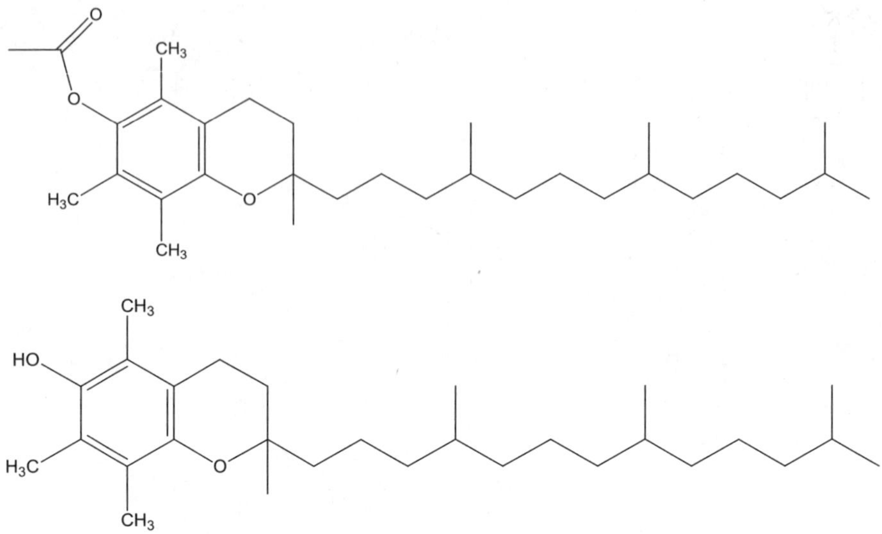

Figure 12.12 Tocopheryl acetate and tocopherol.

12.3.1.5 Choosing a Natural Oil/Butter/Fat/Wax In most cases, the choice will be based on the consumer preference, that is, what is popular with consumers based on actual or perceived benefits of the ingredients. Often one does not choose a particular natural oil because of aesthetics or cost, since many oils will have a similar feel to the consumer. One may avoid certain oils because of aesthetics. For example, castor oil provides a noticeably "heavy" feel when applied to the skin compared to other natural oils. High molecular weight and more saturated oils will feel "heavier" or "oilier. While some oils do contain natural antioxidants, this may or may not be of great value. If, for example, preventing color change is important, an oil loaded with natural tocopherol may not be your best choice. While this tocopherol will help protect the oil against rancidity, the tocopherol will produce a dark red color as it oxidizes.

12.3.2 Natural Waxes

Waxes are solids at room temperature but are not triglycerides. These are typically esters of fatty acids with fatty alcohols, sterols, terpenoids, or combinations of these [45, 46]. Waxes will have a lower proportion (or none) of the components that are liquid at room temperature, so that overall they are solid. They will tend to have higher molecular weight, more saturation, and higher rigidity of structure.

Very common in waxes are monoesters of saturated long-chain acids of C12–C30 and saturated alcohols of C20–C32, giving total chain lengths of C34–C62. In some cases, the acid is unsaturated [47]. Several "waxes" such as Japan wax and bayberry wax are actually fats, being composed of triglycerides, and they have been mentioned earlier and also included below for comparison [43, 48].

12.3.2.1 Examples of Common Natural Waxes, Composition, and Properties When formulating with these waxes, understanding their compositions is important for predicting compatibility with other ingredients. Waxes provide structure and hardness to stick products. While suitable for anhydrous products or sticks, the higher melting point waxes are really not suitable for oil in water or aesthetic emulsions. In general, they contain compounds of very high molecular weight or not chemically similar enough to the rest of the oils and fats commonly used, and usually over time will "kick out" or lead to undesirable stability or appearance changes in most emulsions. To impart viscosity or body to emulsions, naturally derived ingredients with lower melting points and more favorable aesthetics are usually more suitable.

Carnauba Carnauba wax has the highest melting point and hardness among the natural waxes. It is mainly composed of esters of C18–C30 fatty acids but has a significant amount of free fatty alcohols with 30–34 carbons. It contains some unusual components such as esterified fatty diols (~20%), hydroxylated fatty acids, (~6%), and cinnamic acid (~10%), which may be hydroxylated or methoxylated. Carnauba wax is used in products such as lipstick and mascara, where its film forming and wear properties are important, but really does not lend itself for use in systems such as oil in water emulsions. It finds use in shoe and floor polish formulations [49].

Candelilla Candelilla wax contains mostly hydrocarbons (~50%) with about 30% fatty esters, ~12–14% triterpenoid esters, and 7–9% fatty alcohols and acids. Its melting range is about 67–80°C. It is used in similar cosmetic applications as carnauba wax [49]. Its composition is quite different, however, so we would expect that it will show different compatibilities compared to carnauba wax.

Rice Bran Wax Rice bran wax consists of fatty esters containing C22 and C24 fatty acids and C24–C40 fatty alcohols. The melting point is 75–80°C. The physical properties are reportedly similar to those of carnauba wax [50].

Japan and Bayberry Waxes These are technically fats, composed of triglycerides. Japan wax has a melting point of 45–53°C and is mostly palmitic acid triglycerides. Bayberry wax contains mainly myristic, palmitic, and stearic triglycerides, with a melting point of ~45°C [43, 50].

Beeswax Beeswax has been used since ancient times. The main components of beeswax are wax esters. Esters with 36–54 carbons amount to 70–90%. Free fatty acids are 12–15% and hydrocarbons 10–15%. Beeswax is softer than carnauba or candelilla waxes and has a melting point of 64°C and a penetration of 20 dmm at 25°C versus a penetration of 2 dmm for carnauba wax and 3 dmm for candelilla wax [51]. Beeswax is often used in cosmetic emulsions. Beeswax/borax was one of the first emulsifying systems, the borax serving to form a soap with the free fatty acids. There are also derivatives of beeswax available. When applied to the skin, beeswax presents high friction sensation.

12.3.2.2 Jojoba Oil—An Unusual Natural Waxes Jojoba oil is often referred to as a liquid wax. It is not a triglyceride but contains more than 90% esters of long-chain unsaturated fatty acids with unsaturated fatty alcohols [52]. Jojoba oil is relatively stable to oxidation, even though it contains two double bonds, as they are isolated from one another and any radicals formed will be allylic rather than doubly allylic. The remainder consists of free alcohols, free acids, and sterols. Jojoba oil is widely used in cosmetics. There are numerous derivatives of jojoba available such as butters or waxes that are naturally derived [53].

12.4 SUMMARY

Many cosmetic and personal care products, whether they are emulsions or anhydrous, creams, sprays, suncare or lip products, include natural oils, butters, fats or waxes in their composition. Being natural, these ingredients can vary by source and from lot to lot. Processing can produce dramatically different ingredients from the same source, such as separating a wax and an oil fraction from the original natural butter. Natural ingredients offer some challenges when it comes to formulations.

Understanding the chemistry of these compounds plays a crucial role in understanding their solubility and compatibility with other formulation ingredients and in helping to prevent and solve product stability issues both short and long term. Oxidative stability is also frequently a challenge with natural ingredients. Their chemistry can also influence formulations aesthetics. This chapter provided information about the different oils, butters, fats and waxes available as well as their chemistry, properties, advantages and limitations. It discusses key tools in the application of different types of analyses to a day to day laboratory practice.

ACKNOWLEDGMENTS

I would like to thank my wife Felicia and my daughter Rachel for their support during the preparation of this chapter as well as Ingrid Guentert for her valuable help in locating much of the information. —M.G.

REFERENCES

1. Fridd P, editor. *Natural Ingredients in Cosmetics II*. Dorset: Micelle Press, 1996. p. 61.
2. Hasenhuettl GL. Fats and fatty oils. *Kirk-Othmer Encyclopedia of Chemical Technology*. Published online: February 18, 2005, doi: 10.1002/0471238961.0601201908011905.a01.pub2, p. 802.
3. Fridd P, editor. *Natural Ingredients in Cosmetics II*. Dorset: Micelle Press, 1996. pp. 81–82.
4. Gunstone FD, Harwood JL, Padley FB, editors. *The Lipid Handbook*, 2nd ed. London: Chapman & Hall, 1994. pp. 3–5.
5. King JW, Adams EC, Bidlingmeyer BA. Determination of free fatty acids in natural oils and alkyd reins by high performance liquid chromatography. *J. Liquid Chromatogr.* 1982;5(2):275–304.
6. Hasenhuettl GL. Fats and fatty oils. *Kirk-Othmer Encyclopedia of Chemical Technology*. Published online: 2005, doi: 10.1002/0471238961.0601201908011905.a01.pub2, p. 826.

7. O'Lenick AJ, Steinberg DC, Klein K, Lavay C. *Oils of Nature*. Carol Stream, IL: Allured Publishing, 2008. pp. 3–4.
8. Windholz M, editor. *The Merck Index*, 10th ed. Rahway, NJ: Merck & Co., Inc., 1983.
9. Weast RC, editor. *The CRC Handbook of Chemistry and Physics*, 62nd ed. Boca Raton, FL: CRC press, Inc., 1981.
10. Gunstone FD, Harwood JL, Padley FB, editors. *The Lipid Handbook*, 2nd ed. London: Chapman & Hall, 1994.
11. Gunstone FD, Padley FB, editors. *Lipid Technologies and Applications*. New York: Marcel Dekker, 1997.
12. Hasenhuettl GL. Fats and fatty oils. *Kirk-Othmer Encyclopedia of Chemical Technology*. Published online: 2005, doi: 10.1002/0471238961.0601201908011905.a01.pub2, p. 814.
13. Hasenhuettl GL. Fats and fatty oils. *Kirk-Othmer Encyclopedia of Chemical Technology*. Published online: 2005, doi: 10.1002/0471238961.0601201908011905.a01.pub2, pp. 814–815.
14. Gunstone FD, Padley FB, editors. *Lipid Technologies and Applications*. New York: Marcel Dekker, 1997. p. 27.
15. O'Lenick AJ, Steinberg DC, Klein K, Lavay C. *Oils of Nature*. Carol Stream, IL: Allured Publishing, 2008. pp. 108, 110.
16. O'Lenick AJ, Steinberg DC, Klein K, Lavay C. *Oils of Nature*. Carol Stream, IL: Allured Publishing, 2008.
17. Gunstone FD, Padley FB, editors. *Lipid Technologies and Applications*. New York: Marcel Dekker, 1997. p. 27.
18. O'Lenick AJ, Steinberg DC, Klein K, Lavay C. *Oils of Nature*. Carol Stream, IL: Allured Publishing, 2008. p. 110.
19. Hasenhuettl GL. Fats and fatty oils. *Kirk-Othmer Encyclopedia of Chemical Technology*. Published online: 2005, doi: 10.1002/0471238961.0601201908011905.a01.pub2, pp. 811–817.
20. Hasenhuettl GL. Fats and fatty oils. *Kirk-Othmer Encyclopedia of Chemical Technology*. Published online: 2005, doi: 10.1002/0471238961.0601201908011905.a01.pub2, p. 819.
21. Fridd P, editor. *Natural Ingredients in Cosmetics II*. Dorset: Micelle Press, 1996. pp. 76–77.
22. Hasenhuettl GL. Fats and fatty oils. *Kirk-Othmer Encyclopedia of Chemical Technology*. Published online: 2005, doi: 10.1002/0471238961.0601201908011905.a01.pub2, pp. 819–820.
23. Fridd P, editor. *Natural Ingredients in Cosmetics II*. Dorset: Micelle Press, 1996. pp. 77–89.
24. Fridd P, editor. *Natural Ingredients in Cosmetics II*. Dorset: Micelle Press, 1996. p. 85.
25. Gunstone FD, Padley FB, editors. *Lipid Technologies and Applications*. New York: Marcel Dekker, 1997. p. 529.
26. Gunstone FD, Padley FB, editors. *Lipid Technologies and Applications*. New York: Marcel Dekker, 1997. pp. 27–29.
27. Gunstone FD, Padley FB, editors. *Lipid Technologies and Applications*. New York: Marcel Dekker, 1997. p. 37.
28. O'Lenick AJ, Steinberg DC, Klein K, Lavay C. *Oils of Nature*. Carol Stream, IL: Allured Publishing, 2008. p. 7.
29. Gunstone FD, Padley FB, editors. *Lipid Technologies and Applications*. New York: Marcel Dekker, 1997. p. 29.
30. Gunstone FD, Padley FB, editors. *Lipid Technologies and Applications*. New York: Marcel Dekker, 1997. pp. 29–30.
31. O'Lenick AJ, Steinberg DC, Klein K, Lavay C. *Oils of Nature*. Carol Stream, IL: Allured Publishing, 2008. p. 36.
32. Hasenhuettl GL. Fats and fatty oils. *Kirk-Othmer Encyclopedia of Chemical Technology*. Published online: 2005, doi: 10.1002/0471238961.0601201908011905.a01.pub2, p. 804.
33. Hasenhuettl GL. Fats and fatty oils. *Kirk-Othmer Encyclopedia of Chemical Technology*. Published online: 2005, doi: 10.1002/0471238961.0601201908011905.a01.pub2, pp. 804–805.
34. Naughton, FC. Castor oil. In: *Kirk-Othmer Encyclopedia of Chemical Technology*. Published online: December 4, 2000, doi: 10.1002/0471238961.0301192014012107.a01, pp. 1–2.
35. Abramovic H, Klofutar C. The temperature dependence of dynamic viscosity. *Acta. Chim. Slov.* 1998;45(1):69–77.

36. Naughton, FC. Castor oil. In: *Kirk-Othmer Encyclopedia of Chemical Technology*. Published online: 2000, doi: 10.1002/0471238961.0301192014012107.a01, p. 2.
37. Naughton FC. Castor oil. In: *Kirk-Othmer Encyclopedia of Chemical Technology*. Published online: 2000, doi: 10.1002/0471238961.0301192014012107.a01, pp. 7–9.
38. Laguerre M, Lecomte J, Villeneuve P. Evaluation of the ability of antioxidants to counteract lipid oxidation: existing methods, new trends and challenges. *Prog. Lipid Res.* 2007;46: 244–282.
39. Niki E. Lipid peroxidation and its inhibition: overview and perspectives. *J. Oleo. Sci.* 2001; 50(5):313–320.
40. Laguerre M, Lecomte J, Villeneuve P. Evaluation of the ability of antioxidants to counteract lipid oxidation: existing methods, new trends and challenges. *Prog. Lipid Res.* 2007;46:244–282.
41. Wicks ZW. Drying oils. *Kirk-Othmer Encyclopedia of Chemical Technology*. Published online: May 17, 2002, doi: 10.1002/0471238961.0418250923090311.a01.pub2
42. Gunstone FD, Padley FB, editors. *Lipid Technologies and Applications*. New York: Marcel Dekker, 1997. p. 184.
43. Gunstone FD, Padley FB, editors. *Lipid Technologies and Applications*. New York: Marcel Dekker, 1997. p. 186.
44. Hasenhuettl GL. Fats and fatty oils. *Kirk-Othmer Encyclopedia of Chemical Technology*. Published online: 2005, doi: 10.1002/0471238961.0601201908011905.a01.pub2, p. 805.
45. O'Lenick AJ, Steinberg DC, Klein K, Lavay C. *Oils of Nature*. Carol Stream, IL: Allured Publishing, 2008. pp. 1–2, 85–93.
46. Leray, C.Waxes. *Kirk-Othmer Encyclopedia of Chemical Technology*. Published online: September 15, 2006, doi: 10.1002/0471238961.2301240503152020.a01.pub2, p. 1.
47. Leray, C.Waxes. *Kirk-Othmer Encyclopedia of Chemical Technology*. Published online: 2006, doi: 10.1002/0471238961.2301240503152020.a01.pub2, p. 2.
48. O'Lenick AJ, Steinberg DC, Klein K, Lavay C. *Oils of Nature*. Carol Stream, IL: Allured Publishing, 2008. p. 22.
49. Leray C.Waxes. *Kirk-Othmer Encyclopedia of Chemical Technology*. Published online: 2006, doi: 10.1002/0471238961.2301240503152020.a01.pub2, p. 8.
50. Leray C.Waxes. *Kirk-Othmer Encyclopedia of Chemical Technology*. Published online: 2006, doi: 10.1002/0471238961.2301240503152020.a01.pub2, p. 9.
51. Leray C.Waxes. *Kirk-Othmer Encyclopedia of Chemical Technology*. Published online: 2006, doi: 10.1002/0471238961.2301240503152020.a01.pub2, p. 5.
52. Leray C.Waxes. *Kirk-Othmer Encyclopedia of Chemical Technology*. Published online: 2006, doi: 10.1002/0471238961.2301240503152020.a01.pub2, p. 11.
53. Harlow RD, Litchfield C, Fu HC, Reiser R. The triglyceride composition of *Myrica carolinensis* fruit coat fat. *J. Am. Oil Chem. Soc.* 1965;42(9):747–750.

CHAPTER 13

NATURAL ANTIOXIDANTS AND THEIR EFFECTS ON THE SKIN

Anne Pouillot,[1] Luigi L. Polla,[1] Philippe Tacchini,[2] Alice Neequaye,[2] Ada Polla,[3] and Barbara Polla[1]

[1]Alchimie Forever Sàrl, Geneva, Switzerland
[2]EDEL Therapeutics SA, Lausanne, Switzerland
[3]Alchimie Forever LLC, Washington, DC

13.1 INTRODUCTION

Antiaging skin care represents a large segment of the cosmetic products market. Such products typically claim to contain the ultimate "miracle in a jar," that is, a unique active molecule that maintains the youthful aspect of the skin by promoting cellular activity in the epidermis and the dermis. To the best of our knowledge, however, there is no such miracle antiaging compound: the best current approach being to combine antiaging natural antioxidants acting in synergy. "Synergy" means that the global effects of these combined natural antioxidants will be greater than the sum of each one's specific effects. In this chapter, we will develop the concept how antioxidants play an important role in the prevention of premature aging. We will also address

- why have antioxidants become major ingredients in antiaging cosmetics?
- why are natural antioxidants preferred to synthetic ones by so many brands?
- what are the best approaches to measure antioxidant concentration, efficacy, and long-term effects on the skin?
- what may we learn from the large epidemiological studies on nutrition-based antioxidants that have recently been published?
- what is the relevance of topical application of plant antioxidants (phytoantioxidants), in particular with respect to the outer layer of the skin, the stratum corneum, compared to nutritional antioxidants?

The answers we propose are based on available data and may be subject to change in the future, but they provide a basis for further investigation and pave the way for the optimal use of phytoantioxidants in skin aging prevention.

Formulating, Packaging, and Marketing of Natural Cosmetic Products, First Edition.
Edited by Nava Dayan and Lambros Kromidas.

13.2 OXIDANTS AND ANTIOXIDANTS

In this section, we will review the basic data on oxidants and their role and function in skin aging.

13.2.1 Oxidants

13.2.1.1 ROS and RNS Oxygen (O_2) is essential to the life of aerobic organisms. However, its metabolites represent a potential threat to all living organisms. Indeed, O_2 is metabolized in animal tissue by successive reductions in superoxide anion ($O_2^{\bullet-}$), hydrogen peroxide (H_2O_2), and hydroxyl radical ($^{\bullet}OH$). These different metabolites are called reactive oxygen species (ROS). ROS are either free radicals (with an unpaired electron in their outer orbital sphere) ($O_2^{\bullet-}$, $^{\bullet}OH$) or nonradical (H_2O_2, singlet oxygen ($^1O_2$)) [1]. Nitric oxide ($NO^{\bullet}$), some functions of which overlap with ROS, is synthesized from L-arginine and oxygen by enzymes called NO synthases and is part of the reactive nitrogen species (RNS) [1]. In this chapter, we will essentially focus on ROS.

13.2.1.2 Physiological Functions At low concentrations, ROS exert essential intracellular functions, as second messengers, gene regulators, and mediators for cell activation (kinases and transcription factors). They also play a key role in our body's defense against infectious organisms, a role exemplified by the fact that patients with genetic defects of NADPH oxidase, an enzyme involved in ROS production by white blood cells upon infection, are subject to major infections [2]. Furthermore, ROS are modulators of cell death, whether apoptosis or necrosis [1]. In the presence of high concentrations of $O_2^{\bullet-}$, $NO^{\bullet}$ allows the formation of peroxynitrites that alter mitochondrial membrane potential, which is key to the activation pathways leading to cellular apoptosis [3].

13.2.1.3 Oxidative Stress Oxidative stress is defined by an imbalance between ROS and antioxidants, ROS being in excess. Oxidative stress is deleterious to both cells and extracellular matrix, to nuclear and mitochondrial DNA, to membrane lipids, and to proteins. DNA damage (single-strand lesions, deletions of bases, or "cross-links" between DNA and proteins) forms the basis of UV-induced skin carcinogenesis. Lipid peroxidation affects phospholipids both structurally and functionally and results in rigid and permeable membranes. Protein alterations, whether direct or activated by proteases, are reflected in the skin by both reductions in total amount and alterations of collagen and elastin [4]. Moreover, excessive production of ROS or RNS induces mitochondrial damage, leading to a sharp decrease in ATP and cell death by necrosis [5].

13.2.1.4 UV Exposure as the Major Source of Oxidative Stress in the Skin UV exposure (180–400 nm) leads to multiple cellular damages, by generating $^1O_2$, $^{\bullet}OH$, H_2O_2, and other ROS. UVB rays (290–320 nm) are absorbed by epidermal chromophores such as melanin and urocanic acid and lead to direct molecular damages while also generating ROS. In the presence of H_2O_2,

UVB radiation leads to the formation of $^{\bullet}OH$ [6], which causes DNA damages. UVA rays (320–400 nm) penetrate more deeply in the dermis, increase the production of ROS, and contribute to long-term actinic damage. Both UVA and UVB induce the activation of a wide range of transcription factors in skin cells, including NF-κB (transcription factor involved in inflammation and cellular stress responses) [7], which in turn may increase the production of matrix metalloproteinases (MMPs), a family of enzymes that degrade collagen and elastin.

The skin is continuously affected by environmental factors and notably UV radiation [8]. In the skin, free radicals induced by UV radiation cause damage to DNA, to proteins, and destabilize the membranes of keratinocytes, leading to premature aging of the skin cells: "oxidation = aging." When exposed to UV radiation, the skin undergoes alterations resulting in inflammation, photoaging, and various skin disorders [9]. Skin photoaging is accompanied by wrinkling, loss of elasticity, increased skin fragility, and slower wound healing.

13.2.2 Antioxidants

Living tissues have a control mechanism to keep ROS in balance. When ROS are generated *in vivo*, many antioxidants come into play. Their relative importance depends upon which ROS are generated, how and where they are generated, and which target of damage is considered [1]. Our body defends itself from these phenomena via endogenous antioxidants [1, 10, 11]. However, when endogenous antioxidants become insufficient or imbalanced in defense against oxidants, exogenous antioxidants may help restore the balance.

Antioxidants inhibit the production of ROS by direct scavenging, decrease the amount of oxidants in and around our cells, prevent ROS from reaching their biological targets, limit the propagation of oxidants such as the one that occurs during lipid peroxidation, and thwart oxidative stress thereby preventing the aging phenomenon.

13.2.2.1 Endogenous Antioxidants Endogenous antioxidants are essentially enzymes that catalytically remove oxidants. Major endogenous antioxidants are superoxide dismutase, superoxide reductase, catalase, and glutathione peroxidase. These enzymes play a key role in decreasing the content of oxidants and preventing oxidative damage. Other endogenous antioxidant molecules, such as heme oxygenase, minimize the availability of oxidants. This enzyme is strongly induced by oxidative stress and removes an oxidant (heme) while generating a putative antioxidant (bilirubin that is sensible to $^{1}O_2$) and a prooxidant (iron). In addition, high ferritin levels result in an increased iron scavenging capacity that may confer increased resistance to oxidative stress [12].

The levels and composition of endogenous antioxidant molecules differ from tissue to tissue and by cell type. Both embryonic and adult stem cells express high levels of antioxidant enzymes, which decrease as the cells differentiate [13]. The endogenous antioxidant molecules are often increased after exposure to oxidants [1]. However, this "antioxidant pool" is gradually consumed, as oxidant exposure increases over time.

13.2.2.2 Exogenous Antioxidants Exogenous antioxidants include antioxidants that cannot be synthesized by our body such as vitamins, trace elements, and phytoantioxidants. Vitamin E (tocopherol) is the most powerful liposoluble antioxidant. It inhibits the peroxidation of membrane lipids. It reacts with free radicals to form the radical tocopheryl, a stable substance that stops the chain reaction of the membrane lipids. The chain reaction is propagation of free radicals: the molecules destabilized by a single electron in turn become free radicals that remove an electron from another molecule that thus becomes a radical, and so on. Tocopheryl has the ability to stop this type of chain reaction by membrane lipids. It works in conjunction with other antioxidants such as vitamin C and selenium. Vitamin C is a water-soluble vitamin and has a strong antioxidant activity that protects cells against damage by free radicals. This vitamin reacts with the tocopheryl radical to regenerate and restore vitamin E. It then becomes the ascorbyl radical, which is also relatively stable. Trace elements such as selenium are important cofactors of the activity of antioxidant enzymes.

13.2.2.3 Food-Derived Antioxidants In 1992, Serge Renaud, professor at the University of Bordeaux, proposed as "the French paradox" the apparent contradiction between the nutrition of French men and their health [14]. In the Southwest of France, food is fairly rich in fat and wine is readily consumed, yet the overall health of the population is quite good, the prevalence of cardiac infarction lower than in the United States, and life expectancy is higher than in the Northeast of France. The proposed explanation for this phenomenon is the high consumption of polyphenols in red wine consumed by the inhabitants of Southwestern France [15]. Polyphenols in red wine were found to inhibit low-density lipoprotein (LDL) oxidation *in vitro*, and it was suggested that they could exert cardioprotective effects by limiting LDL oxidation *in vivo*.

The SUVIMAX (SUpplementation en VItamines et Mineraux AntioXidants) study examined 7876 French women aged 35–60 years and 5141 men aged 45–60 years (all apparently healthy) for more than 7 years (1994–2002). Daily, they were given either a placebo or a pill containing 120 mg vitamin C, 30 mg α-tocopherol, 6 mg β-carotene, 100 μg selenium, and 20 mg zinc. This low-dose supplementation had no significant effect on incidence of cancer or cardiovascular disease for the group as a whole. However, for men only, there was a preventive effect for skin and lung cancer. It also appears that diets rich in a variety of phytoantioxidants (fruits, grains, and vegetables) are protective against several human diseases; hence the nutritional worldwide program: eat five servings of fruits and vegetables everyday!

Because the risk of oxidative stress increases with age, and because endogenous antioxidants are gradually consumed over time, prevention strategies are essential. Natural antioxidants play key roles in these strategies (Section 13.3). The skin, as a cutaneous barrier, is constantly subjected to damage from the environment. As such, it consumes its endogenous antioxidants that may be replenished by exogenous antioxidants provided topically (Section 13.4). Also, specific nutritional factors were found to upregulate endogenous antioxidants.

13.3 NATURAL ANTIOXIDANTS (PHYTOANTIOXIDANTS)

Plants suffer from oxidative stress induced by UV radiation as much as animals and humans do, but cannot protect themselves as humans do by exogenous means and have therefore developed multiple strategies and highly effective molecules to defend themselves against environmental stress. For example, edelweiss or lichens contain natural substances that absorb UVB and act as a "screen" [16, 17]. Plants contain multiple antioxidants effective in ideal combinations, the so-called phytoantioxidants, capable of both protecting their own cells and extracellular matrix against oxidative stress induced by UV radiation and of conferring protection to other organisms upon ingestion or topical application.

Most phytoantioxidants belong either to polyphenols or terpenes and form a family of multiple factors from multiple plants (Figure 13.1). Polyphenols are synthesized by plants, participate in their metabolism, and contribute to their defense against environmental stresses. Polyphenols are found in roots, stems, flowers, and leaves of all plants. They differ among themselves by molecular weight, polarity, and solubility. Polyphenols contain an —OH group attached to a benzene ring. The number of phenolic —OH groups and their relative positions are key determinants of polyphenols' antioxidant activity: these phenolic groups exert direct antioxidant effects, modulate protein phosphorylation, and inhibit lipid peroxidation by acting as chain-breaking peroxyl radical scavengers. The large family of flavonoids, stilbens, and terpenes (Table 13.1) help to prevent cellular and extracellular oxidative stress and slow the aging of the skin; the carotenoids do so more specifically by quenching singlet oxygen ($^1O_2$). More than 4000 distinct flavonoids have been identified, the most important being anthocyanidins, flavanols, isoflavones, and flavanones (Table 13.1).

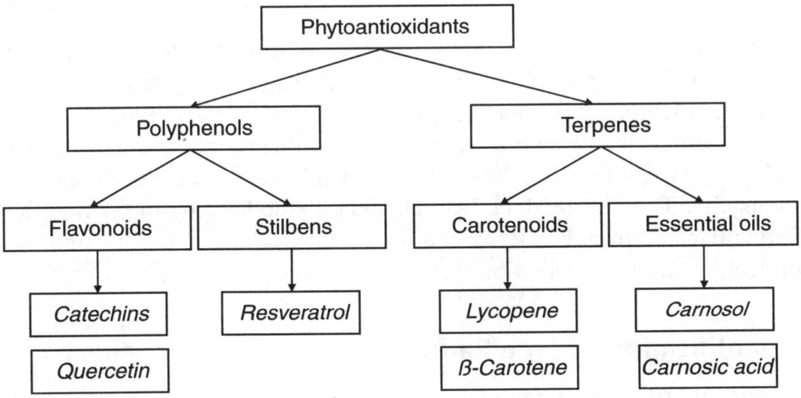

Figure 13.1 Phytoantioxidants: a family. Phytoantioxidants represent large family of molecules: polyphenols and terpenes with subcategories including flavonoids, stilbens, carotenoids, and essential oils. Some examples of phytoantioxidants described in this chapter are shown in italics (adapted from Ref. [50]).

TABLE 13.1 Three Families of Phytoantioxidants with Respective Sources and Effects

Flavonoids	Anthocyanidins	Anthocyans	Present in blueberries, basil grapes; powerful scavengers of free radicals; inhibit lipid peroxidation	Ref. [51]
	Flavanols	Catechins	Present in green tea, grape seeds, litchi; inhibit the production of H_2O_2 induced by UV; protect the endogenous antioxidant systems	Ref. [52]
	Flavonols	Quercetin	Present in green tea; blueberries, grape seeds, apple; protects the antioxidant systems in the skin	Ref. [44]
	Isoflavones	Genistein	Present in gingko biloba, soja; increases the activities of endogenous antioxidants	Refs [52, 53]
	Flavanones	Silymarin	Present in milk thistle; inhibits lipid peroxidation; promotes cell regeneration	Ref. [54]
Stilbens	Resveratrol		Present in grapes, berries; inhibits the production of H_2O_2 and lipid peroxidation	Ref. [54]
Terpenes	Carotenoids	Lycopene	Neutralizes $^1O_2$; scavenges the lipid free radicals; reduces lipid peroxidation	Ref. [55]

13.4 MEASURING ANTIOXIDANT CAPACITY

Phytoantioxidants are comprised of thousands of molecules with different and complementary antioxidant properties, including free radical scavengers, antioxidant defense mechanism elicitors, iron chelators, light and UV physical screens, direct or indirect repair systems. One of their advantages lies in their combined antioxidant properties. Here, we will focus on the methods available for determining their antioxidant capacity. This can be done by analytical methods or by measuring the effects of antioxidants, either directly or indirectly, for example, by analyzing the degree of protection from UV radiation they induce.

13.4.1 Quantification of Free Radicals

As phytoantioxidants scavenge free radicals, the quantification of free radicals can be used for monitoring their protective properties. Such techniques are based on detection of unpaired electrons by spectroscopy. The direct measurement of free radicals is difficult because their half-life is extremely short. $^{\bullet}$OH, for example, has a half-life of 10^{-6} s and a diffusion distance of 10 nm. Electron spin resonance

(ESR), however, enables the direct measurement of radical generation; the use of cold temperatures may contribute to more effective measurements [18]. The direct detection of free radical formation in human and animal skin following exposure to UV radiation has been obtained by low-temperature (−196°C) ESR spectroscopy *in vitro*. These approaches are not used extensively in routine measurements because of their complexity and cost.

13.4.2 Quantification of Damage Markers

The protective role of phytoantioxidants can also be measured by using different techniques of quantification of specific end products resulting from the interactions of free radicals with target molecules such as DNA, lipids, metabolite intermediates, or reporter molecules. Once the quantification of the damage markers is established, the extent of protection provided by antioxidants can be assessed. However, such assays require the use of heavy equipment in a laboratory, as well as expert technicians, are essentially used for indirect measurement of oral uptake of antioxidant-rich food, antioxidant-rich supplements, and/or exposure to hazardous chemicals, and include the following:

- **The 8-hydroxy-2′-deoxyguanosine (8OHdG) assay**. In the case of DNA oxidative stress, guanosine is preferentially oxidized and is used as a biomarker. 8OHdG is detected by standard methods such as high-performance chromatography (HPLC). For example, dermal uptake of polycyclic aromatic hydrocarbons containing oil was shown to increase the urinary production of 8OHdG [19].
- **The Comet assay**. This assay is based on gel electrophoresis of DNA molecules. Intact and damaged DNA display different migrating properties. Damaged DNA displays a "comet"-like trail and the higher the damage the longer the trail. Modifications of the Comet assay have been developed so that the DNA repair capacity can also be monitored. This assay has been used to monitor a variety of conditions that modify DNA, including dietary protective factors [20, 21]. Morley et al. used the Comet assay to investigate green tea's photoprotective effects by comparing DNA damage induced by UV exposure in cultured human cells in the presence or absence of green tea [22].
- Thiobarbituric acid reactive substances (TBARS) and malonaldehyde (MDA) are used as biomarkers of lipid oxidation and were used for measuring the photoprotective role of a rosmarinic acid extract both *in vitro* and *in vivo*, after oral intake [23].
- The isoprostanes assay is considered as one of the most reliable methods for the evaluation of oxidative stress *in vivo* [24]. Isoprostanes are prostaglandin-like products formed *in vivo* from the peroxidation of arachidonic acid [25]. Their dosage has been successfully used to analyze photooxidative UVB-induced skin damage *in vivo*, and a linear correlation between the UVB exposure and the generation of 8-isoprostanes was found [26].

13.4.3 Quantification of Antioxidant Capacity

Phytoantioxidants include several different active principles. Some of these active principles can be purified and measured separately by standard analytical methods. When measuring the effects of an isolated compound, however, the positive or negative modulation of its activity by the biochemical environment is dismissed. Therefore, various assays have been developed to measure the global antioxidant activity, so that these interactions can be taken into account. They can be classified according to the chemical reactions involved, whether electron transfer (ET-based assays) or the hydrogen atom transfer (HAT). These assays, as discussed below, are used for the quantification of the antioxidant capacity of antioxidant-rich food, plant extracts, and biological fluids. Electrochemical-based assays appear promising in the evaluation of complex phytoantioxidants and the measurement of their skin-protective properties.

13.4.3.1 ET-Based Assays ET assays are based on antioxidants' reducing capacity by involving one redox reaction. Such methods include the following:

- The 2,2′-diphenyl-1,1′-picrylhydrazyl (DPPH) assay, where the decrease in absorbance of DPPH is proportional to the concentration of free radical scavengers.
- The trolox equivalent antioxidant capacity (TEAC) assay, where the interactions between antioxidants and 2,2′ azinobis(3-ethyl benzothiazoline-6-sulfonicacid) diamonium salt (ABTS) result in the ABTS change of color (from blue green to colorless).

13.4.3.2 HAT-Based Assays The HAT-based assays involve a synthetic free radical generator, an oxidizable molecular probe, and the antioxidant to be tested. The antioxidant capacity depends on the capacity of donating hydrogen atoms. Such methods include the oxygen radical absorbance capacity (ORAC) and the total radical trapping antioxidant parameter (TRAP) assays. The ORAC assay measures the antioxidant inhibition of free radical damage to a fluorescent probe over time, as compared to Trolox, a water-soluble analogue of α-tocopherol [27]. This is particularly suited to measuring slow and fast acting antioxidants in foods and plant extracts (Figure 13.2) [28]. The TRAP assay is based on the properties of "azo-initiators" resulting in the production of a peroxyl radical that can initiate a peroxidation chain reaction [29, 30].

13.4.3.3 Electrochemical-Based Assays Electrochemical-based assays are used to determine the redox capacity of single or complex compounds. Cyclic voltametry techniques have been adapted to the quantification of the overall reducing capacity of low molecular weight antioxidants in different biological fluids [31]. Such assays do not require the use of reagents and do not depend on absorbance. They do, however, depend on interactions taking place between the tested samples and the electrode surface. The recently developed EDEL assay (EDEL Therapeutics, Switzerland) is based on the titration of an ideal and virtual antioxidant

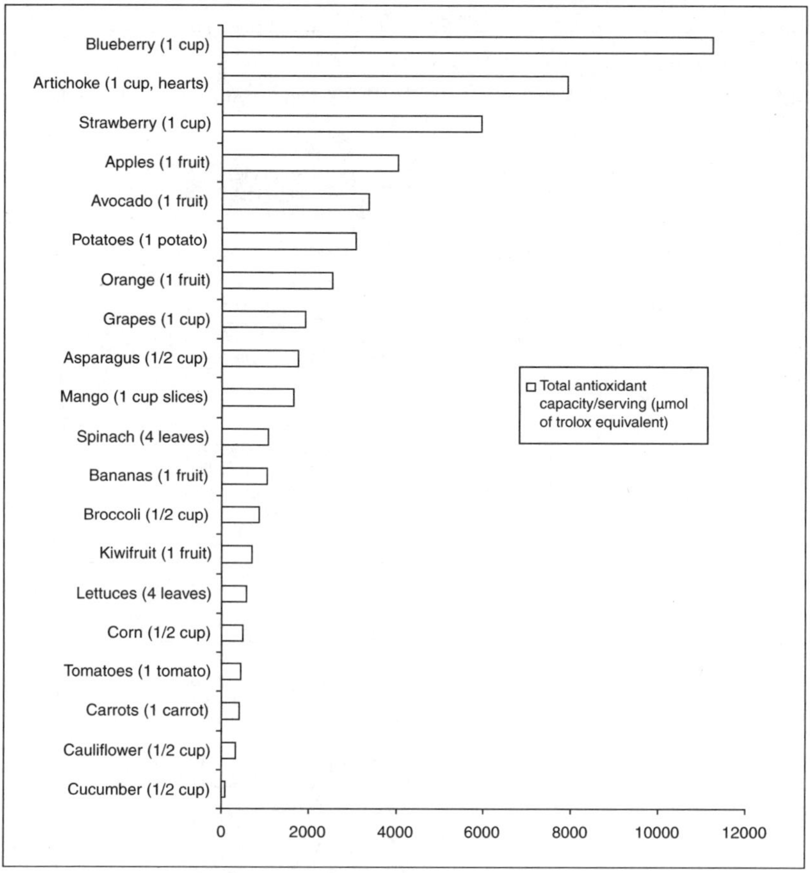

Figure 13.2 ORAC results of different fruits and vegetables. A higher ORAC value represents a more powerful antioxidant activity. Of those tested here, blueberry is the most powerful antioxidant with an ORAC value of 11,400 (adapted from Ref. [28]).

(pending patent number WO/2009/039945), which can be used for measuring a variety of samples (products, cosmetics, biological fluids, and skin) in less than 20 s. This procedure does not require costly or heavy equipment (the electrochemical unit, used in combination with a computer, is portable) and may be applied repeatedly to various biological samples, allowing for direct comparisons. The samples can be analyzed on site upon collection, there is no need for reagents, and single-use strips prevent the risk of cross-contamination. This method allows for *in vitro* antioxidant activity measurements of different cosmetic formulations and their *in vivo* effects on the skin. For *in vivo* measurements, the specially designed well containing PBS is applied to the forehead for 5 min, the required time to extract antioxidants of the skin in the PBS. The antioxidants released in the PBS may then be measured by the electrode in the supernatant. This procedure can be used before and after the application of a protective formulation on the skin (Figure 13.3a and b). The impact of different treatments can also be measured, that is, UV and/or environmental

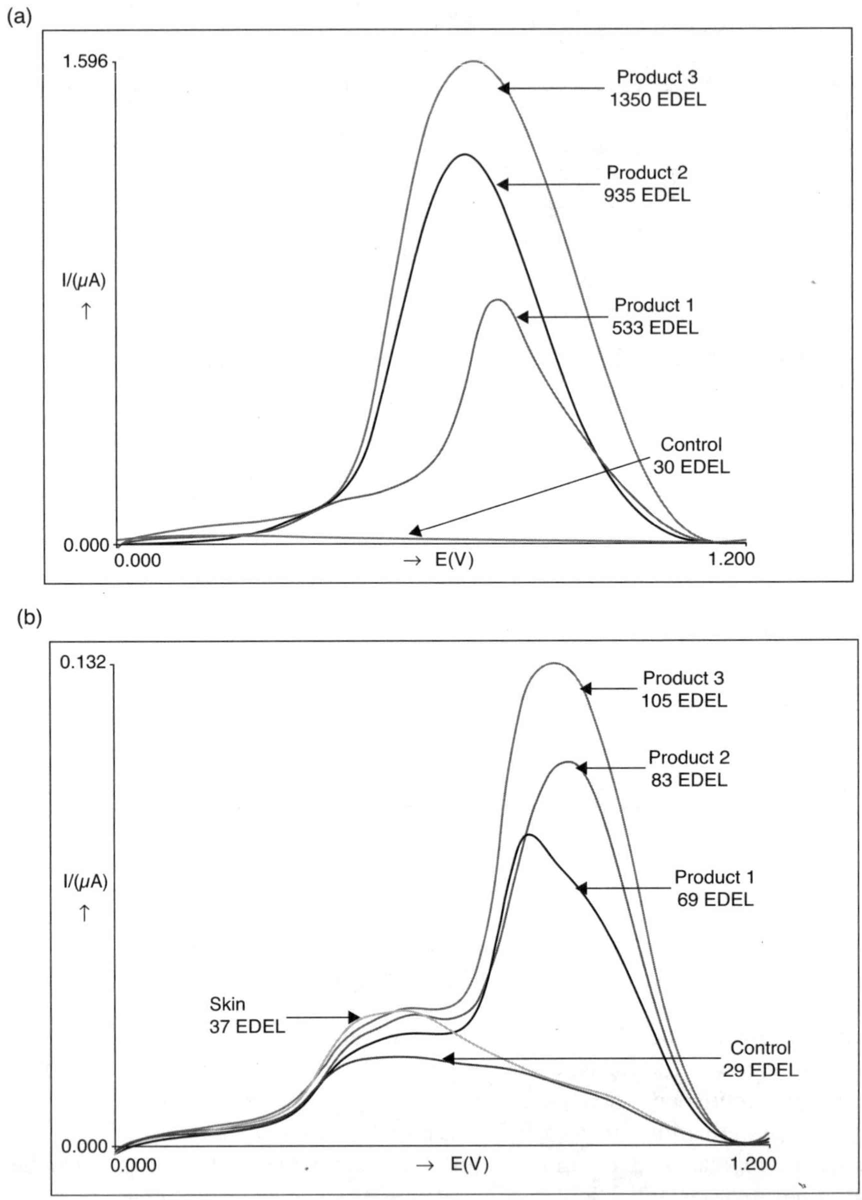

Figure 13.3 Antioxidant activity of three cosmetic products *in vitro* and *in vivo*. (a) Antioxidant results *in vitro* of three cosmetic products in their jars compared to a control solution. Three different cosmetic formulas claiming antioxidant effects were analysed by electrochemistry with Edelscan (Edeltherapeutics, Switzerland). Results are the mean of triplicate measurements. The control is phosphate buffer solution (PBS). (b) Antioxidant results *in vivo* of the same cosmetic products on the skin. 100 μL of PBS was used to extract the skin's antioxidants in 5 min, using an especially made well, on a 314 mm^2 area. Two milligrams of product were applied to the skin for 10 min. The skin's antioxidants were measured by electrochemistry with Edelscan before and after the application of three similar cosmetic formulas. Results are the mean of triplicate measurements. Before product application, the skin contains endogenous antioxidants, and after product application, the level of skin's antioxidants increased.

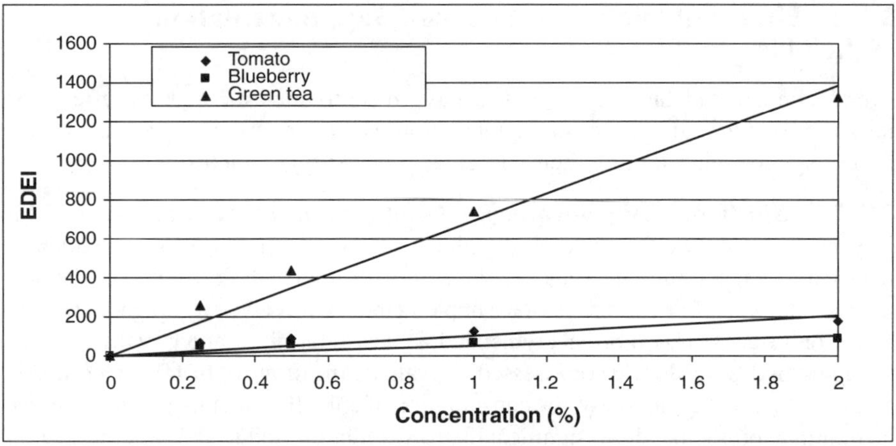

Figure 13.4 Antioxidant activity of three natural extracts *in vitro* as a function of their concentration. A stock solution of 10% w/w of green tea extract, 5% w/w of blueberry extract, and 0.5% w/w of tomato extract each was prepared in phosphate buffer, pH 7.4 (PBS). Each solution was further diluted 50%, 25%, and 12.5%, respectively, into PBS and the antioxidant capacity measured immediately by electrochemistry with Edelscan (Edeltherapeutics, Switzerland). Results are the mean of triplicate measurements. Green tea presents a higher antioxidant activity than tomato or blueberry.

pollution exposure. Figure 13.4 shows the results obtained with three different natural ingredients: green tea, tomato, and blueberry extracts, all contain powerful phytoantioxidants and serve as references. The ability to measure the antioxidant properties of phytoantioxidants and their biological effects will improve the selection of active ingredients, as well as the manufacturing processes, quality control, and packaging of cosmetic natural formulations, and will complement the clinical studies on the effects of antioxidants to determine the most effective antioxidant strategies.

13.5 CLINICAL STUDIES OF NUTRITIONAL AND TOPICAL ANTIOXIDANTS

13.5.1 Effects of Nutritional Supplements on Skin

The effects of antioxidants on cancer and cardiovascular diseases are well described. But what about their effects on the skin? Brosche and Platt showed that consumption of borage oil improves the cutaneous barrier function in elderly individuals, illustrated by a decrease in transepidermal water loss (TEWL) [32]. Specific nutritional factors favor skin hydration, elasticity, and sebum production, and stimulate the physiological properties of the skin [33–35]. Supplementation with vitamin E, vitamin C, and/or carotenoids was shown to provide protection against UV radiation, although the sun protection factor was relatively small compared to topical sunscreens [34].

13.5.2 Effects of Dietary Antioxidant Supplementation on the Skin

Several clinical and laboratory studies have recently investigated the effects on the skin of dietary antioxidant supplementation. Here, we report the results of three of these studies, highlighting their particularly interesting conclusions.

13.5.2.1 Study on Epigallocatechin Gallate Supplementation [36] This study investigated the effects of oral epigallocatechin gallate, a powerful antioxidant in green tea, on the minimal erythema dose and UV-induced skin damage. Female hairless rats were fed a normal diet supplemented with 1.5 ppm epigallocatechin gallate for 8 weeks. The minimal erythema dose was determined and visual scores and transepidermal water loss were assessed to evaluate the severity of UV-induced skin damage. At week 8 of the study, the consumption of epigallocatechin gallate increased the minimal erythema dose significantly. UV radiation-induced sunburn severity alterations in epidermal barrier function were also attenuated by the supplementation of epigallocatechin gallate. The regular intake of epigallocatechin gallate strengthens the skin's tolerance by increasing the minimal erythema dose and thus prevents UV-induced perturbation of epidermal barrier function and skin damage.

13.5.2.2 Study on Supplementation with a Mixture of Various Antioxidants [37] Thirty-nine volunteers with healthy type II skin were divided into three groups ($n = 13$) and supplemented for a period of 12 weeks. Group 1 received a mixture of lycopene (3 mg/day), lutein (3 mg/day), β-carotene (4.8 mg/day), α-tocopherol (10 mg/day), and selenium (75 mg/day). Group 2 was supplemented with a mixture of lycopene (6 mg/day), β-carotene (4.8 mg/day), α-tocopherol (10 mg/day), and selenium (75 mg/day). Group 3 was the placebo group. Skin density and thickness were determined by ultrasound measurements. Skin roughness and scaling were decreased by supplementation with antioxidant micronutrients. In the placebo group, no change was found in any of the parameters.

13.5.2.3 Study on β-Carotene Supplementation [38] Sixteen healthy women were given a dosage of β-carotene (30 mg/day) for 10 weeks. After the 10-week supplementation period, supplementation continued with exposure to natural sunlight for 13 days. During this period, the development of erythema in subjects who had taken β-carotene was much less pronounced than in the placebo group. Supplementation with β-carotene (30 mg/day) before and during sunlight exposure provided protection against sunburn.

These studies indicate that oral antioxidant supplementation may protect against UV-induced skin damage and improves the quality of the skin. How does topical application of antioxidant-rich cosmetics compare to nutritional supplementation?

13.5.3 Effects of Topical Antioxidants on the Skin

Green tea, rosemary, grapes, and tomato are four classical examples of plants that are most studied for their direct antioxidant activity on the skin and skin cells *in vivo* and *in vitro* and serve as a reference for other fruits and vegetables.

13.5.3.1 Green Tea Green tea contains four major flavonoids: epicatechin, epicatechin gallate, epigallocatechin, and epigallocatechin-3-gallate. These molecules have the ability to scavenge ROS: $O_2^{\bullet-}$, $^{\bullet}OH$, H_2O_2, and $^1O_2$. Katiyar et al. have shown that epigallocatechin applied topically on mice exposed to UVB inhibits the production of H_2O_2 both in the dermis and in the epidermis [39].

13.5.3.2 Rosemary Rosemary contains various antioxidants, in particular phenolic diterpenes: carnosol and carnosic acid represent over 90% of the antioxidant properties of rosemary extract [40]. These lipophilic molecules scavenge lipid free radicals, thereby enabling the reduction of lipid peroxidation and inhibiting oxidative damages to skin surface lipids [41]. Pretreatment of human fibroblasts with carnosic acid resulted in the suppression of metalloproteinase-1 messenger RNA elevation caused by UVA irradiation. Carnosic acid also has photoprotective potential [42].

13.5.3.3 Grape Seeds Grape seeds are major sources of resveratrol and quercetin. The stilben resveratrol inhibits lipid peroxidation induced by UVB and significantly decreases UVB-induced skin thickness and edema in hairless mice [43]. The iron chelator flavonoid quercetin maintains and protects the activities of glutathione peroxidase, catalase, and superoxide dismutase after exposure to UV radiation [44].

13.5.3.4 Tomato Tomato is rich in lycopene, a widely studied powerful antioxidant and anticarcinogenic carotenoid with strong reducing ability and the most effective carotenoid in the capture of $^1O_2$. Lycopene scavenges lipid radicals, reduces lipid peroxidation, and prevents erythema caused by UV radiation on the skin [45].

13.6 FROM THE PLATE TO THE JAR TO THE STRATUM CORNEUM

As dietary antioxidant intakes prevent aging effects on various organs including the skin, one could question why it is so important to apply phytoantioxidants to the upper layer of the skin, the stratum corneum. The fact is that the local production of ROS upon exposure to UV radiation and/or other environmental hazards requires topical application of antioxidants in order to optimally prevent ROS-induced local skin damage.

13.6.1 The Stratum Corneum: Structure and Function

The stratum corneum is the outermost layer of the skin. It is 10–20 μm thick, although this thickness varies from a few micrometers to millimeters depending on body parts. It contains only 15% water (whereas the whole body contains 65% and the heart 76% water) and consists of two distinct structural components: the corneocytes and the intercellular lipids. Elias proposed the "brick and mortar model," according to which the stratum corneum is composed of flat cells (bricks) surrounded by a lipid matrix (mortar) [46].

The stratum corneum represents only 10% of the entire skin but contributes to over 80% of the cutaneous barrier function; it prevents the loss of water from the epidermis and provides protection from the outside environment through its antioxidants.

The stratum corneum, due to its critical location at the interface between the body and the environment, is continuously exposed to oxidants, including UV radiation, chemical oxidants, air pollutants, and microorganisms. The stratum corneum is a paradoxical tissue composed of nondividing cells, remaining a metabolically active tissue, able to protect itself, to communicate, and to exchange [47]. The stratum corneum, similar to other human living tissue, contains a number of endogenous antioxidant enzymes, such as catalase, superoxide dismutase, and glutathione peroxidase [48].

13.6.2 The Stratum Corneum as Target for Topical Phytoantioxidants

The stratum corneum is a prime target for cosmetologists for investigating the penetration and efficacy of cosmetic products. Numerous *in vitro* studies have been conducted to demonstrate the penetration of cosmetic ingredients and their effects on the structure and metabolism of skin cells. It is, however, not yet clear as to whether there is and if so to which extent *in vivo* penetration of compounds from cosmetic formulations. Furthermore, the lack of penetration of the products' active ingredients could very well not be a negative characteristic. Indeed, these ingredients are still able to protect the stratum corneum against oxidative and other environmental damages.

The stratum corneum requires antioxidants to protect itself from the environment. α-Tocopherol on the one hand accumulates in the lower part of the stratum corneum, providing protection against lipid peroxidation and allows stabilization of lipid bilayers. Phytoantioxidants on the other hand neutralize UV-induced oxidation of the stratum corneum and provide protection from the environment. The use of phytoantioxidants in cosmetic products may stimulate the stratum corneum to regenerate, to protect itself—and thus the underlying epidermis and dermis—from the harmful effects of UV and other environmental toxins, and nourish the skin in the same way as eating fruits and vegetables nourishes the whole body: "from the plate to the jar."

13.7 MULTIPLE PHYTOANTIOXIDANTS IN COSMETICS: A CASE STUDY

What are the links between all these data and the current world of cosmetics and how does science influence formulation? In order to address this issue, an example of the use of cosmetic formulation and the effects of phytoantioxidants will be discussed.

In 1999, we formulated, according to Swiss quality standards (SwissCos [49]), the serums Diode1® and Diode2® for women (Alchimie Forever, Geneva, Switzerland). These serums contain high concentrations of antioxidant molecules, have antiaging properties, and contribute to the fading and prevention of brown spots. These serums are presented in two bottles: the first bottle, Diode1®, contains water, cellulose gum, chlorphenesin and methylparaben, and 5% of rosemary extract from Morocco (provided by the Swiss supplier Cosmetochem), containing many antioxidant molecules such as terpene phenolics, carnosic acid, and carnosol that have the ability to scavenge lipid free radicals, reduce lipid peroxidation, and have anti-inflammatory properties (Section 13.5.3.2) [41]; the second bottle, called Diode2®, contains water, cellulose gum, chlorphenesin, and methylparaben, and 10% of green tea extract from China (provided by the Swiss supplier Botanica). The main antioxidant polyphenols in green tea are catechins, epicatechins, and their derivatives (Section 13.5.3.1). The antioxidant and healing properties of green tea have been studied in detail in the skin [39]. Measurements performed by EDEL Therapeutics were used to document *in vitro* and *in vivo* antioxidant properties of the serums Diode1® and Diode2®. The results are shown in Figure 13.5a and b. This figure shows that the two serums exert an antioxidant activity that is measurable both in the bottle and after application on the skin, indicating persistent antioxidant activity *in vivo*. These types of measurements should be generalized to confirm the claimed antioxidant effects of skin care products and allow comparisons among them.

13.8 CONCLUSIONS AND PERSPECTIVES

In this chapter, we discussed a number of topics including why antioxidants have become major ingredients in antiaging cosmetics and why natural antioxidants are preferred to synthetic ones by a number of brands. We defined oxidative stress as an imbalance between ROS and antioxidants. In the skin, excess free radicals induced by UV cause damage to cellular macromolecules and destabilize the membranes of keratinocytes, leading to premature aging. Our body defends itself from these phenomena via endogenous and exogenous antioxidants. Natural antioxidants play key roles in these strategies as plants have developed multiple strategies to cope with environmental insults. They generate highly effective molecules to defend themselves against environmental stress. From the large epidemiological studies that have been published recently on nutrition-based antioxidants, we learned that specific nutritional factors favor skin hydration, elasticity, and sebum production, and stimulate the physiological properties of the skin and that oral antioxidant supplementation may protect against UV-induced skin damage.

Current thinking is that a combination of different phytoantioxidants would be the best defense strategy against ROS. We also detailed the best approaches to measure antioxidants and their effects, which will lead to improved cosmetic formulation for the prevention of premature skin aging.

With respect to topical application of phytoantioxidants, we conclude that the stratum corneum is a prime target for cosmetic formulators, as it requires the contribution of antioxidants to protect itself from the environment. Phytoantioxidants

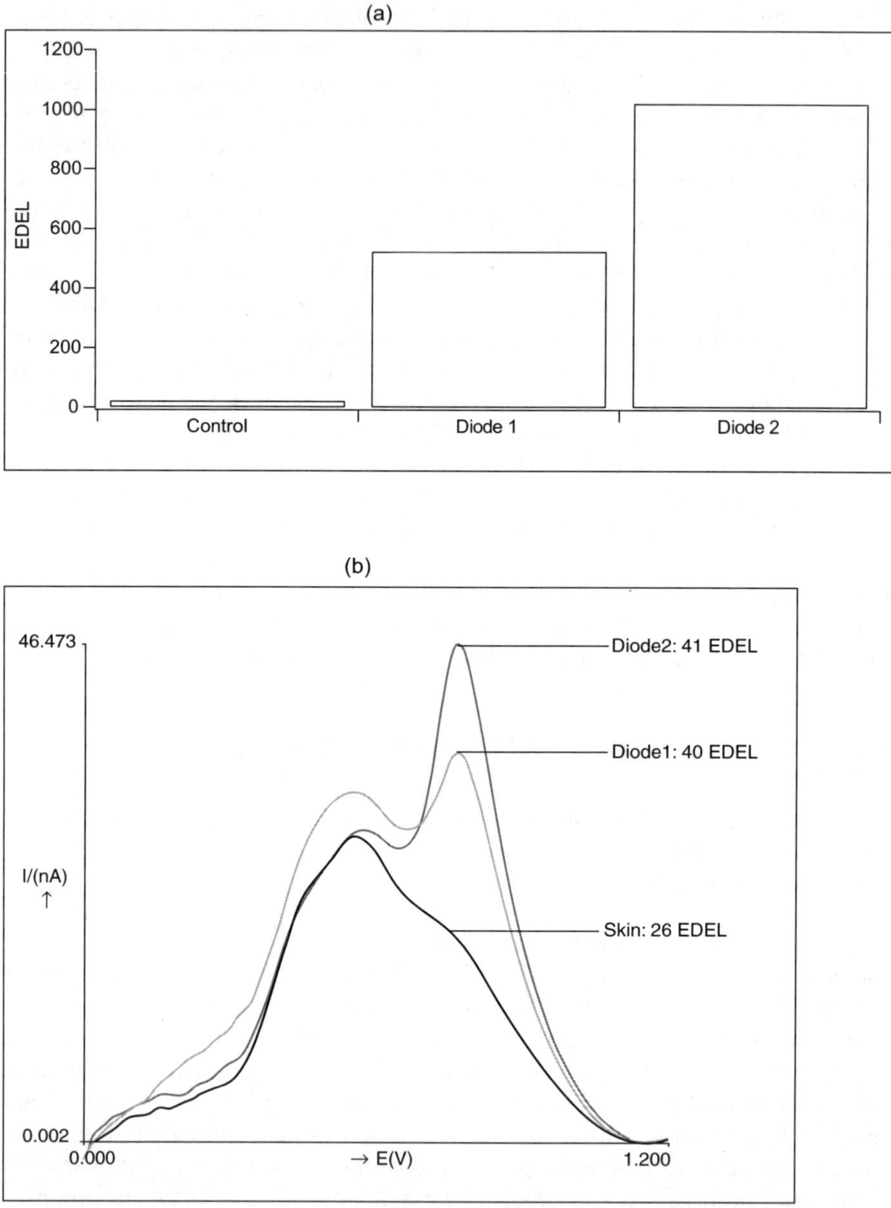

Figure 13.5 Antioxidant activity of Diode1® and Diode2® *in vitro* and *in vivo*. (a) Antioxidant results *in vitro* of Diode1® rosemary and Diode2® green tea serums. The green tea serum displays a higher antioxidant activity than the rosemary serum. (b) Antioxidant results *in vivo* of Diode1® rosemary and Diode2® green tea serums measured directly on the forehead. The antioxidant activity of rosemary and green tea persists once applied on the skin with again a higher activity of the green tea.

neutralize UV-induced oxidation of the stratum corneum, provide protection from the environment, and in cosmetic products may stimulate the stratum corneum to regenerate. Phytoantioxidants are therefore to be used both topically and orally and should be integrated into any antiaging strategy. The use of phytoantioxidants in cosmetic products enables one to nourish the skin and replenish it in antioxidants in the same way as we nourish our body by eating fruits and vegetables.

Further studies along these lines, in particular on the antioxidant capacity of cosmetic formulas *in vivo*, will pave the way for optimal science-based natural health, well-being, and youthful skin.

REFERENCES

1. Halliwell B, Gutteridge JMC. *Free Radicals in Biology and Medicine*. 4th ed. New York: Oxford University Press, 2007.
2. Brown JR, Goldblatt D, Buddle J, Morton L, Thrasher AJ. Diminished production of anti-inflammatory mediators during neutrophil apoptosis and macrophage phagocytosis in chronic granulomatous disease (CGD). *J. Leukoc. Biol.* 2003;73(5):591–599.
3. Poderoso JJ. The formation of peroxynitrite in the applied physiology of mitochondrial nitric oxide. *Arch. Biochem. Biophys.* 2009;484(2):214–220.
4. Zaw KK, Yokoyama Y, Abe M, Ishikawa O. Catalase restores the altered mRNA expression of collagen and matrix metalloproteinases by dermal fibroblasts exposed to reactive oxygen species. *Eur. J. Dermatol.* 2006;16(4):375–379.
5. Pacher P, Beckman JS, Liaudet L. Nitric oxide and peroxynitrite in health and disease. *Physiol. Rev.* 2007;87(1):315–424.
6. Nishigori C, Hattori Y, Toyokuni S. Role of reactive oxygen species in skin carcinogenesis. *Antioxid. Redox. Signal.* 2004;6(3):561–570.
7. Reelfs O, Tyrrell RM, Pourzand C. Ultraviolet A radiation-induced immediate iron release is a key modulator of the activation of NF-kappaB in human skin fibroblasts. *J. Invest. Dermatol.* 2004;122 (6):1440–1447.
8. Mariethoz E, Richard MJ, Polla LL, Kreps SE, Dall'Ava J, Polla BS. Oxidant/antioxidant imbalance in skin aging: environmental and adaptative factors. *Rev. Environ. Health* 1998;13(3):147–168.
9. Scharffetter-Kochanek K, Brenneisen P, Wenk J, Herrmann G, Ma W, Kuhr L, Meewes C, Wlaschek M. Photoaging of the skin from phenotype to mechanisms. *Exp. Gerontol.* 2000;35:307–316.
10. Erenel G, Erbas D, Aricioglu A. Free radicals and antioxidant systems. *Mater. Med. Pol.* 1993;25 (1):37–43.
11. Chaudiere J, Ferrari-Iliou R. Intracellular antioxidants: from chemical to biochemical mechanisms. *Food Chem. Toxicol.* 1999;37(9–10):949–962.
12. Vile GF, Basu-Modak S, Waltner C, Tyrrell RM. Heme oxygenase 1 mediates an adaptative response to oxidative stress in human skin fibroblasts. *Proc. Natl. Acad. Sci. USA* 1994;91:2607–2610.
13. Dernbach E, Urbich C, Brandes EP, Hofmann WK, Zeiher AM, Dimmeler S. Antioxidative stress-associated genes in circulating progenitor cells: evidence for enhanced resistance against oxidative stress. *Blood* 2004;104(12):3591–3597.
14. Renaud S, de Lorgeril M. Wine, alcohol, platelets, and the French paradox for coronary heart disease. *Lancet* 1992;339(8808):1523–1526.
15. Frankel EN, Waterhouse AL, Teissedre PL. Principal phenolic phytochemicals in selected California wines and their antioxidant activity in inhibiting oxidation of human low-density lipoproteins. *J. Agric. Food Chem.* 1995;43:890–894.
16. Vigneron JP, Rassart M, Vertesy Z, Kertesz K, Sarrazin M, Biro LP, Ertz D, Lousse V. Optical structure and function of the white filamentary hair covering the edelweiss bracts. *Phys. Rev. E: Stat. Nonlin. Soft Matter Phys.* 2007;71:011906.

17. Rancan F, Rosan S, Boehm K, Fernandez E, Hidalgo ME, Quihot W, Rubio C, Boehm F, Piazena H, Oltmanns U. Protection against UVB irradiation by natural filters extracted from lichens. *J. Photochem. Photobiol. B* 2002;68(2–3):133–139.
18. Kopani M, Celec P, Danisovic L, Michalka P, Biro C. Oxidative stress and electron spin resonance. *Clin. Chim. Acta* 2006;364(1–2):61–66.
19. Nilsson R, Nordlinder R, Moen BE, Ovrebo S, Bleie K, Skorve AH, Hollund BE, Tagesson C. Increased urinary excretion of 8-hydroxydeoxyguanosine in engine room personnel exposed to polycyclic aromatic hydrocarbons. *Occup. Environ. Med.* 2004;61(8):692–696.
20. Hoelzl C, Knasmuller S, Misik M, Collins A, Dusinska M, Nersesyan A. Use of single cell gel electrophoresis assays for the detection of DNA-protective effects of dietary factors in humans: recent results and trends. *Mutat. Res.* 2009;681:68–79.
21. Valverde M, Rojas E. Environmental and occupational biomonitoring using the Comet assay. *Mutat. Res.* 2009;681:93–109.
22. Morley N, Clifford T, Salter L, Campbell S, Gould D, Curnow A. The green tea polyphenol (−)-epigallocatechin gallate and green tea can protect human cellular DNA from ultraviolet and visible radiation-induced damage. *Photodermatol. Photoimmunol. Photomed.* 2005;21(1):15–22.
23. Sanchez-Campillo M, Gabaldon JA, Castillo J, Benvente-Garcia O, Del Bano MJ, Alcaraz M, Vicente V, Alvarez N, Lozano JA. Rosmarinic acid, a photo-protective agent against UV and other ionizing radiations. *Food Chem. Toxicol.* 2009;47(2):386–382.
24. Montuschi P, Barnes P, Roberts LJ, II. Insights into oxidative stress: the isoprostanes. *Curr. Med. Chem.* 2007;14(6):703–717.
25. Milne GL, Musiek ES, Morrow JD. F2-isoprostanes as markers of oxidative stress *in vivo*: an overview. *Biomarkers* 2005;10:10S–23S.
26. Schneider LA, Bloch W, Kopp K, Hainzl A, Rettberg P, Wlaschek M, Horneck G, Scharffetter-Kochanek K. 8-Isoprostane is a dose-related biomarker for photo-oxidative ultraviolet (UV) B damage *in vivo*: a pilot study with personal UV dosimetry. *Br. J. Dermatol.* 2006;154(6):1147–1154.
27. Cao G, Alessio HM, Cutler RG. Oxygen-radical absorbance capacity assay for antioxidants. *Free Radic. Biol. Med.* 1993;14(3):303–311.
28. Wu X, Beecher GR, Holden JM, Haytowitz DB, Gebhardt SE, Prior RL. Lipophilic and hydrophilic antioxidant capacities of common foods in the United States. *J. Agric. Food Chem.* 2004;52:4026–4037.
29. Wayner DD, Burton GW, Ingold KU, Locke S. Quantitative measurement of the total, peroxyl radical-trapping antioxidant capability of human blood plasma by controlled peroxidation. The important contribution made by plasma proteins. *FEBS Lett.* 1985;187(1):33–37.
30. Ghiselli A, Serafini M, Natella F, Scaccini C. Total antioxidant capacity as a tool to assess redox status: critical view and experimental data. *Free Radic. Biol. Med.* 2000;29(11):1106–1114.
31. Kohen R, Nyska A. Oxidation of biological systems: oxidative stress phenomena, antioxidants, redox reactions, and methods for their quantification. *Toxicol. Pathol.* 2002;30(6):620–650.
32. Brosche T, Platt D. Effect of borage oil consumption on fatty acid metabolism, transepidermal water loss and skin parameters in elderly people. *Arch. Gerontol. Geriatr.* 2000;30(2):139–150.
33. Rampoldi R, Macedo N, Alallon W, Sanguimetti J. Topical vitamin E and ultraviolet radiation on human skin. *Med. Cutan. Ibero Lat. Am.* 1990;18(4):269–272.
34. Boelsma E, Hendriks HFJ, Roza L. Nutritional skin care: health effects of micronutrients and fatty acids. *Am. J. Clin. Nutr.* 2001;73:853–864.
35. Yamamoto Y. Role of active oxygen species and antioxidants in photoaging. *J. Dermatol. Sci.* 2001;27:1S–4S.
36. Jeon HY, Kim JK, Kim WG, Lee SJ. Effects of oral epigallocatechin gallate supplementation on the minimal erythema dose and UV-induced skin damage. *Skin Pharmacol. Physiol.* 2009;22(3):137–141.
37. Heinrich U, Tronnier H, Stahl W, Bejot M, Maurette JM. Antioxidant supplements improve parameters related to skin structure in humans. *Skin Pharmacol. Physiol.* 2006;19(4):224–231.
38. Biesalski HK, Hemmes C, Hopfenmuller W, Schmid C, Gollnick HP. Effects of controlled exposure of sunlight on plasma and skin levels of beta-carotene. *Free Radic. Res.* 1996;24(3):215–224.
39. Katiyar SK, Challa A, McCormick TS, Cooper KD, Mukhtar H. Prevention of UVB-induced immunosuppression in mice by the green tea polyphenol (−)-epigallocatechin-3-gallate

may be associated with alterations in IL-10 and IL-12 production. *Carcinogenesis* 1999;20(11): 2117–2124.

40. Aruoma OI, Halliwell B, Aeschbach R, Loligers J. Antioxidant and pro-oxidant properties of active rosemary constituents: carnosol and carnosic acid. *Xenobiotica* 1992;22(2):257–268.
41. Calabrese V, Scapagnini G, Catalano C, Dinotta F, Geraci D, Morganti P. Biochemical studies of a natural antioxidant isolated from rosemary and its application in cosmetic dermatology. *Int. J. Tissue React.* 2000;22(1):5–13.
42. Offord EA, Gautier JC, Avanti O, Scaletta C, Runge F, Krämer K, Applegate LA. Photoprotective potential of lycopene, beta-carotene, vitamin E, vitamin C and carnosic acid in UVA-irradiated human skin fibroblasts. *Free Radic. Biol. Med.* 2002;32(12):1293–1303.
43. Afaq F, Adhami VM, Ahmad N. Prevention of short-term ultraviolet B radiation-mediated damages by resveratrol in SKH-1 hairless mice. *Toxicol. Appl. Pharmacol.* 2003;186(1):28–37.
44. Kahraman A, Inal ME. Protective effects of quercetin on ultraviolet A light-induced oxidative stress in the blood of rat. *J. Appl. Toxicol.* 2002;22(5):303–309.
45. Stahl W, Heinrich U, Aust O, Tronnier H, Sies H. Lycopene-rich products and dietary photoprotection. *Photochem. Photobiol. Sci.* 2006;5:238–242.
46. Elias PM. Epidermal lipids, barrier function, and desquamation. *J. Invest. Dermatol.* 1983;80:44S–49S.
47. Pouillot A, Dayan N, Polla AS, Polla LL, Polla BS. The stratum corneum: a double paradox. *J. Cosmet. Dermatol.* 2008;7:143–148.
48. Thiele JJ. Oxidative targets in the stratum corneum: a new basis for antioxidative strategies. *Skin Pharmacol. Appl. Skin Physiol.* 2001;14:87S–91S.
49. http://www.swisscos-guarantee.ch/guarantee.html.
50. Liu RH. Potential synergy of phytochemicals in cancer prevention: mechanisms of action. *J. Nutr.* 2004;134:3479S- 3485S.
51. Nakajima JI, Tanaka I, Seo S, Yamazaki M, Saito K. LC/PDA/ESI-MS profiling and radical scavenging activity of anthocyanins in various berries. *J. Biomed. Biotechnol.* 2004;5:241–247.
52. Afaq F, Mukhtar H. Botanical antioxidants in the prevention of photocarcinogenesis and photoaging. *Exp. Dermatol.* 2006;15:678–684.
53. Rigano D, Cardile V, Formisano C, Maldini MT, Piacente S, Bevilacqua J, Russo A, Senatore F. *Genista sessilifolia* DC and *Genista tinctoria* L. inhibit UV light and nitric oxide-induced DNA damage and human melanoma cell growth. *Chem. Biol. Interact.* 2009;180(2):211–219.
54. Baliga MS, Katiyar SK. Chemoprevention of photocarcinogenesis by selected dietary botanicals. *Photochem. Photobiol. Sci.* 2006;5:243–253.
55. Andreassi M, Stanghellini E, Ettorre A, Di Stefano A, Andreassi L. Antioxidant activity of topically applied lycopene. *J. Eur. Acad. Dermatol. Venereol.* 2004;18(1):52–55.

CHAPTER 14

THE USE OF QUERCETIN AND CURCUMIN IN SKIN CARE CONSUMER PRODUCTS

Brian E. Kilfoyle,[1] Diksha Kaushik,[1] Jenna L. Terebetski,[1] Sonali Bose,[1,2] and Bozena B. Michniak-Kohn[1]

[1]Department of Pharmaceutics, Ernest Mario School of Pharmacy, New Jersey Center for Biomaterials, Rutgers, The State University of New Jersey, Piscataway, NJ

[2]Novartis Pharmaceuticals Corporation, East Hanover, NJ

14.1 INTRODUCTION

The cosmetic and pharmaceutical markets have recently seen a significant shift toward more natural ingredients owing to both public demand and the desire to use less synthetic ingredients. This move has been strengthened by more rigorous scientific data on the therapeutic and cellular effects of some of these natural ingredients. Many of these botanical compounds are shown to have a strong potential for use in the treatment of various ailments as antioxidants, anti-inflammatory agents, and anticarcinogenic agents (Table 14.1) [1]. These natural compounds provide a unique class of potentially safe compounds because they have been incorporated for years into the human diet and traditional folk remedies.

Two of the more common botanical compounds are the natural polyphenol curcumin, from the Indian spice turmeric, and quercetin, from apples, onions, and many other plant species. These two compounds have been commonly used throughout history without any real knowledge of their pharmacological properties, but recently they have demonstrated biological activity *in vitro*, *in vivo*, and in human clinical trials. This chapter reviews these two botanical compounds with a focus on their skin-related pharmacological effectiveness.

Formulating, Packaging, and Marketing of Natural Cosmetic Products, First Edition.
Edited by Nava Dayan and Lambros Kromidas.

TABLE 14.1 Common Plant-Derived Compounds and Their Pharmacological Effects [1]

Botanical compound	Pharmacological effects
Resveratrol	Antioxidant, anti-inflammatory, antiproliferative
Vitamin E	Antioxidant, anticarcinogenic, wound healing
EGCG	Antioxidant, anticarcinogenic
Vitamin C	Antioxidant, anticarcinogenic, wound healing
Caffeine	Anticarcinogenic, antioxidant, vasoconstrictor
Pomegranate	Antioxidant, antiviral
Chamomile	Antioxidant, anti-inflammatory
Olive oil	Anti-inflammatory, antioxidant
Ginseng	Anti-inflammatory, antioxidant, anticarcinogenic
Ginkgo biloba	Antioxidant, anti-inflammatory
Rosemary	Antioxidant, anticarcinogenic, anti-inflammatory, antimicrobial
Lycopene	Antioxidant, anticarcinogenic

14.2 CURCUMIN

14.2.1 Background

Curcuma longa, a herb belonging to the Zingiberaceous family, is indigenous to India and Southeast Asia but is also cultivated in locations throughout South America and the Caribbean. The functional components of this herb, when it is ground up, are the rhizomes that yield turmeric. Turmeric has the notable yellow color characteristic of the curcuminoids that it contains [2]. Curcumin, a common ingredient in many curry powders and sauces, has been used for centuries in ancient Indian and Chinese cultures as a medicine, condiment, colorant, and flavoring agent. In addition, curcumin is an established flavor and color additive in foods and has been widely used in the food industry as a preservative. It is added to foods including pickles, margarine, mayonnaise, salad dressing, oil, cheese, and baked goods. The distinct yellow color of curcumin allows it to be used as a coloring agent not only in foods but also in cosmetics, textiles, and pharmaceuticals.

Curcumin has been recognized for its medicinal utility by the Indian traditional health care system, Ayurveda, for thousands of years [3] and was described by Marco Polo in his travels to China during the thirteenth century [4]. Traditional medicinal uses for turmeric include treatment for biliary disorders, anorexia, cough, wounds, hepatic disorders, rheumatism, and sinusitis in Indian medicine and treatment for abdominal pains in Chinese medicine [5]. Turmeric paste mixed with slaked lime has been used as a topical household remedy for the treatment of sprains, swelling, and wounds. These folk remedies are common throughout history and are still commonly used in various parts of the world. Owing to its history of utility, there have been recent attempts to understand the pharmacological functions of curcumin.

14.2.2 Chemistry

Curcumin, alternatively called diferuloylmethane or natural yellow 3, is a polyphenol and the principal curcuminoid present in the spice turmeric. Curcumin is the most active component of turmeric, making up 2–5% of its composition. It was first isolated by Vogel in 1815 [6] and later crystallized and identified as 1,7-bis(4-hydroxy-3-methoxyphenyl)-1,6-heptadiene-3,5-dione by Daybe [7, 8]. The feruloylmethane skeleton of curcumin was elucidated by Lampe and Milobedzka in 1910 [9]. Curcumin has a molecular weight of 368.37 g/mol and a structure with two phenolic rings, as shown in Figure 14.1. Curcumin is an orange-yellow crystalline powder with a melting point of 183°C. It has low solubility in water, but is soluble in organic solvents.

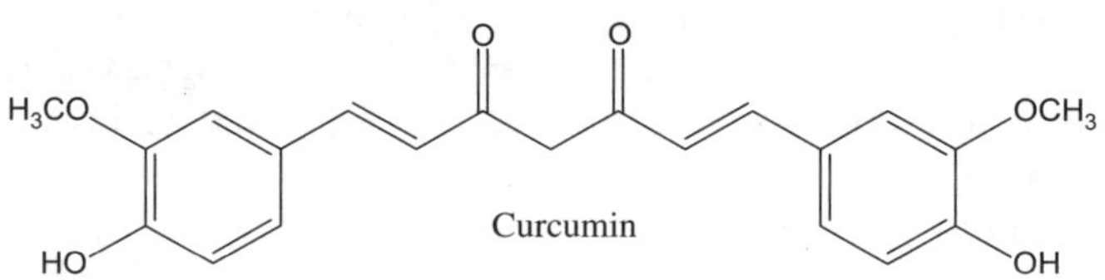

Figure 14.1 Structure of curcumin.

Curcumin is isolated from the rhizomes of *C. longa*. Owing to high cost and significant time requirements, commercial curcumin is typically not available in its pure form (1,7-bis(4-hydroxy-3-methoxyphenyl)-1,6-heptadiene-3,5-dione), but rather it consists of a combination of the naturally occurring curcuminoids often referred to as curcumin I, II, and III. The structures of these curcuminoids are illustrated in Figure 14.2. In general, commercially available curcumin has

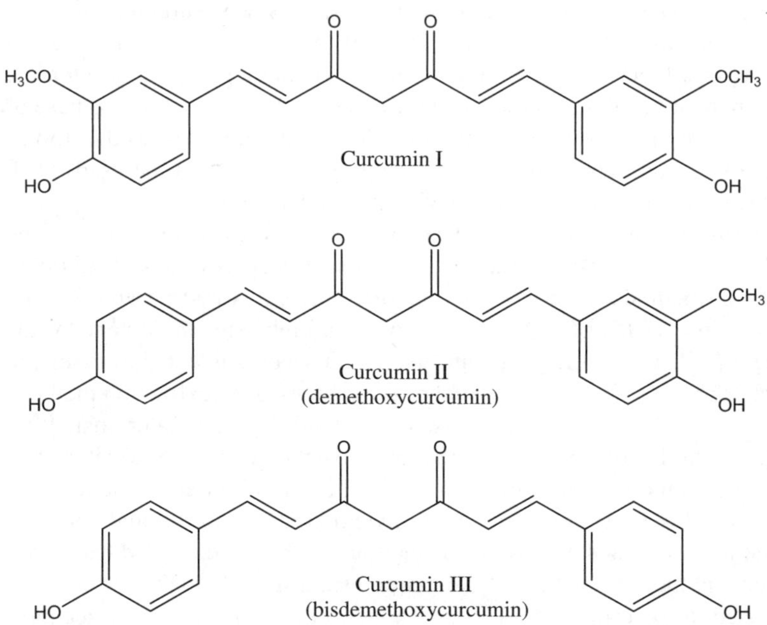

Figure 14.2 Chemical structures of curcuminoids.

a composition of 75–81% curcumin I, 15–19% curcumin II (demethoxycurcumin), and 2.2–6.6% curcumin III (bisdemethoxycurcumin) [10].

14.2.2.1 Cultivation *C. longa* is grown naturally or cultivated in warm rainy regions including India, China, Indonesia, Haiti, Jamaica, and Peru. In general, the cultivation of *C. longa* does not require special practices [11]; however, there are a few procedures that can be utilized for optimal turmeric and curcumin yield. Studies have shown that *C. longa* is a partial shade-tolerant plant and relative light intensity, depending on the region of cultivation, can impact turmeric and curcumin content [12]. Chemical fertilizers can also affect growth, yield, and quality of turmeric with studies showing that nitrogen is the principal nutrient needed for increased vegetative growth of turmeric and that potassium increases curcumin levels, but does not improve the turmeric yield. Akamine et al. suggest that a combination of nitrogen and potassium is best for increasing the yields of turmeric and curcumin [13].

Postharvesting conditions of rhizomes have also been evaluated. Studies suggest that fresh rhizomes yield higher oleoresins compared to dried rhizomes but that dried rhizomes yield extracts containing higher amounts of curcuminoids. In addition, working with dried rhizomes is more convenient and practical from a manufacturing perspective because they have lower moisture content and are less susceptible to microbial infestation [14].

14.2.2.2 Extraction Rhizomes yield the yellow turmeric once boiled, dried, and ground into a powder. To isolate the active curcuminoids, extraction is required. The extraction process yields oleoresin that consists of volatile oils, resin, and approximately 2–8% curcuminoids [15]. Since curcuminoids are insoluble in water, they are commonly isolated by solvent extraction using organic solvents in which they have high solubility (including methanol, ethanol, and acetone). There are limitations on solvent use, depending on the intended use of the turmeric and curcuminoids (e.g., when turmeric extracts are to be used in foods, only acceptable solvents such as ethanol and isopropyl alcohol can be used) [15]. Recovery of the curcuminoids then occurs via solvent evaporation.

The solubility of the curcuminoids depends not only on the composition of the mixture but also on the temperature and pressure of the extraction system. Traditional solvent extraction methods including low-pressure solvent extraction and Soxhlet extraction offer large yields of curcuminoids when a high ratio of solvent to raw material is used [15]. However, yields are decreased when solvent flow rates are increased, limiting process throughput. Supercritical fluid-based extraction methods have also been used for curcuminoid extraction resulting in greater than 90% recovery [16]. In a study comparing low-pressure solvent extraction, Soxhlet extraction, and supercritical fluid-based extraction, supercritical fluid-based extraction with cosolvent produced the maximum curcuminoid yield [15, 17]. Additional concerns with traditional methods such as Soxhlet extraction are that an extended extraction time is required and there is an increased curcumin stability risk due to the elevated temperatures used. Supercritical fluid-based extraction methods are clean technologies that reduce the time and solvent requirements. However, this clean technology is still not commonly used because high operating pressures are required, making scale-up of the process challenging.

Additional extraction methods that have been explored include hydrotropic extraction and microwave-assisted extraction. The hydrotropic process utilizes hydrotropes, water-soluble organic salts, and is free from organic solvents. These hydrotropes penetrate the cell walls of the turmeric plant and make the curcuminoids available for faster dissolution [18]. Water is subsequently added to allow the precipitation of curcuminoids. This is a simple method that results in yields comparable to other extraction processes with much higher purity. Residual hydrotropes can be removed by a simple water rinse, which is beneficial considering the challenge of removing residual solvents in traditional extraction methods. Microwave-assisted extraction utilizes a dual heating mechanism in which the powdered turmeric and extraction solvent are heated simultaneously under microwave energy. At optimal conditions, microwave-assisted extraction allows high yield with a significantly shorter extraction time than other more traditional extraction methods [19].

A purified extract containing a mixture of the three curcuminoids is available on the market. Though it is desirable, due to expense, the isolation of individual curcuminoids is not common practice for marketed material. To date, advances to improve the separation on a processing scale have been minimal. One promising technique involves high-speed countercurrent chromatography (HSCCC) with and without pH zoning. This procedure has been utilized effectively to separate multigram quantities of curcuminoids while maintaining a high level of purity [20].

14.2.2.3 Identification As discussed, there is an assortment of extraction methods used to isolate curcuminoids. To fully assess their utility, it is necessary to accurately detect and quantify the amount of curcuminoids extracted. The most common method for assaying curcumin involves spectrophotometric measurement at 420–430 nm (wavelengths at which curcuminoids exhibit strong absorption). Quantification requires direct absorption measurements and calculations are based on reference values obtained from pure standards. In this method, impurities and other compounds absorbing in the same region are also detected, limiting the utility of spectrophotometric and fluorometric analysis to accurately quantify individual curcuminoids [21].

To determine the exact content of each curcuminoid, separation of the three curcuminoids is necessary. Paper and thin layer chromatographic methods have been used to determine the relative amounts of the three curcuminoids, but these procedures lack the necessary precision and speed [22]. Gas chromatography is also not optimal because curcuminoids have low volatility and are thermally labile. The development of high-performance liquid chromatographic (HPLC) methods has become the standard as they allow more accurate means to separate and quantify the curcuminoids in a reproducible manner [10, 21]. Various detectors have been used in conjunction with HPLC, including mass spectrometers, photodiode array detectors, and UV detectors (most common). Significant progress has been made toward identifying simple, rapid HPLC-UV methods that accurately and precisely quantify the three compounds [23].

Identifying curcumin in the presence of biological fluids is necessary for pharmacokinetic analysis and other *in vivo* methods. This is inherently a more difficult analysis because of the number of potentially interfering biological compounds present. A validated LC/MS/MS assay for curcumin in rat plasma has been reported for this purpose [24].

14.2.2.4 Stability Since curcumin is a bis-α,β-unsaturated β-diketone, it exists in equilibrium with its enol tautomer, as shown in Figure 14.3. Under acidic and neutral conditions, the bis-keto is the predominant form, whereas the enolate form of the heptadienone chain is favored above pH 8 [25]. At pH 1–7, curcumin has a yellow color, and it changes to orange/red at pH above 7.5. These color changes correspond to the pH-dependent instability of curcumin and indicate that curcumin may not be a suitable coloring agent because it loses its hue when stored under alkaline conditions.

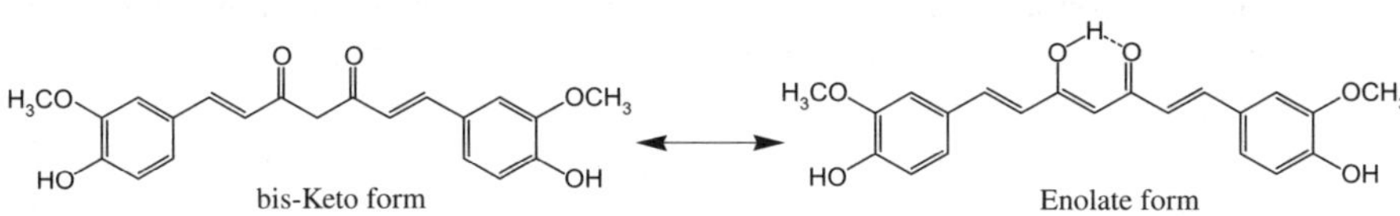

Figure 14.3 Tautomerized structures of curcumin.

For optimal stability, pH should be maintained below 7 because curcumin is highly unstable from pH 7 to 10 [26]. The degradation of curcuminoids under basic conditions corresponds to pseudo-first-order kinetics, with increases in the rates of degradation from pH 7.5 to a maximum at pH 10.2. The rate of degradation then declines from pH 10.2 to 11.95. The main degradation product under basic conditions is *trans*-6-(4′-hydroxy-3′-methoxyphenyl)-2,4-dioxo-5-hexenal, while vanillin, ferulic acid, and feruloylmethane have been proposed as minor degradation products [26, 27]. When evaluating the rates of degradation for each of the curcuminoids, it was noted that bisdemethoxycurcumin was more resistant to alkaline degradation than the other two curcuminoids [28]. Additionally, increased curcumin stability has been demonstrated both in cell culture media containing fetal calf serum and in human blood [27].

Curcumin exhibits strong absorption in the UV/visible range with maxima in the range of 408–430 nm (~420 nm in polar solvents) [29]. It is unstable when exposed to UV/visible radiation both in solution and in solid form. Furthermore, evaluation of the stability of curcumin mixed with microcrystalline cellulose systems revealed that curcumin is sensitive to light with degradation following first-order kinetics. It was also observed that the combined effect of air and light resulted in a significant increase in degradation [11]. Since curcumin acts as a photosensitizer of singlet oxygen, it undergoes self-sensitized decomposition. To avoid photodecomposition of curcumin during storage, it must be protected from light and devoid of any singlet oxygen sources.

14.2.3 Activity

14.2.3.1 Pharmacological Activity Curcumin has a wide array of pharmacological activities that have been the subject of numerous recent reviews, including a comprehensive look at the preventative and therapeutic properties of curcumin by Strimpakos and Sharma [30]. Curcumin has exhibited activity against many diseases, including Alzheimer's, asthma, cardiovascular disease, diabetes, cancer, inflammatory bowel disorder, multiple sclerosis, Parkinson's disease, pancreatitis, psoriasis, depression, and rheumatoid arthritis [7, 8, 31–35]. It has been utilized in wound healing, the prevention of cataract formation, and the inhibition of human immunodeficiency virus (HIV) replication [8]. The overwhelming utility of curcumin in medicine is based largely on its ability to act as an antioxidant and an anti-inflammatory agent.

Curcumin, a known antioxidant, is a potent scavenger of a variety of reactive oxygen species (ROS), including superoxide anion radicals and hydroxyl radicals [36]. It can also inhibit lipid peroxidation, inhibit the generation of ROS by activated macrophages, and lower the production of ROS *in vivo* [37–39]. Since ROS are implicated in the development of various diseases, it is believed that the potent antioxidant activity associated with curcumin helps control these diseases.

The anti-inflammatory effects of curcumin have also proven to be significant. These properties are mediated through the ability of curcumin to suppress inflammatory mediators such as cyclooxygenase 2 (COX-2), lipoxygenase (LOX), inducible nitric oxide synthase (iNOS), tumor necrosis factor (TNF), and nuclear factor κB (NF-κB) [40]. Curcumin's anti-inflammatory activity has been established in the carrageenan-induced edema test in mice and rats [41, 42]. Preliminary clinical trials have also demonstrated the anti-inflammatory activity of curcumin in humans [43].

Curcumin has also been extensively studied for its chemopreventive potential. It is able to suppress the transformation, proliferation, metastasis, and angiogenesis of tumors through the regulation of various transcription factors, growth factors, inflammatory cytokines, protein kinases, caspases, adhesion molecules, and enzymes. Curcumin inhibits the proliferation of cancer cells without causing cytotoxic effects on healthy cells by inducing apoptotic mechanisms including the disruption of p53 stability and the release of cytochrome *c*. Studies have demonstrated that curcumin is effective in inhibiting the proliferation of tumor cells in B-cell and T-cell leukemia, colon carcinoma, epidermoid carcinoma, and various breast carcinomas. Curcumin has protective and therapeutic effects against cancers of the blood, skin, oral cavity, lung, pancreas, and intestinal tract in rats. Clinic trials are ongoing to further explore the anticancer potential of curcumin in humans [44–46].

Curcumin has shown phototoxic effects when examined in numerous bacterial and mammalian cellular models, but, most important, the methods differ on whether curcumin is a generator or scavenger of oxygen [29]. A recent hypothesis suggests that curcumin can exhibit antioxidant and pro-oxidant behavior depending on the surrounding environment. This theory supports the use of curcumin as an antioxidant in numerous applications and also the possibility that it acts as a prooxidant in cancer therapy [29].

14.2.3.2 Treatment of Skin Disease Turmeric has been used as a traditional Indian household remedy with topical administration for several skin diseases, insect bites, wound healing, and rheumatic pain [47, 48]. Numerous studies have been conducted to understand the biological activities of curcumin in relation to skin disease, showing potential for the treatment of skin carcinogenesis, psoriasis, and scleroderma. It also accelerates wound healing, prevents the formation of scars, plays a role in muscle regeneration, and reduces skin inflammation associated with premature aging.

The role of curcumin in the prevention and treatment of skin cancer has been explored extensively. In general, curcumin has significant anticancer potential since it has the ability to suppress proliferation of various tumor cells, downregulate the expression of various inflammatory markers and growth factor receptors, inhibit the activity of kinases, and influence free radical production [40]. Curcumin has shown chemopreventive action in several carcinogen-induced skin cancer models. *In vivo* studies in mice have shown that topical administration of curcumin inhibits 12-*O*-tetradecanoylphorbol-13-acetate (TPA)-induced epidermal inflammation, epidermal DNA and RNA synthesis, and promotion of skin tumors [49, 50]. Tumorigenesis induced by 7,12-dimethylbenzanthracene (DMBA) is also inhibited by curcumin [51], with a reduction in the number of tumors per mouse and decrease in the number of tumor-bearing mice observed [52].

Curcumin has photochemoprotective capabilities, allowing it to counteract the adverse effects of ultraviolet radiation on the skin. Pretreatment with curcumin can alleviate TPA-induced dermatitis by reducing ornithine decarboxylase (ODC) levels induced by ultraviolet A (UVA) irradiation [53]. Studies in mice have shown that topical application of curcumin can also reduce the metallothionein (MT) gene expression levels induced by UVA after topical application of the tumor promoter TPA. It is suggested that this inhibitory effect of curcumin on TPA + UVA-mediated gene expression may be partially attributed to its role as a ROS scavenger [54], as ROS are believed to play a major role in many pathological conditions including skin cancer [55]. Curcumin has also been shown to reduce TPA-induced oxygen generation and intracellular peroxide formation in differentiated HL-60 cells. TPA-induced hydrogen peroxide (H_2O_2) formation was also significantly reduced in ICR mouse skin [56]. These data suggest that curcumin can partially inhibit tumorigenesis owing to its antioxidant properties [57].

The antioxidant properties of curcumin also allow the suppression of apoptosis in specific cell lines by quenching singlet oxygen formed as a result of UV irradiation. Curcumin prevented UV irradiation-induced apoptotic changes in the human epidermal carcinoma A431 cell, with the antiapoptotic effects including inhibition of JNK activation, PAK2 cleavage, and mitochondrial release of cytochrome *c* [58]. Curcumin has also been shown to act as a radiosensitizer by inducing apoptosis in certain cell lines [59]. One study suggests that curcumin pretreatment enhances the effect of radiation on squamous cell carcinoma SSC2, with a decline in cell survival and cell count observed [60]. Curcumin can inhibit the harmful effects induced by irradiation and can also enhance the effect of radiation. These dual modes of action after irradiation depend on dose and system. It is suggested that curcumin may be useful during radiotherapy, enabling the radiation to kill the tumor cells while protecting the normal cells [59].

Studies suggest that curcumin may be a potential treatment option for psoriasis, a skin disease caused by hyperproliferation of cells in the epidermal layer. This disease is characterized by a pink colored plaque covered by loosely adherent silvery white scales (keratinocytes). *In vitro* work suggests that curcumin inhibits keratinocyte proliferation [61]. Reduction in phosphorylase kinase (PhK) level, decreased keratinocyte transferring receptor expression, and decreased severity of parakeratosis and density of epidermal CD8 + T cells have also been observed in curcumin-treated patients [62]. Since curcumin is a selective PhK inhibitor [63], it is proposed that curcumin-induced suppression of PhK activity may result in resolution of psoriatic activity [62].

The impact of curcumin on wound healing has been extensively explored. Wound healing is a complex process involving inflammation, proliferation, and tissue regeneration. Curcumin has been shown to enhance wound healing in rats and guinea pigs in a full thickness punch wound model. In the curcumin-treated animals, a greater number of myofibroblasts, and thus a faster rate of wound closure, were observed. An increase in the infiltration of fibroblasts, neutrophils, and macrophages, along with extensive neovascularization and greater collagen expression, was also noted [64]. The antioxidant properties of curcumin are believed to contribute to enhanced dermal wound healing because extensive oxidative stress generally inhibits tissue remodeling. Curcumin has shown antioxidant effects *in vitro* by inhibiting hydrogen peroxide-induced damage in human keratinocytes and fibroblasts [65]. The incorporation of curcumin in collagen films has been shown to increase wound reduction, enhance cell proliferation, and provide an efficient free radical scavenger [66]. Pretreatment of curcumin in mice exposed to hemibody γ-irradiation led to improved wound contraction and increased collagen, hexosamine, DNA, and nitric oxide synthesis. Enhanced proliferation of fibroblasts and increased vasculature to the wound bed were also observed [67].

14.2.4 Formulation

Based on its numerous demonstrated pharmacological effects, curcumin has been extensively researched in recent years. However, owing to low bioavailability, low solubility, and *in vivo* stability concerns, oral administration of curcumin is unlikely to provide pharmacologically relevant concentrations. Therefore, a plethora of delivery systems have been formulated with curcumin. The literature contains numerous scientific papers containing, among others, novel emulsions, amphiphilic creams, bioresorbable stents [68], colloidal systems [69–74], bioconjugates [75], and transdermal films [76]. A few delivery systems demonstrating utility on the skin will be discussed below, but in no way is this list exhaustive.

For example, a microemulsion system composed of lecithin and Tween 80 as surfactants and ethyl oleate as the oil phase was created. This microemulsion was shown to prevent the degradation of curcumin and to enhance its permeation into the skin. A comparison of different levels of curcumin loading showed the importance of the concentration gradient and particle diameter on skin permeation enhancement [77].

It is common to use polymer films for transdermal and topical drug delivery systems. A study was conducted by Patel et al. varying the ratios of hydrophilic to hydrophobic polymers together with oleic acid as a permeation enhancer in a film-like delivery system. All formulations prepared were found to be stable with the release of curcumin following Higuchi kinetics. The formulation containing hydrophilic polymer with enhancer showed the highest level of *in vitro* permeation and anti-inflammatory activity against carrageenan-induced mice [78]. In a separate study, polymeric films incorporating curcumin were investigated for suppressing inflammation in the treatment of rheumatoid arthritis. Release rates were determined to be constant up to 9 h with a cumulative release of 50.7% curcumin. These polymeric films produced no signs of irritation in a skin irritation study [79]. Another delivery system, curcumin loaded nanofibers, was shown to have antioxidant and anti-inflammatory properties in a diabetic mouse model. In addition, the dressing used in this study provided sustained release of curcumin over 72 h [80].

Based on its wide range of pharmacological effects and its history of use and safety, curcumin has significant potential for use in cosmetic products. As an anti-inflammatory agent, curcumin has been used in antiaging cosmetics because inflammation is often a primary indicator for the aging process [81]. It has also been used in cosmetics targeting the care of sensitive and irritated skin. The skin protecting properties of curcumin have been recognized for many years, with inhabitants of the northern mountain region of India using *C. longa* to protect their skin against sun rays [3]. Research has shown that it is the antioxidant properties associated with curcumin that allow it to function as a photoprotectant by preventing UV irradiation-induced apoptotic changes in humans [55]. The wound healing capabilities of curcumin also offer an opportunity to incorporate curcumin into cosmetics for their skin regenerative activity. In addition, face masks have been used to take advantage of the anti-inflammatory and antioxidant properties of curcuminoids. A skin irritation test showed no irritation in a rabbit model and in healthy human volunteers. The Draize test and skin pigment analyzer were used to score the presence of erythema and edema on the skin [82].

As mentioned earlier, free radicals, generated through exposure to UV radiation and other environmental stresses, have the capacity to significantly damage the skin. Curcumin, in its role as an antioxidant, has the ability to scavenge these free radicals, but its prominent yellow/orange color somewhat limits its general utility in cosmetic applications. However, hydrogenated derivatives of curcuminoids, called tetrahydrocurcuminoids (THC), retain the biological activities of curcuminoids while appearing colorless. This lack of color makes THC more suitable in many cosmetic applications. It has been investigated as a cosmetic ingredient with antioxidant and moisturizing effects [83]. THC has also been used for skin lightening (by inhibiting the synthesis of melanin) to remove age spots, scars, and blemishes [84].

Curcumin is commonly used as a natural colorant [85] but has also found numerous uses as an antiseptic, in the prevention of dry skin, in the treatment of eczema and acne, and in the slowing down of aging and wrinkle formation. Turmeric and curcumin are used in face masks, washes, lip balms, and sunscreens. Clinical data supporting these label claims are limited because curcumin and turmeric are considered natural products and therefore do not always require clinical trials.

An article titled "Turmeric's Therapeutic Benefits in Cosmetics" from the web site for Cosmetics & Toiletries (cosmeticsandtoiletries.com) details some of the specific uses of turmeric and curcumin in the cosmetic field. Briefly, cleansers including Kama Ayurveda Ubtan Soap-Free Body Cleanser and Vegetical Milk

Infusion (by Skeen plus Vegeticals) are said to soften and smooth the skin while cleansing and purifying the skin. Face masks such as that from Juara are used to remove fine lines and blackheads while nourishing the skin and claim that turmeric "detoxifies, brightens, and strengthens the skin." Well-known skincare companies such as Clinique and Estee Lauder have also introduced turmeric containing cosmetics including eye treatments to hydrate, lighten, and protect the skin; a lip balm to rejuvenate the skin; and a face mask to reduce redness and revitalize and restore radiance to the skin. In addition, turmeric is used extensively as an ingredient in sunscreens for its antioxidant benefit. Sunscreens with turmeric are produced by Lierac, Solaire, and Korres Natural Products.

14.2.5 Safety

Curcumin has a long history of human use and is therefore presumed safe. Since curcumin has recently been considered for a variety of clinical applications, the need for formal safety data has become necessary. One of the preliminary toxicology studies was conducted by Shankar et al. in rats, guinea pigs, and monkeys at doses of 2.5 g turmeric per kilogram of body weight. After 3 weeks of oral dosing, no mortality or morphological and histological changes were observed in any species [86]. Srimal and Dhawan also showed that curcumin is well tolerated in mice at doses as high as 2.0 g/kg administered orally [42]. These data provide evidence that turmeric/curcumin is not toxic when administered orally at high doses.

On the basis of a large accumulation of data in animals, indications suggest that curcumin is not genotoxic, not carcinogenic, and does not have reproductive or teratogenic effects [87]. It is also not mutagenic based on the Ames test [43]. However, in a study where mice and rats were fed ethanolic turmeric extract for 14–90 days, some hepatotoxicity was reported. Focal necrosis with and without regeneration was observed, liver abnormalities were noted in mice at 14 days, and necrosis was noted in rats at 90 days [88]. In addition, curcumin has shown phototoxic effects when examined in numerous bacterial and mammalian cellular models [29].

The first phase I human clinical trial with curcumin was not conducted until 2001. In this study, no treatment-related toxicity was observed when curcumin was administered orally to 25 patients at doses up to 8.0 g/day for a 3-month duration [89]. Curcumin has demonstrated safety in at least six human clinical trials, with only a few grade 1 adverse events noted (including diarrhea, headache, rash, and yellowish stool). These effects were not dose related, and the studies conclude that humans can tolerate doses of curcumin up to 12.0 g [90].

14.3 QUERCETIN

14.3.1 Background

Quercetin (3,3′,4′,5,7-pentahydroxyflavone) is a flavonol, a subclass of flavonoids (Table 14.2), that is commonly found in the plant kingdom. Quercetin serves as a defensive mechanism in plants against UV light [91]. Quercetin has a molecular

TABLE 14.2 Various Subclasses of Flavonoids

Class	Examples
Flavonol	Quercetin, kaempferol, morin, rutin, fisetin
Flavones	Hesperitin, naringin, naringenin
Flavanones	Luteolin, chrysin, apigenin
Flavan-3-ols	Catechin
Isoflavones	Genistein, diadzin
Anthocyanidins	Cyanidin, delphinidin

weight of 302.24 g/mol and a molecular formula of $C_{15}H_{10}O_7$. The structural formula of quercetin is shown in Figure 14.4.

Quercetin has been extensively researched in recent years owing to its numerous pharmacological activities, including its anticarcinogenic [92], antiallergic [93] antioxidant, and anti-inflammatory properties [94, 95]. It has shown utility in the prevention of several conditions including arthritis [95], allergies [96], ulcers [97], diabetes-related complications [98], cataracts [99], cancers, obesity [100], cardiovascular disease [101], and microbial infections [94, 102, 103].

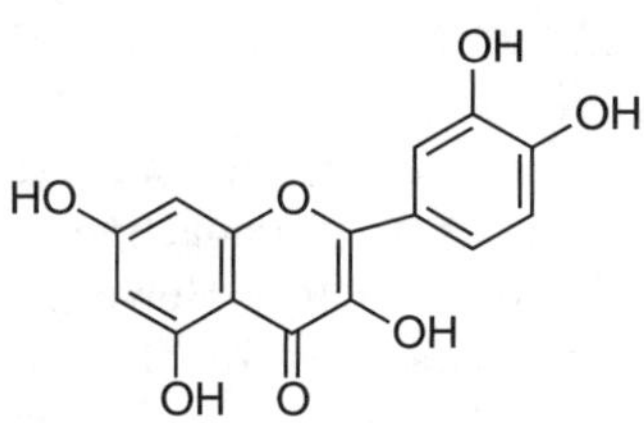

Figure 14.4 Structure of quercetin.

Quercetin is present in numerous edible fruits and vegetables but is predominantly found in onions, teas, apples, berries, and red wine. Hertog et al. [104] reported on flavonol intake (including quercetin) from the Seven Countries Study and reported that tea was the major source of quercetin in the Netherlands and Japan, wine was the major source of quercetin in Italy, and a combination of onions and apples were the chief source of quercetin in the United States, Finland, Greece, and the former Yugoslavia. In this study, the daily intake of flavonoids reported ranged between 3 and 38 mg. A separate study estimated intake of flavonols and flavones in the United States at 20–22 mg/day, of which 73–79.6% was quercetin [105].

14.3.2 Chemistry

Quercetin in fruits and vegetables mainly exists in the form of glycosides (glycone plus aglycone) with quercetin-3-rutinoside (or quercetin-3-rhamnoglucoside or rutin) being the most abundant. Quercetin represents the aglycone part of the glycoside. Other forms of quercetin glycosides include quercetin-4′-glucoside and quercetin-3,4′-glucoside found in onions and quercetin galactosides and quercetin arabinosides

TABLE 14.3 Main Dietary Quercetin Glycosides and Their Sources in the Diet

Quercetin glycoside	Source	Content of quercetin (mg/kg)	Reference
Quercetin-3,4′- glucoside, quercetin-3-glucoside	Onion	284–486	[104]
Quercetin-3-rhamnoglucoside (rutin)	Black tea	10–25	[104]
Quercetin-3-galactoside, quercetin-3-rhamnoside, quercetin-3-arabinoside, quercetin-3-glucoside	Apple	21–72	[104]
Quercetin-3-rhamnoglucoside, quercetin-3-rhamnoside, quercetin-3-galactoside	Blackcurrant	44	[105]

found in apples and berries. Major dietary quercetin glycosides are listed in Table 14.3. The aqueous solubility of quercetin is reported to be as low as 0.55 μM [106]. This can be increased by the addition of lipids and emulsifiers, with co-addition of lipids and emulsifiers yielding the best results. Addition of ethanol has also been shown to increase the solubility of quercetin.

The octanol–water partition coefficient (log P) of quercetin was reported at 1.82 using an adapted shake flask technique [107], 1.20 using an automated continuous flow method [108], and 1.48 using the atom fragment contribution method [109]. All measurements are similar and indicate that quercetin is hydrophobic in nature.

14.3.2.1 Cultivation Factors including the precise part of plant, seasonal variations, agricultural practices, processing conditions, and storage conditions have been shown to affect the yield of quercetin. A comparison of flavonoid content in the flesh and the peel of apples indicated that higher concentrations of quercetin exist in the peel [110]. Seasonal, spatial, and interspecific variations in two traditional varieties of herbal teas showed that the content of quercetin in both varieties of tea reached peak levels in summer and lowest levels in autumn [111]. Significant differences in quercetin content based on geographical location were also observed, with the accumulation of quercetin favored in cooler climates.

A separate study on the common yellow onion (*Allium cepa*) investigated numerous additional variables on flavonol synthesis, including the effect of nitrogen fertilizer level, lifting time, and field curing time [112]. This research showed no effect on quercetin production with variation of nitrogen levels, minor effects due to the lifting times (referring to removing the onions from the ground at different developmental stages), and significant increases in the amount of quercetin after field curing.

Reports suggest that food preparation and storage affect quercetin content in food. Dietary sources of quercetin subjected to boiling and frying showed lower quercetin content compared to their raw equivalent because of thermal degradation and leaching of quercetin into boiling water [91]. However, dietary sources containing quercetin conjugates retain their stability up to temperatures of 100°C. Storage conditions have also been shown to affect quercetin content. As an example, onions lose 25–33% of their quercetin content in the first 12 days of storage [113], while levels in strawberries increased by approximately 32% when stored at −20°C for 9 months.

14.3.2.2 Extraction Quercetin, used in the nutraceutical, cosmetic, pharmaceutical, and food industries, is derived from natural plant sources. To obtain pure quercetin, a series of steps involving extraction, hydrolysis, and purification are performed. Flavonoids including quercetin are commonly extracted from plant materials with solvents including acetone [114], ethyl acetate [115], pure methanol, ethanol, or a combination of methanol/ethanol with water [115–120]. Techniques of extraction commonly include ultrasound–microwave-assisted extraction (UMAE) and Soxhlet extraction. In one study, UMAE was preferred because of a reduction in time required, a higher yield, and better recovery [121]. Quercetin purification from extract is typically performed using various chromatographic techniques, including the use of an octadecyl-functionalized silica flash column, a Sephadex LH-20 column, a silica gel 60 column, and HSCCC [121].

14.3.2.3 Identification Quercetin can be characterized in numerous ways, including ultraviolet spectra, infrared spectroscopy, high-performance liquid chromatography, gas chromatography, mass spectrometry, and nuclear magnetic resonance spectroscopy [121, 122]. Many HPLC methods are found throughout the literature for the determination of quercetin concentration. An example method involves a C18 column with mobile phase of 0.2% phosphoric acid:methanol in a ratio of 50:50 (v/v) and a flow rate of 1.0 mL/min. Column temperature was maintained at 30°C and the UV detection wavelength was set to 360 nm [121].

14.3.2.4 Stability The stability of quercetin at different pH values (2.7, 7, and 10) and different temperatures (room temperature, 4°C, and −20°C) was evaluated [123]. These experiments demonstrated that quercetin is unstable under basic conditions, potentially related to the instability of the center ring structure, resulting in eventual ring fragmentation. Excellent stability was observed under acidic conditions at 4°C and −20°C, although some degradation was observed at room temperature under both acidic and neutral pHs. Quercetin was found to be stable for only short periods at room temperature in human plasma, acetonitrile, and water. Improved quercetin stability is observed in urine samples due to the acidic pH of urine. The stability of quercetin measured by its activity in a lipid peroxidation assay indicated no loss of activity after storage for up to 6 months at 4°C, room temperature, 37°C, and 45°C [124].

The photostability of quercetin (in the range of 290–320 nm) showed no change in the absorption spectrum, indicating quercetin to be sufficiently stable in the UVB range [125]. However, the range of wavelength investigated in this study did not simulate the UVA component (range of 320–400 nm) of natural sunlight. In another study, in which the entire range of wavelength was studied, 23.1% of quercetin content was lost following a 2 h irradiation period [126]. The results of this study compared favorably with another, in which light-induced decomposition of 17–20% occurred after 15 h of exposure to UV radiation [127].

14.3.3 Activity

14.3.3.1 Pharmacological Activity UV rays from the sun are known to deplete the antioxidant systems of the skin and lead to a marked increase of ROS including the superoxide anion, hydrogen peroxide, and hydroxyl radical [128]. The presence of large amounts of ROS subsequently leads to oxidative stress that can (i) damage cell membranes by peroxidation of membrane fatty acids, (ii) lead to DNA breakage, (iii) lead to enzyme inactivation (including free radical scavenger enzymes), and eventually (iv) result in skin damage (e.g., skin aging, tumor promotion, etc.) [129].

Numerous studies have shown that quercetin promotes a wide range of activities, including the scavenging of oxygen radicals [130, 131], protection against lipid peroxidation [132, 133], thereby preventing the free radical chain reaction [134], and complexation of transition metal ions to form inert chelate complexes [135, 136]. A study of the structure–activity relationship of six flavonoids (myricetin, quercetin, kaempferol, luteolin, apigenin, and chrysin) on antioxidant activity in a human dermal fibroblast model revealed that an increase in the number of hydroxyl (OH) groups within the flavonoid was important for efficient free radical scavenging activity. The study ranked the activities as myricetin (six OH) > quercetin (five OH) > kaempferol (four OH) and concluded that flavonoids with a higher number of OH groups are more useful in the prevention of UV stressed skin aging [137].

Quercetin exhibits a higher antiradical activity toward hydroxyl radicals, peroxyl anions, and superoxide anions compared to other flavonoids [138]. These properties of quercetin are due to the presence of three active functional groups in its structure: an *ortho*-dihydroxy or catechol group in the B ring, the 2–3 double bond in conjunction with a 4-oxo function, and the hydroxyl substitution at positions 3 and 5 [130, 139].

Quercetin reduces or prevents photooxidative damage caused by the deeper skin penetration of UVA radiation. It has also been shown to statistically reduce the amount of malondialdehyde (MDA), an end product of lipid peroxidation [140]. Quercetin reduces MDA levels by scavenging the free oxygen radicals, thereby preventing the free radical chain reaction. Similar results have been reported in a separate study where the levels of enzymes involved in lipid peroxidation in the liver of rats exposed to UVA light were measured at different doses and the antioxidant effects of quercetin were determined [141].

14.3.3.2 Treatment of Skin Disease Matrix metalloproteinases (MMPs), a family of zinc- and calcium-dependent endoproteinases, are induced in human skin *in vivo* within hours of exposure to UVB. MMPs are capable of degrading many extracellular matrix components of the skin and cause skin wrinkling and loss of elasticity in both naturally and photoaged skin [142]. Flavonoids, such as quercetin, have demonstrated a reduction in the expression of MMP-1, an interstitial collagenase selective for the degradation of type I collagen, at both the mRNA and protein levels [137]. Quercetin is also effective in protecting fibroblasts, keratinocytes, and endothelial cells from cytotoxic oxidative stress and death. When human dermal cells were exposed to buthionine sulfoximine, which causes cytotoxicity without application of an external stress, quercetin was found to protect the cells from protracted oxidative injury and to reduce cell death in a concentration-dependent manner [143]. The beneficial effect of quercetin in preventing oxidative skin damage has led to its investigation as a potential sunscreen agent. When 10% (w/w) of quercetin was incorporated in an oil/water emulsion, sun protection factor (SPF) values similar to that of homosalate (a reference filter used to establish FDA standards) were obtained [144]. A combination of quercetin and titanium dioxide showed synergistic effects resulting in a product with a SPF of about 30, indicating protection over the whole range of UV [144].

Quercetin has also demonstrated numerous benefits when applied topically to the skin. For one, topical formulations of quercetin have the potential to successfully inhibit UVB-induced skin damage [145]. Studies performed to evaluate the protective effect of two kinds of quercetin formulations (the former being a nonionic emulsion with high lipid content and the latter being an anionic emulsion with low lipid content) on dorsal skins of hairless mice indicated that quercetin was able to reduce UVB-induced oxidative damage by inhibiting increase in myeloperoxidase activity, glutathione depletion, and proteinase secretion/activity [146]. Similarly, Kuniaki et al. formulated a quercetin-derived cosmetic preparation with potent UV absorbent attributes [147].

Like other flavonoids, quercetin possesses excellent antineoplastic properties. However, the use of quercetin as a chemotherapeutic drug has often been marred by its low potency and poor selectivity. To overcome these limitations, Paliwal et al. targeted the delivery of quercetin (50 μM) from a topical formulation into skin cancer cells with pretreatment of the targeted cells with low frequency ultrasound (20 kHz, 2 W/cm^2, 60 s). This treatment not only decreased the LC_{50} of quercetin for skin cancer cells by 80-fold but also destroyed 90% of the skin cancer cells within 48 h [148].

Quercetin plays a curative and preventive role in hypertrophic scarring [149]. Excessive dermal scarring resulting in hypertrophic scars and keloids are serious clinical conditions occurring due to abnormal fibroblast proliferation, collagen overproduction, and contraction following acute or chronic injury [150, 151]. Rapid growth and repair of wounds results in increased histamine synthesis and collagen overproduction indicated by high hydroxyproline levels in the tissue. Experiments performed by El Goweini and Nour [149] on a rabbit ear model showed that quercetin cream applied for 4 weeks resulted in only 40% development of hypertrophic scars compared to 100% in untreated and placebo-treated groups. The quercetin group also showed a significant decrease in histamine and hydroxyproline levels compared to control [149]. The authors concluded that quercetin plays an effective role in preventative treatment of dermal scarring and to a lesser extent for cure of hypertrophic scars due to its antihistaminic and antifibrotic activities.

In another study, *in vitro* evaluations performed on quercetin-treated fibroblasts (cultured from nontreated earlobe keloids and burn hypertrophic scars) showed that quercetin inhibited proliferation and contraction of excessive scar-derived fibroblasts [151]. Similarly, Lim et al. demonstrated dose-dependent inhibitory effects of quercetin on fibroblast overproduction that resulted in keloid and hypertrophic scars [150]. Quercetin has shown an inhibitory effect on fibronectin production by keloid-derived fibroblasts. Immunoassay and electron microscopy techniques indicated an inhibitory effect of quercetin on fibroblast proliferation and production of extracellular matrix fibronectin, suggesting its therapeutic potential in the treatment of excessive scars and other fibroproliferative diseases [152].

Quercetin has also shown additional benefits as a novel dressing material in dermal wound healing. Studies in rats, conducted by Gomathi et al., showed that when quercetin was incorporated into collagenous matrix for dermal wound healing, rats showed better wound healing than control or collagen only treated animals. The results indicated that a quercetin incorporated collagen matrix showed improved wound contraction, increased proliferation of cells with collagen (evident by biomarkers such as hydroxyproline, protein, uronic acid), and increased radical quenching ability [153].

14.3.4 Formulation

Limited permeation of quercetin through the stratum corneum into the deeper skin layers has been observed in *in vitro* permeation studies [154, 155]. Although the liposolubility of quercetin, as shown by a partition coefficient of 1.5–2, is theoretically adequate to permeate the skin, its limited water solubility is believed to hinder permeation and limit bioavailability. Some formulation approaches utilized to improve the solubility and stability of quercetin while increasing permeability into/through the skin are summarized below.

Incorporation of quercetin into microparticles, composed of a lipid matrix stabilized by surfactant molecules, increased the chemical stability and photostability of quercetin [126]. Evaluation of wax-based oil-in-water emulsion systems with varying amounts of lipid (2% and 10%) content showed a faster *in vitro* release of quercetin from the formulation with the lower lipid content, whereas a gradual increase in the quercetin activity with time (measured by antilipoperoxidative activity assay) was detected from the formulation with the higher lipid content. The lipophilic characteristics of quercetin were concluded to decrease its affinity for the components of the formulation, with the lower lipid content resulting in a faster diffusion rate. Although high quercetin activity was detected in the skin samples, no quercetin was detected in the receptor medium after 12 h, indicating the utility of this formulation in topical but not transdermal applications [156].

Incorporation of quercetin into a water-in-oil microemulsion significantly increased the skin penetration of quercetin [157] and required a sixfold lower dose to produce the same *in vivo* effects as obtained with a nonionic emulsion-based system [158]. No quercetin was detected in the receptor phase, indicating only topical delivery. The microemulsion system was also shown to be effective in inhibiting the UVB irradiation-induced decrease in endogenous reduced glutathione (GSH) levels and maintaining GSH levels close to the nonradiated control. Improved skin permeation with the microemulsion system could be attributed to the high thermodynamic activity of the microemulsion system (saturated with quercetin) resulting in a higher penetration rate. The ultralow interfacial tension of the system also ensured excellent contact surface between the skin and the vehicle over the entire application area [159].

Enhancement of quercetin solubility and intradermal delivery using an isopropyl myristate (oil phase), sodium chloride solution (aqueous phase), Tween 80 (surfactant), and ethanol (cosurfactant) microemulsion system was reported [160]. Distribution of quercetin in the deep skin layers was confirmed by skin sectioning using a cryomicrotome.

Kim et al. systematically studied the effect of a wide range of enhancers with different functional groups on the *in vitro* permeation of quercetin and concluded that an optimum balance between the degree of solubilization of drug in the vehicle and the thermodynamic activity of the solution had to be achieved for efficient enhancement [161]. Enhancers with an ester group were shown to promote relatively high enhancement compared to enhancers with ether groups. Among the fatty acid/fatty alcohol enhancers studied, linoleic acid showed the highest enhancement effect [161]. In another experiment, dimethylformamide (DMF) and L-menthol showed L-menthol to be eight times more effective than DMF in improving quercetin penetration [162]. The varying effects of different permeation enhancers can be attributed to the varying influence of enhancers on the biophysical properties of the stratum corneum and their alteration of the skin barrier property [163, 164].

Ester-based prodrugs of quercetin have been synthesized and their water solubility and permeation through excised human skin compared to the parent molecule [165]. The propionyl group was found to have the most optimum balance between water solubility and parent drug lipophilicity required to bring about the highest improvement in skin permeation of quercetin. This prodrug showed an enhancement factor of 12 compared to quercetin. This study concluded that quercetin derivatives with shorter acyl chains are more water soluble than quercetin [165]. This finding was supported by the findings of another study that showed the acetate ester of quercetin to be 450 times more water soluble than quercetin (water solubility of 240.325 μg/mL obtained for quercetin-3-*O*-acetate compared to 0.514 μg/mL obtained for quercetin) [166].

The protective effects of wax-based topical quercetin formulations were evaluated to determine the effect of quercetin in mitigating UVB-induced oxidative stress [158]. UVB irradiation induces prolonged inflammation with an increase in myeloperoxidase (MPO) activity, prompting the use of MPO as a marker for tissue neutrophil content and inflammation [167]. The study results indicated that both quercetin formulations were able to significantly inhibit an increase in MPO activity. Interestingly, the quercetin formulation with the higher lipid content was able to diminish the MPO activity to nonirradiated control levels, indicating a higher retention of quercetin in the epidermis.

In addition, the use of quercetin as an ingredient in the cosmetic field is becoming more and more common. A survey of quercetin containing products has shown that quercetin, typically utilized for its antioxidant properties, is present in creams, face masks, and lip balms. A representative sample of these quercetin containing formulations are discussed here, but it is important to note that clinical data supporting these label claims are limited. Korres (korresusa.com) has produced Quercetin and Oak Age-Reversing day and night creams that claim to support maximum antioxidant protection and create younger, healthier, and more elastic skin. Laura Mercier (lauramercier.com) has produced a Flawless Skin Tone

Perfecting Crème with Quercetin Plus as a moisturizer that incorporates quercetin to diminish skin discoloration, sun damage, and age spots by protecting against free radicals. mdskincare by Dr. Dennis Gross (mdskincare.com) uses quercetin in combination with bisabolol, inhipase, caprylate, and willowherb extract to help soothe redness and reduce inflammation of blemished skin. The formulation is used to treat acne by drawing the blockage from the pore, killing the bacteria, and calming the skin. Ole Henriksen (www.olehenriksen.com) utilizes quercetin in its African Red Tea Exfoliating Lip Salvation to moisturize, nourish, and revitalize the lips.

14.3.5 Safety

A number of *in vitro* studies have been performed with quercetin under different experimental conditions to evaluate its role as an antioxidant and prooxidant. Experiments conducted in mouse thymocytes incubated with 50 μM of quercetin exhibited antioxidant activity with no induction of cytotoxicity [168]. Conversely, when quercetin was tested at concentrations up to 300 μM in human lung embryonic fibroblasts, cytotoxicity was observed with increased levels of ROS [169]. Similarly, in human red blood cells, quercetin oxidized oxyhemoglobin, producing methemoglobin in the presence of extracellular horseradish peroxidase, suggesting its prooxidant nature [170]. Musonda et al. [171] also investigated the antioxidant/prooxidant effects of quercetin on NF-κB and observed that the antioxidant attributes of quercetin surpass its oxidative potential as quercetin inhibited hydrogen peroxide-mediated DNA strand breakage and NF-κB activity. The *in vitro* mutagenic effect of quercetin has been attributed to its prooxidant activity. Reports suggest that in the presence of oxidative conditions (such as hydrogen peroxide, horseradish peroxidase), quercetin degrades to form an *ortho*-semiquinone intermediate, followed by subsequent transformation into *ortho*-quinone accompanied by formation of a ROS. The *ortho*-quinine then undergoes tautomerization to form *ortho*-quinone/quinine methides that bind to DNA and proteins to cause mutagenicity [172].

Despite the discouraging *in vitro* mutagenicity data, quercetin occurs ubiquitously in human diet with a wide distribution in berries, citrus fruits, leafy vegetables, roots, tuber, bulbs, herbs, leaves, legumes, and cereal grains. Daily levels of quercetin range between 3 and 38 mg/day and can be as high as 200–500 mg/day [173]. This diet has been consumed by generations, suggesting no carcinogenic effects in human upon oral administration.

Harwood et al. provides a comprehensive review of quercetin safety and explains the mechanisms for the observed mutagenicity of quercetin *in vitro* and the absence of carcinogenicity *in vivo*. According to the authors, the metabolic fate of quercetin *in vivo*, especially the first-pass effect, significantly reduces the potential for quercetin-mediated adverse effects. The *O*-methylation of quercetin *in vivo* eliminates the potential for formation of potential mutagens. Furthermore, numerous protective mechanisms exist *in vivo* to limit the prooxidant activity of small quantities of systemically available unmetabolized quercetin [173].

To date, no safety data are available for quercetin administration via alternate routes (transdermal, buccal, nasal, etc). These data are critical because these routes avoid first-pass metabolism, thereby avoiding detoxification by phase I and

II enzymes. Since detoxification of quercetin does not occur, there is a possibility of toxicity that may lead to mutation in cells. Moreover, the bioavailability of quercetin aglycone may be increased by nonoral routes of delivery.

Quercetin has been studied for its potential carcinogenicity in numerous long-term experimental animal studies. The majority of these studies are related to the oral administration of quercetin and indicated no evidence of significantly increased incidences of tumor formation or carcinogenicity [173]. One of the few studies involving a skin carcinogenesis model was carried out by Sato et al., in which the tumor initiating activity of quercetin was tested on a two-stage mouse skin model. No apparent tumor initiating activity was observed [174]. In addition, phototoxicity screening of quercetin using a capillary electrophoresis method revealed that quercetin alone was not phototoxic.

14.4 SUMMARY

Quercetin and curcumin, two naturally occurring polyphenols, have been historically recognized to have medicinal utility. Because of this history of utility, recent attempts to understand the numerous biological activities of curcumin and quercetin have been undertaken, resulting in the discovery of their antioxidant, anti-inflammatory, and anticarcinogenic properties. Curcumin and quercetin have been explored for the treatment of numerous skin conditions, including the prevention and treatment of skin cancer, UV radiation damage, psoriasis, wound healing, and hypertrophic scarring. Delivery systems including emulsions, creams, colloidal systems, and transdermal films are currently being developed to increase the bioavailability of these two compounds. Once optimized, curcumin, quercetin, and other similar nutraceuticals should find an even greater utility in the pharmaceutical and cosmetic markets because of the various beneficial effects mentioned above.

REFERENCES

1. Baumann L. Cosmetic natural ingredients: botanical cosmeceuticals. *Household and Personal Care Today*. 2008;14–16.
2. Kapoor L. editor. *Handbook of Ayurvedic Medicinal Plants*. New York: CRC Press, 2001. pp. 149–150.
3. Scartezzini P. Speroni E. Review on some plants of Indian traditional medicine with antioxidant activity. *J. Ethnopharmacol.* 2000;71:23–43.
4. Aggarwal B. Sundaram C. Malani N. Ichikawa H. Curcumin: the Indian solid gold. *Adv. Exp. Med. Biol.* 2007;595:1.
5. Ammon H. Wahl M. Pharmacology of *Curcuma longa*. *Planta Med.* 1991;57:1–7.
6. Vogel E. Pelletier S. Curcumin. *J. Pharm.* 1815;2:50.
7. Goel A. Kunnumakkara A. Aggarwal B. Curcumin as "Curecumin": from kitchen to clinic. *Biochem. Pharmacol.* 2008;75:787–809.
8. Pari L. Tewas D. Eckel J. Role of curcumin in health and disease. *Arch. Physiol. Biochem.* 2008;114:127–149.
9. Lampe V. Milobedzka J. Studien uber Curcumin. *Ber. Dtsch. Chem. Ges.* 1913;46:2235–2240.
10. Jayaprakasha G. Jagan Mohan Rao L. Sakariah K. Chemistry and biological activities of *C. longa*. *Trends Food Sci. Technol.* 2005;16:533–548.

11. Souza C. Osme S. Gloria M. Stability of curcuminoid pigments in model systems. *J. Food Process. Preserv.* 1997;21:353–363.
12. Hossain M, Akamine H, Ishimine Y, Teruya R, Aniya Y, Yamawaki K. Effects of relative light intensity on the growth, yield and curcumin content of turmeric (*Curcuma longa* L.) in Okinawa, Japan. *Plant Prod. Sci.* 2009;12:29–36.
13. Akamine H, Hossain M, Ishimine Y, Yogi K, Hokama K, Iraha Y, Aniya Y. Effects of application of N, P and K alone or in combination on growth, yield and curcumin content of turmeric (*Curcuma longa* L.). *Plant Product. Sci.* 2007;10:151–154.
14. Green C, Hibbert S, Bailey-Shaw Y, Williams L, Mitchell S, Garraway E. Extraction, processing, and storage effects on curcuminoids and oleoresin yields from *Curcuma longa* L. grown in Jamaica. *J. Agric. Food Chem.* 2008;56:3664–3670.
15. Braga M, Leal P, Carvalho J, Meireles M. Comparison of yield, composition, and antioxidant activity of turmeric (*Curcuma longa* L.) extracts obtained using various techniques. *J. Agric. Food Chem.* 2003;51:6604–6611.
16. Sanagi MM, Ahmad UK, Smith RM. Application of supercritical fluid extraction and chromatography to the analysis of turmeric. *J. Chromatogr. Sci.* 1993;31:20–25.
17. Chassagnez-Mendez A, Machado N, Araujo M, Maia J, Meireles M. Supercritical CO_2 extraction of curcumins and essential oil from the rhizomes of turmeric (*Curcuma longa* L.). *Ind. Eng. Chem. Res.* 2000;39:4729–4733.
18. Dandekar D, Gaikar V. Hydrotropic extraction of curcuminoids from turmeric. *Sep. Sci. Technol.* 2003;38:1185–1215.
19. Mandal V, Mohan Y, Hemalatha S. Microwave assisted extraction of curcumin by sample–solvent dual heating mechanism using Taguchi L9 orthogonal design. *J. Pharm. Biomed. Anal.* 2008;46: 322–327.
20. Patel K, Krishna G, Sokoloski E, Ito Y. Preparative separation of curcuminoids from crude curcumin and turmeric powder by pH-zone-refining countercurrent chromatography. *J. Liq. Chromatogr. Relat. Technol.* 2000;23:2209–2218.
21. Taylor S, McDowell I. Determination of the curcuminoid pigments in turmeric (*Curcuma domestica* Val) by reversed-phase high-performance liquid chromatography. *Chromatographia.* 1992;34:73–77.
22. Jayaprakasha G, Rao L, Sakariah K. Improved HPLC method for the determination of curcumin, demethoxycurcumin, and bisdemethoxycurcumin. *J. Agric. Food Chem.* 2002;50: 3668–3672.
23. Wichitnithad W, Jongaroonngamsang N, Pummangura S, Rojsitthisak P. A simple isocratic HPLC method for the simultaneous determination of curcuminoids in commercial turmeric extracts. *Phytochem. Anal.* 2009;20(4):314–319.
24. Liu A, Lou H, Zhao L, Fan P. Validated LC/MS/MS assay for curcumin and tetrahydrocurcumin in rat plasma and application to pharmacokinetic study of phospholipid complex of curcumin. *J. Pharm. Biomed. Anal.* 2006;40:720–727.
25. Sharma R, Gescher A, Steward W. Curcumin: the story so far. *Eur. J. Cancer.* 2005;41: 1955–1968.
26. Tonnesen H, Karlsen J. Studies on curcumin and curcuminoids. VI. Kinetics of curcumin degradation in aqueous solution. *Z. Lebensm. Unters. Forsch.* 1985;180:402–404.
27. Wang Y, Pan M, Cheng A, Lin L, Ho Y, Hsieh C, Lin J. Stability of curcumin in buffer solutions and characterization of its degradation products. *J. Pharm. Biomed. Anal.* 1997;15:1867–1876.
28. Price L. Buescher R. Kinetics of alkaline degradation of the food pigments curcumin and curcuminoids. *J. Food Sci.* 1997;62:267–269.
29. Priyadarsini KI. Photophysics, photochemistry, and photobiology of curcumin: studies from organic solutions, bio-mimetics and living cells. *J. Photochem. Photobiol. C.* 2009;10:81–95.
30. Strimpakos AS, Sharma RA. Curcumin: preventive and therapeutic properties in laboratory studies and clinical trials. *Antioxid. Redox Signal.* 2008;10:511–545.
31. Xu Y, Ku BS, Yao HY, Lin YH, Ma X, Zhang YH, Li XJ. Antidepressant effects of curcumin in the forced swim test and olfactory bulbectomy models of depression in rats. *Pharmacol. Biochem. Behav.* 2005;82:200–206.

32. Aggarwal B, Harikumar K. Potential therapeutic effects of curcumin, the anti-inflammatory agent, against neurodegenerative, cardiovascular, pulmonary, metabolic, autoimmune and neoplastic diseases. *Int. J. Biochem. Cell Biol.* 2009;41:40–59.
33. Shishodia S, Sethi G, Aggarwal BB. Curcumin: getting back to the roots. *Ann. N.Y. Acad. Sci.* 2005;1056:206–217.
34. Wang R, Li YB, Li YH, Xu Y, Wu HL, Li XJ. Curcumin protects against glutamate excitotoxicity in rat cerebral cortical neurons by increasing brain-derived neurotrophic factor level and activating TrkB. *Brain Res.* 2008;1210:84–91.
35. Labbozzetta M, Notarbartolo M, Poma P, Maurici A, Inguglia L, Marchetti P, Rizzi M, Baruchello R, Simoni D, D'Alessandro N. Curcumin as a possible lead compound against hormone-independent, multidrug-resistant breast cancer. *Ann. N.Y. Acad. Sci.* 2009;1155:278–283.
36. Reddy A, Lokesh B. Studies on the inhibitory effects of curcumin and eugenol on the formation of reactive oxygen species and the oxidation of ferrous iron. *Mol. Cell. Biochem.* 1994;137:1–8.
37. Joe B. Lokesh B. Role of capsaicin, curcumin and dietary *n*-3 fatty acids in lowering the generation of reactive oxygen species in rat peritoneal macrophages. *Biochim. Biophys. Acta.* 1994;1224:255.
38. Reddy A, Lokesh B. Effect of curcumin and eugenol on iron-induced hepatic toxicity in rats. *Toxicology.* 1996;107:39–45.
39. Rao M. Curcuminoids as potent inhibitors of lipid peroxidation. *J. Pharm. Pharmacol.* 1994; 46:1013.
40. Menon V, Sudheer A. Antioxidant and anti-inflammatory properties of curcumin. *Adv. Exp. Med. Biol.* 2007;595:105.
41. Mukhopadhyay A, Basu N, Ghatak N, Gujral PK. Anti-inflammatory and irritant activities of curcumin analogues in rats. *Agents Action.* 1982;12:508–515.
42. Srimal R. Dhawan B. Pharmacology of diferuloyl methane (curcumin), a non-steroidal anti-inflammatory agent. *J. Pharm. Pharmacol.* 1973;25:447.
43. Chainani-Wu N. Safety and anti-inflammatory activity of curcumin: a component of turmeric (*Curcuma longa*). *J. Altern. Complement. Med.* 2003;9:161–168.
44. Aggarwal B, Kumar A, Bharti A. Anticancer potential of curcumin: preclinical and clinical studies. *Anticancer Res.* 2003;23:363–398.
45. Duvoix A, Blasius R, Delhalle S, Schnekenburger M, Morceau F, Henry E, Dicato M, Diederich M. Chemopreventive and therapeutic effects of curcumin. *Cancer Lett.* 2005;223:181–190.
46. Leu T. Maa M. The molecular mechanisms for the antitumorigenic effect of curcumin. *Curr. Med. Chem. Anticancer Agents.* 2002;2:357.
47. Maheshwari R, Singh A, Gaddipati J, Srimal R. Multiple biological activities of curcumin: a short review. *Life Sci.* 2006;78:2081–2087.
48. Thangapazham R, Sharma A, Maheshwari R. Beneficial role of curcumin in skin diseases. *Adv. Exp. Med. Biol.* 2007;595:343.
49. Huang M, Smart R, Wong C, Conney A. Inhibitory effect of curcumin, chlorogenic acid, caffeic acid, and ferulic acid on tumor promotion in mouse skin by 12-*O*-tetradecanoylphorbol-13-acetate. *Cancer Res.* 1988;48:5941–5946.
50. Huang M, Ma W, Yen P, Xie J, Han J, Frenkel K, Grunberger D, Conney A. Inhibitory effects of topical application of low doses of curcumin on 12-*O*-tetradecanoylphorbol-13-acetate-induced tumor promotion and oxidized DNA bases in mouse epidermis. *Carcinogenesis.* 1997;18:83.
51. Aburjai T, Natsheh F. Plants used in cosmetics. *Phytother. Res.* 2003;17:987–1000.
52. Huang M, Wang Z, Georgiadis C, Laskin J, Conney A. Inhibitory effects of curcumin on tumor initiation by benzo[*a*]pyrene and 7, 12-dimethylbenz[a]anthracene. *Carcinogenesis.* 1992;13:2183.
53. Ishizaki C, Oguro T, Yoshida T, Wen C, Sueki H, Iijima M. Enhancing effect of ultraviolet A on ornithine decarboxylase induction and dermatitis evoked by 12-o-tetradecanoylphorbol-13-acetate and its inhibition by curcumin in mouse skin. *Dermatology.* 1996;193:311–317.
54. Oguro T, Yoshida T. Effect of ultraviolet A on ornithine decarboxylase and metallothionein gene expression in mouse skin. *Photodermatol. Photoimmunol. Photomed.* 2001;17:71–78.
55. Ahmad N, Katiyar S, Mukhtar H. Antioxidants in chemoprevention of skin cancer. *Curr. Probl. Dermatol.* 2001;29:128–139.

56. Nakamura Y, Ohto Y, Murakami A, Osawa T, Ohigashi H. Inhibitory effects of curcumin and tetrahydrocurcuminoids on the tumor promoter-induced reactive oxygen species generation in leukocytes *in vitro* and *in vivo*. *Jpn. J. Cancer Res.* 1998;89:361–370.
57. Afaq F, Mukhtar H. Photochemoprevention by botanical antioxidants. *Skin Pharmacol. Appl. Skin Physiol.* 2002;15:297–306.
58. Chan W, Wu H. Anti-apoptotic effects of curcumin on photosensitized human epidermal carcinoma A431 cells. *J. Cell. Biochem.* 2004;92:200–212.
59. Jagetia G. Radioprotection and radiosensitization by curcumin. *Adv. Exp. Med. Biol.* 2007;595:301–320.
60. Khafif A, Hurst R, Kyker K, Fliss D, Gil Z, Medina J. Curcumin: a new radio-sensitizer of squamous cell carcinoma cells. *Otolaryngol. Head Neck Surg.* 2005;132:317–321.
61. Pol A, Bergers M, Schalkwijk J. Comparison of antiproliferative effects of experimental and established antipsoriatic drugs on human keratinocytes, using a simple 96-well-plate assay. *In Vitro Cell. Dev. Biol. Anim.* 2003;39:36–42.
62. Heng M, Song M, Harker J, Heng M. Drug-induced suppression of phosphorylase kinase activity correlates with resolution of psoriasis as assessed by clinical, histological and immunohistochemical parameters. *Br. J. Dermatol.* 2000;143:937.
63. Reddy S, Aggarwal B. Curcumin is a non-competitive and selective inhibitor of phosphorylase kinase. *FEBS Lett.* 1994;341:19–22.
64. Sidhu G, Singh A, Thaloor D, Banaudha K, Patnaik G, Srimal R, Maheshwari R. Enhancement of wound healing by curcumin in animals. *Wound Repair Reg.* 1998;6:167–177.
65. Phan TT, See P, Lee ST, Chan SY. Protective effects of curcumin against oxidative damage on skin cells *in vitro*: its implication for wound healing. *J. Trauma.* 2001;51:927.
66. Gopinath D, Ahmed M, Gomathi K, Chitra K, Sehgal P, Jayakumar R. Dermal wound healing processes with curcumin incorporated collagen films. *Biomaterials.* 2004;25:1911–1917.
67. Jagetia G, Rajanikant G. Curcumin treatment enhances the repair and regeneration of wounds in mice exposed to hemibody gamma-irradiation. *Plast. Reconstr. Surg.* 2005;115:515.
68. Nguyen KT, Shaikh N, Shukla KP, Su SH, Eberhart RC, Tang L. Molecular responses of vascular smooth muscle cells and phagocytes to curcumin-eluting bioresorbable stent materials. *Biomaterials.* 2004;25:5333–5346.
69. Kumar V, Lewis SA, Mutalik S, Shenoy DB, Venkatesh, Udupa N. Biodegradable microspheres of curcumin for treatment of inflammation. *Indian J. Physiol. Pharmacol.* 2002;46:209–217.
70. Letchford K, Liggins R, Burt H. Solubilization of hydrophobic drugs by methoxy poly(ethylene glycol)-block-polycaprolactone diblock copolymer micelles: theoretical and experimental data and correlations. *J. Pharm. Sci.* 2008;97:1179–1190.
71. Ma Z, Haddadi A, Molavi O, Lavasanifar A, Lai R, Samuel J. Micelles of poly(ethylene oxide)-*b*-poly (epsilon-caprolactone) as vehicles for the solubilization, stabilization, and controlled delivery of curcumin. *J. Biomed. Mater. Res. A.* 2008;86:300–310.
72. Shaikh J, Ankola DD, Beniwal V, Singh D, Kumar MN. Nanoparticle encapsulation improves oral bioavailability of curcumin by at least 9-fold when compared to curcumin administered with piperine as absorption enhancer. *Eur. J. Pharm. Sci.* 2009;37:223–230.
73. Bisht S, Feldmann G, Soni S, Ravi R, Karikar C, Maitra A, Maitra A. Polymeric nanoparticle-encapsulated curcumin ("nanocurcumin"): a novel strategy for human cancer therapy. *J. Nanobiotechnol.* 2007;5:3.
74. Koppolu B, Rahimi M, Nattama S, Wadajkar A, Nguyen KT. Development of multiple-layer polymeric particles for targeted and controlled drug delivery. *Nanomedicine.* 2010;6(2): 355–361.
75. Kumar S, Dubey KK, Tripathi S, Fujii M, Misra K. Design and synthesis of curcumin-bioconjugates to improve systemic delivery. *Nucleic Acids Symp. Ser.* 2000;75–76.
76. Gopinath D, Ahmed MR, Gomathi K, Chitra K, Sehgal PK, Jayakumar R. Dermal wound healing processes with curcumin incorporated collagen films. *Biomaterials.* 2004;25:1911–1917.
77. Lin C, Lin H, Chen H, Yu M, Lee M. Stability and characterisation of phospholipid-based curcumin-encapsulated microemulsions. *Food Chem.* 2009;116:923–928.

78. Patel NA, Patel NJ, Patel RP. Design and evaluation of transdermal drug delivery system for curcumin as an anti-inflammatory drug. *Drug Dev. Ind. Pharm.* 2009;35:234–242.
79. Verma M, Gupta P, Pokharkar V, Purohit A. Development of transdermal drug dosage formulation for the anti-rheumatic ayurvedic medicinal plants. Available at www.ayurvedam.com/pdf/deverhematic.pdf.
80. Merrell JG, McLaughlin SW, Tie L, Laurencin CT, Chen AF, Nair LS. Curcumin loaded poly(epsilon-caprolactone) nanofibers: diabetic wound dressing with antioxidant and anti-inflammatory properties. *Clin. Exp. Pharmacol. Physiol.* 2009;36(12):1149–1156.
81. Thornfeldt C. Chronic inflammation is etiology of extrinsic aging. *J. Cosmet. Dermatol.* 2008;7:78.
82. Nawanopparatsakul S, Euasathien J, Eamtawecharum C, Benjasirimingokol P, Soiputtan S, Toprasri P, Phaechamud T, Nawanopparatsakul S. Skin irritation test of curcuminoids facial mask containing chitosan as a binder. *Silpakorn Univ. Int. J.* 2005;5:140–147.
83. Arunothayanun P, Wirachwong P, Burananon V, Phisalphong C, Hokputsa S, Suvathi Y, Watnasirichaikul S, Boonkird S. Development of tetrahydrocurcuminoid liposomes as an ingredient for cosmetic products. *Proceedings of the 7th ASCS Conference—Toward a New Horizon: Uniting Cosmetic Science with Oriental Wisdom*, Bangkok, Thailand, 2005.
84. Majeed M, Prakash L.A Lighter Skin Tone and More... With Natural Actives. Sabinsa Cosmetics, 2008.
85. Thornfeldt C. Cosmeceuticals containing herbs: fact, fiction, and future. *Dermatol. Surg.* 2005;31:873–881.
86. Shankar T, Shantha N, Ramesh H, Murthy I, Murthy V. Toxicity studies on turmeric (*Curcuma longa*): acute toxicity studies in rats, guinea pigs & monkeys. *Indian J. Exp. Biol.* 1980;18:73.
87. Hallagan J, Allen D, Borzelleca J. The safety and regulatory status of food, drug and cosmetics colour additives exempt from certification. *Food Chem. Toxicol.* 1995;33:515–528.
88. Deshpande S, Lalitha V, Ingle A, Raste A, Gadre S, Maru G. Subchronic oral toxicity of turmeric and ethanolic turmeric extract in female mice and rats. *Toxicol. Lett.* 1998;95:183–193.
89. Cheng A, Hsu C, Lin J, Hsu M, Ho Y, Shen T, Ko J, Lin J, Lin B, Ming-Shiang W. Phase I clinical trial of curcumin, a chemopreventive agent, in patients with high-risk or pre-malignant lesions. *Anticancer Res.* 2001;21:2895.
90. Lao C, Ruffin M, Normolle D, Heath D, Murray S, Bailey J, Boggs M, Crowell J, Rock C, Brenner D. Dose escalation of a curcuminoid formulation. *BMC Complement. Altern. Med.* 2006;6:10.
91. Aherne SA, O'Brien NM. Dietary flavonols: chemistry, food content, and metabolism. *Nutrition.* 2002;18:75–81.
92. Cragg GM, Newman DJ. Plants as a source of anti-cancer agents. *J. Ethnopharmacol.* 2005;100:72–79.
93. Cheong H, Ryu SY, Oak MH, Cheon SH, Yoo GS, Kim KM. Studies of structure activity relationship of flavonoids for the anti-allergic actions. *Arch. Pharm. Res.* 1998;21:478–480.
94. Hollman PC, Katan MB. Absorption, metabolism and health effects of dietary flavonoids in man. *Biomed. Pharmacother.* 1997;51:305–310.
95. Mamani-Matsuda M, Kauss T, Al-Kharrat A, Rambert J, Fawaz F, Thiolat D, Moynet D, Coves S, Malvy D, Mossalayi MD. Therapeutic and preventive properties of quercetin in experimental arthritis correlate with decreased macrophage inflammatory mediators. *Biochem. Pharmacol.* 2006;72:1304–1310.
96. Anonymous. Quercetin (Monograph). *Altern. Med. Rev.* 1998;3:140–143.
97. Parmar NS, Parmar S. Anti-ulcer potential of flavonoids. *Indian J. Physiol. Pharmacol.* 1998;42:343–351.
98. Shetty AK, Rashmi R, Rajan MGR, Sambaiah K, Salimath PV. Antidiabetic influence of quercetin in streptozotocin-induced diabetic rats. *Nutr. Res.* 2004;24:373–381.
99. Sanderson J, McLauchlan WR, Williamson G. Quercetin inhibits hydrogen peroxide-induced oxidation of the rat lens. *Free Radic. Biol. Med.* 1999;26:639–645.
100. Ahn J, Lee H, Kim S, Park J, Ha T. The anti-obesity effect of quercetin is mediated by the AMPK and MAPK signaling pathways. *Biochem. Biophys. Res. Commun.* 2008;373:545–549.
101. Perez-Vizcaino F, Duarte J, Andriantsitohaina R. Endothelial function and cardiovascular disease: effects of quercetin and wine polyphenols. *Free Radic. Res.* 2006;40:1054–1065.

102. Davis JM, Murphy EA, McClellan JL, Carmichael MD, Gangemi JD. Quercetin reduces susceptibility to influenza infection following stressful exercise. *Am. J. Physiol. Regul. Integr. Comp. Physiol.* 2008;295:R505–R509.
103. Murota K, Terao J. Antioxidative flavonoid quercetin: implication of its intestinal absorption and metabolism. *Arch. Biochem. Biophys.* 2003;417:12–17.
104. Hertog MG, Kromhout D, Aravanis C, Blackburn H, Buzina R, Fidanza F, Giampaoli S, Jansen A, Menotti A, Nedeljkovic S, et al. Flavonoid intake and long-term risk of coronary heart disease and cancer in the seven countries study. *Arch. Intern. Med.* 1995;155:381–386.
105. Sampson L, Rimm E, Hollman PC, de Vries JH, Katan MB. Flavonol and flavone intakes in US health professionals. *J. Am. Diet Assoc.* 2002;102:1414–1420.
106. Azuma K, Ippoushi K, Ito H, Higashio H, Terao J. Combination of lipids and emulsifiers enhances the absorption of orally administered quercetin in rats. *J. Agric. Food Chem.* 2002;50:1706–1712.
107. Rothwell JA, Day AJ, Morgan MR. Experimental determination of octanol-water partition coefficients of quercetin and related flavonoids. *J. Agric. Food Chem.* 2005;53:4355–4360.
108. Brown JE, Khodr H, Hider RC, Rice-Evans CA. Structural dependence of flavonoid interactions with Cu^{2+} ions: implications for their antioxidant properties. *Biochem. J.* 1998;330:1173–1178.
109. Meylan WM, Howard PH. Atom/fragment contribution method for estimating octanol-water partition coefficients. *J. Pharm. Sci.* 1995;84:83–92.
110. D'Abrosca B, Pacifico S, Cefarelli G, Mastellone C, Fiorentino A. 'Limoncella' apple, an Italian apple cultivar: phenolic and flavonoid contents and antioxidant activity. *Food Chem.* 2007;104:1333–1337.
111. Ma M, Hong CL, An SQ, Li B. Seasonal, spatial, and interspecific variation in quercetin in *Apocynum venetum* and *Poacynum hendersonii,* Chinese traditional herbal teas. *J. Agric. Food Chem.* 2003;51:2390–2393.
112. Mogren LM, Olsson ME, Gertsson UE. Quercetin content in field-cured onions (*Allium cepa* L.): effects of cultivar, lifting time, and nitrogen fertilizer level. *J. Agric. Food Chem.* 2006;54:6185–6191.
113. Beecher GR. Overview of dietary flavonoids: nomenclature, occurrence and intake. *J. Nutr.* 2003;133:3248S–3254S.
114. Schieber A, Hilt P, Conrad J, Beifuss U, Carle R. Elution order of quercetin glycosides from apple pomace extracts on a new HPLC stationary phase with hydrophilic endcapping. *J. Sep. Sci.* 2002;25:361–364.
115. Spanos GA, Wrolstad RE. Influence of processing and storage on the phenolic composition of Thompson Seedless grape juice. *J. Agric. Food Chem.* 2002;38:1565–1571.
116. Hertog MGL, Hollman PCH, Katan MB. Content of potentially anticarcinogenic flavonoids of 28 vegetables and 9 fruits commonly consumed in the Netherlands. *J. Agric. Food Chem.* 2002;40:2379–2383.
117. Molnar-Perl I, Fuzfai Z. Chromatographic, capillary electrophoretic and capillary electro-chromatographic techniques in the analysis of flavonoids. *J. Chromatogr. A.* 2005;1073:201–227.
118. Robards K. Strategies for the determination of bioactive phenols in plants, fruit and vegetables. *J. Chromatogr. A.* 2003;1000:657–691.
119. Naczk M, Shahidi F. Extraction and analysis of phenolics in food. *J. Chromatogr. A.* 2004;1054:95–111.
120. Wach A, Pyrzynska K, Biesaga M. Quercetin content in some food and herbal samples. *Food Chem.* 2007;100:699–704.
121. Huang L, Cao Y, Chen G. Purification of quercetin in *Anoectochilu roxburghii* (wall) Lindl using UMAE by high-speed counter-current chromatography and subsequent structure identification. *Sep. Purif. Technol.* 2008;64:101–107.
122. Tokusoglu O, Unal MK, Yildirim Z. HPLC–UV and GC–MS characterization of the flavonol aglycons quercetin, kaempferol and myricetin in tomato pastes and other tomato-based products. *Acta Chromatogr.* 2003;13:196–207.
123. Moon YJ, Wang L, DiCenzo R, Morris ME. Quercetin pharmacokinetics in humans. *Biopharm. Drug Dispos.* 2008;29:205–217.
124. Casagrande R, Georgetti S. R, Verri WA, Jabor JR, Santos AC, Fonseca MJV. Evaluation of functional stability of quercetin as a raw material and in different topical formulations by its antilipoperoxidative activity. *AAPS PharmSciTech.* 2006;7:E10.

125. Vicentini FTMC, Georgetti SR, Jabor JR, Caris JA, Bentley MVLB, Fonseca MJV. Photostability of quercetin under exposure to UV irradiation. *Latin Am. J. Pharm.* 2007;26:119–124.
126. Scalia S, Mezzena M. Incorporation of quercetin in lipid microparticles: effect on photo- and chemical-stability. *J. Pharm. Biomed. Anal.* 2009;49:90–94.
127. Smith GJ, Thomsen SJ, Markham KR, Andary C, Cardon D. The photostabilities of naturally occurring 5-hydroxyflavones, flavonols, their glycosides and their aluminium complexes *J. Photochem. Photobiol. A.* 2000;136:87–91.
128. Wiseman H, Halliwell B. Damage to DNA by reactive oxygen and nitrogen species: role in inflammatory disease and progression to cancer. *Biochem. J.* 1996;313:17–29.
129. Fuchs J, Packer L. *Oxidative Stress: Oxidants and Antioxidants*. London: Academic Press, 1991. pp. 559–583.
130. Bors W, Heller W, Michel C, Saran M. Flavonoids as anti-oxidant: determination of radical-scavenging efficiencies. *Methods Enzymol.* 1990;186:343–355.
131. Bors W, Michel C, Saran M. Flavonoid antioxidants: rate constants for reactions with oxygen radicals. *Methods Enzymol.* 1994;234:420–429.
132. Laughton MJ, Evans PJ, Moroney MA, Hoult JR, Halliwell B. Inhibition of mammalian 5-lipoxygenase and cyclo-oxygenase by flavonoids and phenolic dietary additives: relationship to antioxidant activity and to iron ion-reducing ability. *Biochem. Pharmacol.* 1991;42:1673–1681.
133. Decharneux T, Dubois F, Beauloye C, Wattiaux-De Coninck S, Wattiaux R. Effect of various flavonoids on lysosomes subjected to an oxidative or an osmotic stress. *Biochem. Pharmacol.* 1992;44:1243–1248.
134. Torel J, Cillard J, Cillard P. Antioxidant activity of flavonoids and reactivity with peroxy radicals. *Phytochemistry*. 1986;25:383–386.
135. Afanas'ev IB, Dorozhko AI, Brodskii AV, Korstyuk VA, Potapovitch AI. Chelating and free radical scavenging mechanisms of inhibitory action of rutin and quercetin in lipid peroxidation. *Biochem. Pharmacol.* 1989;38:1763–1769.
136. Cao G, Sofic E, Prior RL. Antioxidant and prooxidant behavior of flavonoids: structure–activity relationships. *Free Radic. Biol. Med.* 1997;22:749–760.
137. Sim G-S, Lee B-C, Cho HS, Lee JW, Kim J-H, Lee D-H, Kim J-H, Pyo H-B, Moon DC, Oh K-W, Yun YP, Hong JT. Structural activity relationship of antioxidative property of flavonoids and inhibitory effect on matrix metalloproteinase activity in UVA-irradiated human dermal fibroblast. *Arch. Pharm. Res.* 2007;30:290–298.
138. Morel I, Lescoat G, Cogrel P, Sergent O, Pasdeloup N, Brissot P, Cillard P, Cillard J. Antioxidant and iron-chelating activities of the flavonoids catechin, quercetin and diosmetin on iron-loaded rat hepatocyte cultures. *Biochem. Pharmacol.* 1993;45:13–19.
139. Saija A, Scalese M, Lanza M, Marzullo D, Bonina F, Castelli F. Flavonoids as antioxidant agents: importance of their interaction with biomembranes. *Free Radic. Biol. Med.* 1995;19: 481–486.
140. Inal ME, Kahraman A, Koken T. Beneficial effects of quercetin on oxidative stress induced by ultraviolet A. *Clin. Exp. Dermatol.* 2001;26:536–539.
141. Inal ME, Kahraman A. The protective effect of flavonol quercetin against ultraviolet A induced oxidative stress in rats. *Toxicology*. 2000;154:21–29.
142. Fisher GJ, Datta SC, Talwar HS, Wang ZQ, Varani J, Kang S, Vorhees JJ. Molecular basis of sun-induced premature skin ageing and retinoid antagonism. *Nature*. 1996;379:335–339.
143. Skaper SD, Fabris I, Ferrari V, Carbonare AD, Leon A. Quercetin protects cutaneous tissue-associated cell types including sensory neurons from oxidative stress induced by glutathione depletion: cooperative effects of ascorbic acid. *Free Radic. Biol. Med.* 1997;22:669–678.
144. Choquenet B, Couteau C, Paparis E, Coiffard LJ. Quercetin and rutin as potential sunscreen agents: determination of efficacy by an *in vitro* method. *J. Nat. Prod.* 2008;71:1117–1118.
145. Gonzalez S, Fernandez-Lorente M, Gilaberte-Calzada Y. The latest on skin photoprotection. *Clin. Dermatol.* 2008;26:614–626.
146. Casagrande R, Georgetti SR, Verri WA, Jr, Dorta DJ, dos Santos AC, Fonseca MJ. Protective effect of topical formulations containing quercetin against UVB-induced oxidative stress in hairless mice. *J. Photochem. Photobiol. B.* 2006;84:21–27.

147. Kuniaki S, Ishimura Y, Nobuyuki N, inventors; Quercetin and rutin derivatives as ultraviolet absorbents. *JP Patent* 01308476, 1988.
148. Paliwal S, Sundaram J, Mitragotri S. Induction of cancer-specific cytotoxicity towards human prostate and skin cells using quercetin and ultrasound. *Br. J. Cancer.* 2005;92:499–502.
149. El Goweini MF, Nour NM. Effect of quercetin on excessive dermal scarring. *Egyptian Dermatol.* 2005;1:1–10.
150. Lim I, Phan TT, Lee ST. Quercetin inhibits keloid and hypertrophic scar fibroblast proliferation and collagen production. *ANZ J. Surg.* 2003;73:A286.
151. Phan TT, Sun L, Bay BH, Chan SY, Lee ST. Dietary compounds inhibit proliferation and contraction of keloid and hypertrophic scar-derived fibroblasts *in vitro*: therapeutic implication for excessive scarring. *J. Trauma.* 2003;54:1212–1224.
152. Phan TT, Lim IJ, Sun L, Chan SY, Bay BH, Tan EK, Lee ST. Quercetin inhibits fibronectin production by keloid-derived fibroblasts: implication for the treatment of excessive scars. *J. Dermatol. Sci.* 2003;33:192–194.
153. Gomathi K, Gopinath D, Rafiuddin Ahmed M, Jayakumar R. Quercetin incorporated collagen matrices for dermal wound healing processes in rat. *Biomaterials.* 2003;24:2767–2772.
154. Bonina F, Lanza M, Montenegro L, Puglisi C, Tomaino A, Trombetta D, Castelli F, Saija A. Flavonoids as potential protective agents against photo-oxidative skin damage. *Int. J. Pharm.* 1996;145:87–94.
155. Saija A, Tomaino A, Trombetta D, Giacchi M, Pasquale AD, Bonina F. Influence of different penetration enhancers on *in vitro* skin permeation and *in vivo* photoprotective effect of flavonoids. *Int. J. Pharm.* 1998;175:85–94.
156. Casagrande R, Georgetti SR, Verri WA, Borin MF, Lopez RF, Fonseca MJ. *In vitro* evaluation of quercetin cutaneous absorption from topical formulations and its functional stability by antioxidant activity. *Int. J. Pharm.* 2007;328:183–190.
157. Vicentini FT, Simi TR, Del Ciampo JO, Wolga NO, Pitol DL, Iyomasa MM, Bentley MV, Fonseca MJ. Quercetin in w/o microemulsion: *in vitro* and *in vivo* skin penetration and efficacy against UVB-induced skin damages evaluated *in vivo*. *Eur. J. Pharm. Biopharm.* 2008;69:948–957.
158. Casagrande R, Georgetti S. R, Verri WA, Dorta DJ, Santos AC, Fonseca MJV. Protective effect of topical formulations containing quercetin against UVB-induced oxidative stress in hairless mice. *J. Photochem. Photobiol. B.* 2006;84:21–27.
159. Kreilgaard M. Influence of microemulsions on cutaneous drug delivery. *Adv. Drug Deliv. Rev.* 2002;54:S77–S98.
160. Kitagawa S, Tanaka Y, Tanaka M, Endo K, Yoshii A. Enhanced skin delivery of quercetin by microemulsion. *J. Pharm. Pharmcol.* 2009;61:855–860.
161. Kim HW, Gwak HS, Chun IK. The effect of vehicles and pressure sensitive adhesives on the percutaneous absorption of quercetin through the hairless mouse skin. *Arch. Pharm. Res.* 2004;27:763–768.
162. Olivella MS, Lhez L, Pappano NB, Debattista NB. Effects of dimethylformamide and L-menthol permeation enhancers on transdermal delivery of quercetin. *Pharm. Dev. Technol.* 2007;12:481–484.
163. Green PG, Guy RH, Hadgraft J. *In vitro* and *in vivo* enhancement of skin permeation with oleic acid and lauric acids. *Int. J. Pharm.* 1988;48:103–111.
164. Yamada M, Uda Y, Tanigawara Y. Mechanism of enhancement of percutaneous absorption of molsidomine by oleic acid. *Chem. Pharm. Bull.* 1987;35:3399–3406.
165. Montenegro L, Carbone C, Maniscalco C, Lambusta D, Nicolosi G, Ventura CA, Puglisi G. *In vitro* evaluation of quercetin-3-*O*-acyl esters as topical prodrugs. *Int. J. Pharm.* 2007;336:257–262.
166. Saija A, Tomaino A, Trombetta D, Pellegrino ML, Tita B, Messina C, Bonina FP, Rocco C, Nicolosi G, Castelli F. 'In vitro' antioxidant and photoprotective properties and interaction with model membranes of three new quercetin esters. *Eur. J. Pharm. Biopharm.* 2003;56:167–174.
167. Bradley PP, Priebat DA, Christensen R. D, Rothstein G. Measurement of cutaneous inflammation: estimation of neutrophil content with an enzyme marker. *J. Invest. Dermatol.* 1982;78:206–209.
168. Lee J-C, Kim J, Park J-K, Chung G-H, Jang Y-S. The antioxidant, rather than prooxidant, activities of quercetin on normal cells: quercetin protects mouse thymocytes from glucose oxidase-mediated apoptosis. *Exp. Cell Res.* 2003;291:386–397.

169. Matsuo M, Sasaki N, Saga K, Kaneko T. Cytotoxicity of flavonoids toward cultured normal human cells. *Biol. Pharm. Bull.* 2005;28:253–259.
170. Galati G, Sabzevari O, Wilson JX, O'Brien PJ. Prooxidant activity and cellular effects of the phenoxyl radicals of dietary flavonoids and other polyphenolics. *Toxicology.* 2002;177:91–104.
171. Musonda CA, Chipman JK. Quercetin inhibits hydrogen peroxide (H_2O_2)-induced NF-kappaB DNA binding activity and DNA damage in HepG2 cells. *Carcinogenesis.* 1998;19:1583–1589.
172. Metodiewa D, Jaiswal AK, Cenas N, Dickancaite E, Segura-Aguilar J. Quercetin may act as a cytotoxic prooxidant after its metabolic activation to semiquinone and quinoidal product. *Free Radic. Biol. Med.* 1999;26:107–116.
173. Harwood M, Danielewska-Nikiel B, Borzelleca JF, Flamm GW, Williams GM, Lines TC. A critical review of the data related to the safety of quercetin and lack of evidence of *in vivo* toxicity, including lack of genotoxic/carcinogenic properties. *Food Chem. Toxicol.* 2007;45:2179–2205.
174. Sato H, Takahashi M, Furukawa F, Miyakawa Y, Hasegawa R, Toyoda K, Hayashi Y. Initiating potential of 2-(2-furyl)-3-(5-nitro-2-furyl)acrylamide (AF-2), butylated hydroxyanisole (BHA), butylated hydroxytoluene (BHT) and 3,3′,4′,5,7-pentahydroxyflavone (quercetin) in two-stage mouse skin carcinogenesis. *Cancer Lett.* 1987;38:49–56.

CHAPTER 15

AYRUVEDIC INGREDIENTS IN COSMETICS

Rajarajeswari Sivalenka and Mangathayaru Putrevu

Celgene Cellular Therapeutics, Warren, NJ

15.1 INTRODUCTION

Ayurveda (ā-yur-ve-da) transpired as a science thousands of years ago in the Indian subcontinent and formed the basis for traditional Indian medicine that exists to this day. The principles of ayurveda are based not only on healing but also on healthy living. Fundamentally, ayurveda means "knowledge of life" ("Ayur" meaning life and "veda" meaning knowledge). The ultimate goal of Ayurveda is best described in the verse from the ancient text *Caraka Samhita* as "ayurvedo amritam," which means the purpose of ayurveda is to achieve immortality. It manifests in the use of natural and herbal wealth for the health and well-being of human kind. Its vast theoretical knowledge base stems from ages of trial and error with the use of a multitude of herbal extracts and traditional practices. However, a lot of these remain yet to be scientifically validated. Ayurvedic medicine has not been explored to such an extent as other cultures' natural remedies like the Chinese or European herbal medicine. The increasing demand for naturals in the Western market drives the need to explore many amazing benefits that ayurveda has to offer.

15.2 ORIGIN OF AYURVEDA AND ITS EVOLUTION

An ancient wisdom of healing system in Indian subcontinent contended in the spiritual tradition or a Vedic religion called the "Sanatana dharma," way before writing was invented. This philosophy used herbs and minerals as a natural cure for healing and treating disease. The principles of this tradition formed parts of the Vedas—the most ancient Hindu scriptures. The principles and essence of the Vedas is divided into four parts—Rig veda, Sama veda, Yajur veda, and Atharva veda. These sections contain the spiritual insights for self-realization and were first inscribed by the great sage Veda Vyasa. The concept of ayurveda is part of the fourth veda the

Formulating, Packaging, and Marketing of Natural Cosmetic Products, First Edition.
Edited by Nava Dayan and Lambros Kromidas.

"Atharva veda." Information regarding plants, minerals, animal products, and their usage for medical purposes has been described in this veda [1]

Origin of ayurveda has been described variously in several legends. According to one school of thought, ayurveda originated from Brahma, who is believed to be the Creator according to Hindu mythology. He preached this wisdom to Daksha who then passed it on to Indra. During adversity, sages prayed to Indra who in turn imparted this knowledge to the sage Atreya. The original Atreya philosophy was documented by his disciple Agnivesha, as *Agnivesha Samhita*. Another legend suggests that the God of healing, Divodasa Dhanvantri, wrote the first text on ayurveda [2]. The above two works were lost in time although passed down the generations via preaching, myths, and legends from sages.

The texts on ayurveda were later documented by sages Caraka, Sushruta, and Vaghbata, which still exist today, evidencing the ancient traditions. Caraka, who was a disciple of Atreya, compiled the first state-of-the-art text on ayurveda as early as 700 BC entitled "Caraka Samhita." Sushruta's text was based on the Dhanavantri school and is named "Sushrutha Samhita." It mainly deals with surgery. A third treatise integrating both texts was compiled by Vaghbata.

Around 1500 BC, ayurveda became the most respected and widely used form of healing practiced in India. During this time ayurvedic medicine diverged into two schools of thought, namely, the Atreya philosophy (school of physicians) and Dhanvantri (school of surgeons) named after the sages who preached them. Ayurveda takes into account the five elements that form the basis of life: earth, water, fire, air, and ether. The three aspects of ayurvedic medicine are described as "Trisutras" that are (1) disease, (2) symptoms, and (3) treatment. These were further elaborated into eight divisions whose practical doctrines of ayurveda are called "ashtanga ayurveda," and include internal medicine, surgery, organic medicine, pediatrics, toxicology, rejuvenating remedy, aphrodisiac remedies, and spiritual healing.

Even today, a vast majority of population around the world still relies on traditional medicine and use of plants and plant extracts for disease treatment and healthy living. Over the past two decades, at least 121 clinically useful prescription drugs have come to be recognized that are derived from higher plants [3]. About a third of these have been adapted from their use in traditional medicine. Many of the plant extracts used in ayurveda are now scrutinized by modern researchers and investigated to prove their asserted benefits. With the coming of age of modern disciplines and techniques in chemistry and biology, many natural products are reinvestigated to prove their biological activities [4]. The use of several ayurvedic crude medicinal plants has found concordance with application in modern medicine. *Asparagus racemosus* (Satavari, asparagus), *Cedrus deodara* (Devadaru), and *Psoralea corylifolia* (Bakuchi, Babchi) are a few good examples of plants that were studied extensively to validate their impending claims. Of the known traditional medicines, approximately one-third are intended for treatment of wounds or skin disorders, as compared to less than 3% of modern drugs [5]. However, the use of such medicinal plant extracts for treatment of skin disorders is based on historical and anecdotal evidence since there has been relatively little data available in the scientific literature on their performance in controlled clinical trials.

15.3 AYURVEDIC BASIS FOR CLASSIFICATION OF SKIN TYPES AND THE TRIDOSHA CONCEPT

The unique holistic approach of ayurveda advocates customized treatments, considering individual constitutions or body types. According to the principles of ayurveda, the human body is known to contain three primal forces called the "tridosha." These three forces are vata, pitta, and kapha that act in tandem. These three factors are represented individually by (1) vata for air or the energy, (2) pitta for fire or the chemical activity, and (3) kapha for earth or material elements. In terms of human body, these would connote as vata for bodily energy, pitta for metabolism (digestion, respiration, etc.), and kapha for material composition of the body. Based upon which of these doshas or combinations dominate in a person, the individual can be classified into various types such as vata, pitta, or kapha or vata-pitta, vata-kapha, or kapha-pitta, and so are their skin types. An imbalance of equilibrium of the three doshas causes diseases indicated by an increase or decrease in one, two, or all of these three doshas.

15.3.1 Vata

The skin of a person with "vata" constitution is described to be characteristically dry, rough, cold, thin with fine pores, and wrinkled. It easily gets dehydrated and aggravation of vata dosha leads to premature aging. Dryness, atrophy, pricking pain, contraction, dilation, hardness, roughness, horripilation, and dusky red coloration have been the findings for skin lesions due to vata imbalance.

15.3.2 Pitta

In case of pitta constitution, the skin is delicate, fair and red toned with freckles, moles, or acne of medium thickness and gets sunburns easily. Pitta dosha of skin is characteristically termed "Bhrajaka Pitta." Skin pigmentation is described to be effected by this dosha. Pitta skin usually results from sun damage, irregular eating or sleep habits, and consumption of certain foods that aggravate this dosha such as spices, pepper, and some kinds of nuts, among others. Rosacea is hypothesized to be a result of pitta dosha. Other skin manifestations due to pitta imbalance are burning sensation, redness, and exudation, softening, and sloughing of affected areas.

15.3.3 Kapha

Kapha skin is described to be thicker and softer. Oily, moist, pale skin with a tendency toward large pores is typically a constitution of kapha dosha. Skin changes due to its imbalance range from whiteness, coldness, pruritus, heaviness, sliminess, and softening. Oil reducing, exfoliating, and clarifying products are therefore suggested to treat this imbalance.

15.4 SKIN AILMENTS AND AYURVEDIC TREATMENTS

Various dermatological conditions and treatment options are elaborated in Ayurveda [6, 7]. Eighteen basic types of skin afflictions are listed to result from changes and imbalances in skin types or doshas described in previous section.

Vitiligo/leucoderma in modern medicine has been described in Atharva veda as disease "Kilasa" along with its prognosis. Several herbal prescriptions are cited for its treatment in ayurveda [8]. One such prescription is powdered seeds of *Psoralea corylifolia* (commonly called Bakuchi). Psoralea is an erect annual found almost throughout India. Besides vitiligo, it is also valued in ayurveda for the treatment of other skin diseases such as psoriasis and inflammatory diseases of the skin [9]. The active principal of this plant called psolaren was isolated in mid-1930s. Its function was later recognized in the formations of skin pigment melanin [10]. It was therefore used to treat leucoderma (white patches/vitiligo). Besides psoralen, meroterpene bakuchiol and other related compounds were isolated from this plant. Bakuchiol has been shown to possess potent antibacterial activity and is also useful for the treatment of psoriasis.

Ocimum sanctum also called holy basil is a plant whose chemical constituents are demonstrated to possess various pharmacological activities [11]. According to ayurveda, *O. sanctum* has vast number of therapeutic applications evidenced in other ancient medicines such as Greek, Roman, Siddha, and Unani. Besides treating many pathological conditions, it is known for treatment of leucoderma (white patches), ringworm (fungal infection), and other skin diseases.

Plants like aloe vera are used for healing of wounds and burn injuries. Tea tree (*Melaleuca alternifolia*) oil is suggested for its antifungal, antiviral, antibacterial, and acaricidal activity against skin infections such as acne, herpes, and scabies [5].

Turmeric (*Curcuma longa*) is proven for its antitumor promoting activity against skin cancer. Curcumin, the active component in turmeric that has found vast use in ayurveda over ages, was isolated two centuries back and its chemical structure was identified as diferuloylmethane. Therapeutic activities of curcumin have been attributed to its antioxidant, anti-inflammatory, antiviral, antibacterial, antifungal properties.

Neem (*Azadirachta indica* ADR) leaf and its constituents have been demonstrated to exhibit anti-inflammatory, immunomodulatory, antihyperglycemic, antiulcer, antimalarial, antifungal, antibacterial, antiviral, antioxidant, antimutagenic, and anticarcinogenic properties [12]. Curcumin and neem are used for healing chronic ulcers and scabies in ayurveda. Its efficacy was proven in studies where scabies patients were treated with "neem" and "turmeric" used as a paste on 814 patients [13]. In 97% of the cases, cure was obtained within 3–15 days of treatment.

The essential oil from sandalwood (*Santalum album*), in the form of emulsion or paste, has been used in India as an ayurvedic medicinal agent for the treatment of inflammatory and eruptive skin diseases. The chemopreventive effects of sandalwood were later demonstrated in a mouse model study [14]. This investigation tested the effect of sandalwood oil (5% in acetone, w/v) in a model where skin papillomas were induced by DMBA (7,2-dimethylbenz(a)anthracene). Sandalwood oil pretreatment was shown to significantly decrease papilloma incidence evidencing its chemopreventive potential against skin cancer.

As exemplified above, therapeutic benefits of various ayurvedic herbs have come to be well recognized and used for various health benefits including delaying symptoms of aging. Current developments and research on pharmacognosy, chemistry, and pharmacology and clinical therapeutics of ayurvedic medicinal extracts have led to such evidences of medicinal use of numerous molecules from ayurvedic experimental base to be beneficial [15]. While the use of such plants and their extracts has been shown to be beneficial in treating ailments, there are also multiple components or chemical constituents of the same plant extract that may function under different conditions or treat varied ailments. Further extracts of different parts of a plant may also be effective in similar or different pathological conditions. Also, certain side effects, or in some cases adverse effects, have been noted and reviewed at several instances in the literature [5]. Among such effects are (1) irritant contact dermatitis caused mechanically (spines or irritant hairs from plant products) or by irritant chemicals in plant sap (especially members of the Ranunculaceae, Euphorbiaceae, and Compositae plant families); (2) phytophotodermatitis resulting from skin contamination by plants containing furocoumarins and subsequent exposure to UV light or phototoxicity (notably members of the Umbelliferae and Rutaceae plant families); and (3) immediate (type I) or delayed hypersensitivity contact reactions mediated by the immune system in individuals sensitized to plants or plant products (e.g., peanut allergy, poison ivy).

While numerous plant-based products, as exemplified above, form the basis for ayurvedic treatments, the main rituals or procedures involved in topical treatments for skin ailments include body massages with such extracts, detoxification by cleansing with clays, and herbal masks. These methods not only are well established in ayurvedic treatments but also have been constantly improvised over the years.

Ayurvedic formulations in general come in various forms such as powders, decoction, fresh juices, tablets, oils, ghee (clarified butter preparation), asavas/arishtas (alcoholic preparations). Most topical applications are powdered extracts from various herbs called "churnas" or oxidized metals or minerals called "bhasmas." These two forms can in turn be incorporated into liquid formulations, which can also be applied topically.

15.5 EXAMPLES OF COMMON AYURVEDIC HERBS FOR COSMETIC USE

In addition to the therapeutic applications of ayurveda, topical cosmetic applications for sensual appeal are also quite well known. Since times immemorial, the use of essential aromatic herbal oils for perfumery, the use of sandalwood, turmeric extracts, and so on for skin care, the use of herbal coloring agents such as henna, and the use of neem extracts in soaps for antimicrobial effects have been very popular. Table 15.1 shows some common ayurvedic herbs that have come to be used most commonly for various cosmetic applications. Most of the herbs and their extracts confer multiple benefits and the use of combination of several herbal extracts for one application is

TABLE 15.1 Some Common Ayurvedic Herbs Most Commonly Used for Various Cosmetic Applications

Herb	Scientific name	Plant part	Application
Cleansing			
Shikakai	*Acacia concinna*	Pods	Shampoos
Myrrh	*Balsamodendron myrrha*	Gum	Soaps and shampoos
Soap wort	*Sapindus trifoliatus*	Fruit	Detergent
Anitmicrobial			
Neem	*Azadirachta indica*	Leaves	Soaps, oral care
Devadaru	*Cedrus deodara*	Stem wood	Soaps
Zedoary	*Curcuma zedoaria*	Rhizome	Soaps
Wound healing			
Turmeric	*Curcuma longa*	Rhizome	Creams, lotions, antibacterial
Gotu kola (Brahmi)	*Centella asiatica*	Plant	Creams
Manjistha	*Rubia cordifolia*	Root	Creams
Garlic	*Allium sativum*	Bulb	Creams
Skin care/moisturization			
Aloe	*Aloe vera*	Leaves	Skin care creams
Marigold	*Calendula officinalis*	Flowers	Skin creams/lotions
Oats	*Avena sativa*	Fruit	Moisturizer creams
Citrus	*Citrus aurantium*	Peel	Creams, lotions
Chandan	*Santalum album*	Stem	Lotions
Wheat germ	*Triticum sativum*	Germ	Lotions
Powders			
Galanga	*Alpinia galanga*	Rhizome	Dusting powder
Orris	*Iris florentina*	Root	Dusting powder
Anti-inflammatory			
Corriander	*Coriandrum sativum*	Plants	Creams/lotions
Chicory	*Cichorium intybus*	Seed	Creams/lotions
Licorice	*Glycyrrhiza glabra*	Root	Creams/lotions
Ginger	*Zingiber officinale*	Root	Creams/lotions
Ashwagandha	*Withania somnifera*	Root	Creams/lotions
Indian frankincense	*Boswellia serrata*	Gum resin	Creams/lotions
Guggul	*Commiphora mukul*	Gum resin	Creams/lotions
Hair care			
Kapurkachir	*Hedychium spicatum*	Rhizome	Oil/conditioning
Benjamin	*Moringa oleifera*	Seed	Oil
Amalaki	*Phyllanthus emblica*	Fruit	Oil/shampoo
Henna	*Lawsonia alba*	Leaves	Shampoo/conditioning
Bhringraj	*Eclipta alba*	Plant	Shampoo
China rose	*Hibiscus rosa sinensis*	Flowers	Shampoo/oils

also well known. Further application of purified isolates from these herbs into modern formulations can indeed be effective in improving efficacy.

15.6 AYURVEDIC HERBS AND MODERN SKIN CARE

Despite the use of numerous herbs and extracts over the ages for various skin care applications, only few have been systematically studied. Nonetheless, modern studies and research are questioning their validity. Efforts are well underway to find the proof of principle for their inconceivable benefits. While research is important for scientific validation, it also provides basis for isolation of active principals from such extracts. On the other hand, they also confer upon formulators a broader array of compounds to play with and boost the efficacy of their formulations for particular claims. Furthermore, the use of these bestows the combined nutraceutical and topical benefits on the end user. The following sections provide an overview of such compounds and reviews studies that demonstrate the potential of various popular ayurvedic ingredients or preparations for specific skin care applications.

15.6.1 Anti-inflammatory

Botanical extracts offer a safe and effective way of alleviating allergy symptoms while avoiding the unpleasant side effects of drugs. Ginger (*Zingiber officinale*) root, neem (*A. indica*) [12], Capsicum (capsacin from *Capsicum annuum*), *Boswellia serrata*, cinnamon (*Cinnamomum cassia*; vasodilator), bavachinin A (*Psoralea corylifolia*), menthol (*Mentha piperita*), eucalyptus oil (*Eucalyptus globules*), clove (*Syzygium aromaticum*), devadaru (*C. deodara*), aswagandha (*Withania somnifera*), guggul (*Commiphora mukul*), *Nyctanthes arbor tristis*, turmeric (*C. longa*) are some of the popular ayurvedic preparations that have been mentioned in texts for their anti-inflammatory effects.

Boswellic acid from gum resin of *B. serrata* (Indian frankincense) is a good example that finds concurrence of use both in ayurvedic system and in Western medicine. Recent studies report safety evaluation [16] of its beneficial compounds particularly 3-*O*-acetyl-11-keto-β-boswellic acid (AKBA) using *in vitro* cytotoxicity tests (HaCaT and NCTC 2544) and fetal dermal fibroblasts (HFFF2).

Muktashukti bhasma (MSB) is an ayurvedic compound prepared by combination of pearl, aloe vera, and vinegar. This preparation was tested for its potential anti-inflammatory activity in studies using acute and subacute inflammation models of albino rats [17]. This was compared with well-known acetyl salicylic acid (ASA) and the mechanism was demonstrated to be by mode of prostaglandin histamine and 5HT inhibition and stabilization of lysosomal membranes.

N. arbor tristis Linn. (Harsingar) is widely used as a decoction in the ayurvedic system of medicine for treatment of sciatica and arthritis. Its anti-inflammatory activity was demonstrated [18] in a study using the alcoholic extracts of leaves of *N. arbor tristis* substantiating its potential.

Antioxidant and anti-inflammatory properties of curcumin, its therapeutic usage against different pathological conditions, are widely discussed in literature

[19, 20]. The anti-inflammatory effect of curcumin is believed to be mediated through its ability to inhibit cyclooxygenase-2 (COX-2), lipoxygenase (LOX), and inducible nitric oxide synthase (iNOS). COX-2, LOX, and iNOS are important enzymes that mediate inflammatory processes. Improper upregulation of COX-2 and/or iNOS has been associated with the pathophysiology of certain types of human cancer and inflammatory disorders. Because inflammation is closely linked to tumor promotion, curcumin with its potent anti-inflammatory property is anticipated to exert chemopreventive effects on carcinogenesis. Devadaru (*C. deodara)* oil has been reported to possess anti-inflammatory and antimicrobial activities and is particularly useful in curing infective wounds.

15.6.2 Antiaging: Wrinkle Treatment

While aging is an inevitable phenomenon and has been part of human life cycle since life originated, it could result by various factors such as extrinsic or environmental and intrinsic such as genetic factors. Skin, being the largest organ and the barrier against external insults, is the first to exhibit any visible signs of bodily aging. Signs of aging in skin are manifested in wrinkles. Human attempts to find therapies to combat and minimize the effects of premature aging are longstanding. Principles of ayurveda describe "jara," that is, "aging" as a result of wearing out. Skin aging according to ayurveda is due to various factors ("doshas" described in Section 15.3) that include (1) loss of moisture balance or "kapha dosha," (2) imbalance of metabolic functions that result from improper chemical or hormonal reactions in skin, called the "pitta dosha," and (3) impairment in circulation of blood or nutrients to skin called "vata dosha." Ayurvedic perspectives and theories on aging and its management in comparison to modern medicine have been reviewed comprehensively [21]. As is with modern medicine, Ayurveda also describes two modes of therapies, that is, curative "vyadhihara" to combat a disease or condition and preventive "urjaskara."

To treat or avoid wrinkles and premature aging from vata imbalance, skin care products to rehydrate and nourish the skin are recommended in ayurvedic practices. Skin massage using natural oils and application of herbal moisturizers are considered. For pitta type of skin that is delicate as described in previous sections, the use of sunscreens is advisable. Hot oil massage and cleansing by gentle exfoliation help rejuvenate skin with kapha imbalance.

Powders of ginger and turmeric roots, rosemary leaves, and pomegranate seeds are recommended for use in oral dosage forms for skin health. A recent article [22] describes a model formulation for complementary use as a topical application to confer antiaging benefits.

15.6.3 Skin Brightening

Skin brightening agents are mostly compounds that function by mechanisms effecting melanin synthesis. Various ayurvedic extracts from herbs such as fennel (*Ferula foetida*), bearberry (*Arctostaphylos uva-ursi* L), licorice (*Glycyrrhiza glabra*), *Sophora angustifolia* (yellow mountain laurel), amla (gooseberry, *Phyllanthus emblica*) fruit, and mulberry (*Morus alba*) have been demonstrated to contain bioflavonoid components that are analogous to L-DOPA, the end product of the first step in melanin synthesis. They thus inhibit by feedback mechanism the conversion of L-DOPA to melanin and therefore function as skin brighteners. The skin brightening effects by white tea, licorice, green tea can be attributed to their antioxidant activity in inhibiting the oxidation steps in melanin synthesis. Curcumin is the other multifunctional ayurvedic gem for which the skin brightening effects are attributed to its antioxidant properties. However, there have been constraints in its use for topical applications due to its colored nature.

15.6.4 Wound Healing

Wounds are defined as cellular or anatomic disruption of tissue [23]. They can result from various factors such as physical, chemical, thermal, microbial, or immunological insults to the tissue. Topical agents used for wound healing include antibiotics, antiseptics, and desloughing agents (that cause chemical debridement). Over the years, various plant products have been used in wound healing. Chemical entities isolated from plants such as flavonoids and phenolic compounds are now often used in formulating treatments and management of wounds. While numerous herbal extracts are subject of scrutiny, there are yet a large number of plants from traditional ayurvedic treasure that could be potential candidates and need to be scientifically validated.

Several plant extracts/decoctions or pastes have been used traditionally in India for treatment of cuts, wounds, and burns. Various traditional methods use plant extracts and preparations according to ayurvedic practices [24]. They discuss the limited validation of Indian medicinal plants in wound healing treatments and provide extensive insights for researchers engaged in validation of the traditional claims and development of safe, effective, and globally accepted herbal treatments for cuts and wounds.

Typically, polyherbal formulations that use combinations of extracts from herbs such as guggul (*Balsamodendron mukul*), maharasanadi kwath (a carrier prepared from various herbs), amla (*P. embica*), guduchi (*Tinospora cordifolia*), manjishta root (*Rubia cordifolia*), licorice (*G. glabra*), and *shankha bhasma* (calcinated conch shell of *Turbinella pyrum*) that are claimed to have wound healing action. They are known for their antimicrobial effects against bacterial infections. These herbal preparations have been reported to promote skin tensile strength and epithelial formation [25].

Recent studies by researchers exemplify the efficacy evaluation of the antiaging activity of ayurvedic preparations [26] by excision and incision wound healing animal models. The authors study the parameters including wound contraction, collagen content, and skin breaking strength, which in turn are indicative of the tissue/cell regeneration capacity, collagen formation, and mechanical strength of skin. They demonstrate the wound healing properties, elevation of collagen content, and better skin breaking strength by treating the skin with antiaging formulations comprising yashada bhasma (a zinc complex) along with *Shorea robusta* resin and flax seed oil.

Neem oil, known for its wound healing properties, is used for dressing foul ulcers, eczema, and skin diseases such as ringworm. Neem leaf extracts have particularly proven useful in wound healing due to their antimicrobial properties, thus preventing secondary infections. Certain clinical studies have also demonstrated the anti-inflammatory efficacy of neem in comparison to known anti-inflammatory compounds such as cortisone acetate.

Helianthus annuus (suryamukhi or sunflower) is a common Indian herb the extracts of which are used in traditional medicine for inflammation in eyes, bite wounds, and bone fractures [27].

Jasminum auriculatum (jasmine) is another herb from south India, which is known for its wound healing properties. Topical application of *J. auriculatum* extracts in ghee accelerated the wound healing process, in a study using rodent model by induction of second-degree burn wounds, up to 6 days. The mucopolysaccharide accumulation was significantly higher in the group treated with medicated ghee [28].

15.6.5 Antiseptic

Sarsaparilla (*Aralia nudicaulis*) is an ayurvedic herb that is claimed to possess antiseptic and antipruritic benefits when applied topically [29]. Saponins and sterols are the components that might contribute to these properties.

Sharpunka (*Tephrosia purpurea*) is an ayurvedic herb well-known for its antibacterial, antiulcer, and anticancer properties. Its therapeutic effects are evidenced in several isolated studies and animal models and its activities are speculated to be due to its composition containing phenolic and flavonoidal compounds such as quercetin.

15.6.6 Acne

Earliest mention of ayurvedic treatment for acne using derivatives of salt, rocksalt, or sea salt called lavana, occurs in Atharva veda [1]. *Charak Samhita* describes acne vulgaris as "Tarunya Peetika," meaning a disease of adolescence. A variety of drugs are used in the ayurvedic system of medicine to treat this condition. *Sharangdhar Samhita* (1984) proposes sookshma triphala, thiostanin, amalakimashi vati, shankhabhasma vati, and sunder vati, among others, to be safe and effective treatment for acne. Sunder vati is an ayurvedic formulation composed of kutaj (*Holarrhena antidysenterica* Linn.), amalaki (*Emblica officinalis*), vidanga (*Embelia ribes* Burm.; Myrsinaceae), and saunth (*Z. officinale* Roscoe; Zingiberaceae). This was tested and proven effective to treat acne in comparison to other known ayurvedic formulations and a placebo control in recent clinical trials [30].

Lately, several such isolated clinical trials have been instrumental in proving the effectiveness of various herbal extracts of the plants described in ayurvedic treatises such as the *Bhavprakasha Nighantu* and *Charak Samhita* in combination therapies of oral and topical applications for treating acne vulgaris [31].

Ayurvedic supplements for acne treatment seem to benefit from the antibacterial properties of neem (*A. indica*). Based on the same rationale, many ayurvedic soaps are formulated using extracts from stems, root, and bark of neem. Haemafine syrup, a natural solution made up of more than a dozen herbs, with blood purifying properties, and Khadir, an extract from herb *Acacia catechu* (heart wood), along with neem are widely prescribed to treat acne, psoriasis, and eczema.

15.6.7 Personal Care

15.6.7.1 Face and Body Treatments The use of sesamum (*Sesamum indicum*) seed and its oil has been well known in Indian traditional medicine. Ayurveda recommends regular topical application of sesame oil to the skin as a health-promoting procedure. Sesame oil and other vegetable oils rich in linoleic acid were shown to contain selective antineoplastic properties, which are similar to those demonstrated for essential polyunsaturated fatty acids and their metabolites [32]. This study proved the potential clinical usefulness of sesame oil recommended for topical application by ayurveda. A recent study describes standardized polyherbal formulations [33] containing sesamum oil as the main ingredient in terms of its active principal or marker compound. Biologically active lignans in sesamum oil are identified as the marker compounds for the oil and its formulations. The authors identify and quantify sesamin and sesamolin, the major lignans of the sesamum oil, and develop methods that could be applied to polyherbal formulations containing the oil for their quantitative estimation.

15.6.7.2 Hair Care Numerous herbs are described in ayurvedic repositories for their activity in promoting hair growth. Analysis of ayurvedic herbal extracts conducted at the research division of Hindustan Lever (Mumbai, India) using chicken chorioallantoic membrane assay was documented [4], employing a test system for screening plant extracts based on the premise that transport of nourishment to the hair root, the site of biochemical activity, would be facilitated by strengthening the hair root system. This model adopted the strategy of angiogenesis (vascularization) that would aid blood supply to the hair papilla. These studies demonstrated *Hedychium spicatum* (shati), *Hemidesmus indicus* (saariva), *Nardostachys jatamaansi* (jatamaansi), *Saussurea lappa* (kushth) roots, among other herbs that were tested, promoted hair growth.

Eclipta alba Hassk is another well-known ayurvedic herb with purported claims of hair growth promotion. A study on albino rats evaluated petroleum ether and ethanol extract of *E. alba* Hassk for its effect in promoting hair growth [34]. This study demonstrated that treatment with 2% and 5% petroleum ether extracts was better than the known positive control 2% minoxidil treatment.

15.6.7.3 Stress Relief: Skin Soothing Abhyanga is a form of traditional ayurvedic massage that uses herbal oils to draw out toxins through the skin, recommended particularly for people with particular skin conditions. Shirodhara, an oil-dripping treatment procedure, is recommended in ayurveda that involves the use of medicated herbal sesame oil and lavender oil [35].

The well-known herbal candidate conferring antistress properties is ashwagandha (*W. somnifera*) [36]. Its main use is as an immunomodulator. Its major phytoconstituents are withanolides. Ashwagandha is one of the best known and most researched ayurvedic herbs and holds a place in the ayurvedic traditions similar to ginseng in Chinese medicine and hence is also called "Indian ginseng." *W. somnifera* is used in several indigenous drug preparations both for maintaining health and for treating several disease conditions. The remarkable immunomodulating properties of *W. somnifera* (ashwagandha) is a very important subject for clinical investigations [37].

15.6.7.4 Oral Care Neem (*A. indica*) bark is the oldest candidate for oral care known in ayurveda. In ancient times, the stems of this tree were used directly for cleansing and protection of the mouth and gums from microbial infections. Several ayurvedic herbal extracts and preparations have been adapted for oral hygiene in several dental and oral care formulations. Among the most well known are the vajradanti (*Barleria prionitis*), peelu tree (*Salvadora persica*), bedda nut (*Terminalia bellarica*), vatadha (*Prunus amygdalus*; almonds), babul (*Acacia arabica*), spearmint (*Mentha viridis*), peppermint (*M. piperita*), cloves, and cardamom (*Elettaria cardamomum*). These have proven effective over the years in acting as astringents, preventing cavities, protecting gums from antibacterial and antimicrobial effects, preventing bad breath, and so on.

15.6.8 Natural Preservatives

Tulsi (*O. sanctum* also referred to as holy Indian basil), known for its antibacterial properties, curcumin from turmeric, and neem (*A. indica*) are well known among numerous other herbs to have preservative properties. *O. sanctum* has been demonstrated for its activity against several Gram-positive bacteria such as *Salmonella typhi* and *Bacillus subtilis* in several isolated studies.

15.6.9 Arthritis Treatments

Using the ayurvedic practice of treating inflammatory disorders with turmeric, Western medicine has also adopted the use of turmeric rhizomes and turmeric extracts in dietary supplements for treatment of arthritis. Data indicating the antiarthritic efficacy of well-characterized curcuminoid-containing turmeric extracts was reported using an animal model of rheumatoid arthritis [20]. The results suggest that three major curcuminoids are responsible for this effect, while the remaining compounds in the crude turmeric extract may inhibit this protective effect.

Guggul, the sticky gum resin from the Mukul myrrh tree, plays a major role in the traditional herbal medicine of India. The primary chemical constituents of guggul include phytosterols, gugulipids, and guggulsterones. It was traditionally combined with other herbs for the treatment of arthritis and skin disease. There has been mention of various oral dosage forms of ayurvedic ingredients for arthritic pain relief. Some of these include kanchnar guggul, goksharidi guggul, trayodashang guggul, yograj guggul, vyousadi guggul, vatari guggul, simhanad guggul, ajamodadi curpa, nimbadi curpa, pancasma curpa, vaisvanara curpa, rasa parpati, anand bhaijrava rasa, mahalaxmi villas rasa, sarvanabhupati rasa, rasnadi kvath curpa, amrat ghrit, and jiraka modka [38]. Based on such well-recognized use of ayurvedic ingredients in pain relief, a recent study described a model topical formulation [22] that uses the ginger oil, *B. serrata* extract, and purified corydalis extract for arthritic pain. This formulation combines the use of well-known pain relief nutraceuticals such as methylmethane sulfate, chondroitin, and glucosamine with the purified ayurvedic compounds described above to increase the efficacy.

15.7 FORMULATION CHALLENGES WITH USING AYURVEDIC HERBS

Exploration into yet unidentified and untapped resources from the wealth of ayurvedic herbs can certainly prove beneficial in discovering new ingredients for skin care. However, there have also been, and will be in future, certain constraints associated with the use of such ayurvedic compounds. Commercial cultivation of many of the wild plant species may pose a hurdle. Herbal extracts used for cosmetics or nutraceuticals are usually mixtures of complex molecules not amenable to synthesis. Purification of chemicals from herbal extracts also compromises their activity. Not only is their functionality but also the effect of various compounds coexisting in the extract is compromised, which may be essential for their stability, and their activity is lost. Thus, their physiological functions could be altered.

While using such novel entities, it is extremely important for the formulators to keep in view certain important aspects [39] such as (1) the use of isolated and purified forms of the ingredients, where possible, from herbal extracts or preparations is necessary to assure good quality and efficacy; (2) chemical identification of each of the active components from extracts and their structure is essential to ensure compatibility for use in a formulation; (3) proper understanding and validation of any changes in formulations is required to ensure stability; and (4) identification and application of a suitable delivery system is vital to ensure consumer desirable benefits.

Application of bioinformatics and data mining systems to herbal texts holds great promise for identifying novel pharmacotherapeutic leads for bioactive compounds [40]. Such technologies make the process of extracting useful information contained within these resources much easier and time effective.

15.8 CONCLUSIONS

Undoubtedly, historical herbal texts and ancient Indian discipline of ayurveda provide a preexisting resource that documents the traditional uses of numerous plant species for therapeutic benefits. Exploration, validation, and documentation of such traditional knowledge are increasingly valuable. The eminent demand for novel natural cosmeceuticals boosts the need to seek out the inconceivable benefits from the rich diversity of Indian herbs and unveil the hidden knowledge that ayurveda has in store. Scientific research and validation of the efficacy of herbal extracts and their isolates holds a great promise in various fields of medicine including skin care.

REFERENCES

1. Prasad PV. Atharvaveda and its materia medica. *Bull. Indian Inst. Hist. Med. Hyderabad* 2000;30(2): 83–92.
2. Murthy AR. Dhanwantari: the God of Hindu medicine. *Bull. Indian Inst. Hist. Med. Hyderabad* 1997;27(1):1–14.
3. Abelson PH. Medicine from plants. *Science* 1990;247(4942):513.

4. Dev S. Ancient-modern concordance in Ayurvedic plants: some examples. *Environ. Health Perspect.* 1999;107(10):783–789.
5. Mantle D, Gok MA, et al. Adverse and beneficial effects of plant extracts on skin and skin disorders. *Adverse Drug React. Toxicol. Rev.* 2001;20(2):89–103.
6. Behl PN. Ayurvedic dermatology in the concept of modern medicine. *Indian J. Dermatol. Venereol.* 1957;23: 2.
7. Patwardhan B. Ayurveda: the designer medicine. *Indian Drugs* 2000;37: 213–227.
8. Prasad PV, Bhatnagar VK. Medico-historical study of "Kilasa" (vitiligo/leucoderma): a common skin disorder. *Bull. Indian Inst. Hist. Med. Hyderabad* 2003;33(2):113–127.
9. Sharma PV. *Dravyaguna Viinan*, vol 2 Varanasi, India: Chaukhamha Bharati Academy, 1986. pp. 691–693.
10. Anderson TF, Voorhees JJ. Psoralen photochemotherapy of cutaneous disorders. *Annu. Rev. Pharmacol. Toxicol.* 1980;20:235–257.
11. Gupta SK, Prakash J, et al. Validation of traditional claim of *Tulsi, Ocimum sanctum* Linn. as a medicinal plant. *Indian J. Exp. Biol.* 2002;40(7):765–773.
12. Subapriya R, Nagini S. Medicinal properties of neem leaves: a review. *Curr. Med. Chem. Anticancer Agents* 2005;5(2):149–146.
13. Charles V, Charles SX. The use and efficacy of *Azadirachta indica* ADR ('Neem') and *Curcuma longa* ('Turmeric') in scabies: a pilot study. *Trop. Geogr. Med.* 1992;44(1–2):178–181.
14. Dwivedi C, Abu-Ghazaleh A. Chemopreventive effects of sandalwood oil on skin papillomas in mice. *Eur. J. Cancer Prev.* 1997;6(4):399–401.
15. Dahanukar SA, Kulkarni RA, Rege NN. Pharmacology of medicinal plants and natural products. *Indian J. Pharmacol.* 2000;32: S81–S118.
16. Burlando B, Parodi A, et al. Comparison of the irritation potentials of *Boswellia serrata* gum resin and of acetyl-11-keto-beta-boswellic acid by *in vitro* cytotoxicity tests on human skin-derived cell lines. *Toxicol. Lett.* 2008;177(2):144–149.
17. Chauhan O, Godhwani JL, et al. Antiinflammatory activity of *Muktashukti bhasma. Indian J. Exp. Biol.* 1998;36(10):985–989.
18. Saxena RS, Gupta B, et al. Study of anti-inflammatory activity in the leaves of *Nyctanthes arbor tristis* Linn.—an Indian medicinal plant. *J. Ethnopharmacol.* 1984;11(3):319–330.
19. Menon VP, Sudheer AR. Antioxidant and anti-inflammatory properties of curcumin. *Adv. Exp. Med. Biol.* 2007;595: 105–125.
20. Funk JL, Oyarzo JN, et al. Turmeric extracts containing curcuminoids prevent experimental rheumatoid arthritis. *J. Nat. Prod.* 2006;69(3):351–355.
21. Datta HS, Mitra SK, et al. Theories and management of aging: modern and ayurveda perspectives. eCAM: nep005, 2009. doi:10.1093/ecam/neq072.
22. Gupta S. Ayurvedic beauty: here's how to formulate beauty products based on ancient Indian discipline. *HAPPI* 2008;September: 46.
23. Bennet RG. *Fundamentals of Cutaneous Surgery.* St. Louis: CV Mosby, 1988. p. 778.
24. Kumar B, Vijayakumar M, et al. Ethnopharmacological approaches to wound healing: exploring medicinal plants of India. *J. Ethnopharmacol.* 2007;114(2):103–113.
25. Nandakarni KM, Nadakarni AK. *Indian Materia Medica.* 3rd ed. Bombay: Popular Prakashan Ltd. 1954. pp. 167–170.
26. Datta HS, Mitra SK, et al. Wound healing activity of topical application forms based on ayurveda. eCAM: nep015, 2009. doi:10.1093/ecam/nep015.
27. Jain SK, Tarafdar CR. Medicine plant love of Sautals (A review of P.O. Bodding's work). *Econ. Bot.* 1970;24: 241.
28. Deshpande PJ. Pathak SN. *Indian J. Med. Res.* 1966;1(1):81.
29. Draelos ZD. Botanicals as topical agents. *Clin. Dermatol.* 2001;19(4):474–477.
30. Paranjpe P, Kulkarni PH. Comparative efficacy of four ayurvedic formulations in the treatment of acne vulgaris: a double-blind randomised placebo-controlled clinical evaluation. *J. Ethnopharmacol.* 1995;49(3):127–132.
31. Lalla JK, Nandedkar SY, et al. Clinical trials of ayurvedic formulations in the treatment of acne vulgaris. *J. Ethnopharmacol.* 2001;78(1):99–102.

32. Smith DE, Salerno JW. Selective growth inhibition of a human malignant melanoma cell line by sesame oil *in vitro*. *Prostaglandins Leukot. Essent. Fatty Acids* 1992;46(2):145–150.
33. Sukumar D, Arimboor R, et al. HPTLC fingerprinting and quantification of lignans as markers in sesame oil and its polyherbal formulations. *J. Pharm. Biomed. Anal.* 2008;47(4–5):795–801.
34. Roy RK, Thakur M, et al. Hair growth promoting activity of *Eclipta alba* in male albino rats. *Arch. Dermatol. Res.* 2008;300(7):357–364.
35. Xu F, Uebaba K, et al. Pharmaco-physio-psychologic effect of ayurvedic oil-dripping treatment using an essential oil from *Lavendula angustifolia*. *J. Altern. Complement. Med.* 2008;14(8):947–956.
36. Muruganandam AV, Kumar V, Bhattacharya SK. Effect of poly herbal formulation, EuMil, on chronic stress-induced homeostatic perturbations in rats. *Indian J. Exp. Biol.* 2002;40(10):1151–1160.
37. Kour K, Pandey A, Suri KA, Satti NK, Gupta KK, Bani S. Restoration of stress-induced altered T cell function and corresponding cytokines patterns by Withanolide A. *Int. Immunopharmacol.* 2009; 9(10):1137–1144.
38. GoI. *Ayurvedic Formulary of India*. 2nd ed. Government of India, 2003.
39. Gupta S. Ayurvedic antiaging. *HAPPI* 2007;46
40. Buenz EJ, Schnepple DJ, et al. Techniques: bioprospecting historical herbal texts by hunting for new leads in old tomes. *Trends Pharmacol. Sci.* 2004;25(9):494–498.

CHAPTER 16

FORMULATING WITH TRADITIONAL CHINESE MEDICINE HERBS

Ina Schlenoff[1] and Robin Choi[2]

[1]Estee Lauder Companies, Melville, NY
[2]New York College of Health Professions, Syosset, NY

16.1 INTRODUCTION

The current focus on youth and health has spurred a quest for the most natural and effective cosmetic ingredients. For centuries, the Chinese culture has utilized herbs to enhance skin treatment and skin appearance. Researching the ancient texts and resources of these Chinese herbs has provided a new pool of ingredients for cosmetic formulations. The herbs used in traditional Chinese medicine are reaching more of a widespread audience within Western society's acceptance and move toward natural products. Both women and men are increasingly more concerned with delaying aging. Furthermore, the market potential for herbs and antiaging is worth over 14 billion dollars and the fastest growing segment of the market [1]. Chinese herbal remedies fulfill the tremendous need for products that are mild and biologically active for today's cosmetic products.

16.2 HISTORY OF TRADITIONAL CHINESE MEDICINE

Traditional Chinese medicine, or TCM, is a science based on over 5000 years of observing the patterns of the human body and nature, and the interaction between them. It is a holistic medicine, preventing, restoring, and maintaining overall health in relation to the surrounding environment. TCM is composed of acupuncture, herbal remedies, including moxibustion and qi gong or tai chi exercise.

The first record of acupuncture is found in the 4700-year-old text *Huang Di Nei Jing (Yellow Emperor's Classic of Internal Medicine)*. This is said to be the oldest medical textbook in the world. The oldest written text on herbs is *Shen Nong Ben Cao Jing (The Divine Farmer's Materia Medica).*

Formulating, Packaging, and Marketing of Natural Cosmetic Products, First Edition.
Edited by Nava Dayan and Lambros Kromidas.

TABLE 16.1 The Five-Element Theory and Associated Correspondences

	Fire	Earth	Metal	Water	Wood
Yin organs	Heart	Spleen	Lungs	Kidneys	Liver
Yang organs	Small intestine	Stomach	Large intestine	Urinary bladder	Gall bladder
Weather	Heat	Dampness	Dryness	Cold	Wind
Tissues	Vessels	Muscles	Skin	Bone	Tendons
Tastes	Bitter	Sweet	Pungent	Salty	Sour
Colors	Red	Yellow	White	Blue/black	Green

16.3 THEORY OF CHINESE MEDICINE

According to TCM, there are 12 pathways or meridians of vital energy (qi) that flows through the body; these are associated with the various organs of the body. When pathways become obstructed, deficient, or excessive, the pathways are thrown out of energy balance; This is what causes illness according to Chinese medical theory. TCM restores this balance and prevents and heals disease.

In the diagnostic theory of TCM, various perspectives are observed to analyze and diagnose skin problems. The eight principal patterns are four pairs of polar opposites, used to define the signs and symptoms presented by the patient; exterior/interior, excess/deficiency, hot/cold, yang/yin. Yang-yin theory is based on the philosophical construct of two polar complements that always exist in the world [2]. The other main diagnostic category is the pernicious influences or evils: wind, cold, heat, dampness, dryness, and fire [2].

In addition, TCM is explained through The Five-Element Theory [3] and ascribes certain phenomena to each. The five elements involved are wood, fire, earth, metal, and water. This system serves as one of the diagnostic and treatment protocols in practice.

The most common correspondences [4] of the five elements are listed in the chart below. This diagnostic information assists in establishing patterns and identifying the TCM disease.

Moxibustion (moxa) is the application of heat to a specific area of the body for the purpose of treating disharmony. The heat is generated from burning the herb *Artemisia vulgaris* (mugwort, Chinese wormwood). Though used often in everyday TCM practice, the skin conditions that are discussed in the next sections are heat conditions and moxa treatments which would be counterintuitive for these skin problems. The warming and invigorating nature of the moxa is used to treat deficiency syndromes and increase circulation.

16.4 THEORY OF HERBAL MEDICINE

Chinese herbal formulas are composed of numerous plants, animals, and minerals. Herbs in TCM extend beyond the botanical world, unlike the Western view of herbal medicine. The medicines come in various forms such as pills, tablets, liquid tinctures, or teas. Patients should inform the practitioners if they have dietary restrictions or ethical concerns since in Western culture, many of the ingredients are frowned upon, such as, squirrel excrement, worms, bones, and so on. Formulas are prescribed after determining the pattern of disharmony and should be taken as prescribed by the practitioner. The chemical constituents begin to alter the body immediately and begin the healing process. As the patient's symptoms begin to change and are alleviated, the formula will be modified. The formula is chosen because of the nature of the ingredient (temperature and taste) and its therapeutic actions [5]. The art of combining herbs should be preferably left to those with clinical experience and training.

According to traditional Chinese medicine, herbs are administered in carefully prescribed formulations. The reason is to assign certain herbs to address the main illness while other may be used to address lesser imbalances in the patient and if need be to counteract toxicities or harsh effects of herbs in the formula. The overall synergistic effect of the herbs in the formula is what makes it more potent than just the sum of the individual herbs.

A well-known saying in TCM theory is "Tong bing yi zhi, yi bing tong zhi—One disease, different treatments; different diseases, same treatment." This philosophy differentiates Eastern medicine from Western medicine, in which TCM differentiates from the Western style of treating symptoms. Eastern medicine focuses on treating the cause of the disease and rarely the symptoms of the disease or its outcome. In the diagnosis of a disease, correct pattern discrimination is vital. For example, two persons with the single symptom of acne may be treated differently to return their health back to balance. The Chinese medicine treatment may be as different as the individuals with different symptoms. A practitioner may assess that one's acne maybe triggered by diet. In another patient, it may be due to stress, and in another it may be due to fluctuations of hormones related to menstruation. These differences dictate the need for different treatment principles for the same disease. In TCM, there is no one universal pill for each disease. In Chinese herbal medicine, different combinations of herbs are prescribed and the formula is tailored to each individual's constitution.

Similar to herbal medicine, choices in nutrition can greatly alter a person's health. Under heat conditions such as acne, inflammation, and fever, cooling foods and herbs should be added to the diet to reduce the heat signs. Nutrition is viewed in TCM as the Five Element Theory. Food is categorized by its nature and its effect on the body. Examples of cooling foods include watermelon, cucumber, soy, and peppermint tea. Foods can also aggravate a skin condition. For chronic heat conditions, that is, rashes or menopause, it is recommended to avoid the hot nature foods. This includes ginger, cinnamon, clove, cumin, and pepper [6]. As with all TCM prescriptions, foods can be added and altered according to the symptoms. TCM recommends to seasonally alter your diet to harmonize your health with the season. This keeps the body balanced and prevents illness.

16.5 APPLICATIONS

The application of traditional Chinese medicine in the field of skin treatment is widely practiced and is a specialty in the education, training, and clinical practice of a TCM practitioner. Dermatological conditions can be very difficult to treat, therefore it is imperative to be able to accurately diagnose the main causal factors that lead to the symptomatic manifestations on the skin. The same herbal formulas traditionally ingested can, in theory, be used topically on the skin. The natural extractions can be added to skin treatment formulations, thereby nurturing the appearance of the skin from the outside.

16.5.1 Facial Rejuvenation

A growing interest in the West is the use of Chinese medicine in the application for facial rejuvenation. Through the use of acupuncture and herbal medicine, practitioners are able to help improve the complexion of the skin. By employing basic strategies such as encouraging the production of blood cells and other vital substances in the body and then stimulating the circulation of blood and qi to the local area, this can help improve the health of the skin. Skin problems such as dryness can be corrected by allowing the blood to properly nourish and moisturize the area; similar strategies can be employed to reduce the appearance of fine lines and wrinkles. Red inflamed skin can be treated with herbs that alleviate heat and detoxify the body. Individuals with any of the above conditions will undergo topical and internal treatments. Herbs can be applied topically to induce local moisturizing, circulating, and anti-inflammatory effects. However, for a longer lasting effect, herbs that are used internally are also recommended. Depending on the patient's constitution different herbal formulas may be employed to address the internal systemic imbalance that may be the chief cause for many skin problems.

A well-known basic herbal formulation useful in promoting the engenderment and circulation of blood and protecting the yin and essence is Si Wu Tang (four-substance decoction) that contains the following herbs:

- Shu Di Huang (Radix Rehmanniae Glutinosae Conquitae) (Chinese foxglove root)
- Bai Shao (Radix *Paeoniae lactiflorae*) (white peony root)
- Dang Gui (Radix *Angelicae sinensis*) (Tang-kuei root)
- Chuan Xiong (Radix Ligustici Chuanxiong) (Szechuan lovage root)

Herbs such as these are good examples of how one can address internal imbalances that can lead to premature aging. These herbs have the ability to enhance and protect those body constituents that are necessary for health and vitality from inside out. Research is beginning to show that the constituents in this formula can nourish blood, regulate immunity, counteract hypoxia, protect against free radical and radiation damage, regulate uterine movement, decrease lipidemia, enhance cardiac function, resist toxicities, and deter cell mutation [7].

Topical application of Chinese medicinal herbs plays an important role in facial rejuvenation. While it is important to take internal herbs to treat systemic internal imbalances, applying the herbs topically is useful for targeting specific problem areas. Often oil or water-based masks and creams blended with Chinese herbs are regularly used in traditional and modern skin care formulas. These herbs can help protect the skin through their antioxidant effects and reduce fine lines and wrinkles by nourishing and hydrating at the cellular level. Popular herbs for topical use include the following:

- Fu Ling (Sclerotium Poriae Cocos) (Tuckahoe)
- Bai Zhi (Radix Angelicae Dahuricae) (Chinese angelica root)
- Dang Gui (Radix Angelicae Sinensis) (Tang-kuei root)
- Chuan Xiong (Radix Ligustici Chuanxiong) (Szechuan lovage root).

16.5.2 Acne Treatment

Acne or acne vulgaris is a common skin problem characterized by inflamed papules, blackheads, whiteheads, pustules, and cysts. According to traditional Chinese medicine, the primary causes of acne involve heat pathogens with an involvement of toxins, dampness, or blood stasis. Heat pathogens can accumulate on the skin that can give rise to redness and inflammation to the local area. Bacteria on the skin that brings forth the redness and inflammation can sometimes be seen as agents of heat pathogens. Plugs of keratin and sebum in hair follicles, which give rise to blackheads and whiteheads, can be explained by the heat that solidifies the fluids of the sebaceous glands. The presence of deep, painful inflamed nodules and pus-filled cysts correlate with the additional presence of toxins, dampness, and blood stasis.

Herbs in Chinese medicine can be used to treat these various pathogens such as heat, dampness, toxins, and blood stasis. These herbs often possess antimicrobial and anti-inflammatory properties.

Three herbs that play an important role in treating heat, dampness, and toxins are commonly known as the "three yellows" or the "three huangs" due to their characteristic yellowish coloration and they are as follows:

1. Chinese pinyin name: Huang qin
   Pharmaceutical name: Radix Scutellariae Baicalensis
   Botanical name: *Scutellaria baicalensis* Georgi; *Scutellaria amoena* C.H. Wright
   English name: baical skullcap root, scutellaria, scute
   Properties: bitter, cold
   Chemical composition: Baicalin, baicalein, wogonin, wogoside, chrysin, oroxylin A, oroxylin-A-glucuronide, skullcapflavone I and II, tenaxin II, koganebanacin, and 7-methoxy-baicalein [8].
   Western therapeutic actions: Antimicrobial effect of Radix Scutellariae Baicalensis (huang qin) and particularly baicalin has shown to demonstrate a broad range of inhibitory effect on many pathogenic bacteria including *Staphylococcus aureus*, *Diplococcus pneumoniae*, *Pseudomonas aeruginosa*, *Bacillus dysenteriae*, *Escherichia coli*, *Bordetella pertussis*, *Vibrio cholerae*, *D. meningitidis*, *Leptospira*, and various species of dermatophytes and influenza viruses. Baicalin has also been shown to potentiate the effect of common antibiotics such as ampicillin, amoxicillin, methicillin, and cefotaxime. Pathogens, resistant to the aforementioned antibiotics,

such as *S. aureus* and methicillin-resistant *S. aureus* (MRSA) were inhibited when these antibiotics were combined with baicalin [9, 10].

Chinese therapeutic functions: One of the huang qin's function is to clear heat and eliminate toxins. It is indicated for disorders of the skin, throat, and eyes marked by lesions, sores, hot swellings, and ulcerations.

2. Chinese pinyin name: Huang lian

   Pharmaceutical name: Rhizoma Coptidis

   Botanical name: *Coptis chinensis* Franch; *Coptis deltoidea* C.Y. Cheng; *Coptis teetoides* C.Y. Cheng

   English name: coptis, coptis root

   Properties: bitter, cold

   Chemical composition: Berberine, coptisine, worenine, palmatine, columbamine, obacunone, obaculactone, jatrorrhizine, and magnoflorine [11].

   Western therapeutic actions: Rhizoma coptidis (huang lian) with berberine as one of its constituents has a very broad antimicrobial action against *Streptococcus pneumoniae, Neisseria meningitidis, S. aureus, Shigella dysenteriae, Mycobacterium tuberculosis*, and various strains of pathogenic fungi and viruses [11].

   Berberine has demonstrated properties to inhibit lipogenesis in human sebaceous glands and is effective in the treatment of acne vulgaris [12].

   Chinese therapeutic functions: Among huang lian's numerous functions is its ability to clear fire toxins. Fire toxins in Chinese medicine can manifests as common dermatological disorders such as abscesses, inflammation, carbuncles, burns, hot lesions, and swellings. Phytochemicals in this herb possess antimicrobial action making it an ideal herb for use.

3. Chinese pinyin name: Huang bai

   Pharmaceutical name: Cortex Phellodendri

   Botanical name: *Phellodendron amurense* Rupr; *Phellodendron chinensis* Schneid

   English name: phellodendron bark, amur cork tree bark, Chinese cork tree bark

   Properties: bitter, cold

   Chemical composition: Berberine, jatorrhizine, magnoflorine, phellodendrine, candicine, palmatine, menisperine, obacunone, and obaculactone [11].

   Western therapeutic actions: Cortex phellodendri (huang bai) has demonstrated antibiotic effects against *S. aureus, D. pneumoniae, Corynebacterium diphtheriae, B. dysenteriae, D. meningitidis, V. cholerae, B. anthracis*, dermatophytes, and *Leptospira* [13].

   Huang bai, according to the Chinese pharmacopoeia, can clear fire and eliminate toxins, in addition to clearing heat and drying dampness. Clinically, it is used to treat abscesses, sores, burns, eczema, redness, and swellings.

These three herbs are often prescribed together and can be used to treat a variety of disorders that are due to dampness, heat, and toxins.

16.5.3 Rosacea Treatment

Rosacea is a chronic inflammatory disorder of the face, particularly the central region. It is characterized by vascular dilation and may or may not accompany papules and

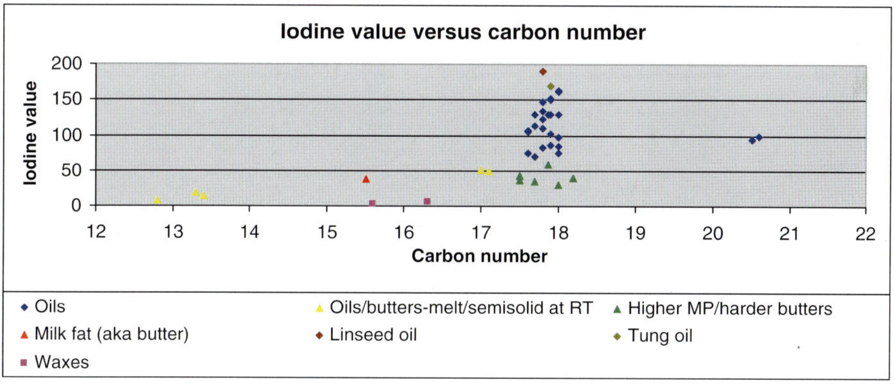

Chart 12.2 Iodine value versus carbon number of natural oils, butters, and fats.

Figure 21.1 Time lapse photographs of PLA bottles exposed for 30 days in industrial composting conditions.

pustules. Patients with rosacea are prone to flushing of the face and having ruddy complexions. Just like acne, rosacea is often described in Chinese medicine as an imbalance involving heat pathogens as well as blood stasis and toxins. The redness and flushing represents the heat trapped in the body, so herbs prescribed for the treatment of acne would be applicable. The deeper the redness the deeper the heat is said to be lodged in the body's interior; thus, in Chinese medicine there are herbs that address heat in the deeper levels of the body. If the bulb of the nose is darker red or even purple in color, then it would suggest the presence of blood stasis and again herbs to help break blood stagnation and to promote circulation would be necessary.

From a Western perspective, the mode of action for a number of these herbs remains unclear. Active compounds can be isolated in many herbs, but that alone does not always explain their effectiveness. A herb's function is not solely based on the sum parts of the active ingredients but rather all the properties of the herb work together in ways that cannot be fully explained from a biochemical approach.

One of the main herbs used in treating rosacea is

Chinese pinyin name: Da huang

Pharmaceutical name: Radix et Rhizoma Rhei

Botanical name: *Rheum palmatum* L; *Rheum officinale* Baill

English name: rhubarb root and rhizome

Properties: bitter, cold

Chemical composition: Chrysophanol, emodin, aloe-emodin, rhein, and physcion [11].

Western therapeutic actions: Radix et rhizoma rhei (da huang) has demonstrated laxative, antineoplastic, anti-inflammatory, and hematological effects, to name a few.

Chinese therapeutic functions: According to Chinese herbal knowledge, this herb exhibits functions, such as clearing internal heat, invigorating the blood, and removing blood stasis, which make it ideal for treating such imbalances that commonly cause rosacea.

16.6 CHALLENGES IN UTILIZING TCM HERBS

Herbs are sold in various forms ranging from whole plants, plant parts, to cut pieces or finely ground powders. It is impossible for a layperson to determine the quality of a plant material or even identify it by visual inspection. The buyer totally depends upon the reputation of the seller having the requisite analytical equipment to establish identity and quality of the botanicals sold. Efficacy and standardization are a major concern and reputable companies advertise the active ingredients.

Consumers are becoming more concerned about the safety of the herbal products they are using. Many Eastern herbal formulas are sold in the Western market and comply with Good Manufacturing Practices and other international quality standards. Packages are stamped with the recognizable symbols to advertise their compliance with international standards.

Other important issues concerning the use of herbal medicine include the development and elicitation of allergies and heavy metal and/or pesticide contamination. Also, suppliers of Chinese herbs acknowledge problems with sustainability and loss of habitat of the species. In almost all cases, appropriate substitutions exist. Many items are listed in Chinese materia medica, but are not used at this time because the substance is either an endangered species or toxic [11]. Manufacturers must adhere to the strict international requirements followed by many industries. These include the endangered species watch lists: CITES [14] and Red List [15]. Knowledge and awareness of the problem are the first steps to overcoming the issues. Most importantly, working with an experienced herbalist will help answer most questions.

As research intensifies in the area of traditional Chinese medicine, practitioners in the West will most likely gain more depth and understanding of the materia medica. For the Western world to fully accept and understand TCM, funding from institutions and corporations is required to increase its accuracy, effectiveness, and reputation. There are continued efforts to investigate and study the mechanisms of action of the active constituents of the herbs. More articles will be written in the Western world giving access to English speakers to understand the medical science. Translation of Asian research continues, yet more will need to occur. TCM treatments will become financially more viable as therapeutic solutions are discovered and treatments are proven effective.

16.7 SUMMARY

Chinese herbal formulas are effective because each constituent contributes to fix the imbalance of the specific condition. The formula is carefully chosen to treat the pattern presented and targets the source of the problem. The activity from each chemical compound works synergistically to bring the body back to homeostasis. The herbal therapy along with suggested lifestyle changes restores harmony to the individual.

In recent years, the demand for natural products has rejuvenated the importance of plant-based raw materials for personal care products. Incorporating Chinese herbs into cosmetic formulas provides new opportunities for business growth in a marketing story based on 5000 years of continual practiced history.

REFERENCES

1. http://www.gcimagazine.com/marketstrends/segments/antiaging/17221979.html?utm_source=Most + Read&utm_medium=website&utm_campaign=Most + Read. (Accessed 22 March 2010.)
2. Kaptchuk Ted J. *The Web That Has No Weaver*. 2nd ed. New York, NY: Contemporary Books, 2000. pp. 215–238.
3. Maciocia G. *The Foundations of Chinese Medicine: A Comprehensive Text for Acupuncturists and Herbalists*. 2nd ed. London: Churchill Livingstone, 2005. pp. 19–40.
4. Dragon V. 2008. http://www.acupuncture.com/education/theory/fiveelemnts.htm. (Accessed 9 January 2010.)
5. Bensky D, Steven C, Stoger E. *Chinese Herbal Medicine: Materia Medica*. Seattle, WA: Eastland Press, 2004.

6. Pitchford P. *Healing with Whole Foods: Asian Traditions and Modern Nutrition*. 3rd ed. Berkeley, CA: North Atlantic Books, 1993.
7. Wang X, Liu H, Guo S. 2008. Clinical application and research of si wu tang a review. Available at http://ajtcvm.org/?p=29. (Accessed 15 January 2010.)
8. Li T. *Chinese and Related North American Herbs: Phytopharmacology and Therapeutic Values*. 2nd ed. Boca Raton, FL: CRC Press, 2009.
9. Chen J, Chen K. *Chinese medical herbology and pharmacology*. California: Art of Medicine Press, 2004.
10. Liu IX, Durham DG, Richards RME. Baicalin synergy with β-lactam antibiotics against methicillin-resistant *Staphylococcus aureus* and other β-lactam-resistant strains of *S. aureus*. *J. Pharm. Pharmacol.* 2000;52(3):361–366.
11. Bensky D, Steven C, Stoger E. *Chinese Herbal Medicine: Materia Medica*. Seattle, WA: Eastland Press, 2004.
12. Seki T, Morohashi M. Effect of some alkaloids, flavonoids and triterpenoids, contents of Japanese–Chinese traditional herbal medicines, on the lipogenesis of sebaceous glands. *Skin Pharmacol.* 1993;6(1):56–60.
13. Chen J, Chen K. *Chinese medical herbology and pharmacology*. California: Art of Medicine Press, 2004.
14. http://www.cites.org/. (Accessed 6 February 2010.)
15. http://www.iucnredlist.org/. (Accessed 6 February 2010.)

CHAPTER 17

THE INSIDE-OUT CONCEPT AS COMPLEMENT TO THE USE OF TOPICAL SUNSCREEN: THE CASE FOR ENDOGENOUS SKIN PHOTOPROTECTION FROM SUNLIGHT BY NATURAL DIETARY ACTIVES SUCH AS TOMATO CAROTENOIDS

Joseph Levy[1,2] *and Yoav Sharoni*[1,2]

[1]*Department of Clinical Biochemistry, Faculty of Health Sciences, Ben-Gurion University of the Negev, Israel*
[2]*Soroka Medical Center of Kupat Holim, Beer-Sheva, Israel*

17.1 INTRODUCTION

Sun exposure is the main cause of many skin defects such as photosensitivity (sensitivity to sunlight that may lead to photodermatosis), sunburn, photoaging, and carcinogenesis. Such defects may arise due to photooxidation of cellular lipids, proteins, and DNA. Therefore, protection from sun radiation is now becoming an important issue due to increased awareness about skin cancer in the developed world. The simplest strategy to minimize skin damage is by exogenous protection using clothing and sunscreens. This chapter addresses the concept of endogenous photoprotection (protection from light) by dietary means as a complementary method to the common use of sunscreens.

Formulating, Packaging, and Marketing of Natural Cosmetic Products, First Edition.
Edited by Nava Dayan and Lambros Kromidas.

17.2 ENDOGENOUS (DIETARY) PHOTOPROTECTION

Exposure to ultraviolet (UV) light from the sun and other man-made sources occurs under everyday circumstances. Extensive exposure to UV can lead to accelerated aging and other defects. It is estimated that additional exposure due to conservative vacationing in warm climates is approximately 30% of the yearly sun exposure. Thus, the skin is exposed to a large degree, about two-thirds of the cumulative erythemal UV dose/year, when no topical or other protection is used [1]. Using protective clothing and sunscreens during regular daily activity have not always proven to be practical. Thus, an important complementary approach is to make photoprotective antioxidants available to the skin before the time of exposure. This notion of endogenous sun protection of the skin came about due to insufficient protection by exogenous means such as the regular use of topical sunscreens [2–5]

Innate endogenous photoprotection of the human skin is provided by skin melanins [6]. Melanins are ubiquitous pigments that scatter and absorb light. Tanning is the process of melanin production in epidermal melanocytes and this process is increased with exposure to sunlight. This natural photoprotection may be enhanced by dietary compounds with photoprotective properties. Photoprotection is based primarily on high absorption efficacy of UV light over a broad range of wavelengths and on intracellular control of oxidative stress and stress-dependent signaling and inflammation. A combination of nutrients, which operate in one of these two mechanisms, may be even more effective than single agents. Nutrients that are important for endogenous photoprotection are found in fruits and vegetables and include carotenoids, tocopherols, flavanols, and other polyphenols, as well as vitamin C. Among these diverse groups of nutrients, those found in tomatoes, which are mainly carotenoids, have elicited special interest, as will be discussed below.

Dietary antioxidants function in skin cells in two major ways. First, they can scavenge reactive oxygen species (ROS) generated in the skin in response to UV exposure. Second, they can activate protective cellular pathways and interfere with signaling pathways altered in the skin in response to direct UV radiation and enhanced ROS formation. Botanical antioxidants are gaining popularity, and an increasing number of skin care products containing such ingredients are introduced into the market. However, antioxidants will be effective only if they are present in the skin at the time ROS are produced as a consequence of radiation [3]. While protection through individual dietary components in terms of sun protection factor (SPF) may be considerably lower than that achieved using topical sunscreens, an increased lifelong overall protection via dietary photoprotective regimens may contribute significantly to skin health.

The endogenous UV protection by dietary means was tested in several human studies and their mechanism of action was examined in various cellular models. Carotenoids were the main dietary ingredients used in these studies with a predominant role allotted to tomato carotenoids. However, other carotenoids, plant polyphenols, isothiocyanates, vitamins, and oils also exerted positive effects.

17.3 ULTRAVIOLET SUN RADIATION

UV radiation is divided into three spectra subdivisions: UVC (200–290 nm), UVB (290–320 nm), and UVA (320–400 nm). Ozone in the Earth's atmosphere absorbs 100% of UVC and consequently has no significant impact on human health. In contrast, approximately 90% of UVB, and virtually no UVA, is absorbed by the atmosphere [7, 8]. Therefore, severe depletion of the ozone layer as a result of climate changes would have a large impact on the amount of UVB exposure. UVB component of the solar spectrum would present the greatest risk for human health. It is well recognized that exposure to solar UVB irradiation is the main cause of skin cancer in humans [7, 8]. This UVB radiation causes sunburns, DNA strand breaks, and pyrimidine dimers ultimately leading to malignant transformation. Exposure to UVA light is the main reason for the process of photoaging to take place, whereas UVB has a minor effect on this process [7, 8]. The epidermis presents a weak resistance to UVA penetration compared to UVB that is mostly absorbed by this layer. UVA penetration into the deeper skin layer is responsible for the signs of photoaging, including wrinkling, pigmentation, dryness, and roughness. Moreover, UVA causes immunosuppression [9] due to a decrease in Langerhans immune cells in the dermis, which parallels the increase in inflammatory cells in this skin layer [10].

The UVB part of solar radiation is highly erythematogenic and is considered a major cause of sunburn, which starts to develop a few hours after sun exposure, culminating in about 18–24 h postirradiation. Sunburn or erythema is a visible dermal damage following excessive exposure to sunlight. This condition is characterized by tenderness, sometimes painful blistering, and second-degree burns [11]. Direct and indirect damage resulting from photochemical reactions leads to many inflammation-related signs including vasodilatation of dermal vessels, edema, and increased blood flow in the affected area. Damage to proteins and DNA induces morphological changes in keratinocytes and other skin cells. When a cell becomes irreversibly damaged by UV exposure, cell death is triggered by apoptosis, a physiological form of cell death that appears to be a protective mechanism, for disposal of damaged cells that may carry potentially harmful mutations [12].

17.4 CAROTENOIDS AND UV PROTECTION

17.4.1 Introduction to Carotenoids and Health

Carotenoids are a family of compounds comprising over 600 fat-soluble plant pigments that provide much of the color we see in nature. Only a small portion of them is found in human blood and tissues. Apart from their aesthetic role, dietary carotenoids, or foods rich in these colorful pigments, are considered to be beneficial in the prevention of a variety of major diseases, including cardiovascular diseases, cancer, and eye diseases. The carotenoids that have been most studied in this regard are β-carotene, lycopene, lutein, and zeaxanthin. Disease prevention by carotenoids is thought to result, in part at least, from their antioxidant activity, but other mechanisms of protection also do exist.

Fruits and vegetables contain in excess of 40 carotenoids that are routinely absorbed and metabolized by humans [13]. Intake of tomato and tomato-based food products contributes to the absorption of a wide range of carotenoids to human serum and tissues. The prominent carotenoid in tomatoes is the red pigment lycopene that is also among the major carotenoids found in human serum. A prospective cohort study by Giovannucci et al. [14] indicated that intake of lycopene-rich foods (e.g., tomato sauce, tomatoes, and pizza) was inversely associated with risk of prostate cancer, whereas the overall intake of fruits and vegetables was not related. Heating tomato juice in oil was found to be associated with an increased lycopene absorption compared to the absorption of uncooked juice [15]. Interestingly, these studies support the findings of Giovannucci et al. [14] showing inverse association between the consumption of tomato paste or sauce and the risk of prostate cancer, but this inverse correlation was not observed with consumption of minimally processed tomato juice that is consumed uncooked and not with oil and thus has lower bioavailability.

Carotenoid actions and their relation to health and disease is an important issue in many studies including human intervention studies [16]. In this chapter, we will concentrate on studies with tomato carotenoids. These intervention studies were performed mainly with tomato extract preparations that, in addition to the major tomato carotenoid lycopene, contain other carotenoids such as phytoene, phytofluene, and β-carotene. A combination of these carotenoids may produce additive or even synergistic effects on many metabolic endpoints. This concept was also tested in a human intervention study, which will be detailed later, that examines the skin protective effect of carotenoid mixtures. The nutritional significance of lycopene for human health has been demonstrated in studies showing its inhibitory effect on tumor growth in animal models *in vivo* [17–20] and on cancer cells *in vitro* [21–23]. We have demonstrated that lycopene inhibits mammary, endometrial, lung, and leukemic cancer cell growth in a dose-dependent manner [24, 25].

17.4.2 Clinical Studies on the Prevention of Erythema

A major endpoint in the studies of UVB-induced skin damage is the formation of erythema. The sensitivity of an individual toward erythematogenic UV exposure is determined by the minimal erythemal dose (MED). MED is defined as the threshold dose required to produce a detectable reddening of the skin 24 h after UVB exposure. MED values differ between individuals and depend on skin type and the endogenous protection by melanin. Melanin levels determine the skin color and are related mostly to the skin type, which is often categorized according to Fitzpatrick scale ranging from type I to type VI [26]. Skin type I is assigned to people with white or freckled skin, green or light blue eyes, red hair, and high sensitivity to sun light; people with skin type VI have black skin, dark brown eyes, and black hair, almost never experience sunburn. Skin response to UV can be predicted to a good approximation by skin colorimetry, a simple, inexpensive, and noninvasive measurement [27].

Among the carotenoids, β-carotene supplements were the first to be used as an oral, endogenous sun protectant. However, studies demonstrating protection of skin from sun exposure by β-carotene treatment are somewhat contradictory. In one study, it was shown that ingestion of 24 mg/day of β-carotene (from the alga *Dunaliella salina*) for a period of 12 weeks increased carotenoid skin levels and this was associated with protection from UV-induced erythema [28]. At each time point (week 0, 4, 8, 12), erythema was detected by measuring skin color before and 24 h after irradiation, by chromametry (analysis of the degree of reddening of the skin). Erythema intensity was significantly diminished from week 8 onward. Interestingly, erythema suppression was more pronounced when a combination of β-carotene and vitamin E was used. In another study, the protective effects of oral α- and β-carotenes on UVA- and UVB-induced erythema in humans were determined [29]. Subjects were supplemented with 30–90 mg/day of natural carotenoids for 24 weeks. Serum levels of both carotenoids were significantly increased and this was associated with a modest increase in MED of solar simulator radiation. Similar results were reported in other studies [2, 28, 30], but not consistently [31–33]. Interestingly, skin protection was noted only when carotene consumption lasted for more than 10 weeks. Apparently, sustained consumption is required to achieve adequate level of the nutrients in the skin for optimal protection against UV-induced erythema.

The use of high doses of β-carotene in supplements for photoprotection has been controversial due to safety concerns that arose in some clinical studies. In two intervention trials with individuals at high risk for lung cancer, supplementation with pharmacologic levels of β-carotene did not prevent lung cancer, but rather increased lung cancer incidence [34, 35]. These results have ignited debate on the protective or potentially harmful role of β-carotene in human skin [36]. The reasons why β--carotene increased lung cancer risk are not completely understood. However, a major reason that is frequently cited is the large dose of β-carotene selected for supplementation, which resulted in high levels of the carotenoid in the participants' blood. In striking contrast to the very high blood levels achieved after β-carotene supplementation, lycopene levels in human serum are kept at lower levels even after supplementation with "mega" doses. Although the mechanism for the lower blood level of lycopene is not known, it provides a safety gauge against deleterious effect of lycopene after supplementation. In such studies, serum levels of β-carotene are around 3 μM and may exceed 5 μM [37], whereas lycopene levels above 1.2 μM are rarely obtained even with high doses of lycopene. For example, quantities as high as 75 mg/day lycopene did not result in serum level that is higher than 1 μM [38].

The relative inconsistency in β-carotene effects and the safety issue stimulated intense interest in the skin protection activity of other carotenoids in particular lycopene. Thus, Heinrich et al. assessed whether, to achieve sun protection, a mixture of different carotenoids can be substituted for high dose of β-carotene [39]. Volunteers received either β-carotene (24 mg/day) or a carotenoid mix containing β-carotene, lycopene, and lutein, 8 mg each. Serum levels of the three carotenoids and total skin carotenoids increased as expected. The intensity of erythema was diminished in both groups and was significantly lower after 12 weeks of supplementation compared to baseline. Similar results were obtained in another study [40] exploring the effectiveness of an antioxidant combination containing lycopene, β-carotene, alpha-tocopherol, and selenium to reduce UV-induced damages. The mixture was administered daily for 7 weeks and a reduction in UV-induced erythema, concomitantly with attenuation of molecular changes related to skin damage were detected. The authors concluded that oral intake of antioxidant mixture could provide a safe, prolonged, and efficient endogenous photoprotection. An additional randomized, double-blind, parallel group, placebo-controlled study [41] was conducted in healthy young female volunteers (skin type II) to investigate the preventive, photoprotective effect of antioxidative combination containing both lipid- and water-soluble compounds: carotenoids (β-carotene and lycopene), vitamins C and E, selenium, and proanthocyanidins. The antioxidant mixture retarded the development of UVB-induced erythema. This was associated with a decreased expression of matrix metalloproteinases 1 (MMP-1) and MMP-9. These proteolytic enzymes appear to mediate the deleterious effect of UV radiation on skin structure leading to premature aging.

Stahl et al. examined whether erythema can be prevented by consuming tomato paste [42], which contains high amounts of bioavailable lycopene [43]. Tomato paste containing 16 mg of lycopene/day was given together with olive oil to increase the carotenoid absorption. The intervention was performed over a period of 10 weeks and caused an increase in lycopene levels in serum and skin. At the end of the study, erythema formation was significantly lower in the group consuming the tomato paste compared to control, but no significant protection was detected after a short treatment of 4 weeks. This study raises the question whether lycopene alone was responsible for the systemic photoprotection against UV damage by tomato and its products because it is known that, in addition to lycopene, other carotenoids such as phytoene and phytofluene, which are precursors in the biosynthesis of lycopene, are present at significant amounts in tomatoes (see structures of the tomato carotenoids in Figure 17.1).

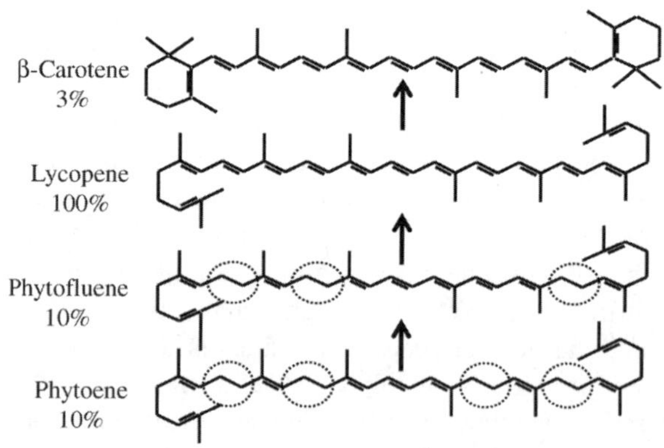

Figure 17.1 Chemical structure of the major tomato carotenoids. The carotenoids are presented in order (from bottom to top) of the biosynthetic pathway, but not all synthetic steps are shown. Circles show the location of single bonds in the backbone of phytoene and phytofluene that are sequentially desaturated to double bonds to get the 11 conjugated double-bond system of lycopene. This conjugated system is responsible both for the red color of lycopene and for its strong antioxidant activity. The percentages shown are the approximate relative content of the presented carotenoids in the tomato, taking lycopene content to 100%.

The photoprotective effects of synthetic lycopene was compared to that of tomato extract and a tomato extract-based drink fortified with phytoene and phytofluene in human volunteers [44]. The amounts of lycopene ingested from these three different sources was similar (about 10 mg/day); however, increasing amounts of other carotenoids present in tomatoes were supplied with the two tomato extract-based preparations. After 12 weeks of supplementation, similar rise in lycopene serum levels and total skin carotenoids was observed in all groups. However, increases in the serum levels of phytoene and phytofluene occurred only in the two tomato extract-supplemented groups and not in the synthetic lycopene-treated group. The protective effect against UV-induced erythema was more pronounced following consumption of tomato-derived products, compared to synthetic lycopene; in this group, the protection from erythema (25%) was not statistically significant. A better and significant reduction in the formation of erythema was evident in the tomato extract group (38%) and the most pronounced protective effect was observed in the fortified tomato drink group (48%). The difference in the efficacy of the tomato products compared to synthetic lycopene might be due to the presence of phytoene and phytofluene. One explanation for the additional protection by phytoene and phytofluene may be due to a sunscreen-like effect of these carotenoids because their absorption maxima are in the ultraviolet light range: 286 nm (UVB) for phytoene and 348 nm (UVA) for phytofluene. Thus, functional foods (like the phytoene plus phytofluene-fortified drink) may be developed to provide endogenous photoprotection more efficiently than that achieved with standard tomato products.

17.4.3 Protection Against Sunburn Cell Formation by Tomato Lycopene

Excessive exposure to the solar UVA and UVB radiation can cause apoptotic skin cell death ("sunburn cells") and suppression of the skin immune response due to depletion of Langerhans cells. During skin infections, local Langerhans cells are involved in immune protection as they become fully functional antigen-presenting cells. In a recent study (personal communication, Zoe Draelos), the efficacy of oral antioxidant supplements in the prevention of UVA/UVB-induced suppression of skin immune response was evaluated in humans. In this study, the effect of tomato extract alone was compared with its combination with natural preparations of lutein and carnosic acid. All ingredients were supplied as oily plant extracts: a tomato extract; a natural carotenoid supplement containing free lutein and lutein esters; a rosemary extract containing mainly carnosic acid (over 35%); soybean oil served as placebo. Subjects with Fitzpatrick skin type I or II were randomized into three balanced groups according to age and sex. After 10 weeks of supplementation, a 3 mm punch biopsy was taken 24 h after exposure to solar simulator (UVB/UVA). The study results show that subjects who consumed the supplements had a significantly lower number of sunburn cells and higher number of Langerhans cells compared to the placebo control group. It is worth noting that subjects who consumed the combination supplement obtained a greater protection than the tomato extract alone in both endpoints measured.

17.4.4 Improvement in Parameters of Human Skin Structure by a Combination of Tomato Lycopene and Other Antioxidants

The studies described above demonstrate that supplementation with tomato lycopene and other phytonutrients is effective in protection against UV-induced skin damage. However, little is known about effects of carotenoids on aesthetic parameters of skin structure. Thus, a study was designed to test the influence of a mixture of antioxidants composed of carotenoids (lycopene, 3 mg/day; lutein, 3 mg/day; β-carotene, 4.8 mg/day), vitamin E (γ-tocopherol, 10 mg/day), and selenium (75 μg/day) on parameters related to skin health and skin aging [45]. Healthy volunteers with skin type 2 were supplemented with antioxidant mixtures for a period of 12 weeks. Upon supplementation, serum levels of the carotenoids increased in the treatment groups but not in the placebo group. Skin density and thickness were determined by ultrasound measurements. A significant increase for both parameters was determined after 6 and 12 weeks treatment with the antioxidant mixture compared to the control group. Roughness, scaling, smoothness, and wrinkling of the skin were determined by surface evaluation of living skin (SELS). Roughness and scaling were improved by supplementation with antioxidant micronutrients. Other SELS parameters, that is, smoothness and wrinkling, were not affected by treatment. Interestingly, in a recent study [46] using modern optical noninvasive *in vivo* methods, both the structures of the furrows and wrinkles and the concentration of lycopene were analyzed quantitatively on the forehead skin of 20 volunteers aged between 40 and 50 years. A significant inverse correlation was obtained between the skin roughness and the lycopene concentration. These findings indicate that higher levels of antioxidants in the skin effectively lead to lower levels of skin roughness.

17.5 ENDOGENOUS SUN PROTECTION BY DIETARY INGREDIENTS OTHER THAN CAROTENOIDS

The use of dietary botanicals other than carotenoids for skin protection from the adverse biological effects of solar UV radiation is drawing considerable attention as improvement in skin structure was achieved by the ingestion of additional phytonutrients. For example, dietary supplementation with flaxseed and borage oil was found to modulate skin appearance in women [47]. Treatment of 12 weeks with both types of oil improved structural parameters of the skin: skin hydration was increased, transepidermal water loss was decreased, and SELS revealed that roughness and scaling of the skin were significantly decreased. In addition, treatment with flaxseed or borage oil reduced skin redness and blood flow, which in this study was induced by nicotinate treatment and not by UV radiation. These data provide evidence that skin properties can be modulated by an intervention with dietary lipid antioxidants.

These investigators also examined the effect of cocoa flavanols on skin surface [48]. In this study, UV-induced erythema was significantly decreased following a 12 week-long treatment with high-flavanol cocoa powder, whereas no change was found in a group receiving low-flavanol powder. Ingestion of the high-flavanol cocoa led to a significant increase in skin density, thickness, and skin hydration. Evaluation of skin surface showed a significant decrease in skin roughness and scaling in the high-flavanol cocoa group. In another study using these cocoa flavanols [49], additional cosmetically relevant parameters of skin surface and hydration were assessed. The findings show that dermal blood flow was increased and oxygen saturation of dermal hemoglobin was elevated following intake of high-flavanol cocoa. The effectiveness of another complex oral nutraceutical preparation containing glycosaminoglycans, various antioxidants, and minerals on cutaneous aging was recently examined in 60 women aged 35–60 years. At the end of the 12 week-long treatment, the oral dietary supplement were shown to improve skin roughness and fine wrinkles, but did not affect skin color [50].

The role of several other antioxidants in photoprotection was studied in animal models. The effect of oral epigallocatechin gallate (EGCG), a powerful antioxidant present in green tea, on UV-induced skin damage was studied in female hairless rats [51]. Intake of EGCG for 8 weeks was demonstrated to increase skin tolerance to radiation by increasing minimal erythema dose. The green tea antioxidant also attenuated UV-induced sunburn severity and alterations in epidermal barrier function. Similar model of hairless mice was used to test the preventive effect of proanthocyanidins from grape seeds in UVB-induced carcinogenesis [52]. Although feeding proanthocyanidins did not show any apparent sign of toxicity, it prevented malignant transformation as evident from a 45% reduction in carcinoma incidence and a 75% reduction in carcinoma size.

17.6 MECHANISMS FOR THE PHOTOPROTECTION BY PHYTONUTRIENTS

Much of the cell injury caused by UV irradiation is associated with oxidative stress and the generation of ROS that have a profound impact on cells in general and on skin cells in particular. Excessive levels of ROS mediate skin inflammation (erythema), may promote cell ageing, cell death, and cancer. Free radical scavenging activity of antioxidant phytonutrients plays a key role in cell protection. However, their low blood and tissue concentrations suggest that other cellular mechanisms, in addition to ROS scavenging, may be more important for the antioxidant effects. A wide variety of protective mechanisms that are associated with the cellular response to oxidative stress have been identified. Among these, the Nrf2 pathway is regarded as most important in protecting cells against chemical- and radiation-induced oxidative and electrophilic stress [53–55].

Nrf2 is a transcription factor that controls the expression and the coordinated induction of a battery of defensive genes encoding detoxifying enzymes and antioxidant proteins named phase II enzymes [53, 54]. Nrf2 is retained in the cytoplasm by an inhibitor protein, Keap1. In response to oxidative/electrophilic stress, Nrf2 is switched on by dissociating it from Keap1 binding. Nrf2 is stabilized and translocates to the nucleus, forming heterodimers with several other transcription factors. These heterodimers bind the antioxidant response element (ARE), a DNA nucleotide sequence residing in promoters of phase II enzymes. This process leads to the coordinated activation of expression of genes that protect cells against free radical damage, prevents apoptosis, and promotes cell survival. Consequently, the induction of the Nrf2 transcription system by UV radiation and its role in protection from UV-induced damage have been studied in detail.

17.6.1 Nrf2\ARE Transcription System Is Involved in Skin Protection from Photooxidative Damage

Kannan et al. demonstrated that low-dose UVB exposure (7.5 J/m^2) led to the nuclear accumulation of Nrf2 and activation of chemoprotective proteins [56]. These results obtained by utilizing low-dose UVB irradiation were similar to those previously observed with natural antioxidants. Intriguingly, high-dose UVB (20 J/m^2) exposure of mouse keratinocyte and human fibroblast cells led to the nuclear exclusion of Nrf2 and the downregulation of ARE-mediated gene expression. This may suggest that skin protection by the Nrf2 transcription system is impaired by a high-dose UV radiation. Not all studies agree with these results. For example, UVB irradiation at physiological doses (10 J/m^2) reduced Nrf2 and Nrf2-dependent gene expression in normal keratinocytes and melanocytes *in vitro* and in *ex vivo* skin organ cultures [57]. However, α-MSH, a UV-induced peptide, prevented or even overcompensated the UVB-induced suppression of Nrf2 and Nrf-dependent genes. The role of the α-MSH in protection of UV-induced damage should be clarified in future studies. In addition, Durchdewald et al. [58] reported that although electrophilic compounds transactivated Nrf2 in keratinocytes and induced nuclear translocation of Nrf2 and the expression of known Nrf2 target genes, UVA and UVB irradiation had no effect.

This controversy is evident in another study that has shown that UVA irradiation, but not UVB, causes nuclear translocation and accumulation of Nrf2 in dermal fibroblasts [59]. To evaluate the protective role of Nrf2 against UVA radiation, they examined UVA-induced apoptosis, using dermal fibroblasts derived from Nrf2 or Keap1 gene knockout mice. Disruption of Nrf2, causing inactivation of the ARE protective system, increased the number of apoptotic cells following UVA irradiation. Along the same line, disruption of Keap1, causing activation of ARE, decreased the apoptotic cell number compared to wild-type controls. These findings thus reinforce the concept that the Nrf2–Keap1 pathway plays an important role in the protection of skin against UVA irradiation.

The effects of UVA on global gene expression in human dermal fibroblasts and keratinocytes was examined in another study [60]. A cluster of known ARE-dependent genes such as heme oxygenase 1 and glutamate-cysteine ligase modifier subunit were induced by UVA radiation and by oxidized lipids that are formed in response to the radiation. Both UVA irradiation and treatment with oxidized lipids increased nuclear accumulation and DNA binding of Nrf2. Silencing and deficiency of Nrf2 suppressed all these effects. Taken together, the data show that UVA-mediated lipid oxidation induces expression of the redox-regulated transcription factor Nrf2.

The role of Nrf2–Keap1 system in protection from UV-induced damage was also examined *in vivo* [61]. In this study, a single irradiation with UVB induced stronger and longer lasting sunburn reaction in Nrf2-null mice compared to the wild-type animal. Histological changes, including epidermal necrosis, dermal edema, inflammatory cell infiltration, sunburn apoptotic cell formation, and accumulation of oxidative DNA products after UVB irradiation, were more prominent in Nrf2-null mice. These findings indicate that the Nrf2 pathway plays an important role in protection of the skin against acute UVB reactions, including cutaneous cell apoptosis and oxidative damage.

17.6.2 Consumption of Plant-Based Foods Protect Skin by Activating Nrf2

A promising means for upregulating the levels of cytoprotective enzymes in the human population is the consumption of plant-based foods such as carotenoids, polyphenols, and isothiocyanates [55, 62]. Induction of the Nrf2/ARE protective system by these phytochemicals over the course of a human lifetime might lead to reduced rates of UV-induced damage in skin. In support of this hypothesis, several studies examined the role of Nrf2 in the protective effects of various phytonutrients on UV-induced damage. For example, Kimura et al. [63] have demonstrated that Nrf2 plays an essential role in keratinocyte protection by quercetin. Quercetin is a major natural polyphenol found in fruits and vegetables such as citrus, onions, kale, and broccoli. It was found that quercetin pretreatment strongly suppressed UVA-induced apoptosis in human HaCaT keratinocyte cells, markedly increased protein levels of the transcription factor Nrf2, induced the expression of antioxidative genes, and dramatically reduced the production of ROS following UVA irradiation. Remarkably, Nrf2 knockdown by transfection with Nrf2-siRNA diminishes the protective effect of quercetin. Thus, quercetin protects cells from UVA damage mainly by elevating intracellular antioxidative activity via the enhanced accumulation of Nrf2, the transcription factor for antioxidant genes. Nrf2 was also found to be important for the protective effect of the isothiocyanate sulforaphane against photooxidative damage in nonskin cells, the retinal pigment epithelial cells [55]. A series of cinnamoyl-based electrophilic compounds were identified as potent Nrf2 activators and inducers of phase II enzymes in human Hs27 skin fibroblasts, HaCaT keratinocytes, and primary skin keratinocytes [64]. These results suggest feasibility of developing cinnamoyl-based Nrf2 activators as novel photochemopreventive agents targeting skin cell photooxidative stress.

It is noteworthy that carotenoids activate the Nrf2/ARE transcription system and induce the expression of phase II enzymes [62]. Although such studies were not performed using skin cells, the findings suggest that carotenoid effects in the protection of UV-induced skin damage are mediated, at least in part, by the Nrf2 system. Various electrophilic phytonutrients have been shown to induce this transcription system by disrupting the inhibitory activity of Keap1 on Nrf2. However, hydrophobic carotenoids such as lycopene lack any electrophilic groups and, consequently, are unlikely to directly activate Nrf2. Interestingly, we demonstrated that carotenoid oxidation products such as mono- and diapocarotenals are the active mediators in the stimulation of the Nrf2/ARE system [65].

17.7 CONCLUSIONS: THE POWER OF SYNERGISTIC COMBINATIONS

To conclude, prevention of UV skin damage employing nutritional approaches has been proven to be effective in many human studies. The biochemical processes involved in this protective effect are not fully understood. However, the induction of phase II enzymes and activation of the Nrf2/ARE transcription system play a key role in this biochemical process. We suggest that combinations of various carotenoids, or carotenoids with other micronutrients, will produce significant skin protective effects, most likely by synergistic interactions. Going back to Mother Nature, effectiveness and safety are married together in the whole tomato or, more generally, in diets rich in vegetables and fruits.

REFERENCES

1. Adhami VM, Ahmad N, Mukhtar H. Molecular targets for green tea in prostate cancer prevention. *J. Nutr.* 2003;133;2417S–2424S.
2. Mathews-Roth MM, Pathak MA, Parrish J, Fitzpatrick TB, Kass EH, Toda K, Clemens W. A clinical trial of the effects of oral beta-carotene on the responses of human skin to solar radiation. *J. Invest. Dermatol.* 1972;59:349–353.
3. Sies H, Stahl W. Nutritional protection against skin damage from sunlight. *Annu. Rev. Nutr.* 2004;24:173–200.
4. Stahl W, Heinrich U, Aust O, Tronnier H, Sies H. Lycopene-rich products and dietary photoprotection. *Photochem. Photobiol. Sci.* 2006;5:238–242.
5. Pathak MA. Sunscreens: topical and systemic approaches for protection of human skin against harmful effects of solar radiation. *J. Am. Acad. Dermatol.* 1982;7:285–312.
6. Ortonne JP. Photoprotective properties of skin melanin. *Br. J. Dermatol.* 2002;146 (Suppl. 61):7–10.
7. Ting WW, Vest CD, Sontheimer R. Practical and experimental consideration of sun protection in dermatology. *Int. J. Dermatol.* 2003;42:505–513.
8. Rabe JH, Mamelak AJ, McElgunn PJ, Morison WL, Sauder DN. Photoaging: mechanisms and repair. *J. Am. Acad. Dermatol.* 2006;55:1–19.
9. Seite S, Zucchi H, Moyal D, Tison S, Compan D, Christiaens F, Gueniche A, Fourtanier A. Alterations in human epidermal Langerhans cells by ultraviolet radiation: quantitative and morphological study. *Br. J. Dermatol.* 2003;148:291–299.
10. Fourtanier A, Bernerd F, Bouillon C, Marrot L, Moyal D, Seite S. Protection of skin biological targets by different types of sunscreens. *Photodermatol. Photoimmunol. Photomed.* 2006;22:22–32.
11. Stahl W, Sies H. Carotenoids and flavonoids contribute to nutritional protection against skin damage from sunlight. *Mol. Biotechnol.* 2007;37:26–30.
12. Kulms D, Schwarz T. Molecular mechanisms of UV-induced apoptosis. *Photodermatol. Photoimmunol. Photomed.* 2000;16:195–201.
13. Khachik F, Goli MB, Beecher GR, Holden J, Lusby WR, Tenorio MD, Barrera MR. Effect of food preparation on qualitative and quantitative distribution of major carotenoid constituents of tomatoes and several green vegetables. *J. Agric. Food Chem.* 1992;40:390–398.
14. Giovannucci E, Ascherio A, Rimm EB, Stampfer MJ, Colditz GA, Willett WC. Intake of carotenoids and retinol in relation to risk of prostate cancer. *J. Natl. Cancer Inst.* 1995;87:1767–1776.
15. Stahl W, Sies H. Uptake of lycopene and its geometrical isomers is greater from heat-processed than from unprocessed tomato juice in humans. *J. Nutr.* 1992;122:2161–2166.
16. Krinsky NI, Johnson EJ. Carotenoid actions and their relation to health and disease. *Mol. Aspects Med.* 2005;26:459–516.

17. Nagasawa H, Mitamura T, Sakamoto S, Yamamoto K. Effect of lycopene on spontaneous mammary tumor development in SHN virgin mice. *Anticancer Res.* 1995;15:1173–1178.
18. Sharoni Y, Giron E, Rise M, Levy J. Effects of lycopene enriched tomato oleoresin on 7,12-dimethyl-benz[*a*]anthracene-induced rat mammary tumors. *Cancer Detect. Prevent.* 1997;21:118–123.
19. Narisawa T, Fukaura Y, Hasebe M, Ito M, Aizawa R, Murakoshi M, Uemura S, Khachik F, Nishino H. Inhibitory effects of natural carotenoids, alpha-carotene, beta-carotene, lycopene and lutein, on colonic aberrant crypt foci formation in rats. *Cancer Lett.* 1996;107:137–142.
20. Okajima E, Tsutsumi M, Ozono S, Akai H, Denda A, Nishino H, Oshima S, Sakamoto H, Konishi Y. Inhibitory effect of tomato juice on rat urinary bladder carcinogenesis after *N*-butyl-*N*-(4-hydroxybutyl)nitrosamine initiation. *Jpn. J. Cancer Res.* 1998;89:22–26.
21. Eichler O, Sies H, Stahl W. Divergent optimum levels of lycopene, beta-carotene and lutein protecting against UVB irradiation in human fibroblasts. *Photochem. Photobiol.* 2002;75:503–506.
22. Smit N, Vicanova J, Cramers P, Vrolijk H, Pavel S. The combined effects of extracts containing carotenoids and vitamins E and C on growth and pigmentation of cultured human melanocytes. *Skin Pharmacol. Physiol.* 2004;17:238–245.
23. Pastori M, Pfander H, Boscoboinik D, Azzi A. Lycopene in association with alpha-tocopherol inhibits at physiological concentrations proliferation of prostate carcinoma cells. *Biochem. Biophys. Res. Commun.* 1998;250:582–585.
24. Levy J, Bosin E, Feldman B, Giat Y, Miinster A, Danilenko M, Sharoni Y. Lycopene is a more potent inhibitor of human cancer cell proliferation than either α-carotene or β-carotene. *Nutr. Cancer.* 1995;24:257–267.
25. Amir H, Karas M, Giat J, Danilenko M, Levy R, Yermiahu T, Levy J, Sharoni Y. Lycopene and 1,25-dihydroxy vitamin-D3 cooperate in the inhibition of cell cycle progression and induction of differentiation in HL-60 leukemic cells. *Nutr. Cancer* 1999;33:105–112.
26. Fitzpatrick TB. The validity and practicality of sun-reactive skin types I through VI. *Arch. Dermatol.* 1988;124:869–871.
27. Andreassi L, Flori ML, Rubegni P. Sun and skin: role of phototype and skin colour. *Adv. Exp. Med. Biol.* 1999;455:469–475.
28. Stahl W, Heinrich U, Jungmann H, Sies H, Tronnier H. Carotenoids and carotenoids plus vitamin E protect against ultraviolet light-induced erythema in humans. *Am. J. Clin. Nutr.* 2000;71:795–798.
29. Lee J, Jiang S, Levine N, Watson RR. Carotenoid supplementation reduces erythema in human skin after simulated solar radiation exposure. *Proc. Soc. Exp. Biol. Med.* 2000;223:170–174.
30. Biesalski HK, Hemmes C, Hopfenmuller W, Schmid C, Gollnick HP. Effects of controlled exposure of sunlight on plasma and skin levels of beta-carotene. *Free Radic. Res.* 1996;24:215–224.
31. Wolf C, Steiner A, Honigsmann H. Do oral carotenoids protect human skin against ultraviolet erythema, psoralen phototoxicity, and ultraviolet-induced DNA damage? *J. Invest. Dermatol.* 1988;90:55–57.
32. Garmyn M, Ribaya-Mercado JD, Russel RM, Bhawan J, Gilchrest BA. Effect of beta-carotene supplementation on the human sunburn reaction. *Exp. Dermatol.* 1995;4:104–111.
33. McArdle F, Rhodes LE, Parslew RA, Close GL, Jack CI, Friedmann PS, Jackson MJ. Effects of oral vitamin E and beta-carotene supplementation on ultraviolet radiation-induced oxidative stress in human skin. *Am. J. Clin. Nutr.* 2004;80:1270–1275.
34. Heinonen OP, Huttunen JK, Albanes D, et al. Effect of vitamin E and beta carotene on the incidence of lung cancer and other cancers in male smokers. *New Engl. J. Med.* 1994;330:1029–1035.
35. Omenn GS, Goodman GE, Thornquist MD, Balmes J, Cullen MR, Glass A, Keogh JP, Meyskens FL, Valanis B, Williams JH, Barnhart S, Hammar S. Effects of a combination of beta carotene and vitamin A on lung cancer and cardiovascular disease. *New Engl. J. Med.* 1996;334:1150–1155.
36. Biesalski HK, Obermueller-Jevic UC. UV light, beta-carotene and human skin: beneficial and potentially harmful effects. *Arch. Biochem. Biophys.* 2001;389:1–6.
37. IARC. *Carotenoids: IARC Handbook of Cancer Prevention*, vol. 2. WHO Press, 1998.
38. Paetau I, Khachik F, Brown ED, Beecher GR, Kramer TR, Chittams J, Clevidence BA. Chronic ingestion of lycopene-rich tomato juice or lycopene supplements significantly increases plasma concentrations of lycopene and related tomato carotenoids in humans. *Am. J. Clin. Nutr.* 1998;68:1187–1195.

39. Heinrich U, Gartner C, Wiebusch M, Eichler O, Sies H, Tronnier H, Stahl W. Supplementation with beta-carotene or a similar amount of mixed carotenoids protects humans from UV-induced erythema. *J. Nutr.* 2003;133:98–101.
40. Cesarini JP, Michel L, Maurette JM, Adhoute H, Bejot M. Immediate effects of UV radiation on the skin: modification by an antioxidant complex containing carotenoids. *Photodermatol. Photoimmunol. Photomed.* 2003;19:182–189.
41. Greul AK, Grundmann JU, Heinrich F, Pfitzner I, Bernhardt J, Ambach A, Biesalski HK, Gollnick H. Photoprotection of UV-irradiated human skin: an antioxidative combination of vitamins E and C, carotenoids, selenium and proanthocyanidins. *Skin Pharmacol. Appl. Skin Physiol.* 2002;15:307–315.
42. Stahl W, Heinrich U, Wiseman S, Eichler O, Sies H, Tronnier H. Dietary tomato paste protects against ultraviolet light-induced erythema in humans. *J. Nutr.* 2001;131:1449–1451.
43. Gartner C, Stahl W, Sies H. Lycopene is more bioavailable from tomato paste than from fresh tomatoes. *Am. J. Clin. Nutr.* 1997;66:116–122.
44. Aust O, Stahl W, Sies H, Tronnier H, Heinrich U. Supplementation with tomato-based products increases lycopene, phytofluene, and phytoene levels in human serum and protects against UV-light-induced erythema. *Int. J. Vitam. Nutr. Res.* 2005;75:54–60.
45. Heinrich U, Tronnier H, Stahl W, Bejot M, Maurette JM. Antioxidant supplements improve parameters related to skin structure in humans. *Skin Pharmacol. Physiol.* 2006;19:224–231.
46. Darvin M, Patzelt A, Gehse S, Schanzer S, Benderoth C, Sterry W, Lademann J. Cutaneous concentration of lycopene correlates significantly with the roughness of the skin. *Eur. J. Pharm. Biopharm.* 2008;69:943–947.
47. De Spirt S, Stahl W, Tronnier H, Sies H, Bejot M, Maurette JM, Heinrich U. Intervention with flaxseed and borage oil supplements modulates skin condition in women. *Br. J. Nutr.* 2009;101:440–445.
48. Heinrich U, Neukam K, Tronnier H, Sies H, Stahl W. Long-term ingestion of high flavanol cocoa provides photoprotection against UV-induced erythema and improves skin condition in women. *J. Nutr.* 2006;136:1565–1569.
49. Neukam K, Stahl W, Tronnier H, Sies H, Heinrich U. Consumption of flavanol-rich cocoa acutely increases microcirculation in human skin. *Eur. J. Nutr.* 2007;46:53–56.
50. Udompataikul M, Sripiroj P, Palungwachira P. An oral nutraceutical containing antioxidants, minerals and glycosaminoglycans improves skin roughness and fine wrinkles. *Int. J. Cosmet. Sci.* 2009;31:427–435.
51. Jeon HY, Kim JK, Kim WG, Lee SJ. Effects of oral epigallocatechin gallate supplementation on the minimal erythema dose and UV-induced skin damage. *Skin Pharmacol. Physiol.* 2009;22:137–141.
52. Mittal A, Elmets CA, Katiyar SK. Dietary feeding of proanthocyanidins from grape seeds prevents photocarcinogenesis in SKH-1 hairless mice: relationship to decreased fat and lipid peroxidation. *Carcinogenesis* 2003;24:1379–1388.
53. Kaspar JW, Niture SK, Jaiswal AK. Nrf2:INrf2(Keap1) signaling in oxidative stress. *Free Radic. Biol. Med.* 2009;47:1304–1309.
54. Niture SK, Kaspar JW, Shen J, Jaiswal AK. Nrf2 signaling and cell survival. *Toxicol. Appl. Pharmacol.* 2010;244:37-42.
55. Gao X, Talalay P. Induction of phase 2 genes by sulforaphane protects retinal pigment epithelial cells against photooxidative damage. *Proc. Natl. Acad. Sci. USA* 2004;101;10446–10451.
56. Kannan S, Jaiswal AK. Low and high dose UVB regulation of transcription factor NF-E2-related factor 2. *Cancer Res.* 2006;66:8421–8429.
57. Kokot A, Metze D, Mouchet N, Galibert MD, Schiller M, Luger TA, Bohm M. {Alpha}-MSH counteracts the suppressive effect of UVB on Nrf2 and Nrf-dependent gene expression in human skin. *Endocrinology* 2009;150:3197–3206.
58. Durchdewald M, Beyer TA, Johnson DA, Johnson JA, Werner S, Auf dem Keller U. Electrophilic chemicals but not UV irradiation or reactive oxygen species activate Nrf2 in keratinocytes *in vitro* and *in vivo*. *J. Invest. Dermatol.* 2007;127:646–653.
59. Hirota A, Kawachi Y, Itoh K, Nakamura Y, Xu X, Banno T, Takahashi T, Yamamoto M, Otsuka F. Ultraviolet A irradiation induces NF-E2-related factor 2 activation in dermal fibroblasts: protective role in UVA-induced apoptosis. *J. Invest. Dermatol.* 2005;124:825–832.

60. Gruber F, Mayer H, Lengauer B, Mlitz V, Sanders JM, Kadl A, Bilban M, de Martin R, Wagner O, Kensler TW, Yamamoto M, Leitinger N, Tschachler E. NF-E2-related factor 2 regulates the stress response to UVA-1-oxidized phospholipids in skin cells. *FASEB J.* 2010;24:39-48.
61. Kawachi Y, Xu X, Taguchi S, Sakurai H, Nakamura Y, Ishii Y, Fujisawa Y, Furuta J, Takahashi T, Itoh K, Yamamoto M, Yamazaki F, Otsuka F. Attenuation of UVB-induced sunburn reaction and oxidative DNA damage with no alterations in UVB-induced skin carcinogenesis in Nrf2 gene-deficient mice. *J. Invest. Dermatol.* 2008;128:1773–1779.
62. Ben-Dor A, Steiner M, Gheber L, Danilenko M, Dubi N, Linnewiel K, Zick A, Sharoni Y, Levy J. Carotenoids activate the antioxidant response element transcription system. *Mol. Cancer Ther.* 2005;4:177–186.
63. Kimura S, Warabi E, Yanagawa T, Ma D, Itoh K, Ishii Y, Kawachi Y, Ishii T. Essential role of Nrf2 in keratinocyte protection from UVA by quercetin. *Biochem. Biophys. Res. Commun.* 2009;387:109–114.
64. Wondrak GT, Cabello CM, Villeneuve NF, Zhang S, Ley S, Li Y, Sun Z, Zhang DD. Cinnamoyl-based Nrf2-activators targeting human skin cell photo-oxidative stress. *Free Radic. Biol. Med.* 2008;45:385–395.
65. Linnewiel K, Ernst H, Caris-Veyrat C, Ben-Dor A, Kampf A, Salman H, Danilenko M, Levy J, Sharoni Y. Structure–activity relationship of carotenoid derivatives in activation of the electrophile/antioxidant response element transcription system. *Free Radic. Biol. Med.* 2009;47:659–667.

PART V

ANALYSIS OF NATURALS

CHAPTER 18

CHROMATOGRAPHIC TECHNIQUES FOR THE ANALYSIS OF NATURAL PRODUCTS IN COSMETICS

Stefan Gafner and Jacquelyn R. Villinski

Tom's of Maine, Kennebunk, ME

18.1 INTRODUCTION

The analysis of natural products in cosmetics is as fascinating as it is challenging. Extracts from terrestrial or marine sources are inherently complex and can contain hundreds of individual components with a large chemical diversity. The concentrations of the individual compounds in an extract vary enormously. In addition, changes in growing conditions, harvesting techniques, and postharvest processing will lead to noticeable differences from lot to lot. Even with the best techniques, only a limited number of molecules can be simultaneously detected. Considering that cosmetic products often contain a plethora of botanical extracts, which, for aesthetic and financial considerations (contrary to the popular belief, high-quality botanicals tend to be rather expensive), are often present in very small amounts in the product, the analyst has to overcome a number of hurdles to analyze such extracts within the formulation. Nevertheless, we believe that natural products analysis is the most exciting job an analytical chemist can get.

Although the qualitative and quantitative evaluation of a whole botanical extract (chromatographic fingerprinting) has gained more importance as separation and data management technologies improve, the assessment of a botanical based on a restricted number of characteristic compounds (marker compounds) is still the most widespread analytical approach, in particular in finished products. Marker compounds have to comply with the following criteria proposed by the European Medicines Agency (EMEA) [1]:

- Be a characteristic for the botanical extract.
- Constant in the presence of the herbal preparation.

Formulating, Packaging, and Marketing of Natural Cosmetic Products, First Edition.
Edited by Nava Dayan and Lambros Kromidas.

- Chemically defined.
- Analyzable with routine analytical equipment.
- Have to have an adequate content in the herbal raw material and herbal drug preparation.
- Have to be stable.

This chapter will focus on the main chromatographic techniques, thin-layer chromatography (TLC), high-performance liquid chromatography (HPLC), and gas chromatography (GC) to analyze skin care products with natural ingredients. Chromatography was developed from a rudimentary tool in the early twentieth century to a number of techniques capable of addressing some of the most challenging and complex analytical problems. In an overview by Salvador et al. [2], separation techniques (chromatography/electrophoresis) made out 69% of the methods used for cosmetic analysis (see Figure 18.1a). The most popular separation technique is HPLC, followed by GC, TLC, and capillary electrophoresis. Other

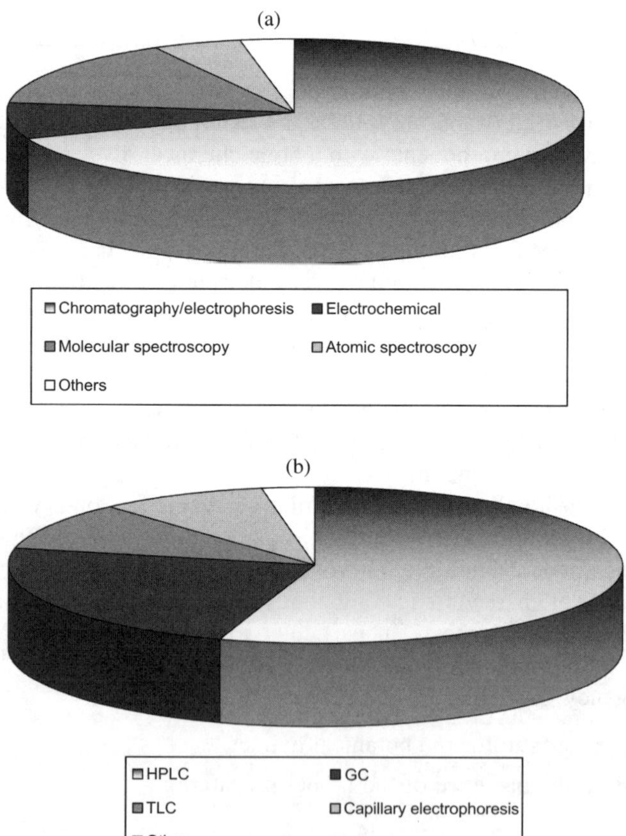

Figure 18.1 (a) Frequency of use of analytical methods in cosmetic analysis. (b) Frequency of use of separation techniques in cosmetic analysis (reproduced with permission from Ref. [2]).

chromatographic techniques (e.g., supercritical fluid chromatography) are less widespread (Figure 18.1b) in cosmetics.

At present, HPLC is the most widely used technique in the cosmetic industry due to its versatility to measure a wide array of compounds of different polarity and the amount of information that can be obtained using different detection systems. Ultraperformance liquid chromatography (UPLC) is a modification of standard HPLC using shorter columns containing a stationary phase of a very small particle size that requires an increase in the flow rate. This technique has recently gained a lot of attention in the community of natural products scientists due to its superior resolution and shorter run times. GC is the gold standard for the analysis of volatile compounds, for example, fragrances. It is also a very useful tool for detection of noxious contaminants such as herbicides, pesticides, or plasticizers, which unfortunately still find their way into botanical raw materials and extracts. Last but not least, TLC is one of the earliest techniques that was used to separate herbal extracts and is still in use due to its low equipment costs (although high-end TLC equipment tend to rival the costs of a HPLC-UV instrument) and versatile applications. Current applications of TLC in the analysis of natural products have mainly focused on correct identification of the material and detection of adulteration. Despite a number of papers reporting good results with quantitative high-performance thin-layer chromatography (HPTLC), the industry is still mainly using HPLC and GC to measure the contents within a given product.

18.2 SAMPLE PREPARATION BASICS

Sample preparation is one of the most crucial steps in the analytical procedure. According to Meyer and Majors [3], approximately two-thirds of the time involved in a typical chromatographic procedure goes into sample preparation. The sample preparation is considered to be slow and labor intensive. Even with the recent improvements in sample automation, the personal care industry is slow in taking advantage of high-throughput sample preparation devices. Majors [4] noticed that the sample load is often not big enough to warrant laboratory robots and that the maintenance labor and costs of such robots may offset the gains in efficiency. In a recent survey on sample preparation techniques in the pharmaceutical industry, the ten most popular techniques were weighing, pH adjustment, filtration, dilution, internal standard addition, column chromatography, evaporation, sonication, centrifugation, and concentration [4]. An overview on analytical procedures for cosmetic ingredients shows that dissolution in an adequate solvent was the most often used solubilization technique, followed by chemical reaction in aqueous media. In terms of purification steps, solid-phase extraction (SPE) turned out to be the most popular, followed by liquid–liquid extraction, distillation, and headspace volatilization. If a chemical reaction was used in the sample preparation step, derivatization was most often used with a relative frequency above 65%, while complexation, saponification/hydrolysis, and redox reactions made up the remainder [2].

The wide variety and the complexity of matrices used in skin care products will inevitably force the analyst to adapt the sample preparation technique according to the formulation. Depending on the matrix in which the extract or isolate has been formulated, the complexity of the sample preparation will vary and thus the overall uncertainty of the final result. Every additional step in sample handling is an additional source of uncertainty. Therefore, whenever possible, the sample preparation should include as few steps as necessary [3]. Another problem with analyzing herbal extracts in skin care formulations is the low concentration of the analyte. Independent of the analytical procedure, the reproducibility of an analysis decreases with decreasing analyte concentration [5]. For example, in a product containing 0.01% of marker compound, the expected relative standard deviation (RSD) is approximately 5%. Decreasing the extract concentration by a factor of 10 will bring the expected RSD close to 8%.

For single-marker compounds, the methods to obtain the analyte in a suitable solution are not different from the methods used to analyze a drug product. However, analyzing a wide variety of components from the botanical may force the analyst to use more complex sample preparation techniques or to resort to a number of different sample preparation techniques for each class of phytochemicals, each of which will be analyzed separately. The Guide to Sample Preparation, published by LCGC magazine in 2000 [6], is an excellent source of information on sample preparation strategies. Due to very small amounts of actives and the difficulties in sample preparation, many companies that produce skin care products will perform the quality control on the incoming botanical raw material before its inclusion in the formulation but will not test the ingredient in the finished product.

18.3 THIN-LAYER CHROMATOGRAPHY AND HIGH-PERFORMANCE THIN-LAYER CHROMATOGRAPHY

Modern HPTLC is an efficient method that uses layers of particles with a smaller diameter (5 μm) compared to 12–20 μm for conventional TLC. The use of HPTLC leads to a greater separation efficiency, faster separations, and improved detection limits [7]. The advantages of TLC lay in the relatively simple methodology, the cost-effectiveness (many samples can be analyzed on the sample plate), the low amounts of solvents used, and the ability to detect many components due to the availability of a large number of reagents for detection. Furthermore, the choice of solvents is less limited than with reverse-phase HPLC or GC. One of the fastest growing areas of application in HPTLC is the presentation of multiple samples as an image to illustrate similarity and difference. This allows for a relatively quick pass/fail decision.

Some of the limitations of HPTLC are pointed out by Reich and Widmer [8]. The use of HPTLC requires manual interaction. Furthermore, some of the parameters influencing TLC are difficult to standardize (e.g., humidity, composition of the gas phase in the TLC chamber). The separation power of HPTLC is lower than that of other chromatographic techniques. Finally, for quantitative analysis, the linear range is lower than the commonly used HPLC-UV or GC-FID and the use of a chemical reagent to make the analyte detectable can lower the precision of the analysis.

18.3.1 Sample Preparation of Raw Materials

Most of the TLC work in natural products chemistry is done on the incoming raw material (whole or powdered plant material, or extract). Extracts are best dissolved in the least polar solvent that allows solubilizing the material. For powdered drugs, Wagner and Bladt [9] give the following recommendations:

- **Anthraquinone Glycosides, Bitter Principles, Flavonoids, and Arbutin**: 1 g of powdered drug is extracted by heating in a water bath for 10 min with 5 mL methanol.
- **Alkaloids**: 1 g of powdered drug is moistened with approximately 1 mL 10% ammonia solution; 5 mL methanol is added and the drug is subsequently extracted by heating for 10 min.
- **Saponins**: an extract prepared the same way as for anthraquinone glycosides is reduced to approximately 1 mL, mixed with 0.5 mL water, and then extracted with 3 mL *n*-butanol. The butanol phase is used for TLC analysis.
- **Terpenes, Coumarins, Phenol Carboxylic Acids, and Valepotriates**: 1 g of powdered drug is extracted by heating under reflux for 15 min with 10 mL dichloromethane. The filtrate is evaporated to dryness and the residue dissolved in 0.5 mL toluene.
- **Essential Oils**: Wagner and Bladt recommend the use of a specific thermo-distillation apparatus for the TLC analysis of essential oils. In case that such apparatus is not available, the same extraction technique as used for terpenes can be utilized.

18.3.2 Sample Preparation of Finished Products

Procedures for preparing samples are generally similar to those for GC and HPLC. In the best case, the sample can be dissolved in the application solvent. Ideally, the application solvent is volatile, free of water, and as nonpolar as possible in order to avoid a circular distribution of the analyte. Unfortunately, most matrices in skin care products cannot be diluted simply in a nonpolar solvent; therefore, a more sophisticated approach has to be taken.

A classic way to prepare samples out of a variety of matrices is based on the distribution of the analyte between an aqueous phase and a nonmiscible organic phase in conjunction with formation of charged molecules based on pH. A typical process is shown in Figure 18.2. Depending on the analyte of interest, this approach will be tailored to suit the needs of the analyst. Nowadays, this traditional approach is often replaced by SPE. Both the liquid–liquid partition and the SPE suffer from a loss of analyte that shows in lower recovery rates. Therefore, it is imperative to use an internal standard of known concentration to evaluate the influence of the sample preparation steps on the accuracy of the final result.

A search of the published literature on the analysis of botanicals in finished products by TLC did not produce any results. However, the technique is used for preliminary data gathering. In an example, Masse et al. analyzed kojic acid and arbutin by TLC. The sample preparation in this case involved mixing 3.5 g of

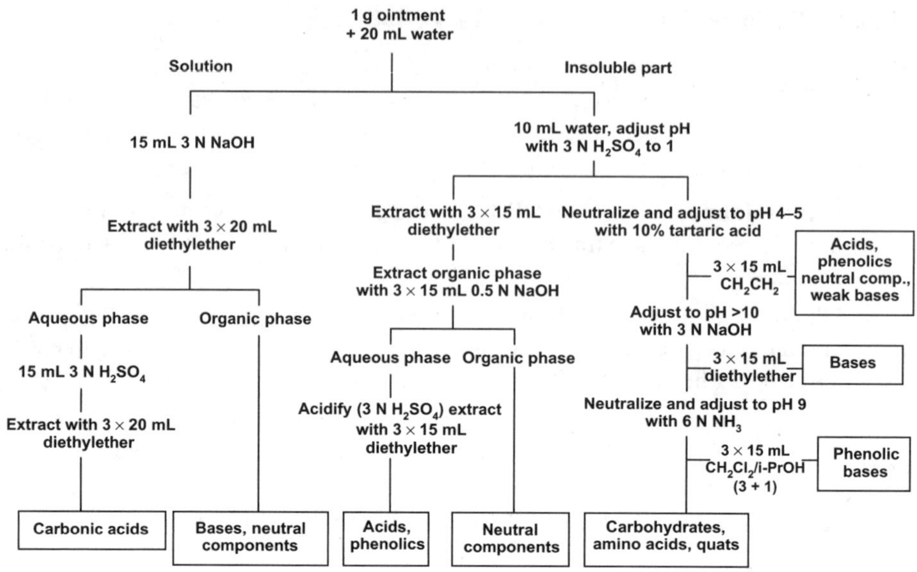

Figure 18.2 Procedure to extract analytes from o/w emulsions (reproduced with permission from Ref. [15]).

commercial product in 10 mL ethanol, which then was sonicated for 20 min. After centrifugation, the supernatant was analyzed by TLC [10].

18.3.3 Analysis

The results depend a lot on the type of instrumentation used. There will be a huge difference between the TLC of our undergraduate years (a scissor "custom-cut" size TLC was placed in a jelly jar and developed with the same over-and-over-used solvent mixture) and the sophisticated HPTLC equipment available nowadays. The application technique is crucial; the diameter of the applied spot should not exceed 1 mm for HPTLC plates and the spot has to be completely dry before the development. Equally important are the geometry of the developing chamber and the composition of the gas phase at the beginning and during chromatography. An equilibrium time of 30 min is recommended for a 20×20 cm plate size. If traditional TLC chambers are used, the plate should be introduced as quickly and regularly as possible. Today, fully automatic developing chambers are available, which allow controlling the saturation and do not disturb the gas phase at any time.

18.3.4 Detection and Documentation

There are a number of nondestructive detection methods available for TLC. The most commonly used is UV detection at 254 and 366 nm. Many components in an extract are not detectable using UV light and the analyst has to use spray reagents to make the products visible. Popular nonspecific spray reagents include the vanillin-sulfuric acid, the anisaldehyde-sulfuric acid, or the natural product reagent (NPR) for flavonoids, carbohydrates, and anthocyanins. There are a large number of specific reagents available as well, such as the Dragendorff reagent for alkaloids or the Berlin blue reaction for hydroquinone derivatives. In order to convert the visual impression into an image, the plates can be scanned with a flat-bed scanner or photographed by a camera. Flat-bed scanners are less expensive, but camera-based systems are more versatile. For qualitative and quantitative evaluations, scanning densitometry is regarded as the standard. Instruments use wavelengths between 190 and 800 nm in single or multiwavelength scans and can even record spectra similar to a photodiode array (PDA).

18.3.5 Applications

18.3.5.1 Identification of Raw Material HPTLC is known to be an excellent choice to compare fingerprints of an extract with an authenticated voucher specimen of the same plant. The 2008 edition of the United States Pharmacopeia (USP31/NF26) [11] lists TLC as an identification tool for a number of plants: black cohosh (*Actaea racemosa*), chamomile (*Matricaria recutita*), chaste tree (*Vitex agnus-castus*), red clover (*Trifolium pratense*), echinacea (*Echinacea angustifolia*, *E. pallida*, and *E. purpurea*), eleuthero (*Eleutherococcus senticosus*), feverfew (*Tanacetum parthenium*), garlic (*Allium sativum*), ginger (*Zingiber officinale*), ginkgo (*Ginkgo biloba*), American ginseng (*Panax quinquefolius*), Asian ginseng (*P. ginseng*), goldenseal (*Hydrastis canadensis*), hawthorn (*Crataegus monogyna* or *C. laevigata*), horse chestnut (*Aesculus hippocastanum*), licorice (*Glycyrrhiza glabra* or *G. uralensis*), maritime pine (*Pinus pinaster*), milk thistle (*Silybum marianum*), stinging nettle (*Urtica dioica*), pygeum (*Prunus africana*), St. John's wort (*Hypericum perforatum*), saw palmetto (*Serenoa repens*), and valerian (*Valeriana officinalis*). In addition, there are a number of excellent sources to find information on TLC and HPTLC conditions for a particular herb, such as the books written by Wagner and Bladt [9], Wichtl [12], and Pachaly [13]. Another source is the monographs of the American Herbal Pharmacopeia (AHP) [14], which has published HPTLC conditions of a number of botanicals important to the North American market.

18.3.5.2 Detection of Adulteration The presence of mislabeled raw material is more frequent than one might think. Adulteration can happen when plant material is harvested in the wild and a plant is mistakenly gathered in place of another, or in an agricultural facility due to human error. A plant with a known history of adulteration is skullcap (*Scutellaria lateriflora*), which has been replaced by material from the same genus *Scutellaria* or by hepatotoxic Germander species (*Teucrium chamaedrys* or *T. canadense*). Distinction of *S. lateriflora* and *Teucrium* species is readily done by HPTLC (see Figure 18.3).

18.3.5.3 Quick Screening Masse et al. used TLC to identify kojic acid and arbutin in skin-lightening products. Trial runs were performed on RP-18, cellulose, and silica gel plates. RP-18 did not provide satisfactory results, and in the system with cellulose plates (using *tert*-butanol–2-butanone–water–formic acid 40:30:15:15 as mobile phase), only kojic acid migrated. Good results were obtained with silica gel

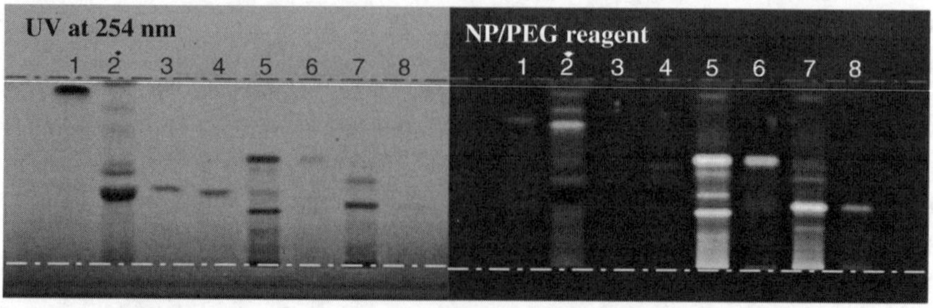

Figure 18.3 HPTLC analysis to distinguish *S. lateriflora*, *T. chamaedrys*, and *T. canadense* [49]. Lane 1, baicalein; lane 3, baicalin; lane 4, ikonnikoside I; lane 6, verbascoside; lane 8, teucrioside; lane 2 is the extract of *S. lateriflora* with a band at Rf = 0.40 corresponding to baicalin and ikonnikoside I of lanes 3 and 4. Lane 5 is the extract of *T. canadense* with two bands at Rf = 0.32 and 0.57 corresponding to the reference standards in lane 6 and lane 8, respectively. Lane 7 is the extract of *T. chamaedrys* with a band at Rf = 0.32 of teucrioside.

and a mixture of ethyl acetate, methanol, and water (85:17:13). The compounds were detected with the Gibbs reagent [10]. Aflatoxins can be detected using the method published by AOAC. The method uses silica gel TLC plates and chloroform–acetone (9:1) as the mobile phase. Aflatoxins are detected by UV at 360 nm. A quick screening for cariogenic sugars in toothpaste flavors by TLC is used in our lab. The method is performed according to a protocol developed by Auterhoff and Kovar [15], with silica plates as stationary phase and a mixture of isopropanol, ethyl acetate, and water (70:20:10). The sugars are detected using the anisaldehyde–sulfuric acid reagent. A similar approach could be easily used to detect potential allergens in fragrances or finished skin care products.

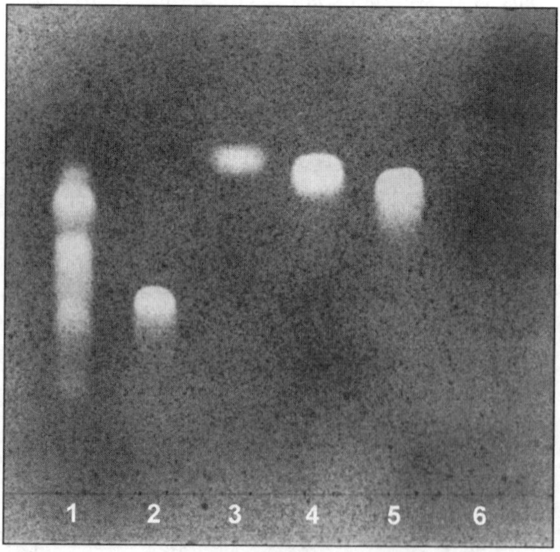

Figure 18.4 Screening for free-radical scavenging activities in extracts and pure compounds using DPPH as a spray reagent (reproduced with permission from Ref. [50]). Lane 1, *Merremia emarginata* extract; lane 2, chlorogenic acid; lane 3, *p*-coumaric acid; lane 4, caffeic acid; lane 5, gallic acid; lane 6, cinnamic acid. The applied amounts were 100 μg for the extract and 10 μg for the pure compounds.

18.3.5.4 Bioactivity Evaluation The concept of a quick screening for bioactive compounds by TLC (TLC bioautography) was initiated in 1970 by Homans and Fuchs [16]. The advantage is that the bioactive compound can be characterized after separation on TLC based on the Rf value and on the detection method, and can be isolated in a very targeted fashion. TLC bioautography has been used to screen for antibacterial and antifungal actives [17, 18], to determine free-radical scavenging properties (Figure 18.4 [19]), or to find enzyme inhibition activities [20, 21].

18.4 GAS CHROMATOGRAPHY

Gas chromatography (GC) is the technique of choice for the analysis of thermally stable volatile and semivolatile components. The most common configuration is gas–liquid chromatography where the mobile phase is an inert gas and the stationary liquid phase is chemically bound to the inner surface of the column. Helium is the most common mobile phase; however, a few applications also use nitrogen, hydrogen, or P-5 (5% methane in argon). GC provides rapid resolution for complex mixtures. A disadvantage is that it is applicable only for compounds that can be volatilized or chemically modified to a more volatile compound. Often compounds exceeding a molecular mass of 500 cannot be analyzed by GC. Also, thermally labile compounds are not suitable to GC analysis. These compounds would decompose in the hot injection port before advancing to the column. Some GC instruments can be modified with a cool-on column injection port where the thermally labile sample can be introduced. Then the injection port can be heated rapidly to bring the remainder of sample material on to the column and clean the injection port in preparation for the next sample.

18.4.1 Sample Preparation of Raw Materials

Often GC analysis of raw materials is performed to assess the purity of a compound or the active ingredient concentration in plant material (whole, powdered, or extract). As in TLC, dissolving the compounds of interest in a compatible solvent is the quickest and simplest sample preparation. For mass spectrometers (MS) and flame ionization detectors (FID), typical solvents include hexane, petroleum ether for nonpolar analytes, and ethanol, methanol, and acetone for polar analytes. Dichloromethane is an excellent solvent for nonpolar analytes, but is increasingly avoided due to toxicity.

18.4.2 Sample Preparation of Finished Products

A variety of sample preparation techniques can be explored to prepare the cosmetic product for analysis. They range from the simple "dilute and shoot" solubilizing of the product to acid/base extraction, concentration, redissolving in a compatible solvent and addition of chemical adjuncts that are better suited to GC analysis (e.g., formation of fatty acid methyl ethers). Table 18.1 gives an overview of sample preparation techniques for GC analysis of natural cosmetic products.

TABLE 18.1 Sample Preparation Techniques Used for GC Analysis of Natural Cosmetic Products

Species	Marker compound(s)	Matrix	Sample preparation	Recovery (%)	Reference
Various	Fragrance allergens (linalool, limonene, citronellol, geraniol)	o/w emulsions	Microdistillation, SDE, SPME	98–103, 91–108, n. det.	[25]
Various	Fragrance allergens (linalool, limonene, citronellol, geraniol)	w/o emulsions	Microdistillation, SDE	94–99, 99–147	[25]
Various	Fragrance allergens (linalool, limonene, citronellol, geraniol)	Surfactant	Microdistillation, SDE	98–103, 91–108	[25]
Various	Fragrance allergens (linalool, limonene, citronellol, geraniol)	Bar soap	SDE	97–123	[25]
Arnica montana	Sesquiterpene lactones	Ointment, gel, strateum corneum	Extraction with acetone, evaporation, solid-phase extraction (NH_2)	98–102, n. det., 100–106	[24]
Aloe barbadensis; *A. arborescens*; *A. ferox*	Barbaloin, aloenin	Lotion, cream	Solid-phase extraction (C-18), evaporation, trimethylsilylation	85, 67	[23]
Aster family (*Arnica* spp.)	Sesquiterpene lactones	Ointment	pH-based extraction, solubilize in dichloromethane	75	[22]

Various	Contact allergens	w/o and o/w emulsions, aftershave, shower gel	Microdistillation, pH based	n. det.	[29]
Matricaria recutita	α-Bisabolol, bisabol oxides A and B, *trans*-β-farnesene, cycloethers	Deodorant stick	Solvent extraction (hexane) by sonication, followed by filtration	n. det.	Villinski (unpublished)
Various	Menthol, camphor, eucalyptol	Ointment	Solvent dilution with petroleum ether	98–101, 95–99, 99–110	Villinski (unpublished)
Various	Synthetic contaminants, including preservatives and plasticizers	Deodorants	SPME headspace	n. det.	[28]

Abbreviations: SDE, simultaneous distillation extraction; SPME, solid-phase micro extraction; n. det., not determined.

If possible, the best approach for sample preparation is to dilute the skin care product in a suitable solvent. In our lab (Villinski, unpublished), muscle balm containing menthol, camphor, and eucalyptol was dissolved in petroleum ether for analysis. Methyl salicylate was used as the internal standard. Recovery of the analytes ranged from 97% to 103%. In cases where sample dilution is not possible, there are a variety of other methods that can be employed to extract the analytes from the matrix. Classical separation techniques such as liquid–liquid extraction with separatory funnels or sonication can be employed when instrumentation is not available. Raddatz et al. [22] exemplified this procedure removing sesquiterpene lactones from an emulsion. The aqueous phase was adjusted to an alkaline pH to open the lactone ring. This makes the analytes soluble in the aqueous phase, providing opportunities to remove the interfering matrix components in the organic phase by liquid–liquid partitioning. Reacidifying the aqueous phase leads to ring closure; therefore, the sesquiterpene lactones are less water soluble and extractable with a nonwater miscible organic phase (e.g., dichloromethane). This method is labor intensive, depends on the expertise of the scientist, and results in less than 90% recoveries.

Another technique that requires minimal instrumentation and is relatively inexpensive is SPE. SPE can be employed to remove matrix interferences, concentrate the analytes, or further clean up an extract before analysis, which will simplify chromatography and improve quantitation. Since the columns are designed to be disposable, there is no concern about cross-contamination between samples. SPE columns are available with many different packing media depending on the application. Nakamura et al. used C-18 SPE columns to clean tape stripping skin samples for the aloe components, barbaloin and aloenin [23]. Wagner and Merfort used NH_2 SPE columns for the analysis of sesquiterpene lactones [24].

One technique that can be done manually or automated is solid-phase microextraction (SPME). SPME is not an exhaustive extraction technique; it involves the use of a fiber coated with an extracting phase. The quantity of analyte extracted by the fiber is proportional to its concentration in the sample as long as the equilibrium between the analyte in the fiber and the sample is reached. The advantages of SPME include minimal solvent use and fast sample preparation time. SPME uses a polymer-coated fused silica fiber on the end of a modified syringe. The fiber can be exposed to the analytes in the headspace above the sample or immersed in the sample for a specified time. After exposure to the sample, it is retracted and the holder is placed in the GC injection port or the HPLC interface desorption chamber. The fiber is exposed inside the injection area to release the analytes into the GC or HPLC. Fiber selection is based on the analyte type and the molecular weight. Leonhardt [25] optimized SPME conditions with two types of fibers in a study of headspace SPME for some fragrance allergens. The carbowax/divinylbenzene (CW/DVB) fiber gave good initial results but was irreversibly contaminated after two injections. The divinylbenzene/carboxen/polydimethylsiloxane (DVB/CAR/PDMS) fiber was recommended for screening of fragrance allergens. Polar analytes such as coumarin and cinnamyl alcohol were better extracted when the sample solution contained phosphate buffer (pH 7). Smaller volumes of the sample solution also increased the concentration of analytes in the vapor phase. The fibers are fragile and are prone to breakage. Fibers used in immersion methods will have a shorter lifetime than fibers that are used in headspace methods due to irreversible polymer contamination. A review of quantitative SPME used to analyze drug products was published by Lord and Pawliszyn [26]. Sample composition (including ionic strength and pH of the sample solution), temperature, and equilibrium times must be carefully controlled for quantitative SPME. The use of an isotopic labeled standard was the preferred method for quantification. Other quantification methods discussed were external standard, standard addition, and internal standard.

There are several types of commercial apparatuses that are available to assist with the extraction of the target analytes from cosmetic products. The benefits of these instruments include less variability and automation and some are able to do multiple samples simultaneously. Of course, instrumentation will be more expensive than traditional extraction/concentration techniques mentioned earlier.

The microdistillation apparatus can be temperature programmed. The sample solution is heated gradually and the volatile components are collected in the receiving vial, which contains water, sodium chloride, and a nonmiscible organic solvent. After the distillation is complete, the organic layer is analyzed by gas chromatography. Leonhardt spiked 29 fragrance allergens into several cosmetic formulations (o/w and w/o emulsions, and a surfactant-based skin cleaning product). All 29 analytes were recovered using microdistillation and the four most prevalent fragrance allergens (limonene, linalool, citronellol, and geraniol) had recoveries between 94 and 104% [25].

Simultaneous distillation extraction (SDE) has been used to remove analytes from a water-soluble matrix and distill them into an organic solvent. However, formation of artifacts is a known limitation of this extraction procedure. A comparison between microdistillation and SDE for identification of fragrance allergens in a variety of cosmetic products showed that microdistillation and SDE had similar recoveries for some of the spiked allergens, but microdistillation provided more precise results, had shorter sample preparation time ($\sim$2 h), and the extract was ready for GC analysis without a solvent concentration step. SDE could not accurately recover some of the allergens in the three formulations because the analyte could not be resolved from the matrix interferences. The matrix interferences caused erroneously high analyte recoveries (up to 140% depending on the fragrance compound) or the analyte was indistinguishable from the background [25].

Sometimes even after matrix interferences are removed and analytes are suitably concentrated, the compounds may still be in a form that is not conducive to GC. Chemical reactions can be utilized to add adjuncts that improve chromatography. A classical example is the preparation of methyl ester derivatives of fatty acids to form compounds that are easily chromatographed. Trimethylsilylation was used to analyze aloe components in skin care products. These reactions, although useful, are difficult to force completion. The recoveries of aloe constituents were 85% in a lotion and 67% in a cream [23].

18.4.3 Analysis

Before performing the actual analysis, a review of GC instrument parameters is required. The injection port temperature, pressure, inlet liner dimensions, and column carrier gas flow rate must be optimized for the best resolution with the shortest run times. The expansion volume of the solvent in the GC injection port must be kept in mind. An expansion volume calculator (web site: http://www.chem.agilent.com/en-US/Support/Downloads/Utilities/Pages/GcPressureFlow.aspx) can be useful to set initial temperature and pressure parameters. The choice of column nowadays depends on the injection port configuration. Natural product analyses reviewed used capillary columns (internal diameter less than or equal to 0.53 mm) although packed columns are still available in the marketplace. Capillary columns provide faster analysis time, greater resolution, and sharper peak shape. There are many phases to choose depending on the type of analysis required. Commercially available columns span the range of nonpolar to polar compounds. Specialty columns can resolve chiral compounds, basic compounds, and organic acids without derivatization.

18.4.4 Detection

Common detectors include the MS and FID. The specific detectors, which include the electron capture detector (ECD) for chlorine and other electrochemical compounds, the flame photometric detector (FPD) for sulfur- and phosphorus-containing compounds, and the nitrogen phosphorus detector (NPD) are used only rarely for the analysis of natural products.

The popularity of MS can be attributed to its increased sensitivity over most other analytical techniques. There are two techniques to ionize analytes as they pass from the GC into the mass analyzer. The first and most popular is electron ionization (EI). As molecules pass into the mass analyzer, they are bombarded with 70 mV and break into characteristic fragments that provide molecular spectral information. The spectra can be searched against commercial libraries and predict with great confidence the presence of desirable or undesirable components even without standard reference materials. The other method to ionize compounds is chemical ionization (CI). CI begins by ionizing methane gas; in turn, the methane ionizes the molecule creating a $[M + H]^+$ radical. This is a milder ionization and there is rarely a significant fragmentation. CI is very useful for determining molecular mass, while EI gives information about the molecular structure. The primary disadvantage of MS is the cost compared to other available GC detectors.

In contrast, the FID is an economical alternative to MS. The FID has high sensitivity and wide range of linearity (often up to seven orders of magnitude). It ionizes compounds that burn in a hydrogen–air flame. The response is proportional to the number of carbon atoms in the molecule. The FID will not respond to water and permanent gases. A major drawback is the need for standard compounds as identification is based on retention time of components.

Historically, classical pesticide and herbicide analysis has been performed utilizing ECD and NPD. The specificity of ECD to electrophilic compounds makes it ideal for the analysis of organochlorinated pesticides, PCB, and herbicides. The NPD is an excellent detector for organophosphorus pesticides. However, as methods are being harmonized around the world, pesticide analysis is trending toward GC/MS or GC/MS/MS for identification and confirmation [27].

The electronic nose is relatively new to the cosmetic industry and is not widely accepted [2]. It differs significantly from other chromatography techniques as it characterizes the cosmetic's aroma with no separation of constituents. The primary limitation of this technique is that the sensor response tends to shift with time. Therefore, odor databases are applicable only for a few months. Data interpretation is reported to be cumbersome, requiring multivariate data analysis methods such as principal component analysis (PCA) or cluster analysis (CA).

18.4.5 Applications

18.4.5.1 Characterization of Raw Materials and Finished Products The 2008 edition of the United States Pharmacopeia (USP31/NF26) has three GC methods for specification of raw materials. The content of β-sitosterol in stinging nettle (*U. dioica* and *U. urens*) is quantified by extracting the powdered herb with chloroform followed by derivation and analysis by GC/FID. Similarly, the contents of α-bisabolol and its derivatives in chamomile (*M. recutita*) are quantified by distilling the volatile oil, dissolving it in cyclohexane, and analyzing the solution by GC/FID. Alpha-bisabolol (levomenol), bisabolol oxide A, and bisbolol oxide B are quantified using a standard solution of α-bisabolol. A Carbowax 20 column is suggested and the temperature program begins at 70°C for 4 min; then it is increased at 4°C/min to 230°C and maintained for 10 min [11]. The fatty acid content of saw palmetto (*S. repens*) plant material is obtained by derivatizing the fatty acids to their methyl esters with boron trifluoride and analyzing by GC/FID on a polyethylene glycol column (Carbowax 20M). Initially, the temperature program is 120°C for 3 min, then increased at 50°C/min to 220°C, where it is maintained for another 12 min [11].

The most widely used method to evaluate the composition of essential oils is GC/MS. Essential oils are often very complex and adulteration is a concern. As an example Santalol isomers were analyzed in commercially available sandalwood oil (reported as *Santalum album*). The chromatography was performed using two columns: ZB-WAX (polar) and DB-5MS (nonpolar). The ZB-WAX column provided better resolution and peak shape of the santalols. The oven temperature program began at 40°C and increased to 200°C at 2°C/min. The majority of these oils failed the internationally recognized standard of 90% santalol content [30].

18.4.5.2 Detection of Contaminants or Allergens Headspace SPME-GC/MS was used to identify the presence of undeclared synthetic preservatives in raw materials and finished skin care products. Some of the natural deodorants were found to contain the synthetic antioxidants BHT and 3-BHA. The analysis of raw materials sold as natural preservatives revealed in some cases the presence of chloromethylbenzenes, toluene, dichlorobenzaldehydes, and 2-phenoxyethanol. Plasticizers such as phthalates could be detected in raw materials that were stored in plastic containers [28]. The amount of allergens in creams, lotions, cleansing products, liquid soap, shower gels, and bar soaps was quantified by GC-MS. Leonhardt compared microdistillation, SDE, and headspace SPME sample preparation for the extraction of fragrance allergens in many types of formulations. Noack utilized pH-based microdistillation to show that contact allergens such as benzyl alcohol, eugenol, and linalool can be formed from their respective precursor acetates (benzyl acetate, eugenyl acetate, and linalyl acetate) under long-term storage conditions [25, 29].

18.4.5.3 Stability of Active Ingredients GC coupled with many detectors can provide quantitative results on cosmetic products. These methods can be used to study the stability of the active ingredient along with its decomposition products. In our laboratory, we monitored the OTC drug components of a muscle balm to comply with FDA regulations and support expiration dating of the product. The method was used to quantify camphor, menthol, and eucalyptol using methyl salicylate as an internal standard. An INNOWax column was chosen for its superior resolution and peak shape. The temperature program began at 75°C for 4 min and increased at a rate of 17°C/min to 200°C and maintained for 4 min. The muscle balm samples were dissolved in petroleum ether. Recoveries of a placebo balm spiked with camphor, menthol, and eucalyptol ranged from 97% to 103%.

Another example is the stability of essential oil components from a clinically tested chamomile extract used in a deodorant stick formulation. Approximately, 1.5 g of deodorant was dispersed in 10 mL hexane. Menthol was used as the internal standard and the sample solution was sonicated for 30 min. The solids were allowed to settle and the supernatant was analyzed on a HP-5MS (30 m × 0.25 mm × 0.25 μm) column. The GC temperature program began at 50°C, held for 2 min, ramp 7°C/min to 300°C, and held for 10 min. The compounds were quantified using the response factor for α-bisabolol as standards for *trans*-β-farnesene, bisabolol oxide A and B, and en-in-dicycloethers were unavailable (Villinski, unpublished).

18.4.5.4 Bioavailability of Ingredients Wagner and Merfort studied the penetration behavior of sesquiterpene lactones in arnica preparations. Two ointments and one gel were evaluated. The products were rubbed onto pig ear skin and incubated for 1–4 h. The stratum corneum (SC) was removed by a series of adhesive tape stripping. The adhesive tapes were extracted in acetone with an internal standard, concentrated, then redissolved in methanol and further purified by eluting with methanol through a Sephadex column. The final extract was analyzed by GC/MS. While both ointment and gel penetrate the SC, ointment preparations appear to continue to penetrate up to 4 h while the gel formula would need more frequent application [24].

18.5 HIGH-PERFORMANCE LIQUID CHROMATOGRAPHY

HPLC has been recognized as the most versatile chromatographic technique for natural products analysis. HPLC has made tremendous improvements in terms of automation, convenience, speed, sensitivity, choice, and stability of column stationary phases, and availability of different detection systems. Despite the large number of different products available, the analysis of natural products is predominantly done using 150–250 mm long octadecylsilyl columns with a particle size between 4 and

5 μm. The mobile phase often consists of methanol–water or acetonitrile–water mixtures with the addition of small amounts of an acid (e.g., trifluoroacetic acid, formic acid, phosphoric acid, or acetic acid) in order to obtain acceptable peak-shapes for the acidic compounds. For an acceptable separation of the complex extracts, the analysts often resort to lengthy run times; durations of 50–80 min are common in the analysis of natural products. However, the introduction of spherical particles with smaller diameters of 3 μm or less, and the introduction of instruments capable of handling higher pressure has led to an increase in separation efficiency and resulting in better separations with shorter columns and shorter separation times.

The advantages of HPLC are its versatility, the good separation power, the possibility to use a number of different detection systems that can be used to obtain a large amount of qualitative and quantitative data, and the option to recover the analyte of interest after the elution. The disadvantages are the relatively (compared to TLC or GC) high costs of instrumentation and maintenance and the larger amounts of solvents used. Also, the accuracy, precision, and linear range of the analysis and the nature of compounds that can be analyzed vary greatly depending on the type of detector used.

18.5.1 Sample Preparation of Raw Materials

Techniques to prepare plant material for HPLC analysis have been reviewed by Romanik et al. [31]. In order to obtain the compounds of interest, extractions with various solvents are carried out. Besides traditional extraction techniques (maceration, percolation, soxhlet extraction, steam distillation), sonication, accelerated solvent extraction, microwave-assisted extraction, ultrafiltration, supercritical fluid extraction, and solid-phase microextraction have been used. Table 18.2 shows a comparison of various liquid–solid techniques used for extracting plant metabolites. Most compendial methods (e.g., United States Pharmacopeia [11], American Herbal Pharmacopoeia [14], and European Pharmacopoeia [32]) use simple techniques such as extraction on a shaker or under reflux conditions, or sonication to get suitable solutions for HPLC analysis. For total flavonoids (e.g., for ginkgo leaves or chamomile flowers), the initial extraction is followed by a hydrolysis step. Most flavonoids in plants are found as a variety of different glycosides; the hydrolysis step cleaves the sugar moieties and leads to a much simpler chromatogram containing only the aglycones.

TABLE 18.2 Comparison of Various Liquid–Solid Extraction Techniques Used in the Analysis of Plant Metabolites

Extraction	Soxhlet	Sonication	ASE	MAE	SFE
Cost	Low	Low	High	Medium	High
Extraction time	6–48 h	<30 min	<30 min	<30 min	<60 min
Solvent use (mL)	200–600	<50	<100	<40	<10

Source: Reproduced with permission from Ref. [31].

ASE, accelerated solvent extraction; MAE, microwave-assisted extraction; SFE, supercritical fluid extraction.

18.5.2 Sample Preparation of Finished Products

According to the literature data, the most popular approach for preparing natural products in cosmetic matrices is to find a suitable solvent in which the compounds of interest and the matrix are soluble, or where the analytes can be extracted with a simple sonication or heating step. Subsequent centrifugation or filtration through HDPE filters leads to "injection-ready" solutions. As mentioned earlier, the fewer steps the analyst can use in the sample preparation, the better the outcome of the analysis in terms of recovery and precision. However, the compatibility of the analyte and possible components of the matrix have to be evaluated carefully. There are a number of cosmetic ingredients that can lead to major headaches. We remember clogging up more than one HPLC guard-column by directly diluting xanthan gum-containing preparations in water. Xanthan gum and other thickeners will precipitate when mixed with a mobile phase containing enough organic solvent. Another example of sample preparation issues we had in our lab is the analysis of licoricidin, an isoflavan from Chinese licorice (*G. uralensis*) in toothpaste. Licoricidin is a very lipophilic molecule; therefore, it seemed to make sense to make a slurry of the toothpaste sample with straight ethanol. Besides suffering from painful arm muscles due to the prolonged vehement shaking of the sample in order to evenly disperse the toothpaste in ethanol, the chromatography was unacceptable—the insolubility of licoricidin in the initial mobile phase led to an important fronting and overlapping with other compounds of the extract. The problem could be solved by preparing a slurry of toothpaste in a small amount of water, which was then filled to volume with ethanol. Particular attention to matrix interference has to be paid when using a mass spectrometric detector. Frauen analyzed Calendula constituents in various preparations by HPLC-MS and noticed a strong interference of matrix components on the ionization of faradiol-esters. He resolved the issue by adding the exact amount of Calendula-free matrix to the external standards [33]. Table 18.3 gives an overview of sample preparation techniques applied to various natural ingredients.

18.5.3 Analysis

It is impossible to list all the different HPLC methods that have been published on botanical extracts. A Google "scholar" search for *E. purpurea* and *E. angustifolia* and "quantitative HPLC analysis" gave more than 400 hits. According to van Beek and Montoro [34], more than 3000 papers on *G. biloba* have appeared since 2001, and about 400 of them pertain to chemical analysis in a broad sense. We think it would be difficult to find a frequently used botanical within Europe or North America for which no HPLC method has been published, yet.

In terms of separation techniques, analysts of natural products seem to suffer from a lack of creativity. Despite the availability of a large number of various stationary phases and detection systems, reverse-phase chromatography using UV or visible detection is predominant. Out of the 25 HPLC methods for botanicals found in the USP31, 22 used packing L1, which means a C-18 column with a particle size between 1.5 and 10 μm. Notable exceptions were the use of a C-8 column for the analysis of lycopene in tomatoes (*Lycopersicon esculentum*) and the determination of

TABLE 18.3 Sample Preparation Techniques Used for HPLC Analysis of Natural Compounds in Cosmetics

Species	Marker compound(s)	Matrix	Sample preparation	Recovery (%)	Reference
Arnica montana	Sesquiterpene lactones	4 emulsions and 2 gels	MeCN	n.det.	[41]
Calendula officinalis	Faradiol esters	o/w emulsion	MeOH-*t*-butyl-methylether (4:1)	99–103	[33]
Camellia sinensis	EGCG	Hydrophilic ointment USP	Liquid–liquid partitioning: H_2O–ethyl acetate	n.det.	[43]
C. sinensis	Catechins	o/w and w/o emulsions	MeOH–H_2O (vitamin C) –THF (9:4:27)	98–105	[33]
Centella asiatica	Asiatic acid, asiaticoside, madecassic acid	Madecassol ointment	MeOH–H_2O (9:1)	n.det.	[51]
Echinacea purpurea	Cichoric acid	o/w and w/o emulsions	MeOH–THF–H_2O (1:2:7)	107	[33]
Glycyrrhiza spp.	Triterpenoids	Anolan cream; vidermina	SPE diol	98–99	[52]
			SPE C-18	90–91	
			MeOH	n.det.	
G. glabra	Glycyrrhizin, glycyrrhetinic acid	o/w and w/o emulsions	MeOH–H_2O (3:2)	97–10	[33]
Humulus lupulus	Bitter acids	Deodorant	EtOH	91–109	[53]
Magnolia officinalis	Honokiol, magnolol	o/w emulsion	EtOH	n.det.	Gafner (unpublished)

(*continued*)

TABLE 18.3 (*Continued*)

Species	Marker compound(s)	Matrix	Sample preparation	Recovery (%)	Reference
Pinus maritima	Pycnogenol	o/w emulsion	Liquid–liquid partitioning: H_2O–dichloromethane	n.det.	[54]
Pinus and other tree spp. (colophonium)	Abietic acid and other resin acids	Wax strips, mascara	MeCN extraction, evaporation, dissolution in MeCN–H_2O (7:3), SPE (MAX phase)	76–97	[55]
Sapindus mukorossi	Saponin	Soap	H_2O	n.det.	[56]
Terminalia sericea	Sericoside	o/w and w/o emulsions	MeOH–H_2O (1:1)	103	[33]
Thymus vulgaris	Thymol	o/w emulsion	EtOH	n.det.	Gafner (unpublished)
Various	Fragrance allergens	o/w emulsion, hair conditioner	MeCN	90–103	[57]

Preparation methods indicating solvents without further explanation: the sample is suspended or diluted in the solvent by shaking or sonication and purified by one or two additional steps (e.g., centrifugation or filtration).

Abbreviations: EtOH, ethanol; H_2O, water; MeCN, acetonitrile; MeOH, methanol; SPE, solid-phase extraction; THF, tetrahydrofuran, n. det., not determined.

pycnogenol in maritime pine, *Pinus* spp., as well as the content of fructose and dextrose in cranberry (*Vaccinium macrocarpon*) preparations with a cation exchange column. The detection system was almost equally uniform, with 21 methods using UV or visible detection. The remaining four methods used an evaporative light-scattering detector (triterpene glycosides in black cohosh (*A. racemosa*) and ginkgo lactones (*G. biloba*)), a refractive index detector (RID) (fructose and dextrose in liquid cranberry preparations), or a fluorescence detector (scopoletin in stinging nettle, *U. dioica*) [11].

There are some advantages of using C-18-based column packing: the material allows a separation of a variety of compounds of different polarity, which is particularly helpful for extracts, the material is stable and can be used in a wide range of pHs (nowadays, specially bonded C-18 columns tend to have a good life span even under alkaline conditions up to a pH of 12), the solvent mixtures are almost always easy to handle, and the retention behavior is very reproducible, especially compared to normal-phase HPLC. In comparison to a C-8 column, the C-18 material tends to have higher retention and better shape selectivity [35].

The recent introduction of UPLC using columns with a narrow size distribution of particles in the range of 1.7 µm in connection with instruments that can handle pressures of approximately 1000 bar has enabled a new level of performance. The advantages of UPLC are a marked reduction in analysis time (up to 10-fold shorter than conventional HPLC) and significant improvements in sensitivity and resolution (Figure 18.5). The trend is toward columns with even lower particle size (1 µm) and pressures of up to 7000 bar [36].

18.5.4 Detection and Documentation

Among all HPLC detectors, the most widely used is UV/Vis. Reportedly, it has the best combination of sensitivity, linearity, versatility, and reliability [37]. Most natural products absorb in the range of 200–500 nm, and even substances with only one or two double bonds can be detected at short wavelengths (e.g., ginsenosides at 203 nm). While many methods use one single wavelength to detect all natural products having a chromophor, the use of a photodiode array detector, which provides whole UV/Vis spectra of any constituent during the separation, provides information about the type of natural product within the extract.

The evaporative light scattering detector (ELSD) has become more and more widely used in the analysis of botanicals in the past decade. The main interest in this detection system is its ability to detect any analyte with a lower volatility than the mobile phase, and the ability to use the ELSD in gradient mode. There are no restrictions in regard to UV absorption of solvents and analyte. Furthermore, as the UV/Vis detector, ELSD can be coupled to HPLC directly. The disadvantages are the lower sensitivity of the ELSD, a lower dynamic range due to the exponential detector response, the restricted use of buffers, and the destructive nature of the method, which does not allow samples to be recovered after the run [38]. Nevertheless, the ELSD is frequently used to analyze molecules without a chromophore such as carbohydrates, lipids, or terpenes. It is also often used as an addition to UV/Vis and MS detection in order to obtain a more complete picture of an extract (Figure 18.6).

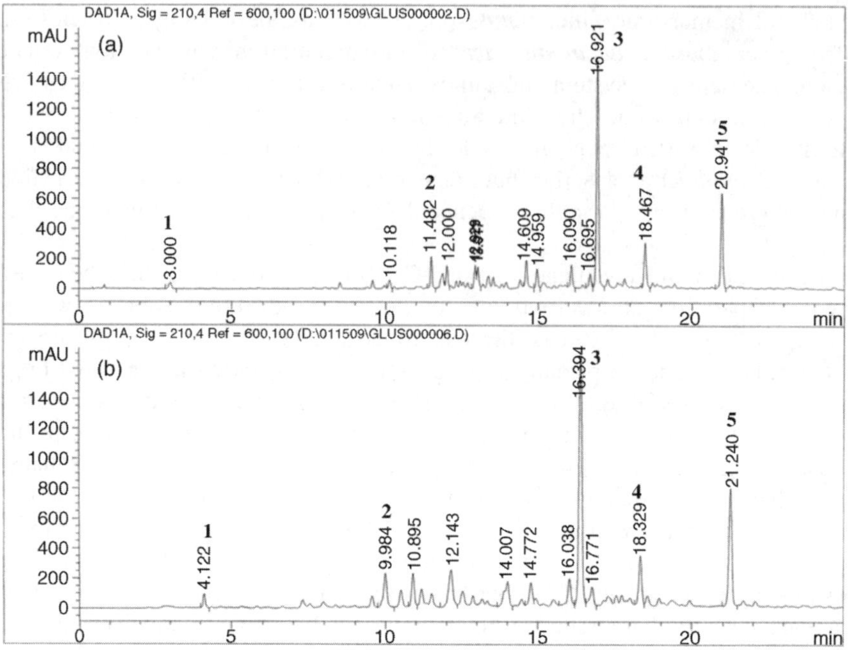

Figure 18.5 Comparison between the separation of a licorice CO_2 extract on C-18 using (a) 1.8 μm and (b) 5 μm particle size (a: Zorbax XDB RR, 4.6 × 50 mm; b: Zorbax Eclipse XDB 4.6 × 250 mm) using the same run time. Selected compounds: 1, liquiritigenin; 2, glycycoumarin; 3, licoricidin; 4, 1′-methoxyficifolinol; 5, licorisoflavan A. Notice the difference in the retention behavior and resolution; some of the larger peaks in the chromatogram (b) consist of several coeluting compounds, which are better resolved using the 1.8 μm particle size column.

The MS detectors used to be restricted to a few high-end laboratories 15–20 years ago, but manufacturers have come out with excellent benchtop models that are much less expensive and use less lab space. This has led to the widespread appearance of HPLC-MS instruments in research and industrial laboratories. Mass spectrometry is a key technique to obtain structural information on the compounds contained in a botanical extract. While MS detectors will be able to provide either nominal or exact mass information for the determination of the molecular formula, fragmentation of the molecular ion using tandem or hybrid instruments gives important structural information and can, in certain cases, help to identify the natural product of interest. Quantitative analysis with MS is still less common than by UV/Vis

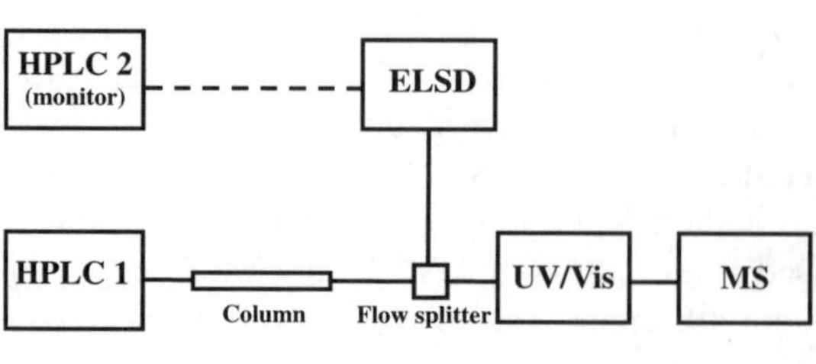

Figure 18.6 Typical setup for an HPLC workstation in natural products analysis.

maybe due to a lack of precision. However, the ability to use shorter run times (peaks rarely have to be separated fully using quantitative MS) and the lower detection limits are advantages of MS and may help to increase the popularity of MS detectors for quantitative results.

The electrochemical detector (ECD) measures electroactive groups, which are present in many natural products. The detector is inexpensive, sensitive, and selective. Functional groups that are sensitive to oxidation include phenols, aromatics, amines, thiols, and quinolines, and those compatible with reduction are olefins, esters, ketones, aldehydes, ethers, and quinones.

Other instruments that are used to detect natural products separated by HPLC are the refractive index (RID), fluorescence (FD), nuclear magnetic resonance (NMR), and chemoluminescence (CLD) detectors. A comparison between the various detectors is given in Table 18.4.

In terms of documentation, chart recorders with quantifications based on cutting out peaks with scissors and weighing them on an analytical balance may still haunt middle-aged chromatographers in their dreams, but fortunately belong to the past. HPLC instruments generally come with fairly user-friendly software. Most companies offer integrated networking solutions, so you can process the data from the computer in your office while having a cup of coffee with a bagel.

Basic parts of the software include several peak integration options (e.g., choice between automatic and manual) and a number of options for your calibration curve settings (important for nonlinear detectors such as the ELSD). Every software has its pros and cons, but the most challenging part is to get used to it. If there is one recommendation to give, we almost exclusively use manual integration (rather than automatic) for every peak due to the complexity of the chromatograms. Many factors (e.g., small peaks, differences in peak size, and resolution) can lead to wrong automatic integration parameters. Such errors can be avoided if the peak is manually integrated in a consistent fashion.

18.5.5 Applications

18.5.5.1 Identification and Characterization of Raw Material In most industrial laboratories, the identification of the raw material is done by HPLC. Besides the well-established identification and quantification of marker compounds (see Table 18.3), some labs look at "fingerprint" chromatograms, in which every constituent in an extract is evaluated. Fingerprint chromatograms are especially helpful for extracts in which either the actives are unknown and no suitable marker has been identified or the actives are composed of an array of different constituents (e.g., black cohosh [39]).

18.5.5.2 Detection of Adulteration Similar to TLC, HPLC is very helpful to detect potential adulterants in a raw material. Adulteration can come either from misidentification or from purposeful addition of ingredients. For a while, citrus-based preservatives have been proposed to the cosmetic industry as alternatives to parabens or phenoxyethanol. Unfortunately, some of the more efficacious products contained the synthetic quaternary ammonium base benzethonium chloride, which indeed will

TABLE 18.4 Comparison of Different HPLC Detectors

	Type	Dynamic range	Detection limit (ng)	Specificity	Chromatographic compatibility	Structural information	Ease of use	Popularity for NP analysis	Cost
UV	Spec.	***	***	**	***[a]	*	****	****	*
DAD	Spec.	***	***	**	***[a]	**	****	***	**
FD	Spec.	**	****	***	****	*	***	*	**
ECD	Spec.	****	****	**	**	—	**	**	*
RID	Univ.	**	**	—	*[b,c]	—	***	*	*
ELSD	Univ.	**[d]	**	—	**[c]	—	****	*	*
CAD	Univ.	****[d]	***	—	**[c]	—	****	—	**
MS	Both	**	****	***	**[c]	***	**	***	***
MS-MS	Both	**	****	****	**[c]	****	*	***	****
NMR	Univ.	**	*	*	*[e]	****	*	**	****

Source: Reproduced with permission from Ref. [37].

Abbreviations: Spec., specific; Univ., universal; Both, can be used for universal or specific detection.
[a] Nonlinear response.
[b] Incompatible with gradients.
[c] Usually incompatible with nonvolatile buffers.
[d] Only solvents with a low UV cutoff.
[e] Only deuterated solvents in the on-flow mode.

provide a good preservation but has nothing to do with a naturally occurring grapefruit seed extract. Benzethonium chloride is easily detected by HPLC-MS [40].

18.5.5.3 Evaluation of Stability This is one of the main applications of HPLC in cosmetics. The stability can be evaluated on the raw material in order to examine potential issues. A widely used approach is both to expose the raw material to various pHs (0.1 N HCl, 0.005 N acetic acid, 0.1 N NaOH) and to 3% hydrogen peroxide and to check the degradation over the course of a week. For this approach, a combined HPLC-UV/MS/MS instrumentation is particularly helpful. The analyst not only can determine the degradation rate but also can get structural information about the degradation products and tentatively identify the mechanisms leading to the breakdown of the parent compound.

The same instrumentation can be used to evaluate the stability of an extract or a marker compound in the finished product. Using HPLC-UV/MS/MS has allowed the characterization of a number of minor degradation products of hops bitter acids in deodorant stick prototypes (Figure 18.7). Frauen [33] has evaluated the stability of green tea catechins in w/o and o/w emulsions for 6 months. He separated the catechins on a GROM-SIL 120 Octyl-5 CP column (60 × 2 mm ID, 5 μm particle size) using a methanol–water (0.1% formic acid) gradient. Without additional stabilizing agents, only traces of catechins were found after 6 months at 40°C in the w/o emulsions and no catechin was present in the o/w emulsion. He noticed that addition of 0.2% ascorbic

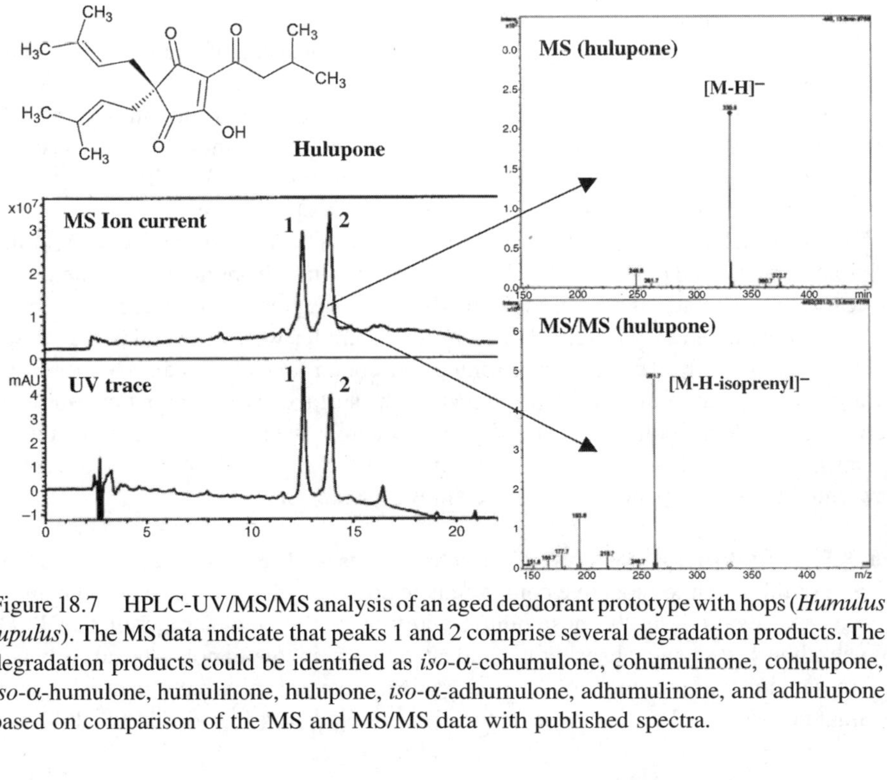

Figure 18.7 HPLC-UV/MS/MS analysis of an aged deodorant prototype with hops (*Humulus lupulus*). The MS data indicate that peaks 1 and 2 comprise several degradation products. The degradation products could be identified as *iso*-α-cohumulone, cohumulinone, cohulupone, *iso*-α-humulone, humulinone, hulupone, *iso*-α-adhumulone, adhumulinone, and adhulupone based on comparison of the MS and MS/MS data with published spectra.

acid did show the best results, while butylhydroxytoluene (BHT) or combinations of BHT with EDTA were less successful in preventing the oxidative degradation. The same researcher also looked at the stability of Calendula faradiol-esters in w/o emulsions. The esters were separated on a GROM-SIL 120 ODS-3 CP column (60 × 2 mm ID, 3 μm particle size) using a gradient of *tert*-butylmethylester and methanol. Again, the addition of antioxidants markedly improved the stability of the compounds. The best results were obtained with 0.5% tocopherol acetate and 0.05% BHT. A publication by Bilia et al. [41] reports that the stability of arnica sesquiterpene lactones strongly depends on the type of matrix used. The sesquiterpene lactones were rather stable in o/w emulsions and anphyphil cream, a cream base of the National Formulary of the Italian Pharmacopoeia. The same compounds, however, degraded rapidly in a w/o emulsion and a hydroxyethylcellulose gel. The HPLC-PDA analysis was done on a Vydac 201 TP 54 C-18 column (254 × 4.6 mm ID, 5 μm particle size) with a gradient comprised of three solvents: methanol, acetonitrile, and water (adjusted to pH 3.2 with formic acid).

***18.5.5.4 In Vitro* or *In Vivo* Bioavailability** Due to the low detection limits with HPLC-UV or HPLC-MS, it is possible to evaluate the bioavailability of the active(s) or a marker compound *in vitro* or *in vivo*. Gebre-Mariam et al. [42] have used a multilayer membrane system to evaluate the *in vitro* release of kaempferol glucosides from creams containing a *Melilotus elegans* extract. Membranes were collected after 0.5, 1, 2, 3, and 4 h and then extracted with 80% methanol using a sonicator. After a hydrolysis step, the total kaempferol glucosides were quantified as kaempferol by HPLC-UV. The kaempferol solutions were injected onto a Nucleosil RP 18 column (125 × 2 mm ID). The mobile phase consisted of water (0.1% formic acid) and methanol (0.1% formic acid), which was run under gradient conditions.

Dvorakova et al. [43] looked at the *in vitro* transdermal penetration of EGCG after topical application of EGCG from a 10% hydrophilic ointment. The maximal concentration of EGCG was reached after 24 h and represented 19% of the topical dose applied to the skin. EGCG was separated on an Absorbosphere HS[7] C-18 column (150 × 4.6 mm ID, 5 μm particle size) with a mixture of 82% water (0.05% H_3PO_4) and 18% acetonitrile and quantified by HPLC-UV at 280 nm. The human skin penetration of cosmetic formulations containing either EGCG from green tea or quercetin from *G. biloba* was assessed by dal Belo et al. [44]. Skin samples were mounted between two diffusion cells and 10 mg/cm^2 of formulation containing 6% green tea extract or ginkgo extract was applied for 6 h and 24 h to the surface. The stratum corneum was removed by tape stripping and extracted by methanol. The molecules of interest were quantified by HPLC-UV again. Results indicated that both EGCG and quercetin presented good skin penetration and retention from cosmetic formulas.

***18.5.5.5 In Vitro* Bioassays** A number of research groups have used various HPLC settings to find bioactive compounds in complex mixtures such as a botanical extract. An example is the postcolumn addition of the stable free-radical DPPH, which shows a strong absorption band at 517 nm in methanol. The absorbance decreases in the presence of a free radical scavenger, which can be easily evaluated using an HPLC-UV detector. The method was introduced in 2000 and is suitable for both isocratic and gradient HPLC runs with mobile-phase compositions ranging from 10% to 90% organic solvent in water or buffer (pH 3–6) and a variety of stationary phases [45]. Other examples describe approaches to find enzyme inhibitors in an extract without the need of time-consuming isolation work. Postcolumn addition of Ellman's reagent can be used to screen for acetylcholinesterase (AChE) inhibitors. The reagent is converted into a yellow pigment by AChE monitored at 405 nm. If an inhibitor is in an extract, the HPLC trace will show negative peaks [46]. Nikolic et al. [47] used pulsed ultrafiltration MS to look for cyclooxygenase 2 (COX-2) inhibitors in a variety of matrices, including plant extracts. The COX-2 enzyme was trapped in solution in an ultrafiltration chamber, to which the extract was added. After incubation, the unbound phytochemicals were washed away by a buffer solution. The bound molecules were later eluted onto the HPLC column after disrupting the enzyme–receptor complex using, for example, methanol or a pH change. The released ligands were analyzed by MS or tandem MS. A good overview of techniques for bioactive natural compounds' screening, coupling HPLC with bioassays, has been published in 2009 [48].

18.6 CONCLUSIONS

The amount of published literature on natural products in cosmetics is very small, despite the fact that the use of natural products in the form of extracts or isolates is growing in popularity and many companies try to take advantage of this trend to make such natural cosmetics available. Some of the industry research is kept private in order to have a competitive advantage or is found in the patent literature, where experimental conditions seldom are described in detail; this leaves very little in terms of good studies available to the public.

Despite the presence of excellent natural cosmetics in the marketplace, the faith of consumers in the efficacy and safety of such products is put to test again and again as media reports unveil the use of extracts of little value (e.g., water that was used to "wash" certain aromatic plants), the presence of potentially harmful synthetic chemicals in natural cosmetics, or the occurrence of fraudulent claims. The characterization of natural products by the techniques presented in this chapter has to be part of the quality control as a basis for consistent safety and efficacy of any cosmetic product. The advancement in analytical techniques beyond quality control presents a wonderful opportunity not only to the few research groups that get funds to study natural products in cosmetics but also to the industry. Hopefully, the natural cosmetics industry will notice that a good reputation can be built by excellent research published in respected journals.

REFERENCES

1. EMEA. Reflection paper on markers used for quantitative and qualitative analysis of herbal medicinal products and traditional herbal medicinal products (EMEA/HMPC/253629/2007). Available at http://www.emea.europa.eu/pdfs/human/hmpc/25362907enfin.pdf.

2. Salvador A, March JG, Vidal MT, Chisvert A, Balaguer A. General review of published analytical methods for cosmetics. In: Salvador A, Chisvert A, editors. *Analysis of Cosmetic Products*. 1st ed. Amsterdam: Elsevier, 2007. pp. 72–82.
3. Meyer VR, Majors RE. Minimizing the effect of sample preparation on measurement uncertainty. *LCGC Europe* 2002;15:2–5.
4. Majors RE. Trends in sample preparation. *LCGC Europe* 2003;16:71–81.
5. Horwitz W, Kamps LR, Boyer KW. Quality assurance in the analysis of foods and trace constituents. *J. AOAC* 1980;63:1344–1354.
6. Majors RE. LC-GC guide to sample preparation. *LCGC North America (Suppl.)* 2000;18:20.
7. Sherma J. Thin-layer chromatography in food and agricultural analysis. *J. Chromatogr. A* 2000;880:129–147.
8. Reich E, Widmer V. Plant analysis 2008: planar chromatography. *Planta Med.* 2009;75:711–718.
9. Wagner H, Bladt S. *Plant Drug Analysis*. 2nd ed. Berlin: Springer, 2001.
10. Masse MO, Duvallet V, Borremans M, Goeyens L. Identification and quantitative analysis of kojic acid and arbutin in skin-whitening cosmetics. *Int. J. Cosmetic Sci.* 2001;23:219–232.
11. USP, 31: United States Pharmacopeial Convention. *United States Pharmacopeia and the National Formulary (USP 31-NF 26)* Rockville: The United States Pharmacopeial Convention, 2008.
12. Wichtl M, editor. *Herbal Drugs and Phytopharmaceuticals*. 3rd ed. Boca Raton: Medpharm Scientific Publishers, 2004.
13. Pachaly P, editor. *DC-Atlas, Dünnschichtchromatographie in der Apotheke*. 5th supplement. Stuttgart: Wissenschaftliche Verlags-Gesellschaft, 2002.
14. Upton R, editor. American Herbal Pharmacopoeia (AHP), http://www.herbal-ahp.org
15. Auterhoff H, Kovar KA. *Identifizierung von Arzneistoffen*. 5th ed. Stuttgart: Wissenschaftliche Verlagsgesellschaft, 1985.
16. Homans AL, Fuchs A. Direct bioautography on thin-layer chromatograms as a method for detecting fungitoxic substances. *J. Chromatogr.* 1970;51:327–329.
17. Rahalison L, Hamburger MO, Monod M, Frenk E, Hostettmann K. A bioautographic agar overlay method for the detection of antifungal compounds from higher plants. *Phytochem. Anal.* 1991;2:199–203.
18. Hamburger MO, Cordell GA. Bioautographic assay for antibacterial compounds. *J. Nat. Prod.* 1987;50:19–22.
19. Cuendet M, Hostettmann K, Potterat O. Iridoid glucosides with free radical scavenging properties from *Fagraea blumei*. *Helv. Chim. Acta* 1997;80:1144–1152.
20. Marston A, Kissling J, Hostettmann K. A rapid TLC bioautographic method for the detection of acetylcholinesterase and butyrylcholinesterase inhibitors in plants. *Phytochem. Anal.* 2002;13:51–54.
21. Wangthong S, Tonsiripakdee I, Monhaphol T, Nonthabenjawan R, Wanichwecharungruang SP. Post TLC developing technique for tyrosinase inhibitor detection. *Biomed. Chrom.* 2007;21:94–100.
22. Raddatz H, Hillenbrand B, Henle, T. Zur Bestimmung von Sesquiterpenlactonen in Naturkosmetischen Emulsionen. *Lebensmittelchemie* 1999;53:60.
23. Nakamura H, Kan T, Kishimoto K, Ikeda K, Amemiya T, Ito K, Watanabe Y. Gas chromatographic and mass spectrometric determination of aloe components in skin care cosmetics. *Pharm. Soc. Japan* 1989;35(3):219–225.
24. Wagner S, Merfort I. Skin penetration behaviour of sesquiterpene lactones from different *Arnica* preparations using a validated GC-MSD method. *J. Pharm. Biom. Anal.* 2007;43:32–38.
25. Leonhardt S.Vergleichende Untersuchungen zur Bestimmung Kontaktallergener Duftstoffe in Kosmetika und Wasch-und Reinigungsmitteln. Diploma thesis. Technische Universität Dresden, 2006.
26. Lord H, Pawliszyn J. Microextraction of drugs. *J. Chromatogr. A* 2000;902:17–63.
27. Codex Alimentarius. FAO/WHO Food Standards: Guidelines on Good Laboratory Practice in Residue Analysis, CAC/GL 40-1993, Rev. 1-2003, 2003.
28. Villinski JR, Gafner S. Comparison of solid phase micro extraction with headspace analysis for screening of volatile components in natural preservatives. Proceedings of the American Society of Pharmacognosy Annual Meeting, Honolulu, HI, 2009.
29. Noack I, Raddatz H, Rockstroh K, Henle T. Bildung von Kontaktallergenen aus Precursor-Duftstoffen bei der Aufarbeitung und Lagerung von kosmetischen Mitteln. *Z. Lebensmittelchem.* 2009;63:123–124.

30. Howes MJR, Simmonds MSJ. Evaluation of the quality of sandalwood essential oils by gas chromatography–mass spectrometry. *J. Chromatogr. A* 2004;1028:307–312.
31. Romanik G, Gilenast E, Przyjazny A, Kamiński M. Techniques of preparing plant material for chromatographic separation and analysis. *J. Biochem. Biophys. Methods* 2007;70:253–261.
32. European Pharmacopoeia Commission, *European Pharmacopoeia*. 6th ed, vol. 1 Strasbourg: Directorate for the Quality of Medicines and HealthCare of the Council of Europe (EDQM), 2008.
33. Frauen M. Analytik kosmetisch wirksamer Pflanzenextrakte mit der Flüssigchromatographie-Massenspektrometrie (LC/MS) Dissertation. Universität Hamburg, 2001.
34. van Beek TA, Montoro P. Chemical analysis and quality control of *Ginkgo biloba* leaves, extracts, and phytopharmaceuticals. *J. Chromatogr. A* 2009;1216:2002–2032.
35. Euerby M, Peterson, P. Chromatographic classification and comparison of commercially available reversed-phase liquid chromatographic columns using principal component analysis. *J. Chromatogr. A* 2003;994:13–36.
36. Ganzera M, Sturm S, Stuppner H. Plant analysis: state of the art and future developments. *Planta Med.* 2009;75:671–682.
37. Wolfender JL. HPLC in natural product analysis: the detection issue. *Planta Med.* 2009;75:719–734.
38. Ganzera M, Stuppner H. Evaporative light scattering detection (ELSD) for the analysis of natural products. *Curr. Pharm. Anal.* 2005;1:135–144.
39. Li WK, Chen SN, Fabricant D, Angerhofer CK, Fong HHS, Farnsworth NR, Fitzloff JF. High-performance liquid chromatographic analysis of black cohosh (*Cimicifuga racemosa*) constituents with in-line evaporative light scattering and photodiode array detection. *Anal. Chim. Acta* 2002;471:61–75.
40. Ganzera M, Aberham A, Stuppner H. Development and validation of an HPLC/UV/MS method for simultaneous determination of 18 preservatives in grapefruit seed extract. *J. Agric. Food Chem.* 2006;54:3768–3772.
41. Bilia AR, Bergonzi MC, Mazzi G, Vincieri FF. Development and stability of semisolid preparations based on a supercritical CO_2 Arnica extract. *J. Pharm. Biom. Anal.* 2006;41:449–454.
42. Gebre-Mariam T, Asres K, Getie M, Endale A, Neubert R, Schmidt PC. *In vitro* availability of kaempferol glycosides from cream formulations of methanolic extracts of the leaves of *Melilotus elegans*. *Eur. J. Pharm. Biopharm.* 2005;60:31–38.
43. Dvorakova K, Dorr RT, Valcic S, Timmermann B, Alberts DS. Pharmacokinetics of the green tea derivative, EGCG, by the topical route of administration in mouse and human skin. *Cancer Chemother. Pharmacol.* 1999;43:331–335.
44. dal Belo SE, Gaspar LR, Maia Campos PMBG, Marty JP. Skin penetration of epigallocatechin-3-*O*-gallate and quercetin from green tea and *Ginkgo biloba* extracts vehiculated in cosmetic formulations. *Skin Pharmacol. Physiol.* 2009;22:299–304.
45. Koleva II, Niederländer HAG, van Beek TA. An on-Line HPLC method for detection of radical scavenging compounds in complex mixtures. *Anal. Chem.* 2000;72:2323–2328.
46. Ingkaninan K, de Best CM, van der Heijden R, Hofte AJP, Karabatak B, Irth H, Tjaden UR, van der Greef J, Verpoorte R. High-performance liquid chromatography with on-line coupled UV, mass spectrometric and biochemical detection for identification of acetylcholinesterase inhibitors from natural products. *J. Chromatogr. A* 2000;872:61–73.
47. Nikolic D, Habibi-Goudarzi S, Corley DG, Gafner S, Pezzuto JM, van Breemen RB. Evaluation of cyclooxygenase-2 inhibitors using pulsed ultrafiltration mass spectrometry. *Anal. Chem.* 2000;72:3853–3859.
48. Shi SY, Zhou HH, Zhang YP, Jiang XY, Chen XQ, Huang KL. Coupling HPLC to on-line post-column (bio)chemical assays for high-resolution screening of bioactive compounds from complex mixtures. *Trends Anal. Chem.* 2009;28:865–877.
49. Gafner S, Bergeron C, Sudberg S, Sudberg ÉM, Kelly L, George M, Gauthier R, Guinaudeau H, Angerhofer, CK. Adulteration of skullcap (*Scutellaria lateriflora* L.): from digital photo-microscopy to HPLC-NMR. First Natural Health Product Conference, Montreal, Canada, 2004.
50. Cuendet M.Recherche de nouveaux composés capteurs de radicaux libres et antioxydants à partir d'une plante d'Indonésie: *Fagraea blumei* (Loganiaceae) et de trois plantes d'altitude: *Bartsia alpina*

(Scrophulariaceae), *Loiseleuria procumbens* (Ericaceae) et *Campanula barbata* (Campanulaceae). Dissertation. Université de Lausanne, 1998.

51. Inamdar PK, Yeole RD, Ghogare AB, De Souza NJ. Determination of biologically active constituents in *Centella asiatica*. *J. Chromatogr. A* 1996;742:127–130.
52. Andrisano V, Bonazzi D, Cavrini V. HPLC analysis of liquorice triterpenoids: applications to the quality control of pharmaceuticals. *J. Pharm. Biom. Anal.* 1995;13:597–605.
53. Dumas ER, Michaud AE, Bergeron C, Lafrance JL, Mortillo S, Gafner, S. Deodorant effects of a supercritical hops extract: antibacterial activity against *Corynebacterium xerosis* and *Staphylococcus epidermidis* and efficacy testing of a hops/zinc ricinoleate stick in humans through the sensory evaluation of axillary deodorancy. *J. Cosmet. Dermatol.* 2009;8:197–204.
54. Vertuani S, Buzzoni V, Manfredini S, Braccioli G. Evaluation of the stability of oligomeric proanthocyanidins from *Pinus pinaster* Ait. in cosmetic formulations. *SÖFW J.* 2001;127:20–23.
55. Nilsson U, Karlberg, AT, Lassen P. Development of an analysis method for quantification of colophonium components in cosmetic products. Environmental Project No. 1272, Danish Ministry of the Environment, 2009. http://www2.mst.dk/udgiv/publications/2009/978-87-7052-927-3/pdf/978-87-7052-928-0.pdf
56. Shiau IL, Shih TL, Wang YN, Chen HT, Lan HF, Lin HC, Yang BY, Ko CH, Murase Y. Quantification for saponin from a soapberry (*Sapindus mukorossi* Gaertn.) in cleaning products by a chromatographic and two colorimetric assays. *J. Fac. Agr. Kyushu Univ.* 2009;54:215–221.
57. Villa C, Gambaro R, Mariani E, Dorato S. High-performance liquid chromatographic method for the simultaneous determination of 24 fragrance allergens to study scented products. *J. Pharm. Biomed Anal.* 2007;44:755–62.

CHAPTER 19

THE USE OF NUCLEAR MAGNETIC RESONANCE SPECTROSCOPY FOR THE IDENTIFICATION OF BIOMARKERS FOR QUALITY CONTROL OF PLANT EXTRACTS

Kan He[1] and Marc Roller[2]

[1]Naturex, Inc., South Hackensack, NJ
[2]Naturex SA, Avignon, France

19.1 INTRODUCTION

A number of plant extracts are formulated into cosmetic products for the consumer's benefit [1–3]. The ability to identify and characterize constituents of interest is essential in the development and quality control of herbal products. Most reported methodologies for the analyses of the plant extracts are mainly high-performance liquid chromatography (HPLC) coupled with various detectors, usually ultraviolet-visible (UV-Vis) and mass spectroscopes (MS). The UV detector measures analytes (i.e., chemical constituents that are determined in an analytical procedure) that possess UV chromophores (the unit of the molecule responsible for UV absorption) as is often the case for conjugated system. UV chromatophore detection is especially useful for identification of polyphenols and flavonoids in plants. MS is a detection system capable of providing more structural information, such as molecular weight, molecular formula, related fragmental ions, and so on. Although both UV-Vis and MS offer high sensitivity, none provides more molecular structure information than nuclear magnetic resonance (NMR). NMR is one of the two techniques used for identification of chemical structures, especially of unexpected or unknown molecules. The other technique is X-ray crystallography, in which a single crystal is required.

Formulating, Packaging, and Marketing of Natural Cosmetic Products, First Edition.
Edited by Nava Dayan and Lambros Kromidas.

The basis for NMR detection is that some isotopic nuclei, such as proton (^{1}H) and carbon-13 (^{13}C), have a characteristic spin ($I = 1/2$) and magnetic moments (a measure of the magnitude and direction of magnetism). When these nuclei are placed in a strong external magnetic field, it results in two spin states, the magnetic moment of low energy $+1/2$ (or α) state and high energy $-1/2$ (or β) state. At equilibrium, the numbers of nuclei in α state are slightly higher than in β state. If an appropriate radiofrequency irradiation is applied, the nuclei in α state absorb energy to jump to the β state. When these nuclei return to equilibrium, they release characteristic energies (signals) detected by the NMR spectrometer. These NMR signals provide a vast amount of structural information based on their different frequency (chemical shift) and signal multiplicity (spin–spin coupling), which reflect the distinct chemical environment of the detected nuclei. NMR spectroscopy has long been a standard tool for organic and natural product chemists for chemical structure determination. It is also an effective technique to study the metabolite fingerprint and profile in plant analysis [4]. Since the majority of organic molecules found in plants are composed of H, C, O, and N, ^{1}H and ^{13}C NMR spectroscopy are the two most commonly used NMR forms in natural products research, while information on O and N is usually deduced from the chemical shift of H and C indirectly. Therefore, ^{1}H and ^{13}C NMR techniques are the focus of this chapter and their applications in the analysis of small molecules in the plant crude extracts are described and demonstrated.

Proton or carbon nuclei in different organic molecules do not experience the same magnetic field when placed in an external magnetic field. The difference in frequency is expressed using a term called chemical shift, denoted by the Greek letter delta (δ), which is given with respect to a reference frequency and expressed as parts per million (ppm). The commonly used reference is tetramethylsilane (TMS) that is set to zero (δ) on the right side in an NMR spectrum, while most nuclei resonate at downfield (low frequency compared to TMS) and appear on the left side of TMS. Chemical shifts are characteristics of the chemical environment of a given nucleus and are affected by the change in electronic structure due to a chemical structural change. Therefore, a proton or a carbon bound to an oxygen atom tend to resonate more downfield from TMS compared to the same nuclei without attached oxygen. For instance, the methyl absorption peak in ethane appears at δ 0.86 ppm (left of TMS) and it shifts to δ 1.22 ppm (further left of TMS) after a change in electronic environment caused by introducing the electron withdrawing oxygen atom in ethanol. Another valuable parameter is peak split caused by vicinal nuclei. When a sample is placed into a magnetic field, the nuclei in the sample molecules may split into two energy levels, aligned with ($+1/2$ state) or opposed against ($-1/2$ state) in the applied field. The two different orientations of nuclear spins induce and generate two small magnetic fields that influence nearby nuclei by adding to or subtracting from the main field. As a consequence, the nearby nuclei resonate in two new fields and the one-peak signal is split into two. This can be described as spin–spin coupling and the distance between lines in a multiplet is denoted by a coupling constant, J (measured in Hz). When the difference in chemical shifts is much larger compared to the coupling constant of two coupled nuclei, the multiplicity of a peak follows the $n + 1$ rule where n is the number of vicinal spin-coupled nuclei. In the example of ethanol, the neighboring methylene protons (2H) split the methyl into three peaks (a triplet) and

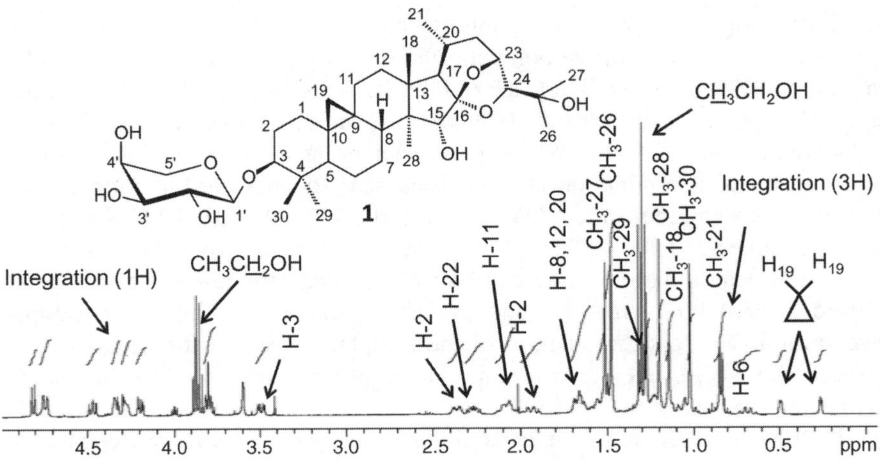

Figure 19.1 ^{1}H-NMR of compound **1**. The spectrum was acquired using 400 MHz spectrometer in pyridine-d_5. Chemical shifts from δ 0–5 ppm are presented. Methyl protons are at about δ 0.8–1.6 ppm; methylene and methine protons are at about δ 1–2.5 ppm; protons attached on the carbons next to oxygen are around δ 3–5 ppm. Compound **1** was recrystallized from EtOH and was not dried completely. Therefore, the signals of EtOH residue (around δ 1.29 ppm triplet for CH_3 and 3.88 ppm, quartet for CH_2) are seen. The signal intensity is shown by integration line on the vertical axis representing one proton per unit.

methyl protons (3H) split methylene into four peaks (a quartet) (Figure 19.1). Spin–spin coupling has played the most important role in NMR, as it shows the fine structure of organic molecules. The coupling constant is often used as a source of the relative stereoconfiguration determination because J is dependent on the chemical bond angle, which is sensitive to the molecular geometry. Nuclear Overhauser effect (NOE) is another powerful technique used to obtain information about molecular structural conformation or configuration of organic molecules. The NOE effect is observed when one resonance is irradiated, the intensity of a neighboring resonance changes if two nuclei are spatially close enough.

Since ^{1}H NMR has a narrow range (0–10 ppm chemical shifts) of resonances, many signals of compounds are overlapped and are too complex to be interpreted. This disadvantage is less severe in ^{13}C NMR owing to its wide chemical shift range (0–200 ppm), but the latter suffers from low sensitivity due to naturally low abundance of ^{13}C nuclei. The introduction of additional spectral dimensions (2D or multidimensional NMR) makes the spectra simplified and the completed ^{1}H and ^{13}C NMR signal assignments become available for complicated molecules. In 2D NMR spectra, the correlations between coupled systems, for example, ^{1}H–^{1}H and ^{1}H–^{13}C, are displayed in a form called "cross-peaks." By exploring these cross-peaks, the connectivity of each nucleus can be established, which leads to the establishment of a structural skeleton and finally the elucidation of the chemical structure. The following 2D NMR techniques are widely used in natural products research: COSY (correlation spectroscopy) and TOCSY (total correlation spectroscopy) are often chosen for ^{1}H–^{1}H correlation; HSQC (heteronuclear single-quantum correlation) or

HMQC (heteronuclear multiple quantum correlation) for ^1H–^{13}C correlation; HMBC (heteronuclear multiple bond correlation) for long-range ^1H–^{13}C correlation; and NOESY (nuclear Overhauser effect spectroscopy) or ROESY (rotating frame NOE spectroscopy) for ^1H–^1H NOE correlation. All these experiments are standard in the setup of modern NMR spectrometers and their application in natural product structural determination has been discussed in detail in the literature [5]. For the analysis of small molecules (MW < 3000), the 300–400 MHz NMR spectrometer (price range \$250,000–\$500,000) provides sufficient sensitivity and resolution for structure determination. Time can be from several minutes for acquiring ^1H NMR to several hours for ^{13}C and 2D NMR spectra depending on sample concentration. The less the quantity of the sample, the longer the accumulation time needed. Usually, several milligrams of a sample are required at least in order to obtain clear spectra.

In order to demonstrate the capabilities of ^1H and ^{13}C NMR for compound identification, the authors selected a triterpene glycoside that was isolated from black cohosh as one example to describe the step-by-step procedure of structural elucidation utilizing the above-mentioned NMR experiments. Other examples, including the applications on extracts of *Cimicifuga racemosa*, *Aloe vera*, *Fraxinus excelsior*, *Ginkgo biloba*, *Smilax glabra*, *Trigonella foenum-graecum*, and *Hibiscus sabdariffa*, are also illustrated in the following sections.

19.2 TRITERPENE GLYCOSIDE FROM BLACK COHOSH

Black cohosh is a common name of *C. racemosa* L. Nutt (Ranunculaceae) (syn. *Actaea racemosa* L.). It is one of the most extensively studied herbal products, used as a herbal drug in Europe and a dietary supplement in the United States for the alleviation of menopausal symptoms [6]. The main constituents in black cohosh are triterpene glycosides (TGs) and are used as markers for standardization of commercial products. TGs have shown inhibitory effects of tumor promotion on mouse skin [7]. Topical application using black cohosh for skin whitening has also been reported, in which the root extracts of black cohosh effectively suppressed the formation of melanin *in vitro* [8]. Most of the TGs contained in black cohosh belong to the 9,19-cycloartanol series and so far about 62 TGs have been identified [6]. Typical adulterants found in marketplace are Asian *Cimicifuga* species and one of them is *C. foetida* that is also rich in TGs. Analytical methods using LC-UV DAD (diode array detectors), LC-ELSD (evaporative light-scattering detector), and LC-MS have been developed for the identification of different *Cimicifuga* species for quality control. All these methods require prior TG structure knowledge. For example, a specific triterpene glycoside, cimigenol 3-*O*-α-L-arabinopyranoside, is found only in *C. racemosa*, not in other Asian *Cimicifuga* species and it has been used to distinguish black cohosh from its adulterants. What follows is a brief description of the procedure using NMR to elucidate the chemical structure of cimigenol 3-*O*-α-L-arabinopyranoside (hereinafter referred to as compound **1**).

Compound **1** was isolated and purified by column chromatography and its purity was detected by HPLC—appearing as a single peak. Its molecular weight

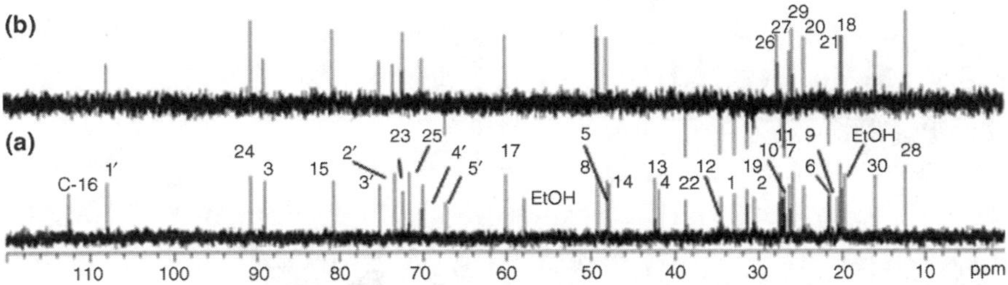

Figure 19.2 (a) ^{13}C-NMR of compound **1** (100 MHz) in pyridine-d_5. It is a proton-decoupled ^{13}C-NMR spectrum in which the protons were saturated by irradiation of a radio frequency, so that the spectrum was simplified by removing ^{13}C–^{1}H splitting and each carbon signal was seen as a single line. The intensities of ^{13}C signals with proton decoupling are also enhanced two–three times owing to NOE effect; (b) DEPT-135 NMR spectrum of **1** obtained using a 135 degree decoupler pulse. Since signals are generated from proton polarization, quaternary carbon signals (C-4, -9, -10, -13, -14, -16, and -25) were not seen. The CH_2 signals (C-1, -2, -6, -7, -11, -12, -19, -22, and -5′) were inversed, while the signals of CH_3 and CH remained the same. The number labeled on each signal is the carbon position of the chemical structure.

(MW 620) was obtained by LC-MS analysis and the molecular formula was determined by high-resolution positive FABMS analysis as $C_{35}H_{56}O_9$ (*m/z* 621.3996 MH^+, calculated 621.3927). A ^{1}H and a proton decoupled ^{13}C NMR spectra were acquired as a starting point of molecular structure assessment. In the ^{1}H NMR spectrum (Figure 19.1), the proton ratio was obtained by integration of each peak, as the peak area under the signals is proportional to the number of protons, so that each functional group can be estimated, for example, methyl group having 3H and methine 1H. The recognizable signals of two very upfield signals at chemical shifts of δ 0.27 and 0.51 ppm, each having one hydrogen by integration, were assigned to be cyclopropane-methylene protons. The upfield signals (δ 0.8–1.6, each with 3H) were seven methyl groups and a sugar moiety appearing at chemical shifts of δ 3.7–4.8. Other characteristic four signals due to connecting to oxo-substituted carbons appeared downfield δ 3.4–4.2. The ^{13}C NMR (Figure 19.2a) displayed 35 single lines (excluding two lines from ethanol residue) representing a total of 35 carbon signals, which suggested a pentosyl (5C) and a triterpene aglycone (30C).

In order to compensate the information lost in proton decoupled ^{13}C spectrum, a DEPT (distortionless enhancement by polarization transfer) experiment is usually performed to find out the protonation levels, where the signals of CH_3 and CH are maintained, the CH_2 inversed, and the quaternary carbons disappeared. In this example, the DEPT spectrum revealed the constitution of all 35 carbons: 7 methyls, 9 methylenes, 12 methines, and 7 quaternaries (Figure 19.2b). Eleven of them were oxo-substituted carbons (downfield chemical shifts δ 66–112). Five of them were ascribed to an L-arabinopyranosyl functionality (δ 107.5, 73.0, 74.7, 69.6, 66.8) and one carbon bearing two oxygen atoms (ketal group) due to its very downfield at 112 ppm (C-16). Since there are no double bonds (1 degree per double bond) or aromatic functional groups observed, 8 degrees of unsaturation can be calculated from $C_{35}H_{56}O_9$ that make up for eight rings in this molecule. Combining all the information above, structure **1** was considered a seven-membered ring triterpene attached with one

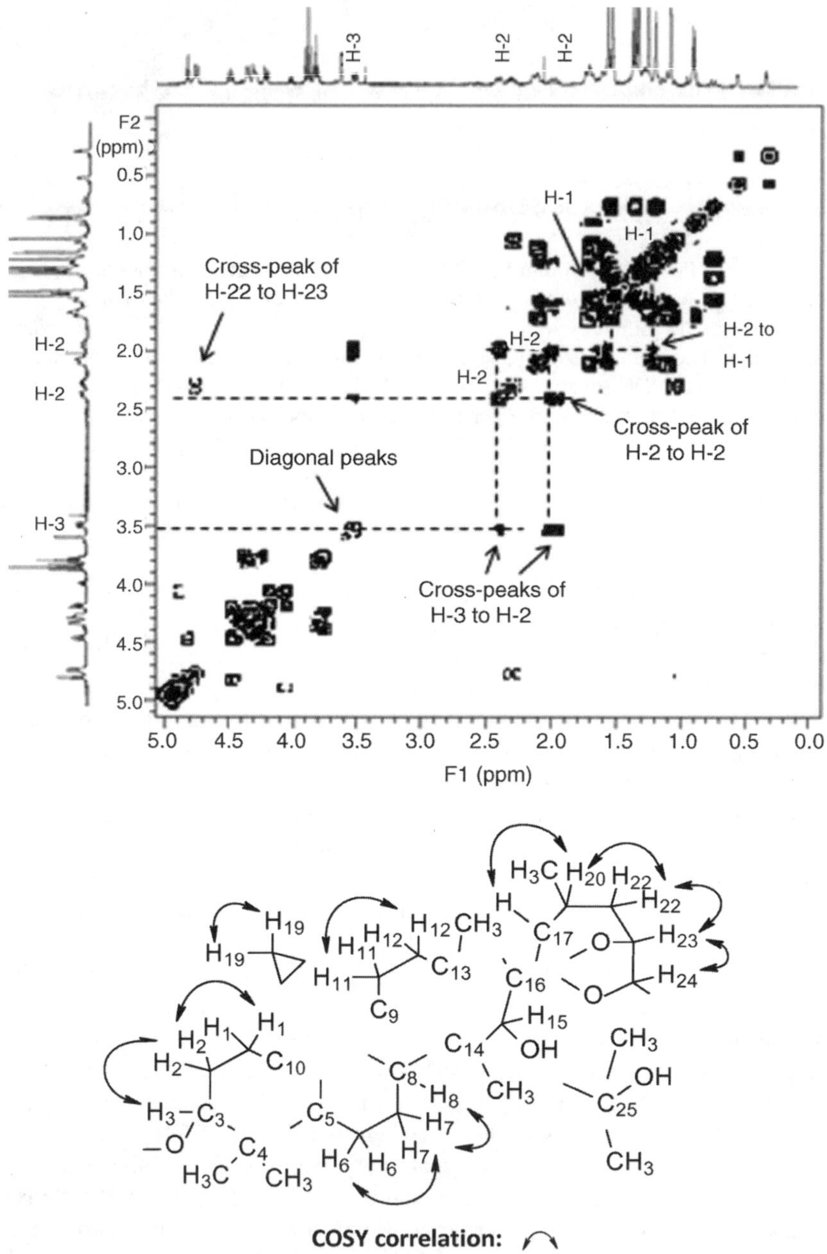

Figure 19.3 ^1H–^1H COSY spectrum of **1**. Contour plot is used in the 2D-NMR spectrum. Both horizontal and vertical axes (F1 and F2) are proton chemical shifts. The diagonal peaks represent a 1D proton spectrum. The presence of off-diagonal peaks (cross-peaks) is the result of correlations between spin–spin coupling. For example, draw a horizontal line from δ 3.5 (H-3) through the diagonal peak and intersect two cross-peaks. Draw two vertical lines from the two cross-peaks, pass through diagonal peaks, and intersect at horizontal axis (F1) at δ 1.94 and 2.36, which are the chemical shifts of the two H-2s being coupled with H-3.

arabinose. Biogenetically, triterpenes are derived from the precursor of 2,3-oxidosqualene; thus, most triterpenes have an oxygen atom at the C-3 position. The observed cyclopropane unit is likely at the positions of C-9 and C-19 that were formed due to the migration of hydrogen from C-9 during the biosynthesis of triterpene. Therefore, 9,19-cycloartenol-type tripterpene can be a prototype in this study and the readily recognizable H-3 can act as one of the entry points to trace out of the coupled spin systems.

Since H-3 and H-2 are coupled to each other, two cross-peaks due to the correlation between H-3 and H_2-2 were observed in the ^1H–^1H COSY spectrum of **1** (Figure 19.3). The H_2-2 were further correlated with H-1 but blocked by a quaternary carbon (C-10). Therefore, only a partial structure -CH(OH)CH_2CH_2– for C-3–C-1 was obtained. As a result of a similar mapping, a partial structure –$CHCH_3CH_2CH(O)$ CH(O)– for C-17 to C-24 could be established by the presence of cross-peaks between H-24 and H-23, H-23 and H_2-22, H_2-22 and H-20, and H-20 further coupled to H-21 and H-17. The two other partial structures -$CHCH_2CH_2CH$– for C-5 to C-8 and -CH_2CH_2– for C-11 to C-12 were mapped out on the basis of the observed correlations from H-5 to H-6, further H-7, and ended at H-8, as well as the correlations between H_2-11 and H_2-12, respectively.

The TOCSY provides ^1H–^1H COSY-like spectra, but is unlike COSY, which displays only cross peaks for directly coupled protons such as H-3 to H-2 and H-2 to H-1, but not H-3 to H-1. In the TOCSY spectrum (Figure 19.4), however, H-3

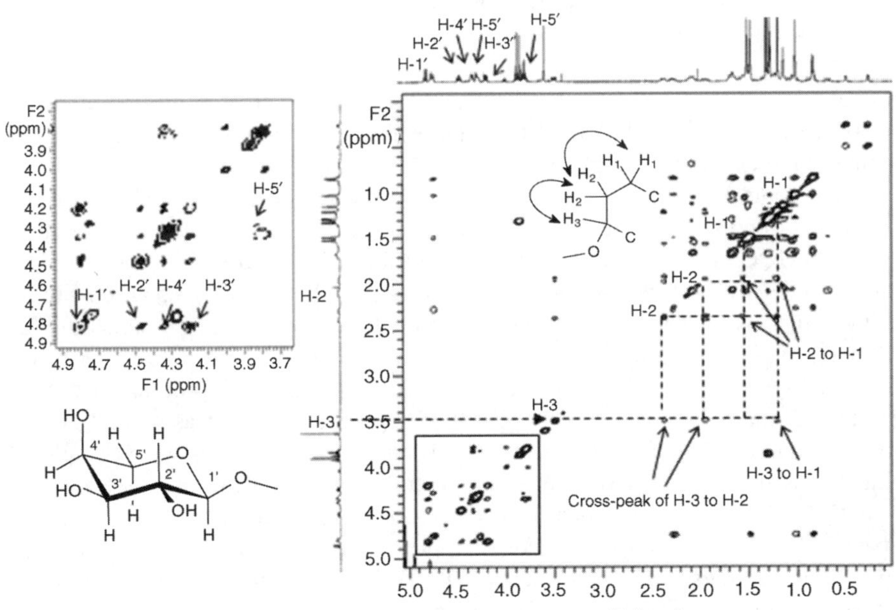

Figure 19.4 TOCSY spectrum (partial) of **1** (mixing time = 80 ms). The left spectrum was expansion for the sugar moiety. The cross-peaks from anomeric proton H-1′ till H-4′ could be seen, while due to small coupling between H-4′ and H-5′, all the cross-peaks to H-5′ were not observed except for two germinal H-5′ protons.

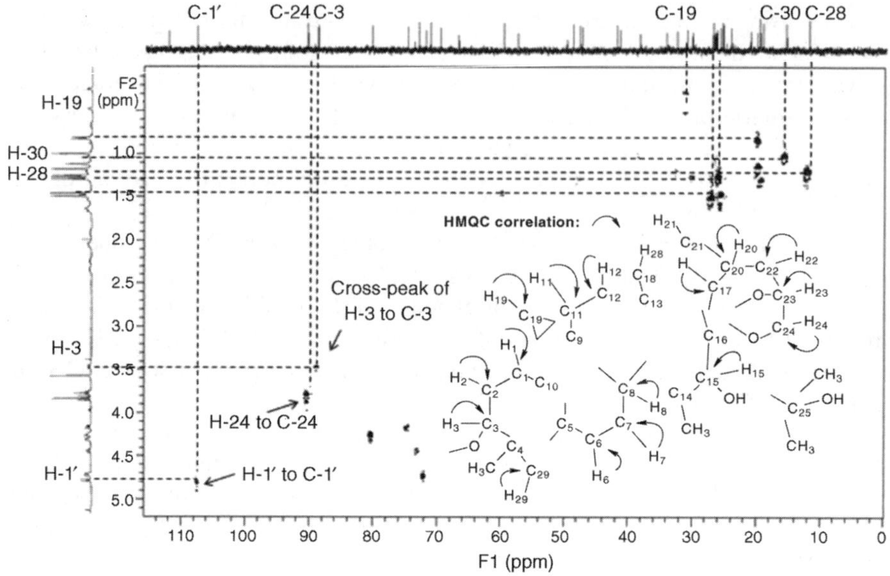

Figure 19.5 HMQC spectrum of **1**. HMQC is used to correlate directly C–H coupled nuclei. Horizontal and vertical axes (F1 and F2) are carbon and proton chemical shifts, respectively. When a proton is bonded to a carbon, for example, the H-3 to C-3, a cross-peak can be observed at the intersection of the horizontal line from H-3 (δ 3.5) and vertical line from C-3 (δ 88.6). Through H assigns C chemical shift or through C assigns H. No cross-peaks are shown for quaternary carbon signals.

correlation can be relayed to H-1 and thus H-3 showed the cross-peaks to both H_2-2 and H-1. This is especially useful for establishing a network of mutually coupled protons, such as sugars and peptides.

When most proton assignments have been finished in COSY and TOCSY, the assignment of corresponding carbon signals can be performed using C-H correlation spectra, HMQC or HSQC. In the HMQC spectrum of **1** (Figure 19.5), the chemical shift at δ 88.9 was ascribed to C-3, when a cross-peak between δ_H 3.5 (H-3) and δ_C 88.9 was displayed. Theoretically, all the protonated carbons can be assigned in this way and it was true for the ^{13}C signal assignment of C-1–C-3, C-5–C-8, C-11, C-12, C-17, C-19–C-24, and C-1′–C-5′, due to their corresponded protons that were determined unambiguously. For those isolated substituents, such as methyl groups at C-18, 26–30 and signals of quaternary carbons, the assignments were completed with the assistance of HMBC and other spectra, such as NOESY or ROESY.

HMBC spectrum of **1** identifies correlations of protons with carbons that are separated by more than one bond. For example, a number of cross-peaks from long-range C-H couplings were displayed in Figure 19.6. Among the seven methyl signals, two $-CH_3$ protons at δ 1.01 and 1.26 showed long-range correlations (three bonds, $^3J_{CH}$) with C-3 and they are assigned to C-30 and C-29, respectively. The CH_3-18, -21, -26, -27, and -28 were assigned in a similar way. More importantly, the observed long-range correlations of CH_3-29 and CH_3-30 with C-4 ($^2J_{CH}$), C-3 and C-5 ($^3J_{CH}$) established the connection of CH_3-29 and CH_3-30 with C-4, C-3, and C-5, linking the

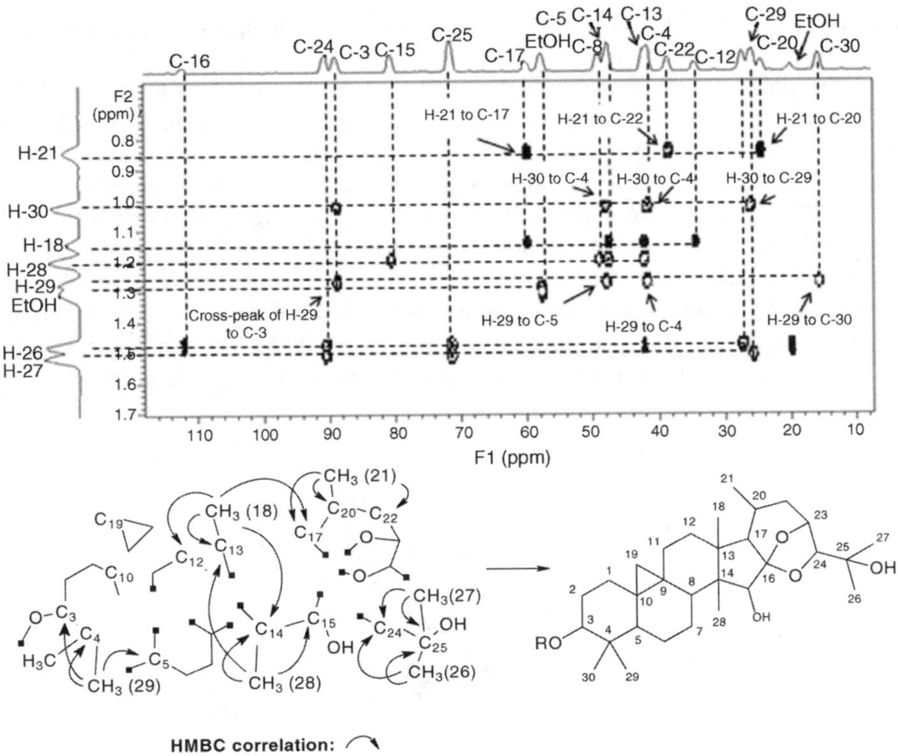

Figure 19.6 Partial HMBC spectrum of **1**, expansion of chemical shifts region at δ 0–1.7 for ^{1}H and δ 10–114 for ^{13}C. HMBC is used to determine proton–carbon correlation in which only long-range (two–four bond) couplings are emphasized. It provides C–H connectivity suitable to all kinds of carbons including quaternary carbons. For example, methyl protons at C-29 have a two-bond coupling with C-4, and three-bond couplings with C-3, C-5, and C-30. Therefore, four cross-peaks can be obtained, when drawing the horizontal line from H_3-29 (δ 1.26) and met with four perpendicular lines from ^{13}C dimension at δ 15.4 (C-30), 41.4 (C-4), 47.5 (C-5), and 88.6 (C-3), respectively.

two fragments of C-1–C-3 and C-5–C-8 and generating a new partial structure of C-1 through C-8. A similar strategy was used: the methyl protons at δ 1.14 (H_3-18) exhibited $^{2}J_{CH}$ correlation with the carbons of C-13 and $^{3}J_{CH}$ with C-12, C-14, and C-17, suggesting that C-18, C-12, C-14, and C-17 all are connected to C-13 quaternary carbon. The correlation of CH_3-28 with C-14, C-13, and C-15 indicated that the quaternary carbon C-14 was the bridge to C-28, C-13, and C-15 and the partial structures from C-11 to C-15 and C-17 were linked. The connection of C-17 to C-20 and through C-20 to C-24 were based on the cross-peaks seen among CH_3-21 to C-20, C-17, and C-22. Finally, the assignment of CH_3-26 and CH_3-27 arose from their long-range correlations with C-25 (δ 71.0) and C-24 (δ 90.2). The long-range correlations through five bonds (not frequently observed) between CH_3-26 (CH_3-27) and C-16 were also observed that further confirmed the connectivity deduced from the above discussion (Figure 19.6). Collectively, structure **1** was assigned to be cimigenol attached with an L-arabinosyl.

The position for the sugar moiety was determined by means of the comparison of the difference of the ^{13}C chemical shift (glycosidation shift) between glycoside and aglycone. Generally, the signal of the glycosidic site (α-carbon) tends to shift downfield (4–11 ppm), while its adjacent positions (β-carbon) have an upfield shift by a small magnitude (0.5–5 ppm). In this example, the glycosidation site of **1** was suggested at C-3 due to a 10.8 ppm downfield shift found in **1** (δ 88.6) compared to its aglycone, cimigenol at 77.8 ppm. The ROSEY spectrum supported this conclusion since a clear correlation between H-3 and H-1′ (because they are spatially close together) was observed (Figure 19.7). A coupling constant ($J = 7$ Hz) for the anomeric proton suggested that the arabinosyl group was present in an α configuration.

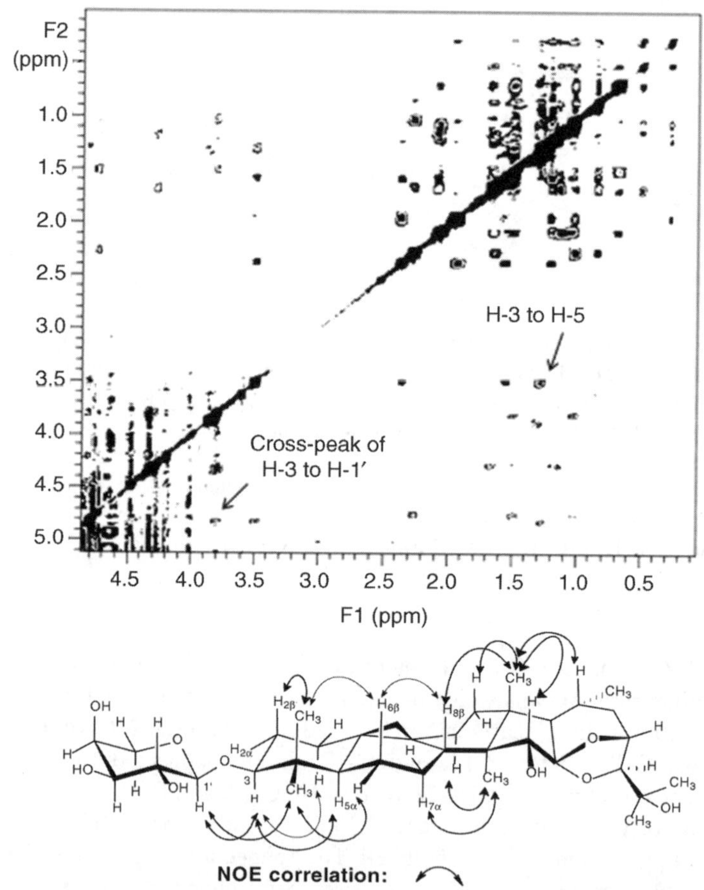

Figure 19.7 ROESY spectrum (partial) of **1**. The presence of cross-peaks is the result of NOE correlations between two nuclei that are close to each other in space. These correlations are used to obtain structural conformation or configuration of organic molecules. For example, the observed correlation between H-3 and H-1′ due to being close to each other indicated that the arabinose was attached at C-3 hydroxyl group.

So far, the chemical structure of **1** was established on the basis of the NMR spectroscopic data. Its relative stereoconfiguration as a last task was resolved by a combination of the proton coupling constants (J) and NOE correlation analysis observed in the NOESY or ROSEY spectrum. The configuration for 3-OH was β oriented, which was deduced from α-oriented H-3 as it appeared as a doublet doublet peak with the coupling constants of 11.6 and 4.4 Hz, respectively. The large J (11.6 Hz) was due to a diaxial coupling between H-3 and axial H-2β when H-3 took an α configuration. Similarly, a distinguishable signal at δ 0.70 ppm with large coupling constant ($J = 12.4$ Hz) was assigned to be H-6β due to two diaxial couplings with H-5 and H-7, respectively. In the ROSEY spectrum of **1** (Figure 19.7), the cross-peak between H-3 and H-5 indicated an α axial configuration for H-5. The β axial configuration of H-8 was derived from the correlation between H-8 and H-6β. The correlations of H-8 with CH_3-18, H-15, and H-19 indicated that all these protons were cofacial. Continuing the NOE mapping using the same strategy, the relative configuration of **1** was obtained and is presented in Figure 19.7.

19.3 NMR PROFILE OF BLACK COHOSH EXTRACT FOR SPECIES IDENTIFICATION AND QUALITY CONTROL

The steps for structure elucidation on a pure compound are briefly described above. When dealing with mixtures, such as in the case of crude plant extract, which usually contains hundreds of compounds, the NMR spectra are complicated by overlapping signals. Due to the limitation of low sensitivity and dynamic range, the signals from minor components will not be seen, which makes it possible to identify the signals arising from major constituents without involving a tedious isolation and purification procedure.

In the course of developing an analytical method for black cohosh extract years ago, the main challenge encountered was that there were no commercial standards available. When tested by HPLC–UV DAD/ELSD/MS, the major peaks found in the extract were suggested to be triterpene glycosides based on their UV and mass spectra. Acquiring the ^{1}H NMR of the extract, signals corresponding to the 9,19-cycloartenol-type triterpene glycosides were observed. The decoupled ^{13}C NMR (Figure 19.8) and DEPT spectra provided more structural information. Although over 200 signal lines appeared, the distribution of these signals exhibited the pattern of the triterpene glycosides: chemical shifts of δ 12–27 for methyl signals, δ 20–40 for methylene, δ 40–60 for methine and quaternary carbons, δ 60–90 for oxo-substituted carbon, δ 100–110 for sugar anomeric carbon, and δ 170–180 for carbonyl groups. Most of the major signals were assignable and could be ascribed to five triterpene glycosides, including cimigenol arabinoside (**1**), actein (**2**), 26-*epi*-deoxyactein (**3**), cimicifugoside (**4**), and 25-*O*-acetylcimigenol xyloside (**5**), by comparing the NMR data with those published in the literature for these compounds. The major signals in the carbohydrate region were attributed to D-xylopyranosyl and L-arabinopyranosyl moieties. The signals around δ 88 were assigned to be C-3, the glycosidic site, for all the major TGs and were confirmed by the cross-peak for H-3 to C-1′ in HMBC spectrum. Other characteristic signals that

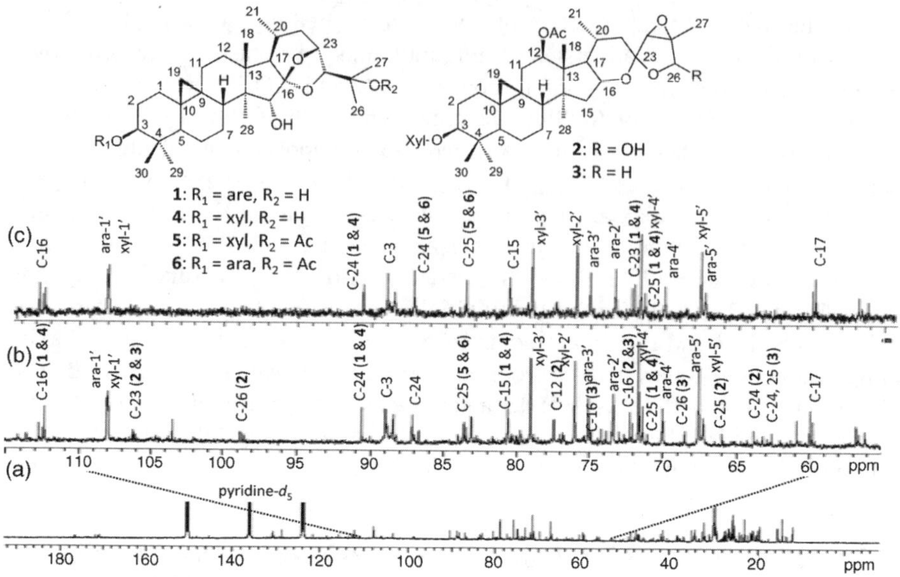

Figure 19.8 Carbon-13 NMR spectra of black cohosh extract and a commercial product. Fifteen grams of *C. racemosa* root powder was extracted with 75% ethanol (2 × 200 mL) at 40°C for 4 h. The combined extract was concentrated to dryness by rotary evaporation *in vacuo* at 40°C. The residue was dissolved in 1 mL of pyridine-d_5 and filtered through cotton into an NMR tube for NMR experiments. The spectra were acquired with broadband proton decoupling and the number of scans was 150,000. (a) Whole spectrum of *C. racemosa* extract; (b) the expanded region of the oxycarbons between δ 55 and 115; (c) the expanded region of the oxycarbons of a commercial product. Labeled signals from cimigenol arabinoside (**1**), actein (**2**), 26-*epi*-deoxyactein (**3**), cimicifugoside (**4**), 25-*O*-acetylcimigenol xyloside (**5**), xyl: β-D-xylopyranosyl; ara: α-L-arabinopyranosyl.

were used to support the existence of these triterpene glycosides were shown in Figure 19.8. All these deduced structures were later confirmed after they were isolated and analyzed by NMR in the pure form.

The value of using NMR has been proven in the identification of compounds, especially for those unknowns in plant extracts. For example, one black cohosh extract was analyzed by HPLC-ELSD in a routine quality control test. It was found that two compounds (**5** and **6**) appeared as major peaks on the HPLC chromatogram (Figure 19.9b), while they were usually minor constituents in regular black cohosh products. Since the LC-ELSD analysis could not provide any further structural information, the extract was then subjected to the NMR analysis and the major signals shown in ^{13}C NMR were identified as those from 25-*O*-acetylcimigenol 3-*O*-xylopyranoside (peak **5**) and 25-*O*-acetylcimigenol 3-*O*-arabinopyranoside (peak **6**) (Figure 19.8c). Although the intensity of ^{13}C signals is not actually proportional to the sample concentration due to the NOE effect, the large signals still reflected relatively high amounts of the components in the mixture. In this case, the major signals seen in the ^{13}C NMR spectrum arose from compounds **1**, **2**, **5**, and **6**, reflecting the major peaks seen on HPLC chromatogram. The observed downfield resonance of the C-25

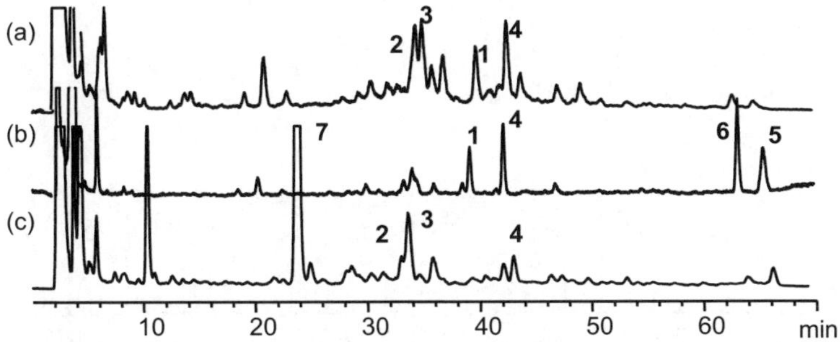

Figure 19.9 HPLC chromatograms of *Cimicifuga* extracts. Chromatography was performed on a Phenomenex Prodigy ODS (3) column (5 µm, 4.6 ID × 250 mm) with a flow rate of 1.0 mL/min and ELSD detector. Mobile phase consisted of a step gradient starting with 30% MeCN in 1% aqueous formic acid with H_2O, increasing to 40% MeCN between 0 and 30 min, then increasing to 60% MeCN between 30 and 60 min. The column temperature was constant at 35°C. Chromatograms of (a) *C. racemosa*; (b) a commercial product; and (c) *C. foetida*.

signal at δ 83.1 due to anisotropic effect of acetyl group on the 25-hydroxyl group compared with C-25 signals at δ 71.0 in **1** and **4**. Its neighboring carbon signal, C-24, showed 3.5 ppm upfield shift. These data suggested that the 25-hydroxyl was substituted by acetyl groups in **5** and **6**. We later obtained a standard from a colleague and one of the two peaks was confirmed to be **5** and another peak was proposed as **6** by the combination analysis of MS and NMR data. When black cohosh extract is prepared under mild conditions, such as at low temperature, compounds **5** and **6** were rarely seen. Early studies had indicated that catalyzed by acid or base, 23-*O*-acetylshengmanol xyloside was converted into 25-*O*-acetylcimigenol 3-*O*-xylpyranoside [9]. Such conversion could also occur under intense heat [10]. The identification of compounds **5** and **6** by NMR led to understanding the mechanism of the formation of these two artificial components. Therefore, harsh heat has been avoided during production in order to maintain the original chemical profile.

Asian *Cimicifuga* species, especially *C. foetida* (Shengma) have been used as adulterants of black cohosh in the US dietary supplemental market. Shengma extract displayed a similar signal pattern to black cohosh in ^{13}C NMR, except for one group of peaks that was dominant (Figure 19.10). These signals were attributed to 30 carbons for a triterpene and a xylose. Three characteristic signals included one double bond at δ 116 and 148 (Δ^{7-8}) and two carbonyl groups above δ 200, which related it to the structures of cimicifugosides H isolated from other *Cimicifuga* species. This major compound was thus identified as cimicifugoside H-1 (**7**) after comparison with literature data [11] and was further confirmed by the 2D correlation spectra. For example, long-range correlations of H-15, H-17 with C-16 and H-22 with C-23 were observed in HMBC spectrum confirming the positions of two carbonyl groups. The cross-peak of H-3 to H-1′ indicated the xylosyl attached at C_3-OH. HPLC-ELSD analysis indicated **7** to be a major peak, accounting for about 30% of the total by peak area. Therefore, cimicifugoside H-1 can be considered as a marker for *C. foetida*.

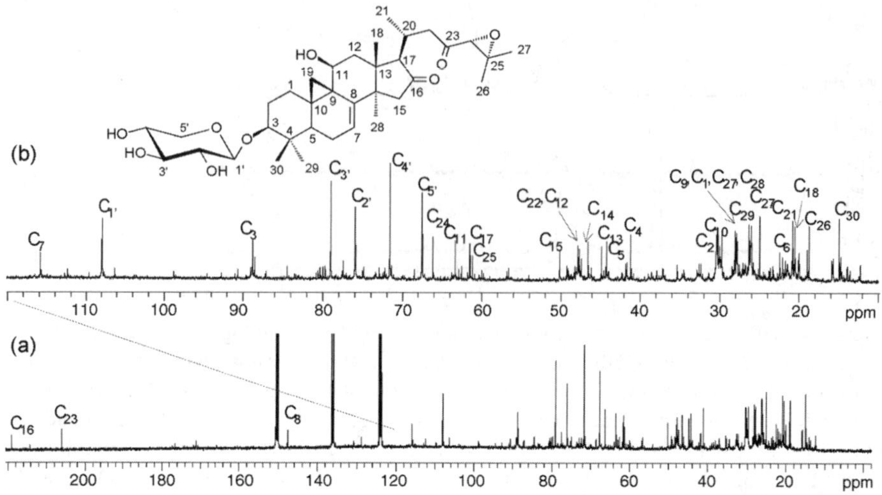

Figure 19.10 Carbon-13 NMR spectra of commercial product (*C. foetida*). The powder (100 mg) was dissolved in 1 mL of pyridine-d_5 and filtered through cotton into an NMR tube for NMR experiments. The spectrum was acquired with broadband proton decoupling and the number of scans was 90,000. (a) Whole spectrum; (b) the expanded region between 10 and 120 ppm.

19.4 PROTON NMR OF ALOE PRODUCT

Aloe gel derived from the inner part of the leave of *A. vera* L. (*Aloe barbadensis* Miller) is used extensively in the cosmetic industry for its anti-inflammatory, wound healing, and other effects [12]. Aloe gel contains >98% water and about 0.5% solids made of polysaccharides (PS), free sugars, proteins, minerals, organic acids, and many other compounds. The PS, especially the acetylated mannans (acemannan), are suggested to be some of the major active components responsible for the biological properties. The difficulty in quality control of *A. vera* products and preparations lies in the lack of major and characteristic markers in the gel. The standards for a good quality of aloe product have been changing from time to time. At present, an aloe product that has high PS content is considered to be good quality. Methods for analysis of PS are usually indirect and time consuming. It requires breaking down PS first and identifying composition of individual monosaccharides, glycosyl linkage, and so on. Diehl and Teichmuller have developed a simple and quick ^{1}H NMR method to examine *A. vera* for purity and quality [13]. In this method, three main components, glucose, malic acid, and PS, can be observed directly in the ^{1}H-NMR spectrum. Other compounds from enzymatic or chemical decomposition, such as lactic acid and acetic acid, as well as the preservatives benzoic acid and sorbic acid can be clearly recognized. Proton NMR has been used on commercial aloe gel powder and provides useful information to assess the identity and quality [14].

The aloe crude PS can be prepared from repeated ethanol precipitation where the alcohol-insoluble PS and minerals are separated from other molecules. The raw PS is dissolved only slightly in water and forms a gel-like solution. Its major NMR signals

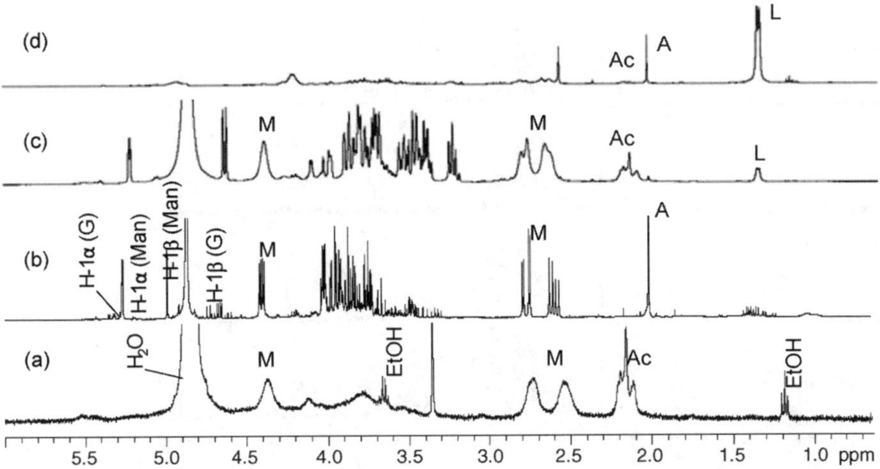

Figure 19.11 Proton spectra of aloe polysaccharides (a) before acid hydrolysis and (b) after acid hydrolysis; typical ^{1}H-NMR spectra of (c) high and (d) low qualities of commercial aloe products. L: lactate; A: acetic acid; Ac: acetate from acemannan; M: malate; G: glucose; Man: mannose.

that can be seen are malate and acetyl groups from acemannan (Figure 19.11a). The NMR peaks were broad due to the long relaxation times of the nucleus, caused by slow movement of the large polymer in the high viscous solution of the mucilaginous PS. The signals at δ 1.9–2.3 arose from the acetyl groups on the polysaccharide backbone and these signals have been considered the fingerprints for aloe PS [13]. When the PS were completely hydrolyzed in acid, the "fingerprint" of acetate signals was converted into a singlet peak (ca. δ 2.0–2.2) corresponding to a free acetic acid. The major monosaccharides were identified as α- and β-mannopyranose, plus minor and approximate equal amount of α-, β-glucopyranose, galactopyranose, and arabinopyranose (Figure 19.11b).

A good quality aloe product should contain high amount of PS (5%–10%) indicated by signals of acetyl groups of acemannan in ^{1}H-NMR with small signals of acetic acid and lactic acid, which are the signs of PS degradation. Other characteristic signals, such as malate, glucose, and mannose, should also be present for identity (Figure 19.11c). The product with a low amount of PS and larger signals of degradation products are considered to be bad quality (Figure 19.11d). Many aloe products were manufactured by treatment with cellulase in order to break the leaf fiber and reduce viscosity [15]. However, the enzymatic hydrolysis will damage aloe PS and increase the amount of acetic acid and lactic acid. In laboratory experiments, it was found that when aloe gel or whole leaf was treated with cellulase, the acetate signals were attenuated over time and eventually disappeared. Also, the signal of acetic acid increased. Cellulase may be contaminated with some other enzymes including mannanase resulting in hydrolysis of acemannan during aloe production (Figure 19.12).

A doublet signal appeared at around δ 1.35–1.45 (changing with pH) arising from the lactate, which is believed to be formed due to fermentation by existing lactobacillus inside the aloe, and is used as an indicator of the freshness of aloe for quality control [13]. We observed in the laboratory experiments that the cellulase makes more of a contribution to the formation of the lactate than the fermentation by

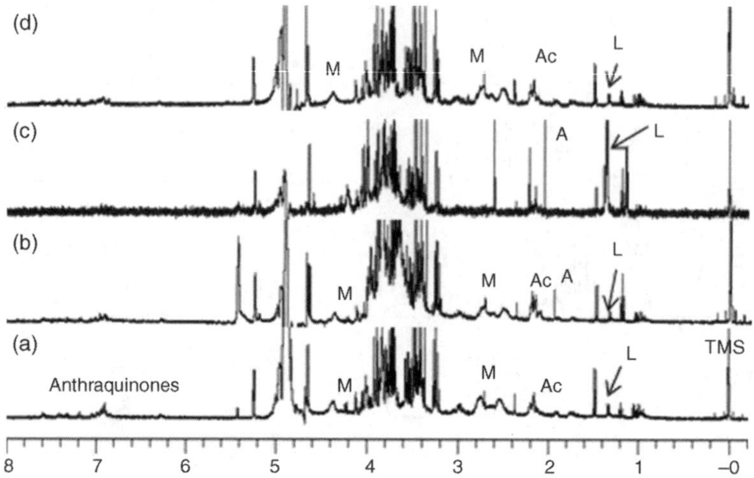

Figure 19.12 Proton NMR spectra of whole aloe leaf extract in D_2O. The H_2O signal was suppressed using presaturation mode and a 32-scan was acquired. (a) Spectrum was obtained immediately after the aloe solution was made without adding cellulase; (b) spectrum after 2 h since adding cellulase (weight of aloe solution to the weight of cellulase is 200:1); (c) spectrum after 48 h since adding cellulase; (d) spectrum after 48 h without adding cellulase. L: lactate; A: acetic acid, Ac: acetate from acemannan; M: malate.

bacteria in the storage period. For example, one aloe sample was prepared by grinding equal amounts of fresh aloe leaf with equal amounts of H_2O and the mixture was centrifuged. The supernatant was subsequently treated with or without cellulase for comparison. The proton signals of lactate were increased slightly in the sample without enzyme, while it significantly increased in the sample with enzyme (Figure 19.12c), and the signals of acetate and malate disappeared as well. A kinetic study indicated an increasing lactate signal with time after cellulase was added to the aloe solution (Figure 19.13). Therefore, on the basis of these observations, controlling the enzymatic hydrolysis should be one of the important steps in aloe production.

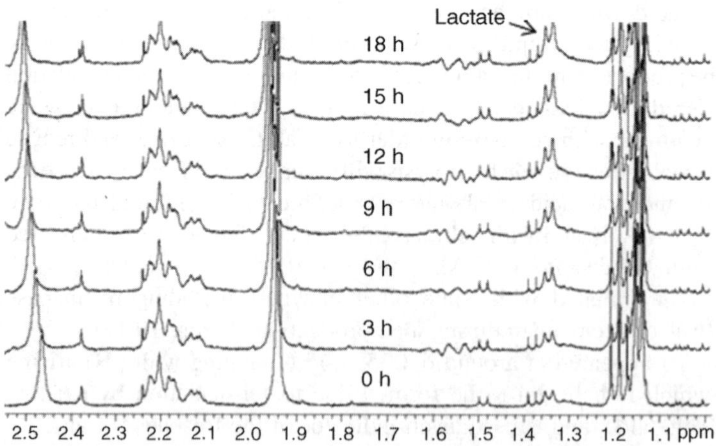

Figure 19.13 Proton NMR spectra of whole aloe leaf extract in D_2O tested at different time period after adding cellulase.

19.5 IDENTIFICATION OF IRIDOIDS IN SEEDS OF *F. EXCELSIOR*

The plant *F. excelsior* L. (Oleaceae) is known as "common ash" or "European ash" in temperate Asia and Europe. The aqueous extract from *F. excelsior* seed (FE) was shown to be highly potent in reducing blood glucose levels without significantly affecting insulin levels [16]. In an *in vitro* bioassay screening, FE was found to activate the nuclear receptor PPARα. PPARα activators have been associated with improving epidermal differentiation and reduce inflammation [17], suggesting that FE has potential to be developed as a cosmetic product.

To investigate the chemical composition, the seed extract of FE was analyzed using LC-UV DAD and LC-MS and two major peaks (**8** and **9**) were observed with DAD and MS detections. Their molecular weights were suggested as *m/z* 686 for **8** and *m/z* 1072 for **9**, respectively, by the MS but were not confirmed because the spectra were complicated by molecular adduct, doubly charged, and fragmental ions. It was reported in the literature that secoiridoids, coumarins, flavonoids, and phenylethanoids were the major components being isolated from the leaves or bark of *F. excelsior*. The UV spectra excluded the possibilities of coumarins, flavonoids, and phenylethanoids for the two peaks. Therefore, secoiridoid molecules were chosen as the focus when the NMR spectra were analyzed.

In the ^{1}H NMR spectrum, several distinctive signals from an oleoside-type secoiridoid glucoside moiety could be assigned: a downfield doublet proton due to attachment to an oxygen at δ 7.5 for H-3, a ketal proton at 5.9 for H-1, a quartet olefinic proton coupled with adjacent CH_3 at 6.1 for H-8, and its neighbor methyl group at 1.7 ppm for CH_3-10 (Figure 19.14a). The ^{13}C NMR signals corresponding for three oleoside aglycone units were observed at approximately δ 155 (C-3, C-3a, C-3c), 95 (C-1, C-1a, C-1c), 125 (C-8, C-8a, C-8c), and 13.5 (C-10, C-10a, C-10c). Besides, five units of glucopyranosyl and two units of phenylethyloxy could be assigned from the major ^{13}C NMR signals (Figure 19.14b). According to the chemical shifts of the anomeric signals (ca. δ 100.7), three of the glucopyranosyl groups were assigned to connect to the three oleoside aglycones at C-1 (**8**), C-1a, and C-1c (**9**), respectively, which were supported by the cross-peaks of H-1′, H-1‴ to the corresponding C-1, C-1a, or C-1c in the HMBC spectrum, leading to the establishment of three oleoside units. In the same spectrum, the long-range correlations of the two anomeric protons (H-1″) with the oxymethylene carbons (δ 71.5 and 72.1) of the phenylethyloxy indicated that the glucopyranosyl groups were connected to the phenylethyloxy moiety at C-1d for **8** and C-1b for **9**. The connections of glucopyranosyl groups through C-6″ to the carbonyl groups of oleosides at C-7 for **8** and C-7a for **9** were established by the HMBC correlations observed among the H-6″ and C-7 and C-7a (ca. δ 172). Considering the molecular weights of 686 and 1072 obtained from MS, structures of nuzhenide for **8** and Gl3 for **9** were proposed by the comparison of ^{13}C NMR data with literature in which the two compounds were previously isolated from the seeds of *Fraxinus americana* [18]. The identification of the two major peaks by NMR analysis allowed a quick analytical method development during FE product development without waiting for time-consuming isolation of **8** and **9**. The final

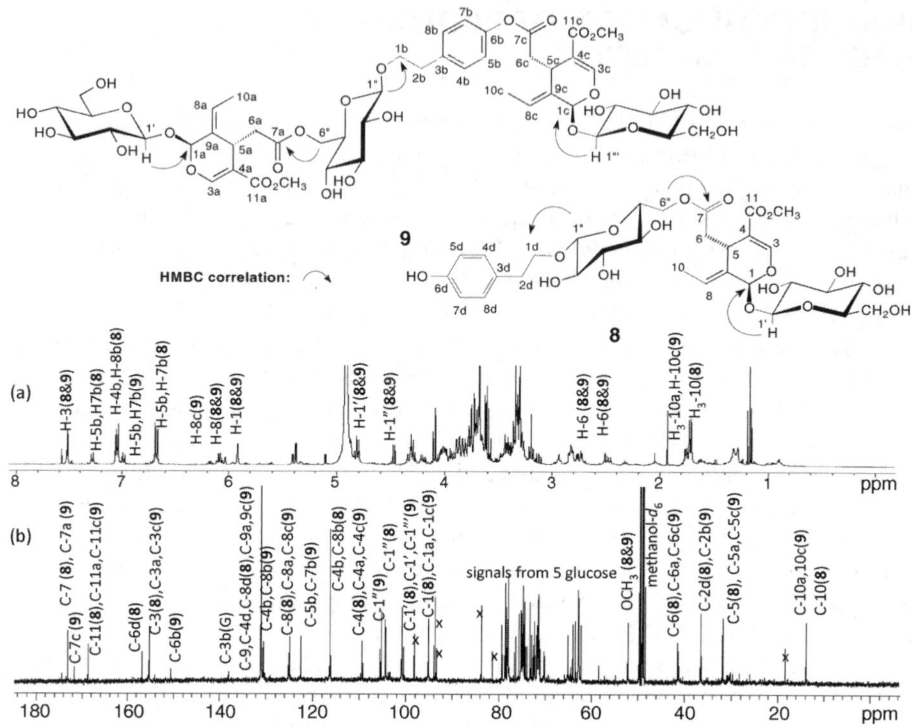

Figure 19.14 NMR spectra of (a) ^{1}H and (b) ^{13}C of *F. excelsior* in methanol-d_4. The tested sample was prepared by directly dissolving the aqueous seed extract of *F. excelsior* into methanol-d_4 and filtered. Compound **8**: nuzhenide; **9**: GI3. Mark "x" indicate the unidentified signals.

confirmation of **8** and **9**, as well as the biological activity test, was achieved through analyzing the pure forms after isolation.

19.6 IDENTIFICATION OF ADULTERATED COMPOUNDS IN *G. BILOBA* EXTRACT

An extract from the leaves of *G. biloba* is known for improving cerebral blood flow. Application of skin care formulations using ginkgo extract was also found in scientific journals [19, 20]. The active constituents in commercial ginkgo products are believed to be ginkgosides and ginkgolides, requiring a minimum of 24% and 6%, respectively. To analyze the ginkgosides, the ginkgo extract is subjected to acid hydrolysis and the contents of flavonol aglycones, quercetin (**10**), kaempferol (**11**), and isorhamnetin (**12**) are measured by HPLC. In analysis of ginkgo extracts obtained from the marketplace, a new peak between quercetin and kaempferol was found, which should not occur in regular *G. biloba* products (Figure 19.15). The LC-UV DAD and LC-MS analyses indicated a possible isoflavone with molecular weight of m/z 270 for this peak. The ^{1}H and ^{13}C NMR spectra showed the signals corresponding to the molecule of genistein (**13**) (Figure 19.16) and it was confirmed by HPLC analysis with genistein

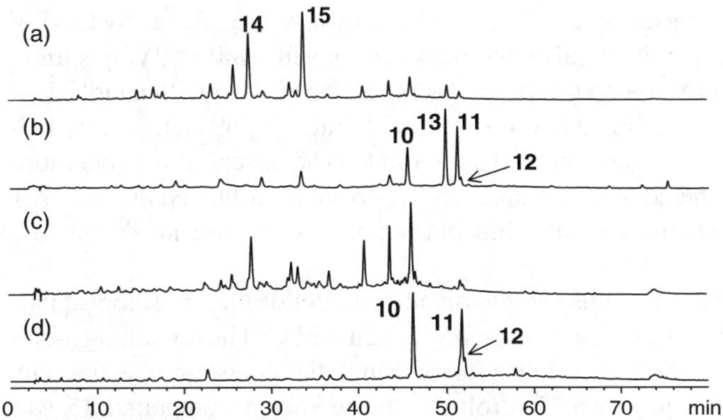

Figure 19.15 HPLC chromatograms of adulterated ginkgo extracts: (a) original extract; (b) hydrolyzed with HCl; unadulterated ginkgo extracts: (c) original extract; (d) hydrolyzed with HCl. Chromatography was performed on a Phenomenex Gemini C18 column (5 µm, 4.6 ID × 250 mm) with a flow rate of 1.0 mL/min and UV detection at 280 nm wavelength. Mobile phase consisted of a step gradient starting with 10% MeCN in 0.1% aqueous acetic acid with H_2O, increasing to 25% MeCN between 0 and 35 min, then increasing to 65% MeCN between 35 and 70 min. The column temperature was ambient. **10**: quercetin; **11**: kaempferol; **12**: isorhamnetin; **13**: genistein; **14**: rutin; **15**: genistein 4'-*O*-α-L-rhamnopyranosyl-(1 → 2)-β-D-glucopyranoside.

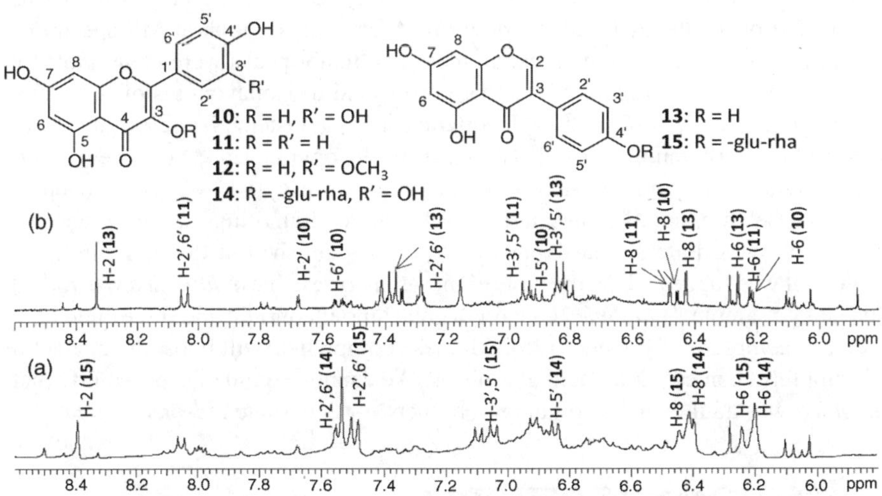

Figure 19.16 Proton NMR spectra of *G. biloba* extract (expansion of chemical shifts in aromatic region of δ 5.6–8.5). (a) Spectrum of ginkgo leaf powdered extract in DMSO-d_6. The major signals in this region were identified as rutin (**14**) and genistein 4'-*O*-α-L-rhamnopyranosyl-(1 → 2)-β-D-glucopyranoside (**15**); (b) spectrum from hydrolyzed *G. biloba* extract by hydrochloric acid. The acid solution was evaporated *in vacuo* and the residue was dissolved in DMSO-d_6. The major signals in this region were identified as quercetin (**10**), kaempferol (**11**), and genistein (**13**). glu: β-D-glucopyranosyl; rha: α-L-rhamnopyranosyl.

standard. The original ginkgo sample before acid hydrolysis was analyzed by LC-UV DAD/MS and a total of five distinguishable peaks displayed similar UV spectra to genistein. Besides rutin (**14**), the major peak was an unknown (**15**) with a molecular weight of *m/z* 578 and two fragmental ions of *m/z* 431 and *m/z* 269 analyzed by LC-MS using electrospray ionization in negative ion mode. The observed fragment ions were corresponding to the loss of a rhamnosyl (146 amu) followed by losing a glucosyl (162 amu), suggesting a possible link of rhamnose to glucose and the glucose to aglycone.

The ^{1}H NMR spectrum of this sample displayed a set of major proton signals that were ascribed to rutin, one of the main peaks seen in HPLC. The remaining major signals were assignable to be a genistein moiety. Since the genistein was the only isoflavone aglycone found after acid hydrolysis, the unknown compound **15** was suggested to be genistein glycoside (with a glucosyl and a rhamnosyl). The chemical shift of H-6 and H-8 appeared at δ 6.24 and 6.41 for genistein, while they were 6.29 and 6.48 for genistin (genistein 7-*O*-glucopyranosyl), shifting downfield compared to genistein due to glycosidation at the C-7 hydroxyl group of genistin. Since the chemical shifts of H-6 (δ 6.25) and H-8 (δ 6.41) in **15** were closer to genistein, it excluded the possibility of glycosidation at 7-hydroxy and the remaining position in genistein was 4′-OH. Therefore, **15** was identified as genistein 4’-*O*-α-L-rhamnopyranosyl-(1 → 2)-α-D-glucopyranoside, which was previously isolated from *Sophora japonica* [21]. Up to this time, no isoflavonoid has ever been found in *G. biloba* leaves [22], except for one report on the isolation of genistein from acid-hydrolyzed ginkgo leaf extract [23, 24]. Likely, the isoflavonoids found in the ginkgo sample was contaminated or adulterated with *S. japonica*. After a closer look at MS spectra of genistein-like peaks, it was found that all these five major peaks were identifiable by molecular weights and fragment ions that correspond to isoflavonoids of genistein, genistin, genistein 7, 4′-di-*O*-β-D-glucopyranoside, genistein 7-*O*-β-D-glucopyranosde-4′-*O*-[(α-L-rhamnopyransoyl)-(1 → 2)-β-D-glucopyranoside], genistein 7-*O*-β-D-glucopyranoside, and genistein 4′-α-L-rhamnopyransoyl-(1 → 2)-β-D-glucopyranoside. These compounds represent the isoflavonoids found in *S. japonica* and the HPLC profile was quite similar compared to those published in the literature [21]. Therefore, it was suggested, on the basis of theses analyses, that *S. japonica* was mixed in this ginkgo sample. *S. japonica* contains a great amount of rutin and some quercetin, as well as kaempferol glycosides [25]. All these compounds will boost the quercetin and kaempferol contents after acid hydrolysis. We cannot exclude the possibility that *S. japonica* was adulterated into ginkgo products for economic reasons.

19.7 PROTON NMR DETECTING FLAVANONOLS IN SARSAPARILLA

For major components in a mixture, using NMR for structural determination has been demonstrated to be as accurate as working on single compound from the above examples. When looking for nonmajor constituents of a mixture, an NMR-based approach is still feasible, where the extract is subjected to a simple purification as seen in the example of sarsaparilla (*S. glabra*) root extract. The aqueous alcoholic extract of sarsaparilla contains flavanonols, including mainly astilbin (**16**) and minor analogues neoastilbin (**17**), isoastilbin (**18**), and neoisoastilbin (**19**), as well as the aglycone taxifolin. (2*R*,3*R*)-(+)-Taxifolin was shown to have an inhibitory effect on melanin biosynthesis by inhibiting the tyrosinase in melanocytes. Its inhibition is more potent than that of the cosmetic agent arbutin and equal to that of kojic acid [26], while astilbin is a mild tyrosinase inhibitory compound [27]. HPLC-DAD/MS analysis of sarsaparilla extract showed a major astilbin peak (compared to standard) and three other peaks with very similar UV spectra and the same molecular weight as astilbin. To confirm the existence of the other three in sarsaparilla extract, the ^{1}H NMR spectrum was studied. The aqueous alcoholic extract of sarsaparilla was partitioned with EtOAc and **16–19** was enriched in the EtOAc phase for NMR analysis. Compounds **16–19** are diastereomers and their chemical shifts were quite similar in ^{1}H NMR except for the H-2, H-3, and rhamnosyl signals. Due to the chirality of C-2 and C-3, the rhamnose can be oriented at a different position toward benzene ring (B-ring) and carbonyl group at C-4. The anisotropic effect of the electrons of benzene and carbonyl resulted in the rhamnosyl signals differing considerably and allowed to distinguish **16–19** through the identification of rhamnosyl groups [28].

TOCSY is a suitable tool for studying sugars, as protons in sugar are mutually coupled, which allows tracking of the proton signals through correlation analysis. The 1D TOCSY is the 2D TOCSY variant and used in the present example. The methyl group (H-6″) was chosen as an entry point by irradiating this signal and 1D TOCSY spectra were acquired (Figure 19.17). When peak at δ 1.15 (H-6″) of **16** was irradiated, the corresponding correlated peaks were detected excepting H-1″ because

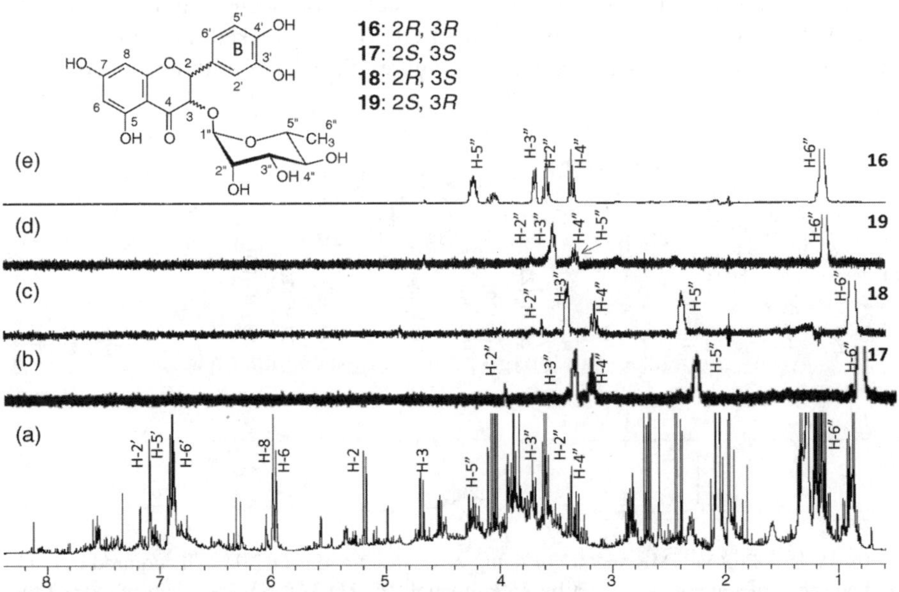

Figure 19.17 Proton NMR spectra of sarsaparilla extract in acetone-d_6. (a) ^{1}H NMR spectrum. The labeled signals were assigned to astilbin; 1D TOCSY spectra of rhamnosyl for (b) neoastilbin (**17**); (c) isoastilbin (**18**); (d) neoisoastilbin (**19**); (e) astilbin (**16**).

of a weak coupling. A similar approach was made by irradiating H-6″ at δ 1.12 for **19**, δ 0.91 for **18**, and δ 0.89 for **17**, and the rhamnosyl signals could be assigned from the congested overlapped peaks. The different rhamnosyl signals were corresponding to different structures of **16**–**19** [29] and these compounds were confirmed to exist in the extract in an indirect way.

19.8 ISOLATION OF (2*S*,3*R*,4*S*) 4-HYDROXYISOLEUCINE FROM FENUGREEK EXTRACT USING ^{1}H NMR AS MONITOR

Amino acid (2*S*,3*R*,4*S*) 4-hydroxyisoleucine (4-OH Ile, **20**) is one of the active components with insulinotropic activity found in the seeds of fenugreek extract (*T. foenum-graecum*). Topical application comprising 4-OH Ile as a cosmetic compound was also reported [30]. To obtain fenugreek extract rich in 4-OH Ile, a purification procedure was needed. Since there was no commercial standard of 4-OH Ile available, and also due to the lack of a simple HPLC detection (amino acids are usually derivatized with chromophores for good detection), the ^{1}H NMR was used to detect 4-OH Ile in each fraction obtained during the purification process. About 1% of 4-OH Ile is found in fenugreek crudes naturally. After extraction, the 4-OH Ile can be as high as 5% in the extract. In the ^{1}H NMR spectrum, the two doublet peaks around δ 0.90 and 1.26 ppm were the characteristic signals from two methyl groups of 4-OH Ile, which were recognizable while the methine signals were buried by overlapped peaks in the extract. The cross-peaks of mutually coupled protons in 4-OH Ile could be detected by TOCSY spectrum. The ^{1}H NMR profile of 4-OH Ile was obtained by 1D TOCSY from the overlapped signals and proved the existence of 4-OH Ile in the extract (Figure 19.18b). Using this technique, the 4-OH Ile was tracked during purification until the pure 4-OH Ile was obtained. In Figure 19.18a and c, three signals in the

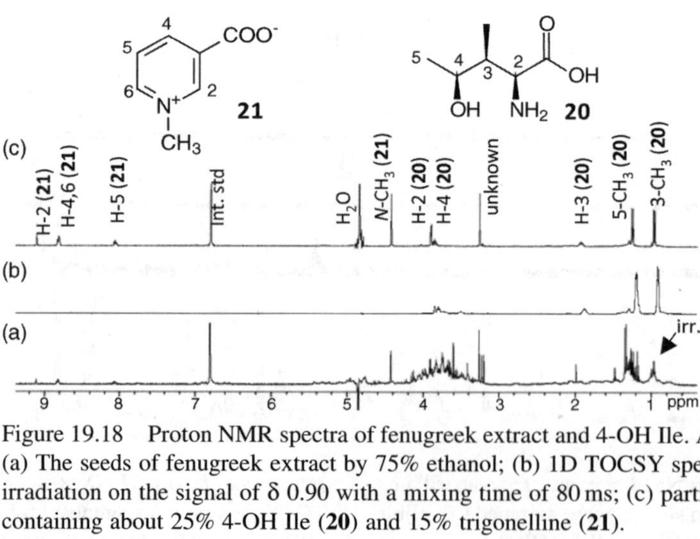

Figure 19.18 Proton NMR spectra of fenugreek extract and 4-OH Ile. All acquired in D_2O. (a) The seeds of fenugreek extract by 75% ethanol; (b) 1D TOCSY spectrum of extract by irradiation on the signal of δ 0.90 with a mixing time of 80 ms; (c) partially purified extract containing about 25% 4-OH Ile (**20**) and 15% trigonelline (**21**).

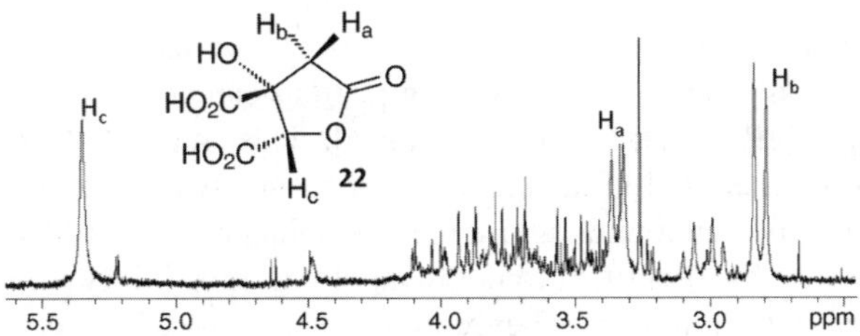

Figure 19.19 Proton NMR spectrum of hibiscus extract in D_2O and structure of **22**.

aromatic region between δ 8.0 and 9.2 were assigned as the signals from alkaloid trigonelline (**21**) and it was confirmed later by HPLC analysis with a standard.

19.9 IDENTIFICATION OF HIBISCUS ACID IN HIBISCUS EXTRACT

H. sabdariffa is a tropical plant and its flower extract is used as folk medicine against many diseases. Hibiscus extract contains anthocyanin pigments and organic acids including citric, malic, tartaric, hibiscus, and ascorbic acids [31]. Using an HPLC method to look for the organic acids of hibiscus flower water extract, one major peak was unidentifiable. The ^{1}H NMR showed two doublet peaks at δ 2.88 and 3.41, respectively, with a large coupling constant ($J = 18.4$ Hz) arising from germinal coupling of the two protons (Figure 19.19). The structure of hibiscus acid was elucidated with the aid of 1D and 2D NMR techniques and its molecular weight was confirmed by HPLC-MS. The major peak was therefore identified as hibiscus acid (**22**).

19.10 CONCLUSIONS

Plants usually contain several major classes of compounds such as alkaloids, phenolics, terpenes, and so on. Unique secondary metabolites of plants are often used as markers for analysis of botanicals and herbal preparation. Quality of botanical product relies heavily on the clear identification of unique constituents so that the correct botanical species and components of interest can be guaranteed. This further warrants clinical results and safety of the product. The majority of available methods used for quality control of plant extracts are HPLC coupled with different detection techniques, including the most common detectors, UV/Vis and MS. These methods are applicable only when the chemical structures of targeted compounds or analytes are identified. For unknown and new compounds, NMR is the only method of choice. Apart from its sensitivity, NMR has been proven to be a powerful tool for structural determination of organic molecules. For characterizing new compounds, the purified form has to be prepared before NMR analysis. When analyzing mixtures, the specific molecule can be identified by extracting its fingerprint from hundreds of others because every single molecule has its own unique NMR fingerprint. As demonstrated by several examples above, the NMR analysis can be applied to crude plant extracts directly for major compound structural determination without undergoing a tedious purification. For minor components, after some simple purification, such as partition or fractionation through a column, and so on, the minor components can be enriched and the same good quality of NMR spectra can be obtained. This makes NMR technology extremely useful to be included in the early analytical method development stage of identifying unknown compounds and in the establishment of NMR profiles of plant extracts for species identification. Owing to its capability for unequivocal chemical structural determination, when combined with HPLC detections, NMR methods can be used to unravel complex metabolites in the plant extracts and significantly expands the range of identified constituents for product development and quality control of botanical extracts.

REFERENCES

1. Thornfeldt C. Herbs in cosmeceuticals: are they safe and effective? In: Draelos ZD, Thaman LA. editors. *Cosmetic Formulation of Skin Care Products*. Cosmetic Science and Technology Series, vol. 30. New York: Taylor & Francis, 2006. pp. 309–350.
2. Draelos ZD. Botanicals as topical agents. *Clin. Dermatol.* 2001;19:474–477.
3. Dureja H, Kaushik D, Gupta M, Kumar V, Lather V. Cosmeceuticals: an emerging concept. *Indian J. Pharmacol.* 2005;37:155–159.
4. Krishnan P, Kruger NJ, Ratcliffe RG. Metabolite fingerprinting and profiling in plants using NMR. *J. Exp. Bot.* 2005;56:255–265.
5. Reynolds WF, Enrýquez RG. Choosing the best pulse sequences, acquisition parameters, postacquisition processing strategies, and probes for natural product structure elucidation by NMR spectroscopy. *J. Nat. Prod.* 2002;65:221–244.
6. Betz JM, Anderson L, Avigan MI, Barnes J, Farnsworth NR, Gerden B, Henderson L, Kennelly EJ, Koetter U, Lessard S, Dog TL, McLaughlin M, Naser B, Osmers RGW, Pellicore LS, Senior JR, van Breemen RB, Wuttke W, Cardellina JH. Black cohosh: considerations of safety and benefit. *Nutr. Today* 2009;44:155–162.
7. Sakurai N, Kozuka M, Tokuda H, Nobukuni Y, TakayasuJ, Nishino H, Kusano A, Kusano G, Nagai M, Sakuraia Y, Lee K-H. Antitumor agents 220. Antitumor-promoting effects of cimigenol and related compounds on Epstein–Barr virus activation and two-stage mouse skin carcinogenesis. *Bioorg. Med. Chem.* 2003;11:1137–1140.
8. Leverett JC, Missler SR, Roh-Schmidt H, Chandra A. Skin whitening compositions containing black cohosh extract. US Patent 7,060,304, 2001.
9. Kusano G. Studies on the constituents of *Cimicifuga* species. *Yakugaku Zasshi* 2001;121:497–521.
10. He K, Pauli GF, Zheng B-z, Wang H-k, Bai N-s, Peng T-s, Roller M, Zheng Q-z. *Cimicifuga* species identification by high performance liquid chromatography-photodiode array/mass spectrometric/evaporative light scattering detection for quality control of black cohosh products. *J. Chromatogr. A* 2006;1112:241–254.
11. Koeda M, Aoki Y, Sakurai N, Nagai M. Studies on the Chinese crude drug "shoma" IX. Three novel cyclolanostanol xylosides, cimicifugosides H-1, H-2 and H-5, from *Cimicifuga* rhizome. *Chem. Pharm. Bull.* 1995;43:771–776.
12. Hamman JH. Composition and applications of *Aloe vera* leaf gel. *Molecules* 2008;13:1599–1616.
13. Diehl B, Teichmuller EE. *Aloe vera,* quality inspection and identification. *Agro Food Ind. Hi Tech* 1998;9:14–16.
14. Bozzi A, Perrin C, Austin S, Arce Vera F. Quality and authenticity of commercial aloe vera gel powders. *Food Chem.* 2007;103:22–30.
15. Waller TA, Pelley RP, Strickland FM. Industrial processing and quality control of *Aloe barbadensis* (*Aloe vera*) gel. In: Reynolds T, editor. *Aloes, the Genus Aloe*. Boca Raton, FL: CRC Press 2004. pp. 149–205.

16. Eddouks M, Maghrani M. Phlorizin-like effect of *Fraxinus excelsior* in normal and diabetic rats. *J. Ethnopharmacol.* 2004;94:149–154.
17. Rosenfield RL, Deplewski D, Greene ME. Peroxisome proliferator-activated receptors and skin development. *Horm. Res.* 2000;54:269–274.
18. LaLonde RT, Wong C, Tsai AI-M. Polyglucosidic metabolites of Oleaceae. The chain sequence of oleoside aglucon, tyrosol, and glucose units in three metabolites from *Fraxinus americana*. *J. Am. Chem. Soc.* 1976;98:3007–3013.
19. Chanchal D, Swarnlata S. Novel approaches in herbal cosmetics. *J. Cosmet. Dermatol.* 2008;7:89–95.
20. Castelli D, Colin L, Camel E, Ries G. Pretreatment of skin with a *Ginkgo biloba* extract/sodium carboxymethyl-beta-1,3-glucan formulation appears to inhibit the elicitation of allergic contact dermatitis in man. *Contact Dermat.* 1998;38:123–126.
21. Qi Y, Sun A, Liu R, Meng Z, Xie H. Isolation and purification of flavonoid and isoflavonoid compounds from the pericarp of *Sophora japonica* L. by adsorption chromatography on 12% cross-linked agarose gel media. *J. Chromatogr. A* 2007;1140:219–224.
22. Singh B, Kaur P, Gopichand Singh RD, Ahuja PS. Biology and chemistry of *Ginkgo biloba*. *Fitoterapia* 2008;79:401–418.
23. van Beek TA, Montoro P. Chemical analysis and quality control of *Ginkgo biloba* leaves, extracts, and phytopharmaceuticals. *J. Chromatogr. A* 2009;1216:2002–2032.
24. Wang F, Jiang K, Li Z. Purification and identification of genistein in *Ginkgo biloba* leaf extract. *Se Pu* 2007;25:509–513.
25. Liu IM, Sheu SJ. Analysis and processing of Chinese herbal drugs. VIII: The study of Sophorae floe. *Am. J. Chin. Med.* 1989;17:179–187.
26. Miyazawa M, Tamur N. Inhibitory compound of tyrosinase activity from the sprout of *Polygonum hydropiper* L. (Benitade). *Biol. Pharm. Bull.* 2007;30:595–597.
27. Hanamura T, Uchida E, Aoki H. Skin-lightening effect of a polyphenol extract from acerola (*Malpighia emarginata* DC) fruit on UV-induced pigmentation. *Biosci. Biotechnol. Biochem.* 2008;72:3211–3218.
28. De Britto J, Mmanickam VS, Gopalakrishnan S, Ushioda T, Tanaka N. Determination of aglycone chirality in dihydroflavonol 3-*O*-α-L-rhamnosides by ^{1}H-NMR spectroscopy. *Chem. Pharm. Bull.* 1995;43338–43339.
29. Kasai R, Hirono S, Chou W-H, Tanaka O, Chen F-H. Sweet dihydroflavonol rhamnoside from leaves of *Engelhardtia chrysolepis,* a Chinese folk medicine, Hung-qi. *Chem. Pharm. Bull.* 1988;36:4167–4170.
30. Potier P, Ouazzani J, Rodelet J-F, Sasaki NA, Zhu Wang Q. Use of (2*S*,3*R*,4*S*)-4-hydroxyisoleucine in cosmetic compounds for topical application. US Patent Application Publication 2005/019357 A1.
31. Ali BH, Al Wabel N, Blunden G. Phytochemical, pharmacological and toxicological aspects of *Hibiscus sabdariffa* L.: a review. *Phytother. Res.* 2005;19:369–375.

PART VI

BIODEGRADATION

CHAPTER 20

BIODEGRADABILITY EVALUATION FOR COSMETIC INGREDIENTS AND FINISHED PRODUCTS

Jennifer K. Saxe

EcoSafety Sciences, Natick, MA

20.1 INTRODUCTION

Biodegradability is a central aspect of the sustainability profile and environmental preferability of formulated products, such as cosmetics. As companies, consumers, and regulators become increasingly interested in evaluating the "green" performance of formulated products, the manner of evaluating biodegradability for these products requires a deeper understanding.

There are three major reasons for evaluating the biodegradability of a cosmetic product:

- To respond to questions from consumers regarding the environmental safety of a product.
- To make a proactive biodegradable claim on a product's packaging or promotional materials.
- To use for internal environmental stewardship initiatives, such as informing the selection of environmentally preferable ingredients when formulating new products.

The evaluation process for biodegradability will differ depending on which of these (or other) purposes is being addressed. This chapter explains how biodegradation can be evaluated using empirical and modeled data for each of these purposes. The proper interpretation and use of data is a fundamental requirement to a sound evaluation, and hence is a focus of this chapter.

Formulating, Packaging, and Marketing of Natural Cosmetic Products, First Edition.
Edited by Nava Dayan and Lambros Kromidas.

20.2 BIODEGRADABILITY EVALUATION AND ITS USE

Biodegradation is the microbially mediated breakdown of organic compounds into smaller organic and inorganic compounds. Frequently, the final products of biodegradation are carbon dioxide and water, or methane. Microbial organisms transform organic chemicals by way of metabolic or enzymatic processes. Biodegradation occurs when microorganisms break chemical bonds, hence transforming the chemicals, in the course of using a chemical as a direct source of energy. Biodegradation can also occur when microbial activity indirectly results in the cleavage of chemical bonds—for example, through the release of extracellular enzymes intended for other purposes—termed cometabolism.

20.2.1 Definition

Biodegradation for a particular chemical can be defined in both qualitative and quantitative terms. Qualitatively, a chemical can undergo either "primary" biodegradation or "ultimate" biodegradation. For a chemical undergoing only primary biodegradation the parent compound degrades, but the end products are different organic compounds (metabolites or daughter products) that can be recalcitrant to further biodegradation. For a chemical undergoing ultimate biodegradation, the parent compound and all complex metabolites degrade to form simple compounds that are ubiquitous in nature, such as carbon dioxide, water, and mineral salts, or are used to synthesize new microbial biomass. Quantitatively, a chemical's rate of biodegradation can be measured and used to describe the chemical's propensity to biodegrade. The difficulty with measuring biodegradation pathways and rates is that both of them depend on the conditions of the test and the particular consortium of microorganisms present, because different microbial populations have different abilities to metabolize chemicals.

There is no unequivocal quantitative threshold that can be used to define a substance as "biodegradable." For instance, the guideline for the most commonly used test to determine biodegradability [1] indicates that the label the test confers—"readily biodegradable"—is an arbitrary classification. The practical definition of "biodegradable" varies depending on the purpose for which the claim is being made. For cosmetics, the three most important audiences for making a claim of "biodegradable" are potential customers in the public at large, formulators within the company, and regulatory authorities. Because there is no unequivocal quantitative definition of biodegradable, it is important to understand the strengths and limitation of the biodegradability data available for a substance used as a cosmetic ingredient, and whenever possible, to make decisions and communicate those decisions to regulators and the public *including* the context and qualification of any claims made.

20.2.2 Degradability Versus Biodegradability

Degradability is of particular interest for substances in cosmetics, because many of these substances are expected to enter the environment via air or water, and it is preferable if they are not persistent in the environment. Persistent chemicals are a concern because they have the potential to travel long distances, to remote and otherwise pristine environments, resulting in long-term, low-level exposure. Such widespread exposure can result in unanticipated adverse effects; for example, chances are higher that commercially significant, endangered, or otherwise important organisms will be exposed, and because of the large variety of organisms exposed, unanticipated effects can occur in organisms more susceptible than those tested in the laboratory. Persistent chemicals also are a concern because any adverse effect they cause cannot be easily reversed, because exposure cannot be mitigated easily after emission and distribution in the environment occur.

The environmental persistence of chemicals is mitigated by degradation processes, which include biodegradation as well as abiotic degradation processes. The most common abiotic degradation processes that occur in the environment are photolysis and hydrolysis. Direct photolysis occurs when a chemical absorbs a photon of light, resulting in the cleavage of one or more chemical bonds. Only chemicals that readily absorb light at wavelengths greater than 290 nm are capable of undergoing substantial direct photolysis in the environment [2]—and then, only in areas where light penetrates, precluding soil and all but the topmost layer of water bodies. Indirect photolysis can occur when a chemical is susceptible to and undergoes attack by a reactive molecule that formed via direct photolysis. For example, reactive hydroxyl radicals (OH•) form in sunlit air and water under certain conditions, and the radical rapidly reacts in the presence of most organic compounds, resulting in chemical transformation (e.g., removal of a hydrogen or addition of OH• across a double or triple carbon bond).

Hydrolysis occurs when a chemical transforms via reaction with water. Water dissociates to form a proton (H^+, or a hydronium ion, H_3O^+, when in water) and a hydroxide ion (OH^-), which add across certain types of bonds, cleaving a chemical into two daughter products. Chemical classes susceptible to hydrolysis include esters, phosphate esters, carbamates, epoxides, halomethanes, and some alkyl halides [3].

A notable distinction between the abiotic degradation processes and biodegradation is that abiotic degradation typically results in incomplete degradation, while biodegradation more often results in complete mineralization. That is, photolysis and hydrolysis do not typically completely degrade a chemical into simple compounds found in nature such as carbon dioxide and water, whereas biodegradation does. However, abiotic degradation can work in concert with biodegradation, to allow a chemical to undergo complete mineralization as shown in Figure 20.1, for acetylsalicylic acid (ASA), or aspirin. ASA undergoes hydrolysis in the presence of moisture when water is added over the ester oxygen, forming salicylic acid and acetic acid (which results in a characteristic vinegar-like odor in aged aspirin containers) [4]. The resulting daughter products are readily biodegradable and are completely mineralized in the presence of microorganisms to form carbon dioxide and water. As a result, ASA would appear to be readily biodegradable in a laboratory test, even though the mechanism of degradation includes an abiotic degradation step.

20.2.3 "Biodegradable" as a Claim on a U.S. Product Label

One use for evaluating biodegradability for a product is to proactively market the product as "biodegradable." Because there is no definitive criterion for categorizing

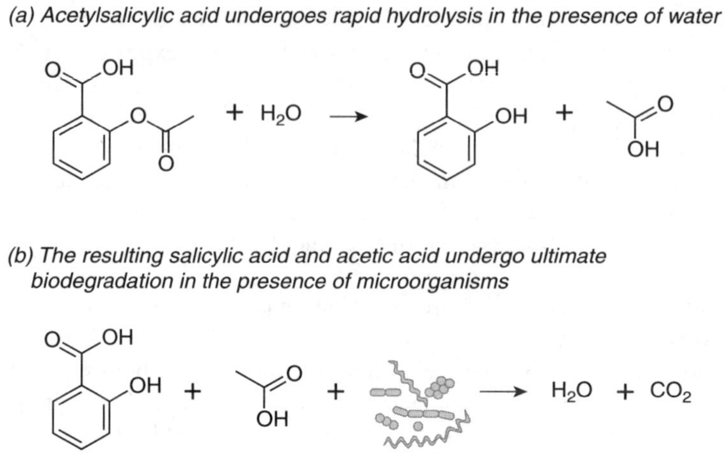

Figure 20.1 Example of abiotic degradation process (hydrolysis) preceding biodegradation: acetylsalicylic acid (aspirin).

a chemical as biodegradable, there is an element of professional judgment involved in the choice to market a product as biodegradable. As an illustration, issues that must be considered when contemplating biodegradability claims for products in the United States are discussed in this section.

In the United States, the Federal Trade Commission (FTC) is the lead agency responsible for ensuring that product marketing claims are not unfair or deceptive. Environmental claims are included within the FTC's purview. To that end, the FTC has issued the "Green Guides" (16 CFR 260[1]) to assist in determining whether environmental claims such as "biodegradable" are legitimate versus deceptive, and, in general, what substantiation would be required to support such claims. In 2007, the FTC began a review process for the Green Guides, to ensure that the Guides adequately address the wide range of claims that proliferated in the marketplace during the early 2000s. As of this writing, the review and any resulting change to the Guides have not yet been completed.

According to the Green Guides, five conditions must be met in order to make an unqualified claim of "biodegradable" for a product. They are as follows: (1) the entire product biodegrades, (2) the residue from biodegradation consists only of elements found in nature, (3) the complete biodegradation of the product must occur within "a reasonably short period of time," (4) biodegradation must occur under conditions of customary disposal, and (5) the previous four points must be substantiated by "competent and reliable scientific evidence."

The first item in the list of conditions indicates that the entire product is biodegradable. As described in Section 20.3.1, biodegradability typically is measured on a chemical-specific basis rather than for a heterogeneous mixture such as

[1] Code of Federal Regulations (CFR) Title 16—Commercial Practices. Chapter I—Federal Trade Commission. Subchapter B—Guides and Trade Practice Rules. Part 260—Guides for the Use of Environmental Marketing Claims.

a cosmetic product. As a result, for formulated products, this condition translates to a need for *each ingredient* in the product to meet the five conditions in order for the finished product to merit the label "biodegradable." However, by definition, the term "biodegradable" applies only to organic compounds. Hence, water and other inorganic ingredients in a product are not subject to this condition.

The fourth item in the list refers to customary disposal, which for most cosmetic products in the United States consists of down-the-drain disposal followed by wastewater treatment. However, a significant portion of some products, such as sunscreens, is expected to enter the environment directly without the benefit of wastewater treatment, such as during recreational swimming. Hence, the final disposition scenario that specifically applies to each product must be considered when determining the conditions to be met for making a biodegradability claim to ensure that the appropriate scenario is selected for substantiation.

The Green Guides provide several examples to illustrate how these conditions would apply for particular products. The example most relevant for making biodegradation claims for formulated cosmetic products follows:

> Example 3: A soap or shampoo product is advertised as "biodegradable," with no qualification or other disclosure. The manufacturer has competent and reliable scientific evidence demonstrating that the product, which is customarily disposed of in sewage systems, will break down and decompose into elements found in nature in a short period of time. The claim is not deceptive. (16 CFR 260.7)

Note that the example does not define a particular rate of biodegradation that would fulfill the criterion for occurrence "in a short period of time." Furthermore, "competent and reliable" scientific evidence is defined as "tests, analyses, research, studies or other evidence based on the expertise of professionals in the relevant area, conducted and evaluated in an objective manner by persons qualified to do so, using procedures generally accepted in the profession to yield accurate and reliable results" (16 CFR 260.5). Hence, the conditions to be met to support an unqualified "biodegradable" claim include aspects that are a matter of professional judgment.

In addition to an unqualified claim of biodegradability, companies can elect to make a qualified claim, to clarify the scope of the conditions under which biodegradation occurs, or the rate and extent of biodegradation, if the conditions of an unqualified claim cannot be met. For example, if an ingredient in a cosmetic product biodegrades only in sewage treatment plants that employ a certain method of advanced or tertiary treatment, which is not common, it would be prudent to qualify the claim to indicate that the product is biodegradable in sewage treatment plants that employ tertiary treatment. Similarly, if an ingredient in a cosmetic product is ultimately biodegradable, but the rate of degradation is relatively slow, it would be prudent to qualify any biodegradable claim to indicate the actual persistence time in the environment, or some similar measure, to avoid the possibility of misleading consumers.

Sections 20.3 and 20.4, respectively, describe empirical test methods and predictive models for determining a chemical's rate and extent of biodegradability. Results from models or studies conducted using the methods described below could potentially constitute reasonable substantiation for unqualified or qualified biodegradability claims, depending on the chemical's ultimate disposition and the results of the study.

20.2.4 Biodegradability as an Environmental Sustainability Metric

Nearly all chemicals are biodegradable. However, not all chemicals can be reasonably labeled as biodegradable, because for some chemicals, degradation occurs only in unusual or highly engineered environments, and over extremely long time periods. A chemical's amenability to biodegradation can be conceptualized as a continuum, where compounds such as glucose are quickly and completely broken down in most environments and compounds such as nonylphenol are more persistent, requiring more time and specific conditions before biodegradation is complete [5]. When evaluating a cosmetic ingredient's biodegradability profile for the purposes of determining the ingredient's environmental safety or sustainability performance, the chemical's position on this conceptual continuum can be considered and compared to other, alternative ingredients, rather than comparing to an arbitrary point on the continuum designated as the threshold for a "biodegradable" label. The flexibility to consider comparative amenability to biodegradation among alternative ingredients allows biodegradability evaluations for the purposes of environmental safety or sustainability assessments to be more informative than a concise "biodegradable" designation on a product label.

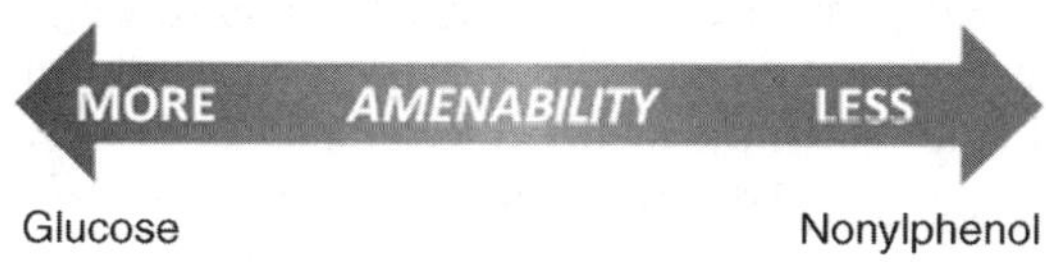

For instance, a cosmetic ingredient that is ultimately biodegradable, but requires significantly more time to break down than the typical retention time in a wastewater treatment plant, would not meet the conditions for an unqualified biodegradability claim. However, it is possible that the alternative ingredients available to fulfill that ingredient's function in a finished product have a less favorable environmental profile, for example, because they are more persistent, or they have a slightly improved biodegradation profile, but unfavorable toxicity characteristics, or they must be used at substantially higher concentrations in a finished product, potentially resulting in greater environmental exposures. Hence, despite the ingredient not fulfilling the conditions for a biodegradable marketing claim, the choice of the cosmetic ingredient reasonably could be deemed to be environmentally preferable to the alternatives. These comparative aspects of biodegradability and the environmental safety aspects that are linked to biodegradability can be explained in responses to questions from the media or the public, or factored into environmentally sustainable formulation programs. In contrast, they cannot be accounted for easily in product labeling because of the prescriptive conditions that must be met and the concise nature of information conveyed on a product label.

In responding to biodegradability questions posed by the media or consumers, companies have the opportunity to use a narrative description of the biodegradability characteristics of the ingredients used in their products, including descriptions of the type of data used, where estimations were made, and when professional judgment was a factor in decision making. In designing ways to incorporate biodegradability into the selection of ingredients when formulating new products, the position of an ingredient on the biodegradability amenability continuum can be represented in any way that is useful to formulators. For example, ingredients can be assigned a numeric score, a color code, or a letter grade that designates its amenability to biodegradation. Factors that can be accounted for in such a system would ideally include the chemical's half-life in wastewater treatment plants and/or the environmental media (i.e., water, sediment, soil, air) in which the chemical is most likely to exist after its use in a cosmetic product, and the likelihood that the chemical will form persistent metabolites after primary biodegradation.

20.3 EMPIRICAL DETERMINATION OF BIODEGRADABILITY

A wide range of test guidelines have been published, describing methods that can be used to determine the biodegradability of a chemical. Test designs range from short-duration, stringent laboratory tests in which only compounds highly amenable to biodegradation will degrade, to long-term field studies in which environmental conditions are well represented. However, experimental biodegradation data are highly variable, because the biodegradation process consists of multiple steps, which depend on chemical structure, laboratory or environmental conditions, and the bioavailability of the chemical. As a result, the outcomes of biodegradation tests are strongly influenced by properties of the chemical, such as solubility, toxicity, and test concentration, because those properties influence bioavailability and the viability of the microbial population in contact with the chemical.

Table 20.1 provides a list of commonly used aerobic freshwater aquatic tests that address a chemical's biodegradability—either alone or, in the case of simulation and field tests, in concert with other competing environmental fate processes, such as sorption to solids and volatilization. The list in Table 20.1 advances in complexity from top to bottom, ranging from screening tests designed to identify substances that biodegrade rapidly and completely (ready biodegradability tests), to tests designed to show that a substance is capable of biodegrading under favorable conditions, and hence is unlikely to be highly persistent in the environment (inherent biodegradability tests), to tests designed to demonstrate biodegradation behavior in environmentally relevant conditions (simulation and field tests). The simpler screening tests at the top of Table 20.1 are cost effective, in part because they measure biodegradation indirectly and hence avoiding the need to analyze for the presence of the test substance or metabolites in the test solution. For chemicals that do not biodegrade swiftly or completely, analysis of chemical residues is undertaken in simulation or field tests. Because residue analysis requires a more substantial effort, including development and performance of extraction, cleanup, and detection techniques, the analysis phase of these studies is significantly more resource intensive compared to

TABLE 20.1 Widely Accepted Test Guidelines Relevant for Assessing the Aerobic Biodegradation of Chemicals Emitted to Freshwater Environments Directly or via Biological Wastewater Treatment Facilities

OECD test guideline	Analogous US EPA test method	Type of analytical determination	Appropriate for test substances that are: Volatile	Appropriate for test substances that are: Low water soluble
Ready biodegradability test guidelines				
301A: DOC Die-Away	OPPTS 835.3110(i)	Indirect, via DOC and optional direct analysis of test substance	No	No
301B: CO_2 Evolution (Modified Sturm Test)	OPPTS 835.3110(ii)	Indirect, via inorganic carbon from trapped CO_2 and optional DOC	No	Yes
301C: MITI[a] (I)	OPPTS 835.3110(iii)	Indirect, via BOD and DOC or direct analysis of residual chemicals	Yes	Yes
301D: Closed Bottle	OPPTS 835.3110(iv)	Indirect, via BOD	Yes	Yes
301E: Modified OECD Screening	OPPTS 835.3110(v)	Indirect, via DOC and optional direct analysis of test substance	No	No
301F: Manometric Respirometry	OPPTS 835.3110(vi)	Indirect, via BOD and DOC or direct analysis of residual chemicals	Yes	Yes
310: CO_2 in Sealed Vessels (Headspace Test)	OPPTS 835.3140	Indirect, via inorganic carbon from trapped CO_2	Yes	Yes
Inherent biodegradability test guidelines				
302A: Modified SCAS[b] Test	OPPTS 835.3210	Indirect, via DOC	No	No
302B: Zahn–Wellens/EMPA[c] Test	OPPTS 835.3200	Indirect, via DOC or COD	No	No
302C: Modified MITI Test (II)	N/A	Indirect, via BOD and DOC or direct analysis of residual chemicals	No	Yes

Guidelines for simulation tests				
303A: Aerobic Sewage Treatment, Activated Sludge Units	OPPTS 835.3240	Indirect, via DOC or COD and direct analysis of the test substance	No	No
303B: Aerobic Sewage Treatment, Biofilms	OPPTS 835.3260	Indirect, via DOC or COD and/or direct analysis of the test substance	No	No
309: Aerobic Mineralisation in Surface Water	OPPTS 835.3190	Indirect, via ^{14}C or direct analysis of the test substance	Yes (slightly)	Yes
314A: Biodegradation in a Sewer System	OPPTS 835.3280	Indirect, via ^{14}C or ^{3}H and direct analysis of residual chemicals	Yes	Yes
314B: Biodegradation in Activated Sludge	OPPTS 835.3280	Indirect, via ^{14}C or ^{3}H and direct analysis of residual chemicals	Yes	Yes
314D: Biodegradation in Treated Effluent-Surface Water Mixing Zone	OPPTS 835.3280	Indirect, via ^{14}C or ^{3}H and direct analysis of residual chemicals	Yes	Yes
314E: Biodegradation in Untreated Wastewater-Surface Water Mixing Zone	OPPTS 835.3280	Indirect, via ^{14}C or ^{3}H and direct analysis of residual chemicals	Yes	Yes
308: Aerobic Transformation in Aquatic Sediment Systems	OPPTS 835.4300	Indirect, via residual ^{14}C and direct analysis of residual chemicals	Yes (slightly)	Yes
Guidelines for field studies with a biodegradation component				
N/A	OPPTS 835.6200 Aquatic (Sediment) Field Dissipation	Indirect, via residual ^{14}C and direct analysis of residual chemicals	Yes[d]	Yes

[a]MITI: Ministry of International Trade and Industry (Japan)

[b]SCAS: semi-continuous activated sludge procedure

[c]EMPA: Swiss Federal Laboratories for Materials Science and Technology

[d]In a field study, the mechanism of disappearance of a volatile test substance is difficult to determine, as volatilization and biodegradation occur in tandem, and may occur in a heterogeneous fashion over the area used to conduct the study.

the simpler tests. In addition, simulation and field tests are performed for longer periods of time—ranging from several months to a year—compared to ready biodegradability tests.

An overview of the different types of aerobic freshwater aquatic biodegradability tests listed in Table 20.1 is provided in the following sections, with an emphasis on the ready biodegradability test, because in most instances it is the test that a company would commission for a cosmetic ingredient, if empirical biodegradability data were sought.

20.3.1 Ready Biodegradability Tests

The simplest and least expensive biodegradability test widely accepted by scientists and regulators is the ready biodegradability test. The goal of the test is to measure mineralization of the test chemical—ultimate biodegradation—under stringent conditions. In an aerobic aqueous mineral salts medium, a relatively large concentration of a test substance is mixed with a relatively small population of microorganisms obtained from a wastewater treatment plant or the natural environment. Mineralization is measured indirectly via depletion of dissolved organic carbon (DOC) from the test solution, depletion of oxygen from the test solution (i.e., biochemical oxygen demand, or BOD), or carbon dioxide evolution. Chemical residues can be measured optionally, but such measurements are not standard protocol. Mineralization over the course of the 28-day test is measured and compared to a target mineralization level deemed the "pass" level. If a test chemical meets the pass level within specific time limits as described below, the chemical can be labeled "readily biodegradable." Often, ultimately biodegradable chemicals do not pass a ready biodegradability test, because the rate of biodegradation is not sufficient to meet the stringent standards of the test. Substances that are toxic to microorganisms or are poorly water soluble can be tested, but require special handling to obtain valid test results.

A ready biodegradability test does not mimic environmental conditions or wastewater treatment facilities. The biodegradation rates that occur in a ready biodegradability test cannot be directly extrapolated to degradation rates in other environments. Rather, the test is useful as a screening tool for chemicals, to determine those for which environmental persistence concerns can be ruled out. The test is stringent in several respects: (1) the ratio of test substance to biomass is high, (2) the biomass is not acclimatized to the presence of the test chemical before the test begins, (3) the test measures mineralization of the test substance, hence a pass level cannot be reached if there are more than trace levels of persistent metabolites generated during the biodegradation process, and (4) the biodegradation rate must be sufficiently rapid to reach a pass level within a relatively short window of time. Because of these factors, "[a] positive result in a test for ready biodegradability can be considered as indicative of rapid and ultimate degradation in most environments including biological STPs[2]" [6].

[2] STP: sewage treatment plant.

Ready biodegradability tests must be conducted on one distinct chemical, and not heterogeneous mixtures, such as finished cosmetic products. In narrow instances, a ready biodegradability test can be performed for mixtures of chemicals, but only for mixtures that are structurally related. The OECD [6] indicates that "homologues of surfactants composed of fatty alcohols of varying chain length or poly(oxyalkylene) polyol materials having defined molecular weight distributions" are specific examples of chemical mixtures for which the ready biodegradability test may be performed. The test is not valid for heterogeneous mixtures because of the manner in which biodegradation is measured, as well as the assumptions inherent in the pass levels. For example, for glycerin (i.e., glycerol), the theoretical oxygen demand (ThOD) and theoretical inorganic carbon (ThIC) production in a ready biodegradability test are determined by the stoichiometry of the mineralization reaction:

$$C_3H_8O_3 + 3\tfrac{1}{2}O_2 \rightarrow 4H_2O + 3CO_2$$

For a test substance concentration equal to 20 mg/L of dissolved organic carbon, the glycerin concentration in the test solution would be 51 mg/L (i.e., glycerin consists of 39% carbon) or 0.56 mmol/L. Hence, ThOD would be 62 mg/L (i.e., 0.56 mmol glycerin × $3^1/_2$ mmol oxygen × 32 mg oxygen/mmol) and ThIC evolved would be 73 mg/L (i.e., 0.56 mmol glycerin × 3 mmol carbon dioxide × 44 mg carbon dioxide/mmol). The pass level for a ready biodegradability test depends on the analyte measured; for OECD test guidelines 301A and 301E, DOC removal must equal at least 70% of the added test substance DOC to pass; for OECD test guidelines 301B and 310, the inorganic carbon evolution must equal 60% of the ThIC to pass; for OECD test guidelines 301C, 301D, and 301F, the oxygen demand must equal 60% of the ThOD to pass. While the pass level is set at only 60–70% of the theoretical maximum values for the various analytes, those pass levels represent virtually complete biodegradation. The 30–40% of the theoretical maximums that are apparently "missing" at the pass level are "assumed to be assimilated by the biomass or present as products of biosynthesis" [6]. This assumption fails for a mixture of heterogeneous chemicals, in which a persistent residue could be present among other readily biodegradable components of the mixture, leading to the requirement that a single chemical or a homologous series be tested.

In addition to requiring the test substance to completely mineralize in order to pass a ready biodegradability test, the biodegradation rate must be rapid. While the duration of the study is 28 days, the majority of degradation must occur during a 10-day window within that 28-day period in order for the test chemical to be considered readily biodegradable. When a microbial population is exposed to a novel, biodegradable carbon source, there is sometimes a lag phase, during which the microbes attempt to synthesize appropriate enzymes to allow utilization of the novel energy source. Negligible biodegradation of the test substance occurs during the lag phase. In a ready biodegradability test, the lag phase is accommodated by starting the clock on the 10-day window only when DOC or oxygen depletion or inorganic carbon production has reached 10% of the theoretical maximum value. Once the clock begins on the 10-day window, the pass level (i.e., 60% or 70%, as described above) must be reached within that window in order for the test substance to be considered readily biodegradable, even if the window closes before day 28 of the test.

The value of using a ready biodegradability test is the ability to assert with confidence that a chemical that has passed the test does not pose a concern for environmental persistence. This concept is borne out by OECD's [6] statement that

> Ready biodegradability tests must be designed so that positive results are unequivocal. Given a positive result in a test of ready biodegradability, it may be assumed that the chemical will undergo rapid and ultimate biodegradation in the environment. In such cases, no further investigation of the biodegradability of the chemical, or of the possible environmental effects of transformation products, is normally required.

Hence, a "pass" result on a ready biodegradability test provides unequivocal substantiation for unqualified biodegradability marketing claims.

If a chemical does not pass a ready biodegradability test, the data can still be useful for other types of biodegradability evaluations. Chemicals for which there is unequivocal evidence of primary or ultimate biodegradation, using any type of test, can be labeled as "inherently biodegradable." If a ready biodegradability test is used to show that a compound is inherently biodegradable, this indicates that the compound ultimately biodegrades and is not persistent in the environment. For instance, Figure 20.2 shows ready biodegradability results for a compound that does not meet the "pass" criteria, but by extending the duration of the study, the results showed that the compound is inherently biodegradable. These data provide strong evidence that the compound will not persist in the environment.

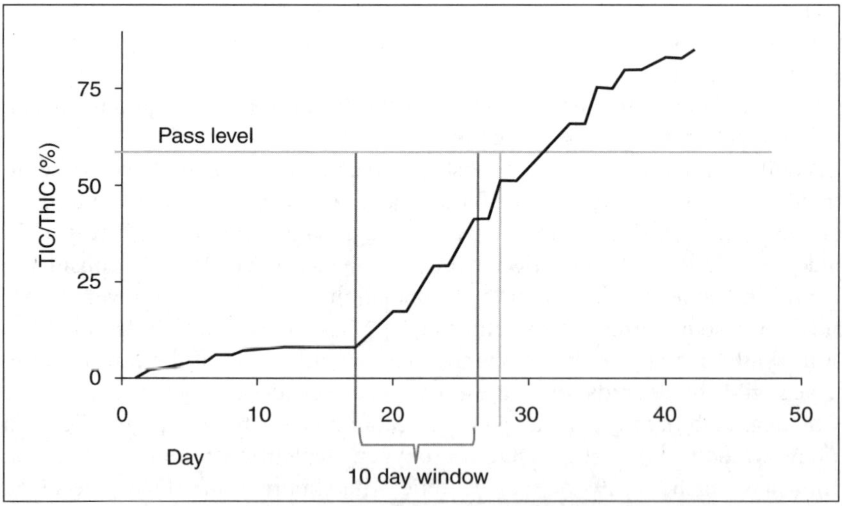

Figure 20.2 Ready biodegradability results for a compound having a long lag phase, failing the 10-day window, but shown to be inherently biodegradable by extending the normal study duration beyond 28 days.

20.3.2 Inherent Biodegradability Tests

Similar to ready biodegradability tests, inherent biodegradability tests are relatively simple and inexpensive. The goal of an inherent biodegradability test is to identify whether a chemical undergoes primary or ultimate biodegradation, without imposing stringent conditions to screen out substances less amenable to rapid and complete mineralization. The experimental design of an inherent biodegradability test (1) uses a low ratio of test substance to biomass, (2) allows for the biomass used in the test to be acclimated to the test compound prior to testing, and (3) allows a prolonged test duration (beyond 28 days) in which biodegradation is monitored. These factors increase the likelihood that the test compound will biodegrade; however, the OECD [6] states that "[because] of the favourable conditions employed in these tests, rapid biodegradation in the environment of inherently biodegradable chemicals cannot generally be assumed." The inherent biodegradability test may be of limited utility for use as substantiation for a biodegradability claim for marketing purposes. However, its results can be useful for determining where a chemical fits on the continuum of amenability to biodegradation, which is valuable for comparing the environmental attributes of alternative ingredients when formulating new products.

An inherent biodegradability test is also a valuable screening tool, but unlike the purpose of the ready biodegradability screen, its objective is eliminating the possibility that the test substance is highly persistent in the environment. In a 2007 ECETOC workshop on biodegradability and persistence, participants criticized the ready biodegradability test for its lack of ability to identify persistent compounds (i.e., not meeting the "pass" criteria does not indicate that a chemical will be persistent in the environment). The modifications to the ready biodegradability test design identified by workshop participants as most likely to yield data that would differentiate slowly biodegradable chemicals from truly persistent chemicals included (1) increasing the test duration (beyond 28 days), (2) enhancing the biomass used, (3) allowing biomass to acclimate to the chemical before testing begins, and (4) increasing the test volume [7]. These suggested modifications to the ready biodegradability design track closely with the inherent biodegradability design, supporting the notion that the biggest asset in using an inherent biodegradability test is to disprove that a chemical will be highly persistent in the environment, rather than proving it is easily degraded, as is the strength of the ready biodegradability test design.

20.3.3 Simulation and Field Tests that Include a Biodegradation Component

Laboratory simulation tests and field tests are the most sophisticated and resource-intensive types of studies in which a substance's biodegradation rate and extent can be evaluated. These tests simulate "degradation in a specific environment by use of indigenous biomass, media, relevant solids (i.e., soil, sediment, activated sludge or other surfaces) to allow sorption of the chemical, and a typical temperature which represents the particular environment" [6]. These types of studies can last on the order of 100 days or longer. The concentration of the test substance is typically low—in the range of <1 to 100 μg/L—in order to allow realistic estimates of kinetics, applicable to environmental conditions. Simulation and field study methods typically require that the test substance be radiolabeled in the portion of the molecule believed to be most recalcitrant to biodegradation in order to facilitate the detecting parent molecule mineralization, the presence of metabolites, and bound residues in the solids present in the test environment. Quantification of the parent compound, and in some cases, the metabolites, is also typically required in simulation and field tests.

Simulation and field tests are not routinely performed for chemicals in formulated cosmetics. Because of the time and expense involved in these tests, they are usually conducted only for chemicals having potent biological activity (e.g., active pharmaceutical ingredients, pesticides) in response to regulatory requirements. When the result of a simulation or field test is available, it provides valuable information about biodegradability. However, it can be difficult to interpret the relevance of these data. Because of the presence of solids in the test environment, biodegradation occurs in tandem with sorption, which can potentially reduce the bioavailability, and hence the biodegradability of the test substance. The implications of simulation and field study data for assessing biodegradability of cosmetic ingredients should be evaluated on a case-by-case basis, because of the variability in test designs and the potential for complex results. For instance, it is not possible to determine whether bound, radiolabeled residues, which cannot be extracted from the solids in a test system, consist of the parent compound, metabolites, or mineralized parent compound that has been incorporated into biomass that is recalcitrant to extraction.

The consensus opinion among participants in a 2007 ECETOC workshop on biodegradability and persistence was that the biodegradability results of laboratory simulation tests could be difficult to interpret due to the following reasons: (1) the environmental compartment tested (e.g., sediment, soil) is not always the most appropriate for assessing overall environmental biodegradability or persistence, (2) the interpretation of bound residues (i.e., ^{14}C unextractable from the solid phase) as they apply to the calculation of biodegradation rate and extent is unclear, (3) the interpretation of chemical dissipation rates from water or sediment with respect to degradation half-lives is unclear, and (4) the necessity for and manner of making temperature corrections is unclear [7]. Nevertheless, the results of a simulation test provide more insight on the likely overall environmental fate of a chemical—including its biodegradation kinetics at realistic environmental concentrations—than do the simpler tests that solely address biodegradability. Their results can sometimes be used to substantiate an unqualified biodegradable marketing claim (e.g., in the case of rapid mineralization in a simulation of aerobic wastewater treatment), and can always be used to help define where a chemical falls on the continuum of amenability to biodegradation, for the purposes of comparative evaluation among alternative ingredients considered when formulating a new cosmetic product.

20.4 PREDICTIVE MODELING OF BIODEGRADABILITY

Scientists and regulators have developed predictive tools to estimate the biodegradability attributes of chemicals for which no empirical data exist. These models are most often used for evaluating new chemicals and, in particular, for screening chemicals to determine which ones are likely recalcitrant to biodegradation and hence potentially persistent in the environment. Such screening processes allow testing resources to be prioritized toward expenditure on chemicals more likely to pose environmental risks.

Models that predict biodegradability differ in the parameters they address, ranging from predicting whether primary or ultimate biodegradation will occur, to biodegradation rates in certain environments, to the expected results of specific empirical tests, such as a ready biodegradability study. The range of models available, the scientific bases for their predictions, and their reliability and limitations have been reviewed elsewhere [8–11].

The concept behind most predictive models for biodegradability is that certain physicochemical characteristics of molecules have been identified as corresponding to an increased likelihood of biodegradability or a decreased likelihood of biodegradability, with examples listed in Table 20.2. The structural moieties listed in Table 20.2, such as the presence of unbranched chain structures suggesting enhanced biodegradability versus multiple aromatic ring structures suggesting inhibition of biodegradability, are used as the basis of many predictive models. While formulators can use the information in Table 20.2 qualitatively, to suggest the likelihood that a substance will biodegrade completely, given its molecular structure, computer models have the advantage of taking into account larger numbers of moieties, interactions between moieties, and other physicochemical factors that are not easily intuited by manual inspection of the molecular structure.

Perhaps the most widely recognized predictive model for biodegradation is the US EPA BIOWIN model, which is a module in the EPI Suite computer modeling application [3]. For example, it is relied upon by US EPA across numerous programs, by Environment Canada, for determining a chemical's persistence under the Canada Environmental Protection Act (CEPA), and in the European Union's

TABLE 20.2 Some Structural Characteristics that Influence a Chemical's Biodegradability [12, 13]

Enhances biodegradability	Retards biodegradability
Low molecular weight	Single or multiple aromatic rings
Presence of only C, H, N, and O	Halogen
Acyclic structures	Quaternary carbons
Acid anhydrides	Tertiary and aromatic amines
Nonbranched alkyl chains	Presence of atoms other than C, H, N, O, and S
Hydroxyl group	Branching terminal nitro group (in noncyclic portion of structure)
Carbonyl group	
Ester group	
Acid attached to chain *or* ring structure	
Ethyl/ethylene attached to chain structure	
Double bond (excluding –N= and –C=S)	

Technical Guidance Document on Risk Assessment [14], which recommends the use of the Biodegradation Probability Program (which is now incorporated into BIOWIN). Users can run the BIOWIN model by entering a substance's molecular structure, using SMILES[3] notation. No other empirical information about the chemical is necessary to obtain biodegradability predictions.

BIOWIN consists of seven separate models, which predict different aspects of a chemical's biodegradability, using different modeling techniques. BIOWIN 1 and 2 are described by Howard et al. [15] and use linear (BIOWIN 1) and nonlinear (BIOWIN 2) methods, based on the fragments that comprise a chemical's structure, to predict whether or not the chemical biodegrades fast. BIOWIN 3 and 4 are described by Boethling et al. [16] and estimate the time required for ultimate and primary biodegradation to occur. BIOWIN 3 and 4 are chemical fragment models based on a survey of experts, in which the time to *complete* chemical degradation (i.e., not the half-life) was semiquantitatively rated on a scale from 1 to 5, where 5 = hours, 4 = days, 3 = weeks, 2 = months, and 1 = longer. The results were translated into chemical fragment models, which are used to predict time for ultimate (BIOWIN 3) and primary (BIOWIN 4) biodegradation to take place for new chemical structures in the BIOWIN software. BIOWIN 5 and 6 are described by Tunkel et al. [17] and use a dataset from the Japanese government developed for their Chemical Substances Control Law (CSCL) consisting of 385 chemicals determined to be "readily degradable" and 499 chemicals determined to be "not readily biodegradable." Like BIOWIN 1 and 2, BIOWIN 5 and 6 use linear (BIOWIN 5) and nonlinear (BIOWIN 6) methods, based on the fragments that comprise a chemical's structure, to predict whether or not the chemical biodegrades fast. BIOWIN 7 predicts the probability of fast biodegradation under methanogenic, anaerobic conditions. Combinations of the results of individual BIOWIN models (BIOWIN 3 and 5 or BIOWIN 5 and 6) can be used to predict whether a chemical would "pass" a ready biodegradability test [3].

US EPA [3] uses the results of the ultimate biodegradation model (BIOWIN 3) to estimate the half-life of chemicals, using the scale shown in Table 20.3. Hence, BIOWIN can be used to predict whether a chemical is readily biodegradable, and if so, the rate of degradation in the environment can be estimated.

A number of limitations to the BIOWIN models are described in the model's documentation [3]. Model predictions tend to be poorer for substances that contain chemical moieties or combinations of chemical moieties that are not adequately represented in the database used to develop the BIOWIN models. For example, the ethoxylate ether fragment does not have a unique coefficient in BIOWIN, and as a result, the biodegradability of (poly)ethoxylate chains is underestimated. Also, BIOWIN does not account for stereochemistry in its predictions, which is important for cosmetic ingredients originating in nature, because of the high frequency of chiral molecules in natural products. In cases such as these, where a model may have shortcomings, it would not be prudent to rely on its results for decision-making purposes.

One issue that is frequently the focus of questions posed by members of the public is the possibility that toxic or otherwise harmful metabolites are formed as

[3] SMILES: Simplified Molecular Input Line Entry System

TABLE 20.3 US EPA [3] Scheme for Translating BIOWIN Ultimate Biodegradability Model Results to a Half-Life in Water

Numerical rating from BIOWIN 3 model	Narrative description of degradation time	Assigned half-life in water (days)
>4.75 to 5	Hours	0.17
>4.25 to 4.75	Hours to days	1.25
>3.75 to 4.25	Days	2.33
>3.25 to 3.75	Days to weeks	8.67
>2.75 to 3.25	Weeks	15
>2.25 to 2.75	Weeks to months	37.5
>1.75 to 2.25	Months	60
<1.75	Recalcitrant	180

a result of primary biodegradation. This question can be answered with high confidence for substances that have attained a "pass" level in a ready biodegradability test, due to the test design. However, for chemicals that are not readily biodegradable, that question can be answered definitively only when more sophisticated simulation tests are conducted to determine if persistent metabolites are formed. If so, then those metabolites would need to undergo separate testing to determine their potential environmental effects. This degree of testing is typically only performed when there is a suspicion that adverse effects are occurring in the environment due to incomplete degradation of a chemical (e.g., due to detection of a metabolite in impaired waters).

There is not a large number of known instances where a chemical biodegrades to form a more harmful metabolite. One well-known instance where this does occur for a class of chemicals that was used in personal care products is for nonylphenol ethoxylate surfactants. Nonylphenol ethoxylate parent compounds undergo relatively rapid primary degradation, because the ethoxylate chain is susceptible to both chemical oxidation and biodegradation. The resulting nonylphenol is less amenable to degradation, undergoing ultimate biodegradation in a short time frame only under certain conditions [5]. Nonylphenol is a concern because it is hydrophobic, bioaccumulative, and elicits endocrine effects [18], which has led to the severe restriction or outright banning of nonylphenol ethoxylates in some jurisdictions [19]. Cases such as this receive media attention and can raise concern among members of the public regarding other chemicals used in consumer products. Because it is extremely resource intensive to rule out the possibility that harmful metabolites are formed from some inherently biodegradable chemicals (i.e., those that undergo primary biodegradation only) through the use of empirical studies, the use of biodegradation modeling can be a valuable tool to address this concern if it is raised for cosmetic ingredients.

The results from reputable, validated models such as BIOWIN can be used to corroborate empirical data of unknown or questionable quality, or can be used as a surrogate for empirical data when empirical studies have not been conducted on a chemical. Model results alone usually would not provide strong substantiation for unqualified biodegradability marketing claims, but they can provide valuable information to allow the environmental sustainability attributes of alternative ingredients to be compared when formulating new products, and they can also provide scientifically reasonable information regarding the relative environmental performance of a particular ingredient or product, for use in responding to questions from the media or the public.

20.5 OBTAINING AND APPLYING BIODEGRADABILITY DATA FOR COSMETIC INGREDIENTS

A biodegradability evaluation should be a central feature of responses to questions from the media or members of the public regarding the environmental safety of a formulated product. In addition, any company's internal initiative for formulating products with an improved environmental sustainability profile should include consideration of biodegradability. The concepts of environmental safety and sustainability can include many facets, such as a chemical's toxic potency, its ability to bioaccumulate in the food chain, its propensity for depleting stratospheric ozone levels, and its ability to form ground level smog. However, if a chemical biodegrades rapidly and completely, these other potential adverse environmental effects are mitigated. A chemical or its metabolites must be present in order to elicit adverse effects in the environment. If a chemical is rapidly biodegradable, it will not be present to bring about adverse effects. In essence, persistence—or the lack of degradability—is a prerequisite for many environmental concerns. Hence, the evaluation of biodegradability is a key factor in determining a chemical's environmental safety or sustainability profile.

A limitation in using biodegradability as a decision-making tool during formulation is that there is a lack of empirical data available for many chemicals. As a result, empirical data typically must be supplemented with model predictions to enable formulators to compare across a wide range of potential ingredients. One way to standardize biodegradability data across several data sources is to translate test or model results to an implied biodegradation half-life. Table 20.4 shows a recommended scheme for performing that translation for commonly used empirical tests that do not derive a half-life as their result (e.g., ready and inherent biodegradability tests). Table 20.3 shows the scheme used to translate BIOWIN model predictions to a half-life in surface water. By using both translation techniques, model results and empirical data can be compared directly when making formulation decisions that account for biodegradability estimates.

In addition to—and likely, more often than—using studies that a company directly selects, commissions, and interprets, biodegradability can be assessed by using data generated by other sources. Sources of biodegradability data range from the peer-reviewed scientific literature, to governmental databases, to study summaries or conclusions directly provided by a cosmetic ingredient manufacturer (e.g., in their promotional literature). In each case, the available information should be critically reviewed for reliability before using. In some cases, the amount of information provided is so little (e.g., a statement that a chemical is "biodegradable" on a material

TABLE 20.4 Biodegradation Rate and Half-Life Estimates in a Sewage Treatment Plant and in Surface Water Recommended by ECB [14], on the Basis of Results from Commonly Used Biodegradability Tests

Test result	First-order rate constant	Corresponding half-life
Sewage treatment plant		
Readily biodegradable	$1\,h^{-1}$	0.69 h
Readily, but failing 10-day window	$0.3\,h^{-1}$	2.3 h
Inherently biodegradable, fulfilling criteria[a]	$0.1\,h^{-1}$	6.9 h
Inherently biodegradable, not fulfilling criteria[a]	0	∞
Not biodegradable	0	∞
Surface water		
Readily biodegradable	$0.047\ day^{-1}$	15 days
Readily, but failing 10-day window	$0.014\ day^{-1}$	50 days
Inherently biodegradable, fulfilling criteria[a]	$0.0047\ day^{-1}$	150 days
Not biodegradable[b]	0	∞

[a]In Zahn–Wellens test, pass level must be reached within 7 days, lag phase should be no longer than 3 days, and percentage removal in the test before biodegradation occurs should be below 15%; *or* in MITI-II test, pass level must be reached within 14 days and lag phase should be no longer than 3 days.

[b]Includes inherently biodegradable substances that do not fulfill the criteria mentioned in footnote *a*.

safety data sheet) that the result should be corroborated through the use of additional empirical data or modeled predictions before relying upon it for decision-making purposes.

Databases that can be consulted for third-party biodegradation data range widely in the type of information provided and the reliability of that information. The European Chemical Substances Information System (ESIS), which contains a link to the International Uniform Chemical Information Database (IUCLID) report for a chemical, will often contain biodegradation study summaries submitted by chemical manufacturers. These summaries range in the level of information presented, sometimes including details of the study design, and sometimes listing only the outcome, such as "readily biodegradable." One caution is that IUCLID data sheets can report data for studies conducted with surrogate compounds rather than the chemical listed in the title of the report. The Australian Government Department of Health and Aging (NICNAS) Index of New Chemical Assessments (Use) Summary Reports contains risk assessment reports for new chemicals used in products including cosmetics. These reports contain a summary of environmental fate and effects data used to perform the assessment, and empirical biodegradability data can be included. The Chemical Risk Information Platform (ChRIP) maintained by Japan's National Institute of Technology and Evaluation (NITE) contains summary results for ready biodegradability tests conducted by the government. The U.S. National Institutes of Health maintains the Hazardous Substances Data Bank (HSDB) that includes short summaries of environmental biodegradability data, with citations to the original reports from which the data were obtained. The U.S. Environmental Protection Agency maintains the High Production Volume Information System (HPVIS), which can contain empirical biodegradation data for HPV chemicals. The European Chemicals Agency (ECHA) maintains a database containing environmental fate and effects data for chemicals registered under Registration, Evaluation, Authorisation and Restriction of Chemical substances (REACH) legislation, which can include empirical biodegradation data. The OECD maintains the eChemPortal database, which is searchable by a substance's Chemical Abstracts Service (CAS) registry number or name, and returns a list of databases in which the chemical is found. The databases participating in the OECD eChemPortal include ESIS, ChRIP, HSDB, HPVIS, and others that potentially contain reliable biodegradability data.

Besides Internet-available databases, empirical biodegradability data can be obtained from the scientific literature. In the peer-reviewed scientific literature, it is common for biodegradability to be evaluated using nonstandard test methods or standard test methods that have been modified to the specifications of the researcher conducting the work. As a result, published studies must be reviewed closely to determine the most appropriate interpretation of the data—especially if the data are to be used to substantiate marketing claims.

For instance, in a peer-review biodegradability study for an anionic surfactant, Hales [20] found that the parent compound, dialkyl sulfosuccinate, would undergo primary degradation, forming several different initial metabolites. One of the initial metabolites was readily biodegradable, whereas the others were not. This biodegradation pathway led to unpredictable results in a series of studies conducted on the compound, as shown in Table 20.5. The compound nearly met the conditions to be considered readily biodegradable, yet it did not degrade completely even in the highly favorable conditions of the inherent biodegradability test. In a porous pot test, which is meant to simulate behavior in an aerobic biological wastewater treatment plant, dialkyl sulfosuccinate ultimate biodegradation was highly variable. These results would not provide strong substantiation for an unqualified marketing claim of biodegradability. However, they do show that the compound is sufficiently amenable to biodegradation to avoid being labeled as environmentally persistent. This type of information is labor intensive to interpret, but can be helpful for communication with the public and for internal product stewardship initiatives.

For naturally derived cosmetic ingredients, testing or modeling biodegradability can be more difficult, because these ingredients are often complex mixtures. While it is likely that ingredients originating in nature are biodegradable, it is not necessarily

TABLE 20.5 Data for Biodegradability Studies from Hales [20] for Dialkyl Sulfosuccinate, Which Forms Several Different Metabolites upon Primary Biodegradation, Only One of Which Is Readily Biodegradable

Test type	Ultimate biodegradation achieved	OECD test guideline
Modified OECD screening	51–62%	301B
Modified Sturm	45–55%	301E
OECD coupled unit	70 ± 5%	303A
Modified SCAS	85–94%	302A
Porous pot at 9°C	62 ± 65%	N/A

the case that they are rapidly or readily biodegradable. For example, plant extracts can have antimicrobial qualities that can reduce the rate of biodegradation. Natural oils can be of low bioavailability because of their low water solubility, slowing the rate of biodegradation. Hence, natural ingredients require scientifically rigorous evaluation to determine their biodegradability characteristics to substantiate marketing claims or to evaluate them against alternative ingredients. One option to overcome the dilemma of testing complex mixtures is to use a purified form of the natural ingredient, which contains a limited number of well-characterized chemicals, which can each be tested individually. Another option is to obtain a chemical analysis of the mixture and test or model the biodegradability behavior of its major components.

When using biodegradability information as part of an environmental preferability assessment for cosmetic ingredients or products, the standard to be met for data quality is highest when the goal is labeling a product as "biodegradable," with no qualifications. The most compelling substantiation for such a claim would be a "pass" in an empirical ready biodegradability study conducted for each organic chemical component of the product. Any other substantiation package (e.g., results of models or other empirical studies) would need to be evaluated on a case-by-case basis by an environmental risk assessor to determine whether the data are sufficiently robust to withstand scientific peer review. When answering questions from the media or the public, or for internal environmental stewardship initiatives, all available data can and should be considered to support decision making. Model results and empirical data that cannot be reviewed for quality can be given less weight in decision making or in explaining the environmental sustainability profile for a cosmetic ingredient or product. Synthesizing a conclusion regarding the biodegradability profile for a cosmetic ingredient or product can be challenging, but making use of all available data provides the most robust support for scientifically defensible decisions.

REFERENCES

1. Organisation for Economic Co-operation and Development (OECD). Ready Biodegradability. OECD Guidelines for the Testing of Chemicals. Test Guideline 301. Adopted July 17, 1992.
2. Organisation for Economic Co-operation and Development (OECD). Phototransformation of Chemicals in Water—Direct Photolysis. OECD Guidelines for the Testing of Chemicals. Test Guideline 316. Adopted October 3, 2008.
3. US EPA. Estimation Programs Interface Suite™ for Microsoft® Windows, v 4.00. United States Environmental Protection Agency, Washington, DC, USA, 2009.
4. van Heijst APM, van Dijk A.Poisons Information Monograph. Acetylsalicylic Acid. International Programme on Chemical Safety (IPCS). PIM 006, 1990. Available at http://www.inchem.org/documents/pims/pharm/aspirin.htm, accessed on July 3, 2010.
5. Bradley PM, Barber LB, Kolpin DW, McMahon PB, Chapelle FH. Potential for 4-*n*-nonylphenol biodegradation in stream sediments. *Environ. Toxicol. Chem.* 2008;27(2):260–265.
6. Organisation for Economic Co-operation and Development (OECD). Revised Introduction to the OECD Guidelines for Testing of Chemicals, Section 3. Adopted March 23, 2006.
7. European Centre for Ecotoxicology and Toxicology of Chemicals (ECETOC). *Workshop on Biodegradation and Persistence*, June 26–27, 2007, Holmes Chapel. Workshop Report No. 10, Brussels, September 2007.
8. Raymond JW, Rogers TN, Shonnard DR, Kline AA. A review of structure-based biodegradation estimation methods. *J. Hazard. Mater.* 2001;84(2–3):189–215.

9. Jaworska JS, Boethling RS, Howard PH. Recent developments in broadly applicable structure–biodegradability relationships. *Environ. Toxicol. Chem.* 2003;22(8):1710–1723.
10. Arnot J, Gouin T, Mackay D. Development and Application of Models of Chemical Fate in Canada. Practical Methods for Estimating Environmental Biodegradation Rates. Report to Environment Canada, Canadian Environmental Modelling Network, CEMN Report No. 200503, Trent University, Peterborough, Ontario, March 31, 2005.
11. Pavan M, Worth AP. Review of QSAR Models for Ready Biodegradation. EUR 22355 EN. European Commission Directorate General Joint Research Centre, Institute for Health and Consumer Protection, Toxicology and Chemical Substances Unit, European Chemicals Bureau, Ispra, Italy, 2006.
12. Boethling RS, Sabljic A. Screening-level model for aerobic biodegradability based on a survey of expert knowledge. *Environ. Sci. Technol.* 1989;23:672–679.
13. Loonen H, Lindgren F, Hansen B, Karcher W, Niemelä J, Hiromatsu K, Takatsuki M, Peijnenburg W, Rorije E, Struijś J. Prediction of biodegradability from chemical structure: modeling of ready biodegradation test data. *Environ. Toxicol. Chem.* 1999;18(8):1763–1768.
14. European Chemicals Bureau (ECB). Technical Guidance Document in support of Commission Directive 93/67/EEC on Risk Assessment for New Notified Substances, Commission Regulation (EC) No. 1488/94 on Risk Assessment for Existing Substances. Directive 98/8/EC of the European Parliament and of the Council concerning the placing of biocidal products on the market. Part III, 2003.
15. Howard PH, Boethling RS, Stiteler WM, Meylan WM, Hueber AE, Beauman JA, Larosche ME. Predictive model for aerobic biodegradability developed from a file of evaluated biodegradation data. *Environ. Toxicol. Chem.* 1992;11:593–603.
16. Boethling RS, Howard PH, Meylan W, Stiteler W, Beaumann J, Tirado N. Group contribution method for predicting probability and rate of aerobic biodegradation. *Environ. Sci. Technol.* 1994;28:459–465.
17. Tunkel J, Howard PH, Boethling RS, Stiteler W, Loonen H. Predicting ready biodegradability in the Japanese ministry of international trade and industry test. *Environ. Toxicol. Chem.* 2000;19:2478–2485.
18. OSPAR. Nonylphenol/Nonylphenol Ethoxylates, Hazardous Substances Series. 2001 (2004 Update). OSPAR Commission. The Convention for the Protection of the Marine Environment of the North-East Atlantic, 2004.
19. European Commission (EC). Commission Regulation (EC) No. 777/2006 of 23 May 2006 amending Annex I to Regulation (EC) No. 304/2003 of the European Parliament and of the Council concerning the export and import of dangerous chemicals, 2006.
20. Hales SG, Biodegradation of the anionic surfactant dialkyl sulphosuccinate. *Environ. Toxicol. Chem.* 1993;12:1821–1828.

CHAPTER 21

OVERVIEW OF BIODEGRADABLE PACKAGING, METHODS, AND CURRENT TRENDS

Gaurav Kale

Chiquita Brands International, Franklin Park, IL

21.1 INTRODUCTION

According to the Environmental Protection Agency (EPA), packaging and container waste is a combination of metal, paper and paperboard, plastics, and wood packaging [1]. Plastic packaging is widely used in a number of food, consumer goods, and pharmaceutical applications due to its versatility and low cost compared to other packaging materials. In 2007, 13.6 million tons of plastic packaging waste was generated, out of which only 1.59 million tons was recovered through recycling [1]. The rest of it ended up in landfills. One of the reasons why plastic packaging may end up in the landfill is the complexity involved in separating and cleaning contaminated packages from food and other biological substances. Considering a volume to weight ratio in case of thermoplastics used in packaging applications, the generated waste is likely to occupy a significantly large amount of landfill area. In efforts to reduce this waste, initiatives are being taken by organizations such as Sustainable Packaging Coalition that defines sustainable packaging as the one that can be “effectively recovered and utilized in biological and/or industrial closed loop cycles” [2]. One of the recovery techniques is composting using the biodegradation process. As a result, a number of polymers are being commercialized that are either biodegradable or compostable. A compostable polymer (plastic) is one that undergoes degradation by biological processes and produces water, inorganic compounds, biomass, and carbon dioxide at a rate consistent with other known compostable materials without impacting the quality of the standard compost and leaves no traces of toxic residues, whereas a biodegradable polymer (plastic) is one that undergoes decomposition due to the action of naturally occurring microorganisms such as fungi, algae, and bacteria. Biodegradation can be either aerobic (composting) or anaerobic digestion (in which the microorganisms break biodegradable material in the absence of oxygen). Considering this difference it can be said that all biodegradable polymers are compostable, but the reverse is not true [3–5].

Formulating, Packaging, and Marketing of Natural Cosmetic Products, First Edition.
Edited by Nava Dayan and Lambros Kromidas.

TABLE 21.1 List of Some Commercially Available Compostable Polymers Used in Packaging Applications

Polymer	Derived from	Type of applications	Company	Brand name
PLA	Biobased	Flexible and thermoformed	NatureWorks, LLC	NatureWorks® PLA
Poly(butylene adipate-*co*-terephthalate) (PBAT)	Petroleum	Flexible packaging	BASF	Ecoflex®
Poly(hydroxyalkanoate) (PHA)	Biobased	Thermoformed, injection molded, flexible	Mirel™	Telles®
PLA and PBAT	Biobased and petroleum	Flexible and thermoformed	BASF	Ecovio®

Compostable polymers could be biobased (i.e., derived from renewable resources), petroleum based, or a combination of both. Some of the commercially available compostable polymers that are used in packaging applications are listed in Table 21.1 [3, 4].

Poly(lactic acid) (PLA) is a polymer that is derived from renewable material such as corn. One of the distinct advantages in using the polymers made from renewable resources such as PLA is that the carbon cycle replenishment speed is significantly higher. For example, at the end of a product life cycle, the package made from renewable resources can be composted to obtain the biomass and it can be utilized as a feedstock to grow resources such as corn. Conversely, when polymers made from petroleum resources are disposed in natural environment, it may take more than a million years for them to be converted again into the fossil fuels [4].

21.2 COMPOSTING PROCESS

Polymer decomposition can typically occur under two biodegradation conditions, aerobic or anaerobic. Composting is an aerobic process, where thermophilic microorganisms use oxygen and organic matter (such as food waste, compostable packages, yard trimmings, etc.) to produce a soil-like substance called humus and generate carbon dioxide as a by-product [3]. The composting process takes place in two steps. In the first step, there is a rapid growth in microbial activity, thereby increasing temperature. This increase in the temperature helps in breakdown of the complex structures in the organic materials. In the second step, the organic materials continue to decompose at a slower rate and at lower temperatures until all remaining carbon is converted to carbon dioxide by remaining microorganisms. The commonly used waste materials for the composting process in the United States are food and yard and manure waste, out of which yard waste composting is more prevalent mainly due to the established infrastructure and ease of backyard home composting [3, 6]. There are several variables that can affect the composting process of a compostable package. They can be classified into the exposure conditions and the material chemical structure as summarized in Table 21.2 [3].

TABLE 21.2 Factors that Affect the Rate of the Composting Process

Factor	Effect on composting process
Exposure	
Moisture	Moisture helps in enhancing the microbial activity and reproduction. It also helps in improving the hydrolysis reaction in polymers, which improves the rate of biodegradation in its early phase
pH	Variability in the pH levels of the compost due to certain acidic or basic materials can impact the microbial activity and thereby decomposition of the organic matter
Temperature	Temperature plays an important role in enhancing and maintaining thermophilic microbial activity in the compost. Higher temperatures help in hydrolysis of polymers as well as speedy decomposition of the organic matter
Oxygen	In the composting process, microorganisms consume carbon from organic matter and oxygen to generate CO_2 and water. If there is no oxygen present, then the process can turn into an anaerobic digestion
Enzymes	Different enzymes found in the compost consume specific organic matter at different rates
Organic material (polymer)	
Molecular weight	The rate of biodegradation changes with the molecular weight of the polymer. Some of the low molecular weight polymers can be easily consumed by the microorganisms commonly found in compost. However, in case of polymers with higher molecular weight (>20,000 Da), the polymer first needs to go through hydrolysis reaction, which leads to a reduction in the molecular weight. Biodegradable polymers typically contain ester, amide, or carbonate hydrolyzable bonds in the polymer backbone. The presence of these hydrolyzable functional groups increases the susceptibility to biodegradation. During hydrolysis, a random nonenzymatic chain scission (breakage of large polymer chains) of ester groups takes place, which leads to reduction in molecular weight
Chemical structure	The polymer is more susceptible to hydrolysis if the chemical structure (i.e., the polymer chain) is more conformationally flexible. This flexibility indicates how much energy it takes to rotate molecules around bonds and how easy it is to move atoms closer to or further away from others
Crystallinity	The crystallinity of a polymer plays an important role in the hydrolysis reaction. Hydrolysis reaction takes place first in the amorphous region. Therefore, polymers that are 100% crystalline are less susceptible to biodegrade at a quicker rate
Size and shape	The size and shape also make a big difference in the biodegradation. For example, a compostable package equally exposed to compost from all sides will biodegrade at a faster rate compared to the one that has more density and less exposure to compost
Copolymer composition	In the case of copolymers that utilize one or more resin together with the parent resin, complications can result and affect the rate of biodegradation due to differences in the chain length and crystallinity

21.3 STANDARD METHODS

Simulated composting methods have been developed and established across the world to evaluate and determine the compostability of a material. All those methods deliver information on compostability of a material based on the amount of carbon dioxide generated from a given mass of organic matter. As per ASTM D6400-04 "Standard specification for compostable plastics" [5], for a material to be certified as compostable it must meet requirements established in the standard such as satisfactory disintegration (less than 10% dry weight of material remaining), biodegradation (greater than 60% carbon conversion to CO_2), and promoting overall safety based on toxicity tests. The disintegration and mineralization values are obtained by the simulated composting test in which the organic matter along with test material is added into an airtight bioreactor that is connected to a continuous air supply (oxygen) and the output air can be captured using a number of methods to monitor CO_2 generation [3, 4, 7, 8]. The carbon dioxide generated in this process can be captured using three different methods: (a) cumulative measurement respirometric (CMR) system in which CO_2 is trapped into a chemical solution (such as sodium hydroxide (NaOH) or barium hydroxide ($BaOH_2$)) and later the amount is determined by another chemical reaction such as titration; (b) direct measurement respirometric (DMR) system in which the output air is directly analyzed using gas chromatography (GC) and the data can be simultaneously logged into a computer to give a cumulative output; and (c) gravimetric measurement respirometric system in which CO_2 from the output air is captured in the CO_2 absorption columns (consisting of pellets of sodium hydroxide) and the evolved amount is measured by the weight increase in those columns [4, 7, 9].

Similar to ASTM standards, methods and guidelines have also been established by ISO (International Standards Organization) and European organizations. EN 13432 "Packaging—Requirements for packaging recoverable through composting and biodegradation—Test scheme and evaluation criteria for the final acceptance of packaging" [10] establishes guidelines to certify a compostable package based on (1) characterization of a material (i.e., presence of heavy metals and volatile solids within established constraints), (2) biodegradability based on the method established in ISO 14855 with a minimum biodegradation level of 90%, (3) disintegration similar to the criteria as in ASTM D6400-04 (i.e., less than 10% dry weight allowable), and (4) compost quality or ecotoxicity to ensure that there is no negative impact on the quality as it will have negative effects on the plant growth or other compost applications [3, 10]. Similar guidelines are also established by Germany, Canada, and Japan in regard to the limit concentrations for the elements that can be present in the packaging materials and detailed information can be found elsewhere [4]. Research studies are in progress to evaluate the performance of biodegradable/compostable packages in real composting conditions. Kale et al. conducted a test on biodegradable PLA bottles in an industrial composting environment consisting of cow manure and wood shavings and under ideal composting conditions of an average temperature of 58°C and 60% relative humidity. The PLA bottles were disintegrated completely into the compost at the end of 30 days. This disintegration trend was also

Figure 21.1 Time lapse photographs of PLA bottles exposed for 30 days in industrial composting conditions. (See the color version of this figure in Color Plates section.)

corrected with the changes in the molecular weight, crystallinity, and glass transition temperature of the polymer. The disintegration trend is illustrated in Figure 21.1 [6].

Certification opportunities for the compostable materials and packages are also possible through some organizations such as Biodegradable Products Institute (New York, USA), US Composting Council (Holbrook, NY), DIN CERTCO (Germany), and AIB Vincotte (Belgium). The certification of a package for use in respective countries is based on the standard test method requirements and established allowable heavy metal and volatile solids limits. Based on all those factors, the compostability of a material or a package is determined and a certification is granted. More information can be found on the respective web sites of those organizations [3].

21.4 COMPOSTABLE PACKAGING AND ITS APPLICATIONS AND CHALLENGES

As shown in Table 21.1, there are a number of commercially available compostable materials that could be further processed to make a package for desired applications. PLA is one of the versatile materials that can be processed to support a wide variety of applications. It can be extruded to make a blown or a cast film as well as blown or injection molded. Typically in food packaging applications, there are some restrictions related to the desired shelf life of the product. However, in the case of consumer goods such as cosmetics, there is no such restriction unless the chemical

TABLE 21.3 Some Commercially Available Compostable Packages and Their Applications

Brand	Company	Base material	Applications
Ecotainer cups	International Paper	Paper with PLA lining	Hot and cold filling
Nature-Tec bags	Northern Technologies International Corporation	Ecoflex™	Kitchen bags
Bare cups	Solo Cup Company	PLA-coated paper	Hot and cold filling
Cereplast Compostables	Cereplast Inc.	PLA with biodegradable additives	Tubs for margarine
Mirel	Metabolix	PHA	Compost bags, durable goods, consumer retail
ECOSPACK™	Adairo Plastic Packaging	Polyethylene, polypropylene, polystyrene, or polyethylene terephthalate with added biodegradable additives	Cosmetic packaging such as tubes

contents of the product react with the package and impact its functionality. Table 21.3 shows some of the commercially available compostable packages, manufacturers, and their applications.

Some companies involved in the manufacturing and distribution of cosmetics have started incorporating environment-friendly materials in the packaging. For example, Cargo Cosmetics has designed a PLA-based packaging system for the lipstick containers that costs similar to the one made using conventional petroleum-based polymers. At the end of life cycle, those containers can be disposed of in the commercial composting facilities [11]. Similarly, packaging material suppliers such as Adairo Plastic Packaging have developed a biodegradable cosmetic packaging ECOSPACK™ (Ecological Cosmetic Packaging) that could be customized for cosmetics manufacturers based on their brand. According to Adairo, the conventional polymers are combined with an organic biodegradable additive that accelerates biodegradation rate without compromising the shelf life. However, those packages are claimed to biodegrade in the microbial environment such as landfill in 1–5 years, and no information has been provided related to their performance in composting environment [12]. Organic Essence is distributing some of its products in a biodegradable jar made from compostable paper. Organic Essence makes only USDA Certified Organic body care products and all the packaging components used in the package system are biodegradable [13]. Thus, in this case, it is very likely that the entire package system combined with its product can be certified as biodegradable and it will be safe to dispose it in industrial composting facilities provided that it meets all the heavy metal limits requirements. Keihl's distributes an "Aloe Vera biodegradable liquid body cleanser" that is packed in a post-consumer recycled bottle. However, this package is not biodegradable and the bottle will need to be resourced through proper channels for recycling at the end of life cycle to meet its cradle to cradle design [14].

If this same product was to be packed in a biodegradable container such as that made from PLA, then it can add more value to the complete package being biodegradable and it will be easier to compost it in industrial composting facilities without any potential risks.

All the test methods and standards discussed above focus on testing material or a part of a package to determine its compostability. Therefore, the time frame of disintegration obtained from those tests is not an actual representation of the performance in industrial or backyard composting environment for a complete package. Also, the tests are carried out on the brand new packages or materials. However, in real-time situations, the package is likely to end up in the compost at the end of its life cycle, which varies from product to product. Typically in case of perishable foods that are packaged in the compostable packaging, it is likely to end up faster in the compostable waste compared to the consumer goods such as cosmetics [3, 4]. The standards and guidelines establish limits for heavy metals, volatile solids, and certain chemicals that are allowable in the compostable package. However, there is no restriction on the actual product content and any allowable limits on those. Since it is not practical to wash down and clean each compostable packaging unit before it goes to the compostable waste, limits or restrictions on packaged product should be established before the package is certified as compostable. In United States, as of 2003, there were 151 industrial composting facilities and 3200 yard waste composting facilities mostly located in California, North Carolina, Maine, and Massachusetts. Not all facilities are currently accepting the compostable packages due to limited awareness and information about the approximate time frame it will take for decomposition [9]. If there is a potential contamination to the compost due to the root cause of the package or its contents, then all the compost and nondecomposed organic matter is likely to end up in the landfill. Thus, this will not complete the "cradle to cradle" cycle for compostable packages, which is the most important factor and motivation in its design and development. Also, since the composting facilities are limited, it is not always worth to transport the packaging waste to those facilities if the benefit of using those materials is offset by greenhouse gas emissions caused by the transportation process. Thus, additional work and research is required for developing standards and guidelines that will allow the end users and each system owner to utilize the "cradle to cradle" feature in compostable packages to its full extent.

21.5 SUMMARY

Plastic packaging has a wide variety of applications in the food and consumer goods industry due to its versatile features and low cost. Based on the municipal solid wastes (MSW) data from Environmental Protection Agency, plastic packaging generates a significant volume of waste every year and a small portion of that is recovered through recycling. The rest of it usually ends up in the landfill and it may take more than million years to be converted back to the fossil fuels. In an attempt to provide replacement/resolution to this problem, biobased and petroleum-based biodegradable polymers are introduced to the marketplace that can decompose in composting environments, so that the "cradle to cradle" cycle is completed in the fastest manner. As those compostable polymers for packaging applications started emerging, standards and guidelines were established to certify them so as to make sure that they are properly decomposed at the end of life cycle. However, there are certain gaps identified in those standards and guidelines related to the consideration of complete compostable package versus an individual compostable material. Research is in progress to address those gaps and to establish a correlation between compostability of a package and other variables. So far, there is no formal agreement between companies, governments, and consumers as to how the composting of those packages will take place; therefore, the main drivers for their use have been green marketing and pseudo-environmental consciousness related to high fuel prices and reduction of greenhouse gases. There is also a lack of infrastructure to properly dispose of those packages in the composting environment. Thus, majority of commercially available compostable packages still end up as municipal solid waste that finally ends up in the landfill. To prevent this from happening, there is a need of harmonization of the standards, certifications, and methods, as well as education and awareness among end users across the world. As there will be an increase in the awareness and the major portion of the compostable waste volume gets disposed of in compostable MSW stream, it is likely to trigger a need for additional infrastructure for proper and complete organic recovery of the waste.

REFERENCES

1. Environmental Protection Agency. Municipal Solid Waste in United States: 2007 Facts and Figures. EPA, Washington, DC, 2008.
2. Sustainable Packaging Coalition. Definition of Sustainable Packaging. 2010. Available at http://www.sustainablepackaging.org/content/?type=5&id=definition-of-sustainable-packaging (accessed on March 2010).
3. Kale G, Kijchavengkul T, Auras R, Rubino M, Selke S, Singh SP. Compostability of bioplastic packaging materials: an overview. *Macromol. Biosci.* 2007;7:255–277.
4. Kijchavengkul T, Auras R. Perspective compostability of polymers. *Polym. Int.* 2008;57: 793–804.
5. ASTM. ASTM D-6400: Standard Specification for Compostable Plastics. ASTM, West Conshohocken, PA, 2004.
6. Kale G, Auras R, Singh SP. Comparison of degradability of poly(lactide) packages in composting and ambient exposure conditions. *Packag. Technol. Sci.* 2007;20:49–70.
7. Kale G, Auras R, Singh SP, Narayan R. Biodegradability of polylactide bottles in real and simulated composting conditions. *Polym. Test.* 2007;26:1049–1061.
8. ASTM. Standard Test Method for Determining Aerobic Biodegradation of Plastic Materials Under Controlled Composting Conditions. ASTM, West Conshohocken, PA, 1998.
9. Kijchavengkul T, Kale G, Auras R. Degradation of biodegradable polymers in real and simulated composting conditions. In: Celina MC, Wiggins JS, Billingham NC, editors. *Polymer Degradation and Performance. ACS Symposium Series*, Vol. 1004. American Chemical Society, 2009.
10. European Committee for Standardization. European Standard EN 13432: Packaging—Requirements for Packaging Recoverable Through Composting and Biodegradation—Test Scheme and Evaluation Criteria for the Final Acceptance of Packaging. European Committee for Standardization, Brussels, Belgium, 2000.
11. Cargo Lipstick Packaging Biodegradable; Turns into Wildflowers. 2007. Available at http://www.sustainableisgood.com/blog/2007/08/cargo-lipstick-.html (accessed on June 2, 2010).

12. NEW! Biodegradable Cosmetic Packaging. Available at http://www.adairo.com/ (accessed on June 2, 2010).
13. Organic Essence. Biodegradable Jars! Oh My!. 2008. Available at http://www.greenecoservices.com/organic-essence-biodegradable-jars-oh-my/ (accessed on June 2, 2010).
14. "Aloe Vera" Biodegradable Liquid Body Cleanser. Available at http://www.kiehls.com/_us/_en/body/aloe-vera-biodegradeable-liquid-body-cleanser.htm (accessed June 2, 2010).

INDEX

2,20-diphenyl-1,10-picrylhydrazyl assay, see DPPH
8-hydroxy-20-deoxyguanosine assay, see 8OHdG
8OHdG, 245
adulteration, 131–133, 333, 337, 345, 353,
advertise (all endings), 37, 40, 43, 53, 54, 58, 67, 68, 70–71, 73, 74, 77, 80, 83–84, 87, 107–122, 177, 310, 393
AET, 198
AFSSAPS, 176
agricultural ingredient(s), 33, 34, 71, 72, 76, 78
agricultural product(s), 33, 57, 58, 59, 71–72, 78, 79
agro ingredient(s), 21, 78
alcohol, 21, 22, 93, 172, 216, 228, 234–236, 262, 276, 291, 293, 342, 346, 374, 380, 381, 399
allergen, 151, 152, 156, 157, 160, 162, 338, 340–343, 345, 346, 350,
allergy, 101, 149–164, 291, 293
American National Standards Institute, see ANSI
amino acid(s), 20, 62, 114, 127, 132, 141, 154–157, 162, 336, 382
animal(s), 19, 22, 23, 24, 25, 61, 62, 63, 64, 79, 81, 82, 84, 94, 103, 118, 120, 129, 134, 138, 141, 143, 163, 216, 228, 240, 243, 245, 267, 269, 275, 278, 288, 295, 296, 304, 316, 321, 323
ANSI, 28, 34–35, 44, 76,
antimicrobial, 170, 174–175, 192, 197–198, 200, 260, 291, 294–295, 298, 307–308, 409
Antimicrobial Effectiveness Testing, see AET
antioxidant capacity, 244–249, 255
antioxidants, 63, 101, 113–115, 118, 128, 229–230, 232–234, 239–255, 259–260, 265–270, 273, 276–278, 290, 293, 295, 307, 314, 316–323, 345, 356
aromatherapy, 17
ARPP, 177
assessment, analytical, 163
assessment, dermatologic(al)/dermal, 111, 159–160, 162
assessment, environmental, 93–94
assessment, exposure, 157, 163
assessment, risk, 93, 97–98, 101, 104, 127, 131, 137, 140, 152, 157, 159, 162, 182, 404, 407
assessment, safety, 68–69, 90, 94, 98, 126–139, 141–143, 156–157, 160–164
assessment, scientific, 98
assessment, systemic , 159–160, 162–163
Autorite' de Regulation Professionelle de la Publicite, see ARPP
Ayurvedic, 287–299
Ayurvedic herbs, see herb(s), Ayurvedic

bacteria, 170–171, 175, 191–194, 200, 202, 265, 269, 277, 298, 307, 376, 411
BDIH, 5, 8, 21, 23, 31, 33, 36, 39, 64, 171–172
Better Business Bureaus, 71, 107
bioactive(ity), 299, 339, 356–357
bioassay, 129, 356–357, 377
bioavailability, 136, 267, 275, 278, 316, 346, 356, 395, 402, 409
bioavailable, 318,
bioburden, 199, 207
biocide(s), 200,
biodegradable, 25, 63, 85, 86, 120, 179, 198, 206, 216, 389–394, 398–409, 411, 414, 416–418

Formulating, Packaging, and Marketing of Natural Cosmetic Products, First Edition.
Edited by Nava Dayan and Lambros Kromidas.

biofilm, 180, 193–194, 209, 397
BioForum, 8, 21, 36
biological substance(s), 411
biomarker, 245, 275, 361–384
biomaterial(s), 200
biopolymers, 200
botanical antioxidant(s), 314
botanical assessment, 151
botanical compound(s), 259–260
botanical drug(s), 142
botanical extract(s), 24, 152, 157–161, 293, 331, 348, 352, 356, 384
botanical ingredient(s), 24, 142, 149
botanical material, 150, 162
botanical preparation, 150
botanical product(s), 24, 149, 383
botanical source, 150
botanical substance, 150, 157–159
botanical(s), 4, 18, 24, 92–93, 97, 127, 131, 142, 149–164, 259–260, 293, 304, 307–309, 314, 320, 331, 333–337, 348, 351–352, 356, 383–384

calendula, 141–142, 292, 348, 349, 356
California Green Chemistry Initiative, 6
California Health and Safety Code, 45
California Organic Products Act, 32, 75
Campaign for Safe Cosmetics, see CSC
CAN, 28–30, 38, 40, 43, 44
Canada Organic Biologique, 79
Canada Organic Logo, 61, 78
Canadian Organic Standards, see COS
carotene, 25, 242, 243, 250, 315–318, 320
carotenoids, 127, 226, 243, 244, 249, 251, 313–324
CARU, 109, 110, 116, 121,
CEN, 27–28
certification, 5–8, 13, 17, 20–21, 23, 30, 33–35, 37–38, 40, 43, 51–52, 55, 57–59, 62–65, 71–72, 74–79, 83–84, 108,120, 125, 142, 171–173, 177, 415, 418
certified, 3, 6–10, 16–17, 25, 31–32, 33, 35–39, 41, 43–44, 47, 55, 57–58, 60–61, 64, 71–80, 83–84, 125–126, 171, 173, 216, 414, 416–417
certified natural, 6, 9, 16, 126
certified organic, 3, 6–10, 31, 58, 73, 75, 80, 84, 125, 216, 416
CFR, 25, 392, 393,
chemical group(s)(ing), 137–139, 142
Chemical Substances Control Law, see CSCL
Chinese herb(s), see herb(s), Chinese
Chinese medicine, 20, 260, 297, 303–310
chromatography, 155–156, 218, 245, 263, 272, 331–357, 361, 364, 373, 379, 414
churnas, 291
cinnamic aldehyde, 91, 101
CIR, 69, 170
CITES, 24, 310
citronella oil, 95–96
claim(s), advertising, 68, 108–111, 122
claim(s), biodegradable, 389, 393
claim(s), cosmetic, 60, 67–71, 74–75, 77, 79, 84, 107
claim(s), deceptive/misleading/fraudulent/false, 70, 79, 109, 112, 177, 357
claim(s), environment(al), 36–37, 85–87, 108, 121, 392, 394
claim(s), ethical, 126
claim(s), green, 67, 85–87
claim(s), health, 126
claim(s), natural, 6, 20, 32, 37, 42–43, 51, 57, 59, 60, 67, 79–85, 87, 121, 173
claim(s), organic, 20, 30, 35, 57–60, 61, 64, 65, 68, 71–80, 84, 87, 107–108, 110, 173
claim(s), product, 5–7, 13, 18, 30, 32, 34–35, 44, 52, 110, 171
Code of Federal Regulations, see CFR
color additives, 25, 54–56, 61, 63, 68, 81
comet assay, 245
Coming Clean Campaign, 84
comparative approach, 133–139, 142
compost, 179, 206, 411–418
compostable package(s), 412, 413, 414, 416, 417–418
consumer, natural, see natural consumer
consumer, organic, see organic consumer
contaminant(s)/contamination, 23, 126, 128–129, 131–132, 134, 137, 150, 170, 175, 180–183, 186–200, 202–208, 247, 291, 310, 342–343, 417
Convention on International Trade in Endangered Species of Fauna and Flora, see CITES
conventional(s), 12, 26, 30, 76, 84, 134, 143, 180–181, 197–199, 201, 203, 205, 206–207, 334, 351, 416
corporate social responsibility, 19
COS, 77
CosmeBio, 31–33, 36

cosmeceuticals, 113–117, 299
Cosmetic Ingredient Review, see CIR
cosmetic standard(s), 27–45, 77, 84
Cosmetics Directive, 20, 85, 100, 163
COSMOS, 7–8, 21–22, 32–33, 36, 44, 77–78
Cramer class, 140
CSC, 85,
CSCL, 404
CTPA, 176
curcumin, 260–269, 278

dermal sensitization, see sensitization, dermal
DGCCRF, 176–177
Dietary Supplement Health and Education Act, see DSHEA
DPPH, 246, 338, 356
Drifters, 12, 19
DSHEA, 82, 129

EC Scientific Committee on Food, 140
ECOCERT, 8, 21, 23, 31–33, 36, 39, 62–64, 77, 108, 171, 172
eco-friendly, 7, 85, 179–208
eco-label, 7, 34, 61, 64, 74
EDEL, 246, 248, 249, 253, 254,
electron spin resonance, see ESR
environmental claim, 36, 37, 85, 86, 108, 392
Environmental Working Group, see EWG
epigallocatechin, 250, 251, 321
ESR, 245
essential oil, 53, 90, 94, 96, 97, 101, 127, 129, 139, 159, 170, 172, 227, 243, 290, 335, 345, 346
ethical(ly), 5, 13, 19, 125–126, 305
European Committee for Standardization, see CEN
European Food Safety Authority, 140, 156
EWG, 85,
extract(s), extractable, extracted, extraction(s), extractive, 21, 22, 24, 25, 52, 53, 55, 59, 60, 62, 64, 76, 81, 82, 89–93, 100, 101, 104–106, 125–129, 132, 133, 138, 139, 141, 142, 149–163, 166, 174, 194, 216, 221, 245–249, 251, 253, 262, 263, 269, 272, 277, 287, 288, 291, 293–299, 306, 316, 319, 331, 333–343, 345–348, 350–353, 355–357, 361, 362, 364, 371–373, 376–384, 395, 402, 409

Fair Packaging and Labeling Act, 60, 68
fatty acid(s), 20, 21, 62, 114, 127, 132, 141, 216–225, 229–231, 234–236, 273, 276, 297, 339,343, 345,
FDA, 20, 22, 23, 31, 33, 52–57, 60, 67–70, 74, 77, 80–82, 86–87, 99, 108, 113, 122, 126, 129–130, 187, 274, 346
FDCA, 51, 55, 68–70, 80, 82,
Federal Trade Commission, see FTC
fermentation/fermenting, 22, 52, 53, 54, 56, 58, 59, 63, 81, 84, 149, 172, 375
Five-Element Theory, 304
flavor, artificial, 52–56, 59, 82
flavor, natural, 53–54, 56, 81–82, 158, 163
flavour, food, see food flavour
food additive(s), 68, 129, 134, 137, 158,160
food allergen(s), 152, 156, 157
food allergy(ies), 152, 156
Food and Drug Administration, 5, 54, 128, 130, 173, see also FDA
food chain, 406
food claim(s), 52–53, 56–57
food contaminant(s), 129, 134, 137
food extract(s), 156
food flavour, 103
food ingredient(s), 81, 126, 130–131, 133–135, 142, 181
food preservative(s), 173
food regulation(s), 39, 130
food safety, 81, 82, 133–135
food standard(s), 34, 39, 42–43, 54, 59, 81, 84
Food Standards and Labeling Policy Book, 59, 81
food supplement(s), 64, 126, 128, 130, 135, 142
food toxin(s), 127
food(s), functional, 4, 319
food(s), health(y), 3, 4, 9, 15
food(s), natural, 17, 53, 173, 181
food(s), organic, 7, 15–16, 30, 33, 39, 43, 57, 71–72, 78, 84, 120
food(s), traditional, 52, 131
food, novel/new, 126, 130, 133–135
foods, cooling, 305
FPLA, see Fair Packaging and Labeling Act
fragrance, 5, 11, 17, 20, 23, 51, 57, 61, 63, 64, 89–105, 129, 171, 190, 227, 333, 338, 340, 342, 343, 345, 350

FTC, 36, 37, 43, 45, 53, 67, 68, 70, 74, 77, 80, 86, 87, 107–109, 110, 111, 121,392

Generaly Recognized as Safe, see GRAS
Geneticaly Modified Organism, see GM(O)
glycerin, 21, 173, 176, 217, 399,
glycol, 18, 92, 174, 176, 229, 345
GM(O), 22, 134,
grape seeds, 222, 244, 251, 321,
GRAS, 23
Green (variations there of), 12–13, 24, 67, 70–71, 86, 108, 120, 170, 179–183, 187, 188, 189, 190, 192, 194, 199, 200, 201, 204, 206, 216, 389, 418
Green Certifier(s), 171, 173
Green claims, see, claims, green
Green Guide(s), 36–38, 43, 45, 86, 87, 108, 121, 392, 393
green packaging, 179–180, 206
green tea, 128, 244, 245, 249, 250, 251, 253, 254, 295, 355, 356
green wash(ers)/washing, 13, 179, 206

Health Canada, 20, 61, 79,
herb(s)(al), 20, 53, 54, 56, 127, 128, 130, 133, 136, 174, 260, 271, 277, 287, 290–299, 303–310, 331, 332, 337, 345, 361, 364, 383
herb(s), Ayurvedic, 291–293, 296–299
herb(s), Chinese, 303–306, 309–310
herb(s), organic, see organic herb
herbal drug(s), 126, 130, 131, 135, 161, 332, 364
herbal ingredient(s), 24, 126
herbal medicine(s), 126–127, 128, 130, 131, 133, 136, 160–162, 287, 298, 304–305, 306, 310
herbal preparation, 295, 331, 383
herbal remedies, 20, 303,
herbicide(s), 7, 333, 344
homeopathic medicine(s), 20, 62
Hotlist, 79

IBD, 8
ICEA, 8, 21, 23, 31, 33, 36, 78
IFIC, 81
IFRA, 92, 97–105
Industry Associations, 5, 41–42
ingredient(s), agricultural, see agricultural ingredient
ingredient(s), botanical, see botanical ingredient
ingredient(s), herbal, see herbal ingredient
ingredient(s), natural, see natural ingredient
ingredient(s), organic, see organic ingredient
inorganic, 53, 54, 56, 62, 63, 194, 390, 393, 396, 399, 411
Inside-out-concept, 313–324
Instituto Biodinamico, see IBD
International Food Information Council, see IFIC
International Fragrance Association, see IFRA
International Standardization Organization, see ISO
ISO, 27–28, 33, 414

JECFA, 127, 128, 139, 140, 163
Joint Expert Committee on Food Additives, see JECFA

Kapha, 289, 294

LDPE, 201
lemongrass, 94–96,
LLNA, 94, 96, 103
logo(s), 59, 61, 64, 70, 78–79, 108,
LOHAS, 11–12
low-density polyethylene, see LDPE
lycopene, 114–115, 128, 243, 244, 250, 251, 260, 315–320, 323, 348

made with, 6–7, 23, 32–35, 54, 57–59, 72–73, 75–76, 83–84,
MBC, 174–175
MED, 316–317
MIC, 174–175
microbial risks, 179–208
microbiology, 181, 187
mineral, 19–22, 25, 35, 54, 55, 62–64, 76,82, 84, 127, 132, 141, 186, 287–288, 291, 304, 321, 374, 390, 398,
mineralization, 391, 397, 398–399, 401–402, 414
minimal erythemal dose, see MED
minimal process(ing), 5, 52–53, 59, 81
minimal processing , 5, 52, 53, 59, 81
minimum bactericidal concentration, see MBC
minimum inhibitory concentration, see MIC

NAD, 71, 83, 107–113, 115–122,
NARC, 107, 121
National Advertising Division, see NAD
National Advertising Review Council, see NARC
National List, 72, 80
National Organic Program, see NOP
National Organics Standards Board, see NOSB
National Sanitation Foundation, see NSF
NaTrue, 23, 31, 32, 33, 36, 39, 44, 172, 173
natural claim(s), 6, 20, 32, 37, 42, 43, 51, 57, 59, 60, 67, 79–85, 87, 121, 173
natural claims, see claims, natural
natural consumer, 11–12
natural cosmetic(s), 3, 5, 8, 9, 12, 15–17, 30, 31, 32, 33, 36, 38, 52, 56, 63, 84, 142, 169, 171–173, 216, 339, 340, 357
natural food claims, 56
natural fragrance, see fragrance
Natural Health Product, see NHP
natural ingredient(s), 3–6, 9, 13, 18, 20, 23, 24, 25, 32, 36, 40, 44, 45, 51, 64, 83, 84, 89, 91, 93, 100, 108, 120, 122, 127, 142, 169, 170, 171, 173, 177, 184, 207, 215–236, 249, 259, 348, 409
natural marketing guidelines, 6
Natural Marketing Institute, 9, 17, 169, also see NMI
Natural Medicines Comprehensive Database, 24
natural personal care, 3, 9–13, 17, 18, 20, 27, 30, 31, 38–43, 45, 125, 170
natural preservative, 169–177, 189, 298, 345
natural product(s), 3- 14, 19, 20, 24, 30, 32, 35, 38, 40, 41, 43, 44, 54, 59, 65, 83, 93, 97, 177, 189, 268, 288, 303, 310, 331, 332, 333, 335, 336, 344, 346, 347, 348, 351, 352, 353, 357, 362, 363, 364, 404,
Natural Products Association, 5, 7, 20, 31, 32, 35–36, 41, 82–83, 120
natural regulation(s), 6
natural substance(s), 63, 81, 126, 149–156, 158, 159, 160, 162–164, 170, 243
Naturalites, 12, 19
naturally derived, 6, 33, 121, 169, 170, 182, 207, 215–216, 226, 227, 235, 236, 408
nature-identical, 172
NGO, 27, 44
NHP, 20, 22, 61, 62,
NMI, 11, 12
NMR, 353, 354, 361–384
nongovernmental, 19,-20, 27–45, 62, 70
Nonstate Market-Driven, see NSMD
nontoxic, 25
NOP, 7, 31–35, 39, 41, 42, 44, 57, 59, 60, 71–78, 80, 82, 84, 87
NOSB, 34, 39, 71, 74, 77
NPA, see Natural Products Association
Nrf2, 322–324
NSF, 7, 31, 32, 34–35, 36, 44, 76, 77,
NSMD, 28–34, 37–40, 42–45
nuclear magnetic resonance, see NMR

oakmoss, 102–104,
OASIS, 7, 31, 32, 35, 41, 42, 44, 76,-77, 172–173,
OCA, 31, 84, 173
OCA, 31–32, 84
OFPA, 57, 58, 59, 71, 74, 82
OPR, 78–79
ORAC, 246, 247
organic agriculture, 4, 8, 63,
Organic and Sustainability Industry Standards, see OASIS
Organic Body Care Standards, 71, 75, 87
organic certification(s), 7, 17, 21, 58, 59, 72, 74, 75, 76, 78, 79, 171, 173
organic consumer, 11–12, 42,
Organic Consumers Association, see OCA
organic cosmetic(s), 8–12, 15, 16, 18, 20, 21, 30, 32, 35, 51, 61, 62, 63, 67, 74–75, 76, 77, 78, 79, 84, 108, 125, 126, 142, 171, 177
Organic Farmers and Growers, 8
organic food(s), 7, 15, 16, 30, 33, 39, 43, 57, 71, 72, 78, 84, 120
Organic Foods Production Act, see OFPA
organic herbs, 73
organic ingredient(s), 3, 9, 19, 21, 22, 26, 31, 32, 33, 34, 35, 40, 41, 57, 58, 59, 72, 73, 75, 76, 110, 125, 393
Organic Monitor, 15, 16, 125,
Organic Personal Care Companies, 10
organic personal care industry, 3, 9, 12
organic personal care market(s), 10, 34, 171
Organic Personal Care Standard(s), 7, 34, 35
organic preservative, 170, 190

organic product(s), 3, 7–13, 15, 16, 18, 20, 21, 24, 26, 30, 32, 33, 35, 39, 58, 59, 60, 61, 71, 72, 73, 75, 78, 120, 171, 176, 177
Organic Products Regulations, see OPR
organic regulation(s), 4, 6
organic sale(s), 9–10
organic seal(s), 7, 19, 34, 59, 60, 71, 73, 74, 76, 79
organic standard(s), 8, 31, 32, 33, 34, 36, 71, 72, 74, 75, 77, 79, 80, 84, 143, 172
organic(s), 6–9, 11–13, 16–19, 21, 22, 25–27, 30–35, 38–45, 52–62, 64, 67, 70–80, 82, 84, 87, 108, 110, 120, 122, 125, 126, 131, 141, 142, 150, 171–173, 177, 179, 184, 190, 192, 194–198, 200, 206, 207, 216, 229, 261, 262, 263, 288, 335, 336, 342–344, 348, 357, 362, 363, 370, 374, 383, 390, 391, 393, 398, 399, 409, 411–414, 416–418
organic, 100%, 18, 32–36, 57–60, 72–73, 76, 79–80, 173
organic, 70%, 32–33, 35, 58–59, 72–73, 75,
organic, USDA, 7–8, 68, 72, 73
OTC, 198, 346
oxidants, 240–241
oxidative stress, 118–119, 240–243, 245, 253, 267, 273, 276, 314, 321, 323,
oxygen radical absorbance capacity, see ORAC

paraben(s), 18, 36, 170–172, 174–177, 253, 353
PCPC, 175–177
PCR, 193, 198
Personal Care Products Council, see PCPC
pesticide(s), 7, 30, 150, 310, 333, 344, 402
PET, 175, 181, 189, 197–198, 202, 206
petroleum, 36, 64, 76, 83–84, 120–121, 198, 201, 297, 339, 341–342, 346, 412, 416–417
Pfizer factors, 109–110
photoprotection, 313–324
phytoantioxidants, 239–255
phytochemical, 132, 133, 135–136, 141, 142, 308, 323, 334, 357
Phytonutrients, 320, 321, 323
Pitta, 289, 294
PLA, see PLA
plant origin, 21
Plant-Based, 120, 121, 184, 198, 207, 291, 310, 323
Plant-derived, 125–143, 226, 260
Poison Prevention Packaging Act, 69
polylactic acid, 198
polypropylene, 201, 416
postconsumer recycled, see PCR
preservative efficacy test, see PET
private standard(s), see standard(s), private
probiotic(s), 20, 62

QAI, 76
QRA, 101–102, 104–105
Quality Assurance International, see QAI
Quantitative Risk Assessment, see QRA
Quercetin, 243, 244, 251, 259,269–278

REACH, 94, 138, 408
reactive nitrogen species, see RNS
reactive oxygen species, see ROS
Research Institute for Fragrance Materials, see RIFM
RIFM, 90, 92–93, 98–99, 103–104
risk assessment, see assessment, risk
RNS, 240
ROS, 240, 241, 251, 253, 265, 273, 277, 314, 321, 323
Rosemary, 73, 173–174, 250, 251, 253, 254, 260, 294, 319

safety assessment, see assessment, safety
sensitization, dermal , 94–97, 159
Soil Association, 8, 21, 23, 31, 33, 36, 77
Spectroscopy, 244–245, 272, 332, 361–384
standards, private , 33, 34, 39, 74, 75, 77, 79
supplement(s), 3,4, 20, 21, 61, 64, 70, 82, 111, 117–119, 126, 128, 129, 130, 134, 142, 143, 160, 161, 245, 249, 296, 298, 317, 319, 321, 364
SUpplementation en VItamines et Mineraux AntioXidants, see SUVIMAX
sustainable, 21, 24, 30–31, 35, 45, 132, 172, 179, 182, 185, 192, 194, 201, 206, 208, 394, 411
SUVIMAX, 242

TEAC, 246
The Federal Food, Drug, and Cosmetic Act, see FDCA
The International Federation of Organic Agriculture Movements, 8

Third-Party Standards, 82–83
threshold of toxicological concern, see TTC
tomato, 128, 247, 249–251, 313–324, 348
total radical trapping antioxidant parameter, see TRAP
TRAP, 246
TRIDOSHA, 289
trolox equivalent antioxidant capacity, see TEAC
TTC, 139–142, 152, 154, 158–161, 163,

U.S. Department of Agriculture, 6, see also USDA
Unconcerned, 12
USDA, 6–8, 30, 31, 32, 33, 34, 35, 39, 41, 42, 43, 44, 52, 57, 58, 59, 60, 67, 68, 71, 72, 73, 74, 75, 76, 77, 78, 79, 81, 84, 120, 173, 416
UVCB substances, 138

Vata, 289, 294, 298,
vegetable oil(s), 21, 56, 160, 297
vitamin(s), 3, 20, 22, 30, 62, 113, 114, 115, 118, 127, 132, 226, 229, 232, 233, 242, 249, 260, 314, 317, 318, 320, 349

WHO, 127, 133, 140, 158, 161, 163
Whole Foods, 4, 10, 23, 129, 134, 137,
World Health Organization, see WHO